Warship 2024

AA005496

Future Surface Combatants

Adelaide, Australia
18-19 June 2024

ISBN: 979-8-3313-0375-4

Printed from e-media with permission by:

Curran Associates, Inc.
57 Morehouse Lane
Red Hook, NY 12571

Some format issues inherent in the e-media version may also appear in this print version.

The Royal Institution of Naval Architects
8-9 Northumberland Street
London, WC2N 5DA
United Kingdom

Phone: 020 7235 4622
Fax: 020 7259 5912

publications@rina.org.uk

Additional copies of this publication are available from:

Curran Associates, Inc.
57 Morehouse Lane
Red Hook, NY 12571 USA
Phone: 845-758-0400
Fax: 845-758-2633
Email: curran@proceedings.com
Web: www.proceedings.com

160 YEARS OF HISTORY, KNOWLEDGE AND EXPERIENCE

THE ROYAL INSTITUTION OF NAVAL ARCHITECTS

Warship 2024:

Future Surface Combatants

18-19 June 2024

Adelaide, Australia

TOGETHER WE WILL SHAPE OUR FUTURE

Sponsored by:

BMT babcock™ SOUTH AUSTRALIA THE DEFENCE STATE

Ansys SH DEFENCE THE UNIVERSITY of ADELAIDE 150 YEARS ASC

THE ROYAL INSTITUTION OF NAVAL ARCHITECTS

Warship 2024: Future Surface Combatants

18-19 June 2024

Adelaide Convention Centre, Adelaide, Australia

THE ROYAL INSTITUTION OF NAVAL ARCHITECTS
8-9 Northumberland Street
London WC2N 5DA

Telephone: 020 7235 4622
Fax: 020 7259 5912

ISSN No: 2752-6569

CONTENTS

ENHANCING NAVAL AUXILIARIES THROUGH HUMAN-CENTRIC DESIGN

A Harris and **C Greenbank** BMT Australia, **P Schumacher** and **S Modra** University of South Australia, Australia

SUMMARY

This paper introduces a novel multidisciplinary approach to naval auxiliary vessel design through the implementation of human-centric modularity. Naval auxiliary vessels are crucial to maritime operations, yet they in turn face constraints in many areas including optimisation of space, durability, and long-term adaptability. By drawing on lessons from submarine habitability, containerisation, and server room design, we view the challenges of designing contemporary auxiliaries as opportunities for innovation.

Leveraging extensive experience in space optimisation and expertise in naval architecture, we propose a dynamic, adaptable, and sustainable approach that drives ship design to anticipate evolving maritime Defence needs. Our approach enhances vessel capabilities and ensures long-term operational effectiveness across the ship's lifespan.

1. INTRODUCTION

Hospital ships have played a crucial role in Naval operations since ancient times; the Ancient Greek and Roman Naval forces recognised the importance of medical care at sea, with ships like the *Therapeia* in the Athenian fleet and designated "immunes" in the Roman Empire fulfilling hospital duties [1]. This tradition persisted through the centuries. Despite their historical significance, the tactical employment of hospital ships received limited attention from naval strategists until World War II. Casualty handling ships initially serving primarily as floating ambulances, but their role expanded during World War II to adopt tactics like the "scoop-and-sail" approach, crucial for maximising medical efficacy during amphibious operations [1].

In the modern era, hospital ships like the USNS *Mercy* and USNS *Comfort* have been pivotal in providing medical support during disasters, conflicts, and epidemics These vessels are equipped and staffed to match the capabilities of onshore hospitals, capable of admitting up to 1000 patients depending on available resources [2]. The adaptability of hospital ships is exemplified by instances like the conversion of a long-distance ferry into a hospital ship during the COVID-19 pandemic in Italy, demonstrating the viability of repurposing non-medical vessels for emergency medical care.

Although almost any ship could be converted into a hospital ship, some designs are inherently better suited than others. The choice of vessel for a hospital ship depends on various factors, including the nature of the mission and the available resources. If some ships are better suited than others for conversion, then this in turn links to exploration of the effective use of space within an existing design. The ability to use space for different purposes leads to the ongoing discussions on modularity in Defence more broadly, but also specifically the concept of modular hospital ships, offering flexibility and cost-effectiveness for military and humanitarian missions [3]. These ships are designed to be rapidly deployable and adaptable to different operational scenarios, essential for meeting evolving demands within constrained budgets. Modular Adaptable Ship (MAS) technologies represent a strategic advancement for the U.S. Navy, enabling the transformation of vessels over their service life to respond to emerging threats and technologies effectively [3].

2. MODULARITY

The concept of modularity in ship design, particularly its application in Naval operations and healthcare logistics, offers significant potential for innovation, offering multifaceted advantages in terms of operational flexibility, economic efficiency, and adaptability. Notably, modularity enables stakeholders to defer investment decisions and adapt to new information and technology, thereby enhancing value robustness under uncertainty [4]. Expanding into naval operations, Minjoo Choi et al [5] illustrate how operational flexibility, enabled by modularity, can mitigate risks associated with early design decisions, thereby emphasising the economic value it provides. Similarly, M. Egilmez et al [6] highlight the potential of modularity to reduce total lifecycle costs and improve adaptability. The combination of these benefits and many others means that there is significant opportunity for modularisation to deliver value to multi-purpose Naval platforms including hospital ships, by providing flexibility to adapt to various mission requirements, and technological advancements, supported by their ability to transform and reconfigure in response to changing operational contexts not foreseen at the time of initial design [4]. This adaptability not only enhances economic efficiency but also contributes to overall mission effectiveness and safety.

The concept design for future warships and auxiliary vessels needs to consider the impact of modularity on the cost-capability trade space, whilst exploring the implications of integrating different roles within a single vessel. This addresses the concurrent requirements of many capabilities, including considerations of naval manpower as a significant through-life cost driver and the potential for autonomy and unmanned systems in future designs [7]. The broader application of modularity and flexibility in future ship designs is a strategic response to the challenges of acquiring and supporting a diverse fleet in an era of rapid technological change and constrained defence budgets [3]. These design concepts can mitigate the risks associated with uncertain future missions and technologies, potentially reducing both modernisation costs and initial acquisition costs [3].

The evolution of modularity in warships demonstrates the practical and efficient optimisation of time and resources in shipbuilding. The concept of mission modules allows warships to adapt to various operational capabilities, optimising the use of defence systems and economic resources [8]. The approach outlined in this paper not only enhances vessel capabilities but also ensures that naval auxiliaries can meet the ever-evolving demands of maritime defence throughout their operational lifespans.

The design of naval auxiliaries presents unique challenges due to the need to accommodate personnel within confined spaces while also meeting modern demands for increased functionalities. Recent advancements in submarine habitability and adaptive design help propose a comprehensive methodology that balances warship design requirements with the need for long-term flexibility.

The concept of human-centric modular design in naval auxiliaries builds on the established principles and addresses the challenges of modern maritime defence. This human-centric modular design philosophy is rooted in the principles of Human System Integration (HSI) which aims to enhance human performance, reduce errors, and improve safety while also reducing through-life cost. Modular design further extends these benefits by allowing for the flexible arrangement of internal spaces to meet the changing needs of shipowners and operators (Braidotti et al., 2020).

This proposed approach leverages the holistic methodologies discussed to address the complexities of naval auxiliary design. By considering the entire life cycle and the systemic interactions between different resources, this approach aims to augment naval auxiliary design to consider sustainment, redundancy and consider future reconfiguration.

Modular hospital units within naval auxiliaries is a concept that draws from the principles of containerisation and service-based procurement to enhance adaptability and efficiency in military medical services. Although the direct impact of modularity on hospital performance is not conclusively positive, it does provide a number of benefits as it addresses issues such as the diverse range of medical support missions and resource constraints, and can lead to improved personnel organisation, equipment utilisation, and support capabilities [4]. In addition, modularity opens the door to more innovative solutions to the problem space. For example, service-based approaches to delivering the capability become possible; the capability delivered by the set of modules could now be managed or owned by different entities or specialised units. That in turn offers new ways to deliver the diverse capabilities required of modern Naval Auxiliaries within a more cost-effective framework, leading to increased efficiency and innovation.

Yet despite the long discussion of modularity and the opportunities it brings, there are fundamental problems with the concept. In practice it is much easier to adopt the mantra of modularity as means to manage the inherent uncertainty over what a ship will be required to do during its service life. Over a nominal 30-year life, a ship might well be required to undertake numerous different roles as technology and the world itself changes around it. It is difficult to specify a design that will remain relevant, so in practice it is easier to manage that uncertainty by designing elements to be changed easily in future. While compelling, this comes at the cost of design inefficiency, often with weight and vessel length penalties to allow for a design that allows the modules to be changed. The trick to unlocking the true potential of modularity is therefore to understand how to use space efficiently because it is that which will provide the most effective means to adapt the ship to different roles between different missions or different decades. Simply leaving volume and allocating it to modularity does not necessarily provide an efficient outcome, as illustrated with the case study below.

3. PROVIDING MEDICAL FACILITIES ABOARD A NAVAL AUXILIARY – A CASE STUDY

An illustrative example of the benefits and drawbacks of the modular approach is the provision of medical spaces within BMT's concept ELLIDA 180 Multi-Role Support Ship (MRSS) naval auxiliary.

Figure 1: ELLIDA 180 Multi Role Support Ship

This ship is intended to primarily operate with Role 1 medical provision, for the treatment of injuries and ill-health among the crew, or for medical emergencies relating from a small-scale rescue operation, such as treating survivors from a small maritime casualty. The capability may need to be extended to Role 2 (casualty stabilisation) or Role 3 (casualty treatment), based upon envisaged operational tasking.

Taskings requiring enhancement of medical capability to Role 2 or 3 could include Humanitarian And Disaster Relief (HADR) and Littoral Strike. The ELLIDA designers therefore explored the possibility of providing a permanent Role 1 capability, enhanced by a combination of dual-role spaces and modular spaces.

Dual role spaces are those that can be temporarily reassigned to a second role, chiefly by the removal and replacement of portable outfit items, furniture, and stores.

Modular spaces are those that can be provided by modules that can be bought onto the ELLIDA and arranged within the "Flex Deck", a 1.5 deck high space suited to the embarkation and use of Twenty foot Equivalent Unit containers (TEUs). TEUs are loaded onto the flight deck, then translate through the hangar, through a roller door into the flex deck.

Table 1. Deck area provision requirements for the various Role 1, 2 and 3 spaces:

Capability Level	Scale (Ward Beds)	Area Required (m^2)
1	5	180
2	10	550
2	20	730
2	40	1080
3	20	1250
3	50	1850

Table 2. Various approaches to providing the capability were considered as follows:

Space	Required in Role 1 baseline?	Build In & mothball?	Dual Use Space?	Modularise?
Operating Theatre	No	Yes	Challenging to arrange due to equipment needs and cleanliness	Requires expanding units and cleanliness is challenging
Scrub Area	No	Yes	Difficult due to outfit needs	As above, part of same area.
Sterilisation and Prep Area	No	Yes	Difficult due to outfit needs	As above, part of same area.
Intensive Care	No	Yes	Possible but could be challenging due to additional equipment and isolation needs	Possible but size of space as needs multiple units; cleanliness could be an issue
High Dependency	No	Yes	Yes, provided sufficiently large space and ease to clean (demonstrated on LSV)	Possible but size of space as need multiple units
Triage and Cas. Reception	Limited scale	Yes	Yes, provided sufficiently large space and ease to clean	Requires multiple units for size plus access issues
Trauma / Resus	Limited scale	Yes	Difficult due to outfit needs	Requires multiple units for size plus access issues
Laboratory	No	Yes	Difficult due to outfit needs	Possible
Pharmacy	No	Yes	Difficult due to outfit needs	Possible
Dental Consulting room	No	Yes	Difficult due to outfit needs	Possible
Doctors Consulting room	Required	n/a	n/a	n/a
General / Low Intensity Ward	Yes, but not at scale	Yes	Yes, provided sufficiently large space and ease to clean (demonstrated on LSV)	Possible but size of space as need multiple units
Med Rest Rooms	No	Yes	Yes	Yes
Med Staff Changing rooms	No	Yes	Possibly	Yes
Mortuary	No	Yes	Requires refrigeration so limited options	Yes, refrigerated containers
Medical Laundry	No	Yes	Difficult due to outfit needs	Yes, can containerise
Medical Stores	Yes but limited	Yes	Possibly but depends on compartment type and location	Yes, can containerise

The resulting arrangement on ELLIDA's flex deck is shown below. Ward facilities are provided by temporary re-role of Embarked Military Forces spaces higher in the ship, served by a medical elevator.

Figure 2: ELLIDA 180 Flex Deck Configured in Medical Role

Such use of TEUs as a basis for military modularisation is becoming more widespread. The UK Royal Navy has demonstrated commitment to modular space, establishing a proposed system of Persistent Operational Deployment Systems (PODS) [9]. PODS are a system of containers similar to TEUs, considered as "plug and play" modules.

Development of modular medical facilities in TEUs is being undertaken, for forward deployed field hospitals. An example of a cardboard mock-up of a Role 2 operating theatre housed within half a hi-cube TEU is shown below.

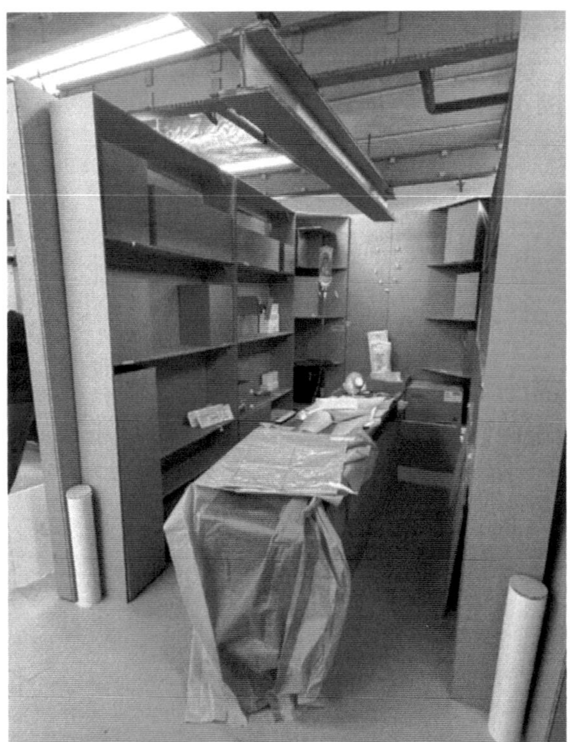

Figure 3: Mock-Up of Role 2 Operating Theatre Within Hi-Cube Container

4. AN EXEMPLAR SOLUTION

The Ellida, while serving as an example to show the advantages of the modular solution, also displays some of the issues that would benefit from further consideration. A limited number of containers (18 in total) dominate the arrangement of the vessel and fill up the available modular space. This results in a less than optimal arrangement and use of space with much empty space above the TEUs and in particular to the starboard side of the flex deck. The deck utilisation factor (including the central passageway) is 0.61 from the point of entry to the complex, where deck utilisation is defined as the footprint area of containerised spaces over gross deck area provision. If considering this as a volume the metric would be even lower.

As well as the space utilisation considerations, there is a potential weight penalty, as the tare weight of a container is usually far greater than the weight of equivalent minor structural or joiner's bulkheads used for compartment divisions in conventional ships. The modular vessel must be sized for the modular payload, including ship's own lifting appliances and load handling systems.

The inevitably greater size and displacement of a highly modular vessel compared to a single role vessel is a comparison that is often made [7], but this fails to account for the savings that can be against the Whole Life Costing of an entire fleet. The modular multi-role vessel can replace more than one single role vessel, with overall fleet size dictated by the concurrency of roles and scale of effect required. The WLC across the whole fleet may demonstrate substantial savings, and this is advantageous for many navies, challenged with the need to meet increasing commitments with limited funds.

The global trends within the research programs of many navies towards TEU-sized modularisation, demonstrates that the possible cost penalties of an individual larger ship are considered to be more than offset by the utility and flexibility of the solution, with many citing the ability to pre-deploy mission packages before ships, allowing for rapid re-role in theatre, or close to an area of operations.

How then, to maximise the benefits of the TEU modular solution in an individual ship design? An approach to the general arrangement of a modular vessel could take the form of maximising the modular space, in dimensions of length x breadth x height, which favour the installation of TEU-based mission packages. The naval architect provides a modular "utility ship" sized around the payload allowance, with defined interfaces for electrical supply, HVAC, water, and other services.

In this concept, the vessel would contain the minimum of propulsion, structure, accommodation, command and control spaces to function as a seagoing ship. A large volume would be dedicated to large, enclosed decks containing the mission pack. The performance characteristics necessary for the ship itself would be determined from operational analysis of all the missions to set speed, range, endurance and other requirements in the traditional manner.

As a practical example, if one of the roles is as a hospital ship then challenge a team to define the necessary space, based on working closely with Navy medical personnel and refine the arrangement until it is clear how many decks are required and what needs to go where. Parametric studies can optimise the space utilisation, until a minimum value is found, below which capability is unacceptably impacted. In parallel area studies for other identified mission packs can be undertaken, for example scientific research, UxV command and control, etc. The footprints for each can be compared for maximum area / dimension requirements when this activity has been completed for all mission packs. If all packs show a similar optimum area, then this validates that the selection of roles are compatible for a single vessel. Each of these activities can be completed in a matter of weeks, not months or years.

It is then for other design teams to work within the payload boundary and its interfaces. The naval architect is now in the consultative role for the mission pack designers, to ensure that regulatory requirements are met at the whole of ship level. The modular payload would still need to comply with normal warship regulations, as applied by the national Naval Administration, such as ANEP 77 Naval Ship Code, SOLAS, MARPOL as applicable.

The ideal form of vessel will vary with the roles to be provided. The roles of ASuW and AAW for frigates and destroyers can require high speed and manoeuvrability on occasion. This is not true of medical, MCM or HADR missions. Therefore, there will still not be a "one size fits all" solution to modular multi-role vessel design. It is likely that one set of missions can be fulfilled by fast and manoeuvrable surface combatants, with appropriate specification of survivability and signatures control. Other missions could conceivably by undertaken by less complex utility ships. These would have much more freedom to move away from traditional surface combatant architecture, allowing for maximisation of modular space.

To consider what type of architecture may suit the utility ship, it can be instructive to look at previous successful ship designs. An example of a ship which could be re-roled from civilian use to either military sealift or a hospital ship was the SS *United States*. This ship pre-dated widespread modularisation in transportation or ship design, so a lengthy refit would have been required for conversion, removing outfit and installing new equipment on an individual item basis.

Another remarkable example of a ship which has been re-roled into a naval auxiliary is RFA *Argus*. Launched as container ship MV *Contender Bezant*, she was temporarily converted into a helicopter support ship when taken up from industry by the UK government to participate in the 1982 Falklands conflict. In 1984 she was purchased and permanently converted into an Aviation Training Ship and, later, with the additional role of a Primary Casualty Receiving Ship (PCRS). The ship motions constraints of aviation meant that several thousand tonnes of concrete ballast were added to raise *Argus*'s centre of gravity and lengthen her natural roll period, but she retains the ability to embark a deck load of containers in lieu of helicopter operations, albeit much reduced from her former capability as a containership. The original containership design with a forward superstructure and a clear deck formed of hatch covers has provided good scope for conversion. She remains in service with the Royal Fleet Auxiliary.

The architecture of *Argus* made her easy to modify and re-role, even though her conversions have not been modular. The high-level characteristics of the ELLIDA 180 design bear some resemblance to a smaller *Argus*. For a utility ship, designed for a modular payload including medical and other complementary missions, a similar arrangement is likely to be well suited, with superstructure forward, accommodation, access routes, machinery and command and control spaces all pushed outboard, below or forward of generous modular open spaces, in place of the holds of a containership design.

Subdivision and watertight integrity of the vessel would be built around these open spaces, with the U-shaped midship section of the ship chosen so that machinery spaces are situated below the modular decks, with the bulkhead deck running along either side of the mission pack before rejoining at the stern behind the pack.

5. DISCUSSION

The total number of hulls within a fleet can be reduced by modularity, where roles are not expected to be concurrently required. Many navies pursue TEU-sized modules. It may be possible to benefit from larger modules. Reference [10] discusses the potential for standardisation across NATO navies and the development of interface standards. An extension of the modular medical facility study on the ELLIDA design could be to further explore the impact of the interface standards of the TEU-based solution, as well as considering the possibility of non-compliant larger modules. This would quantify the benefits and drawbacks of different approaches to modularisation, particularly to identify the ship design characteristics that could improve space utilisation when arranging TEUs in modular multi-role ships.

6. CONCLUSIONS

This review illustrates the potential benefits and considerations of employing modular hospital units in naval auxiliaries, highlighting the importance of adaptability, efficient resource allocation, and integration of technology to enhance military medical support capabilities.

It is concluded from literature review and by illustration of the application to a modern auxiliary ship design, that TEU container modules provide a good basis for the provision of Role 1 and Role 2 medical spaces.

There remains scope for further investigation of space-efficient ship design solutions to suit modular re-role of surface warships,

7. ACKNOWLEDGEMENT

The authors would like to thank Andy Kimber and Alex Aitken of BMT for their support in the preparation of this paper.

8. REFERENCES

1. LANE, D., 'Hospital Ship Doctrine in the United States Navy: The Halsey Effect on Scoop- and-Sail tactics', *Military Medicine,* 1997.

2. BOREN, D., FORBUS, R., BIBEAU, P D., McKENZIE, R., & McKINSEY, K., 'Managing critical care casualties on the Navy's hospital ships', *Critical Care Nursing Clinics of North America, 15 2,* 2003.

3. SCHANK, J., SAVITZ, S., MUNSON, K., PERKINSON, B., McGEE, J., & SOLLINGER, J., 'Designing Adaptable Ships: Modularity and Flexibility in Future Ship Designs', *RAND Corporation,* 2016.

4. KUNTZ, L., & VERA, A., 'Modular organization and hospital performance'. *Health Services Management Research, 20(1)*, 2007.

5. CHOI, M., REHN, C., & ERIKSTAD, S O., 'A Hybrid Method for a Module Configuration Problem in Modular Adaptable Ship Design', *Ships and Offshore Structures, 13*, 2018.

6. EGILMEZ, M., PARK, J M., BAYRAK, A E., EPUREANU, B., & PAPALAMBROS, P., 'Adaptability of Modular Vehicle Fleets to Changing Supply Route Characteristics', *The Journal of Defense Modeling and Simulation: Applications, Methodology, Technology, 17*, 2020.

7. SMITH, D., 'Multi-Role Vessels: Design, Manning & Cost Drivers', *RINA Warship 2019: Multi-Role Vessels*, 2019.

8. PIÑEROS BELLO, L A., & SEGOVIA FORERO, C E., 'Evolution and Present of Modularity in Warships', *Proceedings of the VI International Ship Design & Naval Engineering Congress (CIDIN) and XXVI Pan-American Congress of Naval Engineering, Maritime Transportation and Port Engineering (COPINAVAL),* Springer International Publishing, 2020.

9. HINE, N., 'Royal Navy Keynote Speech: When is a Box Not a Box? The Need to Think Differently', *Defence and Security Equipment International*, 2021.

10. MANLEY, D., 'The NATO Drive to Mission Modularity', *RINA Warship 2018: Procurement of Future Surface Vessels*, 2019.

9. AUTHORS' BIOGRAPHIES

Andrew Harris is Technical Lead, Asia-Pacific region, for Naval Architecture at BMT. He is responsible for maintaining and developing BMT's regional naval architecture capability. Prior to joining the Australian business, Andrew had spent 20 years working in sustainment and design of Royal Navy surface vessels in the UK.

Dr Peter Schumacher is the lead for the Studio for Complex Confined Human Environment Design (SCHED). His expertise is in Human Centred Design methods and processes focused on working with users to understand their environments, work procedures and problems and develops practical solutions that address their needs.

Simon Modra is an experienced design practitioner based in South Australia, with expertise in interior architecture, furniture design, building design, and project management. He holds multiple degrees in architecture and design and is a part-time PhD candidate at The University of South Australia. Simon is a member of the Studio for Complex Confined Human Environment Design (SCHED).

Chris Greenbank is Principal Human Factors scientist for BMT's Asia-Pacific region, and a Chartered Human Factors Specialist with over 20 years of experience supporting a range of high-profile Defence programmes in the UK and Australia. He specialises in the practical application of human factors theory and practice to develop innovative solutions for complex problems.

DESIGNING TO SURVIVE BEYOND 2040 – OPPORTUNITIES AND CHALLENGES OF LUSV

A A Martin, QinetiQ UK

SUMMARY

Parallel developments in uninhabited small surface vessels (USV) and 'loyal wingman' concepts for combat aircraft suggest that large, uninhabited surface vessels (LUSV) will form a significant element of many naval forces in the future.

Such LUSV could function as distributed sensor nodes and additional magazines in a network of linked combat ships. This form of vehicle offers both opportunities and challenges, one of which is the survival of the LUSV in combat. This paper explores the implications of ultra-low crewing and uninhabited LUSV for vessel security, vulnerability, damage control, repair and recovery. In doing so the opportunities for radically different operating methods are explored alongside the naval architecture issues that will arise as LUSV are designed to break the dependence on human limitations.

NOMENCLATURE

BLEVE	Boiling Liquid Expanding Vapour Explosion
CIWS	Close-In Weapons System
HPM	High Power Microwave
LUSV	Large Uncrewed Surface Vessel
MSI/MII	Motion Sickness Incidence/Motion Induced Interruptions
UAV	Uncrewed Aerial Vehicle

1. INTRODUCTION

The future of combat ships could look very different from the fleets of today. There is a general trend towards smaller numbers of crew in both merchant and naval vessels and the first steps are being made towards uncrewed cargo vessels and small naval craft. The most advanced naval experiments are represented by the US Navy's 'Overlord' programme which has so far seen the introduction into service of four standardised Large Uncrewed Surface Vessels (LUSV) (Figure 1) and several experimental types. These are employed on trials to determine how they will integrate with fleet operations, their capabilities, maintenance, and other issues that may only become clear in service. It is recognised that the Royal Australian Navy (RAN) are closely following the US example with plans recently announced to acquire six optionally crewed vessels of the same type operated by the US Navy [1].

It can be seen that these vessels represent a first generation of LUSV, based on commercially-available ship designs. As experience in the operation of these vehicles is gained, it is foreseen that a second generation will evolve, characterised by being purpose-designed to operate with minimal concession to the needs of personnel onboard. One of the key drivers to their design will be their survivability against a variety of threats. Major design decisions will need to be faced including whether such vessels are expendable, whether their uncrewed status allows operation in areas where the threat is too large for crewed ships, and where the balance will lie between susceptibility, vulnerability and recoverability in the design.

We are currently in an era where experimentation will start to reveal both the questions we, as naval architects, should be asking, and some of the solutions. However, the next generation of LUSV are going to need to survive in the combat environment of the 2040's and as designs are prepared now, questions of survivability must be explored now.

Figure 1: 'Nomad' – a US Navy first generation LUSV in 2022 - image via US Navy

3. CHARACTERISTICS OF SECOND GENERATION LUSV

3.1 SIZE

The size of a LUSV may be driven by a number of factors including payload, fuel storage, speed, endurance and seakeeping. In the latter case, the motion environment may be limited by the need to provide radars and other sensors with a stable platform with a predictable motion in a seaway. These factors together may produce LUSV with a size perhaps two-thirds of current crewed equivalents. A big change is that the volume of the ship will not be taken up with accommodation, stores, potable water etc. and there may be significant amounts of void, unused space internally.

Ship size also impacts survivability providing all other factors are common. Smaller size brings reduced signatures at visual and radar wavelengths and also reduced pressure signature in a displacement hull. In parallel, smaller size tends to result in increased vulnerability as the blast/fragmentation effect of a threat warhead will spread over a broadly 'fixed' volume, thus damaging a smaller proportion of the volume of a large ship than a smaller one.

3.2 COMPARTMENT SIZE

Compartment dimensions on most vessels are driven partly by flooded stability concerns, but mostly by human factors. Deck heights for example must account for the height of crew members, with additional allowance for lighting, air-conditioning, pipes and cables. For spaces that have no need for regular human access, these restrictions disappear. It could be that the future will see LUSV that have compartment dimensions that could much taller than at present, or subdivided by additional strength members that would otherwise impede human movement.

3.3 HULLFORM

Accepting that hull motions can be decoupled from human limitations such as Motion Sickness Incidence (MSI) and Motion Induced Interruptions (MII), means that not only can the size of the LUSV be driven by other factors, but also the hullform. This may open up a wider variety of hullforms then may be possible than with crewed warships

3.4 CONTAINERS AND PODS

LUSV are likely to be characterised by role-flexibility, with that flexibility brought by the use of containerised equipment. The Royal Navy has, from 2021, been experimenting with PODS (Persistent Operational Deployment Systems) [2,3] based on the ISO TEU format. Initial efforts have focussed on accommodation, command and control, medical and UAV launchers (Figure 2) although many other variants are being investigated.

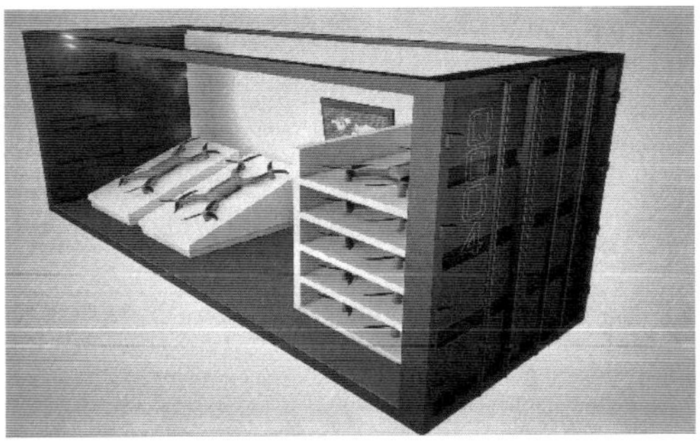

Figure 2: PODS containerised UUV launcher (image via Royal Navy)

4. SURVIVABILITY ISSUES

4.1 EXPENDABLE?

LUSV are expected to be less expensive to purchase than comparable crewed vessels with the same capabilities. However if the LUSV is both a significant sensor node and a magazine for missiles, then the purely financial value of the vessel will still be large. Even more significant may be the functional value of the vehicle; without it the fleet operation will be jeopardised. In this case the LUSV cannot be considered to be expendable and some effort will be needed to ensure that it survives in a combat situation.

4.2 SHOCK/ARMOUR

For any vessel that is not considered expendable, there will be an ongoing need for protection against underwater shock. Although a LUSV, particularly a multihull design may not respond to shock in exactly the same way as a traditional monohull frigate, the approaches used for protecting the equipment on board will be very similar including the use of ductile materials and shock mounting. Protection will also be needed against ballistic attack. For armour solutions for LUSV, these will likely be focussed on protecting any area where a bullet or energetic fragment could pose the threat of catastrophic failure. This includes magazines and fuel systems.

4.3 SYSTEMS SURVIVAL

The fundamental principles of concentration of critical equipment, duplication of major capabilities and separation of duplicate capabilities will apply to LUSV in the same way as crewed warships. Achieving adequate separation will be more difficult in LUSV if they are smaller than fully crewed warships, but will prove of value against smaller threats such as medium calibre gun shells, even if larger anti-ship threats weapons are overmatching the survivability of the LUSV.

4.4 FIRE FIGHTING

Without crew, damage control and fire fighting must be fully automated for LUSV. This will represent a new philosophy for firefighting; some spaces may be sacrificial, others must have rapid extinguishing systems. Re-entry to a fire-damaged compartment will not be possible. In each case early and reliable detection of fire will be critical and it may be necessary to link this to an intelligent damage control command system to ensure both early intervention and minimisation of false alarms. As the damage control command system will be safety critical, it should be robust and able to survive initial weapon effects.

There is the potential for uncrewed spaces to be inerted while at sea, using hypoxic atmosphere control, either by processing the air to increase the nitrogen content at the expense of oxygen, or by filling the space with another gas such as carbon dioxide or argon gas. This will require the compartment to be gas-tight and such a technique is well suited to void spaces. Maintaining a hypoxic environment will be very difficult when faced with penetrating and explosive damage to a compartment and may need to be back-stopped with other measures such as exclusion of flammable materials or use of automatic fire fighting systems. These might be particularly important in machinery spaces where hypoxic atmospheres are more difficult to control as combustion air is drawn in, heat must be exchanged and fuel and lubricant gasses may be found.

4.5 FLOOD CONTROL

Traditional methods of flood limitation such as wedges, plugs, boxes etc. all require human intervention to fit, particularly as any damage tends to be hard to predict and difficult to access. LUSV will not have this option so most flood limitation must be passive. Buoyant syntactic foams will play some part, bonded to the hull, or filling void spaces. Such materials are regularly used to provide buoyancy in free-flood underwater vehicles. Their performance under shock and ballistic attack is not usually a driving factor in their design but some research has been conducted on which to make estimates of their vulnerability. [4,5]

Self-healing materials do not currently exist to cope with the size of damage that an explosion will create [6] but for small scale cracks and tears they could form part of a composite layer in a structure that might limit the permeability of a bulkhead and thus reduce the longitudinal spread of floodwater.

4.6 REPAIR

One of the key functions of crew is to conduct maintenance and repair of equipment. For a LUSV these must be conducted alongside; there is little scope for battle damage repair while underway. An exception to this might be the employment of intelligent fluid systems which use 'smart valves' to sense pressure drops caused by leaks or damage and automatically re-route a flow network to maintain supply, avoiding the damaged area [7]. This is a technology at a high readiness level having been demonstrated through the Royal Navy DINCS project [8].

4.7 MANAGING FUTURE FUELS

Looking into the future, new design warships, including LUSV, could be required to use alternatives to traditional hydrocarbon fuels. The more common suggestions include Methanol, Hydrogen and Ammonia. Considerable work is ongoing to investigate the issues of using these fuels in warships, but their adoption will have several negative implications for survivability some of which are summarised in Table 1.

Table 1. F76 Dieso compared to some alternative marine fuels

	F76 Dieso	Methanol	Ammonia	Hydrogen
Pressurised/chilled	No	No	Either	Yes
Explosive limit	0.6-6.5%	7-36%	15-28%	4-75%
Boiling point	282-338 degC	65 degC	-33 degC (1 atm)	-253 degC
Flash point	54 degC [9]	12 degC	-33 degC	-240 degC
Toxicity	Hazardous by ingestion/inhalation	Lethal by ingestion/inhalation [10]	Vapour lethal to human life at very low concentrations	
Other notes		Burns with clear flame – heat detectors needed	Corrosive to copper	

Ammonia is exceedingly toxic. If it is to be used, then an uncrewed vessel may be the best option for exploring this technology. If the vessel is occasionally crewed, then housing the crew away from any possible fuel leaks may be necessary.

It can be seen that hydrogen must be chilled to a very low temperature and pressurised to store. Any system that employs pressurisation of gas will also be a particular problem for vulnerability; loss of containment could result in a Boiling Liquid Expanding Vapour Explosion (BLEVE). Materials for the fuel tanks are likely to be brittle at very cold temperatures and impact from a ballistic threat could result in either rapid cracking of the container and release of the fuel, or a BLEVE event. Hydrogen will readily leak through any gaps in pipe joints and this could result in gas release and the creation of an explosive environment. This becomes a particular issue for pipes under shock loading as they flex and elastically stretch, or under ballistic attack. As Table 1 indicates, the three alternatives to F76 each have a broad explosive limit so that the atmosphere can be potentially explosive at a wide variety of gas concentrations. Ballistic damage to fuel pressure vessels must be anticipated and it may be necessary to consider armouring both the vessels themselves and any pressurised pipework.

5. THREAT ENVIRONMENT FOR 2040

5.1 TRADITIONAL EXPLOSIVE THREATS

Traditional explosive threats will remain relevant for the foreseeable future. Anti-ship missiles, ballistic missiles, mines and torpedoes will remain potent threats and accessible to non-state organisations, as evidenced by the use of advanced missiles by Hezbollah (2006) and the Houthi group (2023-24). Weapons designed to disable destroyer-sized targets are likely to overmatch smaller LUSV.

5.2 AI AERIAL SWARM THREATS

A subset of traditional explosive attacks is likely to be the AI-enabled swarm threat, particularly when the LUSV is in harbour, or operating close to shore. The wars in Nagorno-Karabakh (2020) and Ukraine (2022-) have been characterised by extensive use of UAV. In parallel the widespread employment of 'drone swarms' for spectacular light-shows indicates the growing potential for UAV swarm attacks, particularly when allied to artificial intelligence. This would tend to indicate that there will be a need for LUSV to employ some form of self-defence, either with laser, HPM or CIWS. This may raise questions about the necessity for a human-in-the-loop for self-defence, particularly in harbour or close to shore, or whether such defence can be automated.

5.3 PHYSICAL SECURITY

As LUSV are in harbour, or close to shore, they may be boarded and damaged. One broad interpretation of the International Convention on Salvage (1989) might even suggest that boarding an uncrewed vessel with a view to bringing it into a safe condition might actually count as salvage! In these situations the physical security of the LUSV might require the temporary accommodation of small crew (perhaps flown out to meet the vessel in port). This would change much about the arrangement of a vessel. A possible solution to this would be to employ containerised accommodation modules that are easily added or removed.

5.4 CYBER SECURITY

LUSV will be heavily dependent of communications to operate, both with secure military data sharing, but also communications with port facilities to arrange maintenance and with operators to continuously monitor machinery and other onboard systems. Interception of low-level machinery control signals could allow cyber-attacks on more sensitive systems, or force shutdown of essential machinery. A possible counter to this will be to limit the amount of communications so that the majority of operations, such as navigation and collision avoidance are conducted autonomously without external control.

6. QUANTIFYING SURVIVABILITY PERFORMANCE

6.1 EFFECTIVENESS

There will be many possible configurations for a second-generation LUSV and a variety of different possible combinations of survivability measures might be employed. However, it is not simple to determine which combination of measures might be most effective against the range of threats envisaged. This is where there is a need to quantify vulnerability and recoverability performance. The software that QinetiQ currently use, and UK MOD have regularly used for this purpose is *Survive UK*.

Survive is a suite of software, initially active from 1998, but continuously developed, expanded, modernised and maintained since then. It calculates the effects of conventional weapon attack, both above and underwater on a ship, submarine or uncrewed marine vehicle. Both vulnerability (the initial effects of blast, fragmentation, shock etc.) and the longer duration effects of fire, flood and smoke can be modelled. This can be combined with both automated and crewed damage control, firefighting and repair to create a timeline of vessel recoverability for a given incident. Typically the output is a probability that the vessel will have certain functions operating initially after an attack. This can reveal when equipment is critical or easily damaged, or where redundancy has been compromised through poor arrangement. Crew casualties may also be predicted. When repeated over many scenarios an average probability of vessel or system survival is calculated. In this way the effects of dissimilar protection strategies (such as adding armour, blast hardening or duplication) can be compared and the most beneficial solution found.

Recent developments in simplified recoverability modelling also allow for the comparison of different damage control approaches that could be applied to a LUSV. Thus the effect of raising or reducing the fire hazard in a compartment could be compared to the effect of using different automated firefighting systems.

6.2 APPLICATION

A notional LUSV has been modelled using *Survive* Software and is illustrated in Figure 3. This model represents a vessel 62m in length with facilities for acting as a sensor node and missile magazine. A small flight deck aft would be able to receive rotorcraft transporting temporary crew members. Accommodation is provided for a team of up to 12 for several days to perform security or other short-duration functions. For routine operations the vessel is autonomous but with a 'human-in-the loop' for combat functions, either based ashore, or afloat as part of a fleet. Space is reserved for either PODS ISO containers or anti-ship missiles.

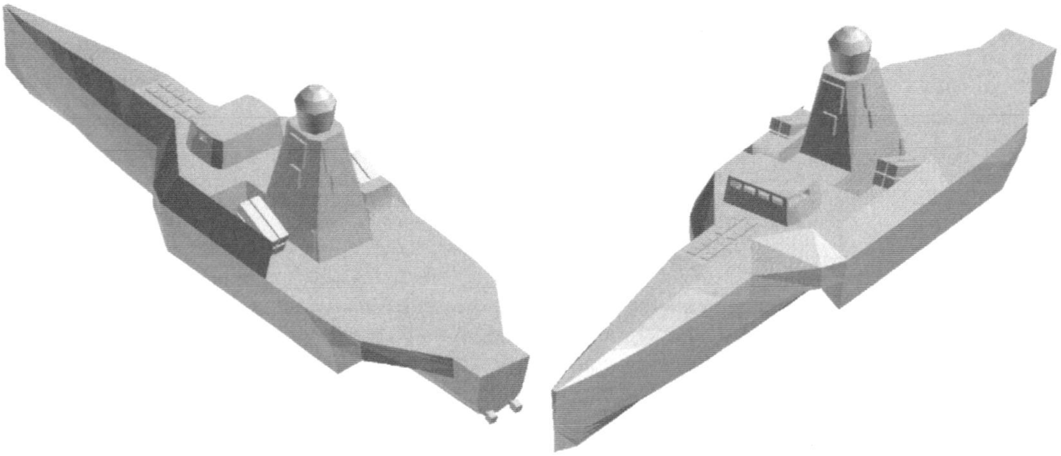

Figure 3: *Survive* sketch design for exploring LUSV survivability

QinetiQ are currently studying how a LUSV of this size can best be configured to survive weapon attack. Investigations that *Survive* can be used for include identifying the level of threat that can be reasonably managed, the degree to which redundancy and separation can be achieved in smaller warships and the effect of creating buoyant and/or hypoxic void spaces on recoverability. The sketch design also forms the basis for investigating how new fuel systems can be integrated into purpose-designed LUSV.

7. CONCLUSIONS

The advances in ship autonomy, AI and threat severity are likely to result in the widespread adoption of purpose-designed Large Uncrewed Surface Vessels as a major component of naval fleets beyond 2040.

These LUSV offer design opportunities for breaking the architectural dependence on human performance and physical characteristics

Such LUSV will also come with challenges, some of which will be associated with survivability.

The effect of removing the crew from many threat scenarios may be met with radically different approaches to vessel survival

Investigations using *Survive* will guide the potential benefits to be gained from different technologies applied to LUSV.

8. ACKNOWLEDGEMENTS

Many thanks to Josh Neill (QinetiQ) for his assistance with this paper.

9. REFERENCES

1. 'Enhanced Lethality Surface Combatant Fleet', ISBN: 978-1-925890-81-5, *Commonwealth of Australia*, February 2024

2. 'Thinking inside the box – the Royal Navy's containerised capability concept | Navy Lookout

3. 'Royal Navy tech office showcases latest deployment units' (mod.uk)

4. DOLESKI, R., PLUNKETT, S., TUCKER, W., 'The rate sensitivity of high strength syntactic foam' via *DTIC May,* 2003

5. ROUSSEAU, C-E., et al. 'Behaviour of syntactic foam under plate impact' Mechanics Research Communications 83 (2017) 1-5 via Elsevier Press 2017

6. SAHIE-POUR, R., BERENBAUM, D., 'Impact of flinch technology on damage control and survivability', *INEC*, 2018

7. VAN OSCH, J M., WORTHINGTON, J P W., et al., 'Distributed Intelligent Networked Control Systems (DINCS) applied to a whole ship integrated Platform Management System', *12th International Naval Engineering Conference and Exhibition* (INEC 2014)

8. WANG, Y., et al, 'Self-healing composites: a review' *Cogent Engineering volume 2, Issue 1,* 2015

9. CITGO F76 Safety Data Sheet

10. Compendium of Chemical Hazards (publishing.service.gov.uk)

9. AUTHORS BIOGRAPHY

Andrew Martin is a naval architect with over 30 years of experience in ship safety, structures and warship survivability research. He is currently the Principal Consultant for marine survivability within QinetiQ Ltd., involved with projects for T31, T45 and future submarines. His role encompasses advancing the modelling of weapon effects in the *Survive* software and exploring issues for the design of future warships. He is a Chartered Engineer and Fellow of the Royal Institution of Naval Architects.

TOWARDS OPTIMISED WARSHIP STRUCTURAL MANAGEMENT DURING FLEET TRANSITION

B Cartwright, University of Newcastle, Pacific ESI Pty Ltd, Australia
M Renilson, Australian Maritime College, University of Tasmania, Pacific ESI Pty Ltd, Australia

SUMMARY

To ensure a surface combatant capability, particularly during the transition between aging and new warships, rigorous assessment of structural service life of naval ships is required. However, such assessment is often based on linear and deterministic methods; the uncertainties regarding the seaway loading, structural configuration, and age-related degradation are implicit. This can lead to unnecessarily conservative maintenance and ship withdrawal decisions. Non-linear and probabilistic analyses enable uncertainties to be taken into account but tend to be time or cost prohibitive. Thus, tailored approaches, including those employed in digital solutions, are receiving attention.

One solution to characterise the structural response of an aging warship in a seaway is coupled Smoothed Particle Hydrodynamics – Finite Element Analysis (SPH-FEA). Initial validation of a novel implementation of SPH-FEA to predict the complex interaction between surface waves and a frigate, by comparing experimental heave and pitch motion responses and vertical bending moments at forward speeds in regular waves, has been completed. To mature the SPH-FEA approach, refined modelling of the fluid (smaller SPH particles) and analyses at a limited number of critical operational conditions are needed.

Although thorough validation of the SPH-FEA techniques is ongoing, this presentation will provide examples of typical analysis results that can provide data-driven decisions in the maintenance and life extension of a fleet. To that end, we propose that knowledge of the exact motions, bending moments and stresses within a ship as it experiences specific large-wave or high-speed conditions can increase the confidence of decisions made about the ship structure. These decisions could be such as maintenance deferral – thus saving money and improving ship availability – f stresses are shown to be within a specific range, or a change of operational envelope to ensure such stresses are not encountered in a particular mission.

Fundamental responses such as hydrostatics, still water bending moment distributions, and dry and wet natural frequencies of the ship structure as developed from the mesh-free technique are essential to build confidence in the analysis technique. We will present results from these fundamental responses and a progression of more complex conditions. An example is Figure 1, which shows non-dimensionalised midship bending moment from numerical simulation and experiment at one speed, and one wavelength as a function of wave steepness [1].

To aid in the understanding of how corrosion might influence the stress in a dynamic condition, a localised area of corrosion was artificially created by reducing the original plate thickness in a narrow athwartship strip in the midship keel region of the frigate. The frigate was then driven through large waves and the stress observed.

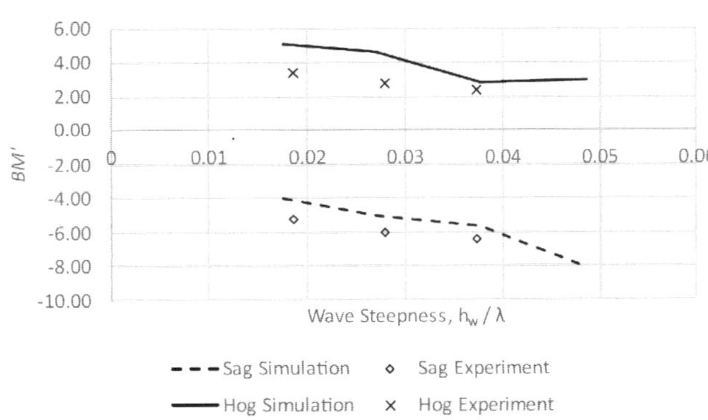

Figure 1: Non-dimensionalised midship bending moment with increasing wave steepness for one speed and one wavelength.

The non-dimensionalised stress in the various degrees of thinned hull plate as functions of time is shown in Figure 2 [2]. Of particular interest is the asymmetry of the hog and sag stress values at this location.

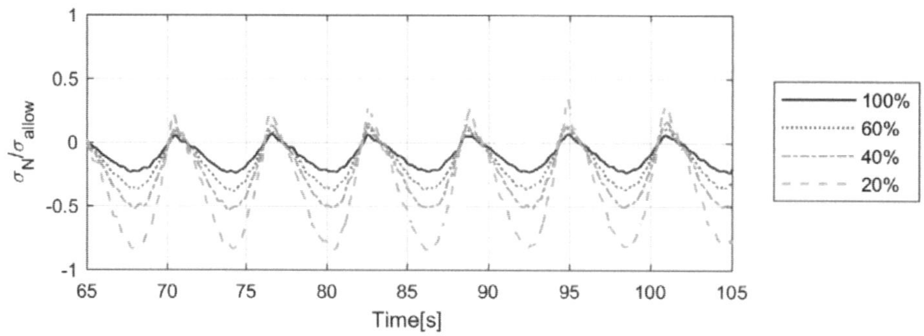

Figure 2: Normalised stress, in longitudinal direction, at reference structural location, for 100%, 60%, 40% and 20% remaining thickness of plate (for Fn=0.18, L/λ=1.0, and H/L=0.028).

Figure 3 [2] illustrates the change in the normal stress range at the reference structural location with different thickness diminutions, for the given load case. Also indicated in the plot are the minimum plate thickness of 80% specified in the class rules, the allowable (design) stress, and the fatigue strength reference value for the given structural detail.

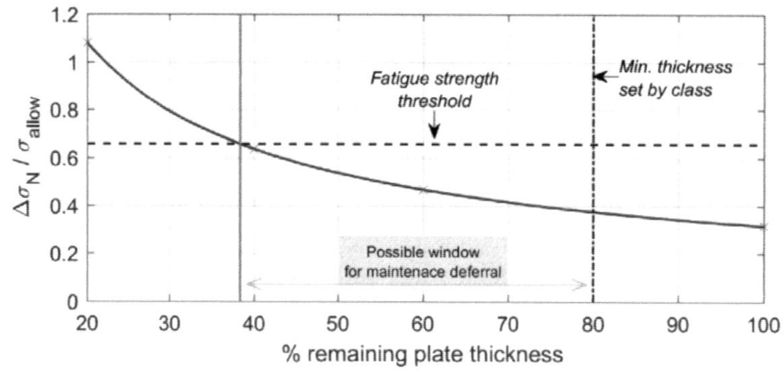

Figure 3: Normal stress range at reference structural location with different plate thickness, for Fn=0.18, L/λ=1.0, and H/L=0.028)

Figure 3 represents just one location at one operating condition. Similar plots at other high-stress locations of the ship potentially subject to corrosion, and at other operating conditions would be needed to determine the criticality of structural defects due to corrosion in the hull. From this suite of results, decisions about maintenance prioritisation, remaining hull life or retirement could be made.

2. REFERENCES

1. CARTWRIGHT, B., MAGOGA, T., RENILSON, M., "Validation of a Smoothed Particle Hydordynamics Approach to Model the Interaction between Surface Waves and a Ship", *Pacific International Maritime Conference*, Sydney, 2023.

2. CARTWRIGHT, B., RENILSON, M., MCGUCKIN, D., MAGOGA, T., MULCAHY, L., MELCHERS, R., ALLEN, C., "Towards a Cost-Effective Approach to Extending the Life of Naval Surface Ships", *Pacific International Maritime Conference,* Sydney, 2022.

3. AUTHORS BIOGRAPHY

Bruce Cartwright is a Research Associate at University of Newcastle, Australia, and is completing his PhD on meshfree analysis of ships in waves. He is also a Senior Simulation Engineer at Pacific ESI Pty Ltd in Sydney.

Martin Renilson is an Adjunct Professor at the Australian Maritime College, Launceston, and works as a consultant with Pacific ESI Pty Ltd in Sydney.

THE APPLICATION OF DISRUPTIVE TECHNOLOGIES TOWARD IMPROVED MISSION CAPABILITIES FOR SAFE LAUNCH AND RECOVERY EVENTS IN HIGHER SEA CONDITIONS

Dr. B Ferrier, Syntek Technologies, Dynamic Interface Laboratory, USA, **Dr. M. R. Belmont** and **Dr. J T Christmas,** University of Exeter, UK

SUMMARY

In piloted aviation, a deck officer normally guides an arriving air vehicle to a safe deck landing. The Deck Officer uses the combination of visual cues and deck behaviour by "feeling the deck" to anticipate the deck's condition. In automated operations, the approach is different.

One of the most challenging tasks in the deployment of basic ship-based VTOL Uncrewed Air System (UAS) is the launch and recovery and deck handling performance on smaller ship platforms. This is owing to the random nature of seaway. The time needed to set-up an air vehicle from the perch to its recovery may be longer than the deck motion quiescent windows may persist. Quiescent Period Prediction (QPP) objective is to provide ship motion conditions and motion predictions with sufficient forecasted time (over a minute) to launch, recover and complete other motion sensitive tasks regardless of the seaway. The wave radar measured seaway is combined with an artificial intelligent (AI) machine.

Owing to the advance of AI techniques and other technologies, this challenging task is achievable. The initial research demonstrated that owing to the random nature of the sea, AI by itself would not suffice. To furnish the AI machine a stable starting system, the fundamental concept is to measure remote sea surface profiles by wave radar to predict the future wave forces acting upon a vessel. In this research, wave measurements were combined with AI models to predict the next few seconds in near real-time. Test results, which are characterised as encouraging, are discussed.

NOMENCLATURE

DI	Dynamic Interface Study
DOF	Degree-of-freedom
EI	Energy Index (calculated by LPD)
HLA	High Level Architecture
LPD	Landing Period Designator
NATOPS	Naval Air Training and Operating Procedures Standardization
RAO	Response Amplitude Operator
RN	Royal Navy (UK)
RNAS	Royal Navy Air Station (UK)
SHOL	Ship-Helicopter Operational Limit
SMP	Ship Motion Program
SMS	Ship Motion Simulation (NAV routines)
TCS	Tactical Control Station
UAV/S	Uncrewed Aerial Vehicle/System
USV	Uncrewed Surface Vessel/System
WOD	Wind-over-deck

1. INTRODUCTION

Uncrewed Air Systems (UAS) are seen by many to be force multipliers and the capabilities they present, once integrated into a digital battlespace will, in time, change the face of maritime warfare. The area of Ship Air Integration is a familiar domain for crewed aircraft but without the pilot for almost instantaneous decisions and workload assessments the operation and integration of UAS on board is in its infancy. To improve mission effectiveness and deck activity tempo, the RN is investing in is the development of Quiescent Period Prediction (QPP). This is an application that has utility across all motion sensitive systems and one that can not only improve safety but increase capability.

Awareness of the ship's position and movement in the ocean is key to air operations. To be able to predict this and target the right time to deploy and recover UAS/USV vehicles, will give Command greater utility and flexibility (figure 1).

Figure 1: Early UAS – Ship Launch and Recovery

The operational challenge is to define in a timely manner periods of low deck motion activity (quiescence). This is particularly critical when involving other robotic vehicles as found in the integrated Expeditionary Strike Group plan **[1]**, MoD/University of Exeter's Quiescent Period Prediction (QPP) purpose of is to forecast (<u>more than a minute(s) depending on the ship and recovery platform</u>) acceptable deck motion periods for **safe and predictable** launch and recovery of piloted and remotely piloted air and sea vehicles. With a view of conducting automated UAS launch and recovery, the specific objectives are: (i) to both give the operator guidance on the optimum ship's heading and speed to initiate launch/recovery vehicular activity and (ii) to predict periods of limited flight deck motion (for safest recovery) over timescales of typically 60 to 120 seconds ahead. The goal is to expand ship UAS/USV operating deck limits to approximately Sea State +6.

From the measured sea surface, a short-term deterministic wave model can be constructed allowing the wave system to be propagated to the ship's location. The ship's response to the wave spectrum is calculated to determine the level of quiescence at the time of encounter. The results are mapped dynamically on speed-polar visual graphics providing the ship's best course to steer to ensure encountering the quiescent zone in a timely manner. This procedure defines the ship motion aspect of the deck environment. A user advice system displays the predicted quiescent periods as the suggested best periods in which to execute an operation.

The QPP application goal is to predict, a few tens of seconds ahead in time, low intensity, i.e., quiescent periods (QPs) within the measured sea waves system. Such quiescent periods are natural feature. of all sea wave systems and occur on a continuous basis (figure 2).

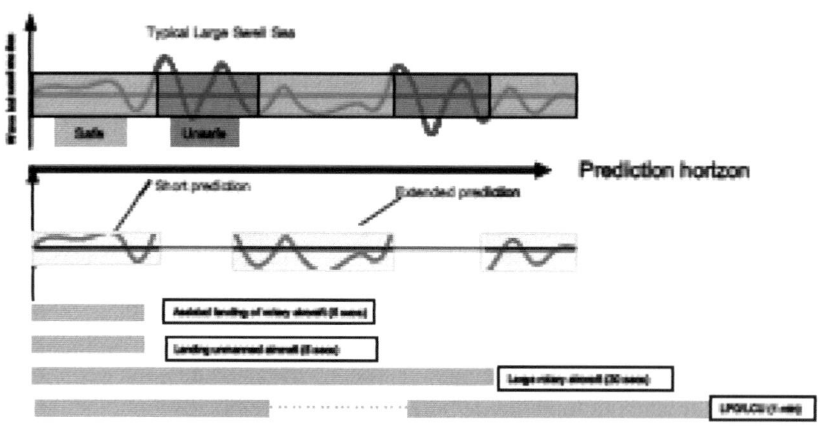

Figure 2: Wave Groups and Quiescent Periods

The UK Ministry of Defence has had an on-going interest in the development of QPP. This has taken the form of both supporting specific work on different technical aspects and in providing sea trials to prove the principal of QPP.

This article describes the MoD/Royal Navy sponsored development of the UK's QPP System with discussions on test procedures, timely seaway mapping and ship motion characteristics specifically to complete a motion sensitive task like UAS launch and recovery. Operating over deck whilst the ship is experiencing a quiescent period has the secondary effect of minimized air wake confusion. This is owing to fewer ship structure excursions into and out of the air flow. A dynamic diagram is shown that is used to both create the "best course to steer" and the best time to execute an operation.

2. QUIESCENT PERIOD PREDICTION

Quiescent Period Prediction (QPP) is a class of technologies that use remote sea surface profile measurements to predict the future wave forces acting on a vessel in space and time. Such technologies have prediction time horizons measured in minutes and offer predictions of the actual motions of the vessel (figure 3). The purpose of this type of technology is to identify so called "Quiescent Periods" in the next few minutes. Knowing by how long a low motion period will begin and how long that period will last, makes it possible to safely execute a motion sensitive operation. These launch and recovery event may be possible even when the mean prevailing conditions exceed the traditionally considered safe deck limits.

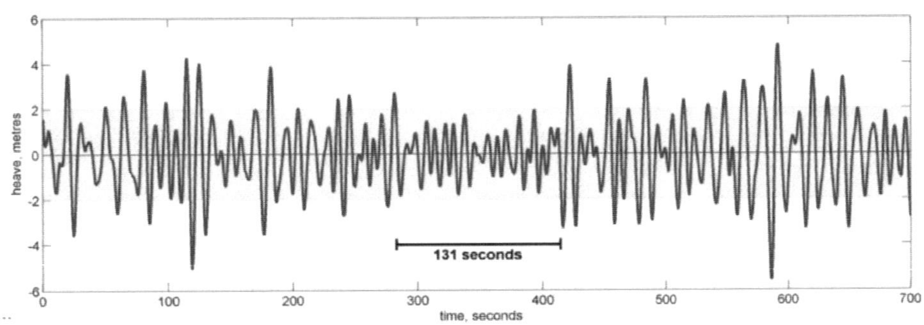

Figure 3: Quiescent Period recorded motions.

A QPP system achieves this by using a remote wave sensor system to measure the sea surface several hundred meters in around the ship **[2]**. From the measured sea surface, a short-term deterministic wave model is constructed allowing the wave system to be propagated to the ship's location **[3]**. The ship's response to the wave spectrum is calculated to determine the level of quiescence at the time of encounter.

The QPP purpose is to provide ship motion conditions and motion predictions with sufficient time to launch, recover and complete other motion sensitive tasks regardless of the seaway. Based on lessons learned from the use of a pre-prototype QPP on trials in the North Atlantic and Irish Sea, it is conceivable, that once integrated, shipboard communities could operate in much higher sea conditions recovering the air/sea vehicles with elevated confidence and reliability. The caveat is that SHOL can be constructed meeting basic deck limit x vehicle interface criteria **[4]**. As stated, the system is designed to safely expand ship operating and deck recovery envelopes. The computational flow diagram corresponding to QPP system components is displayed in figure 4.

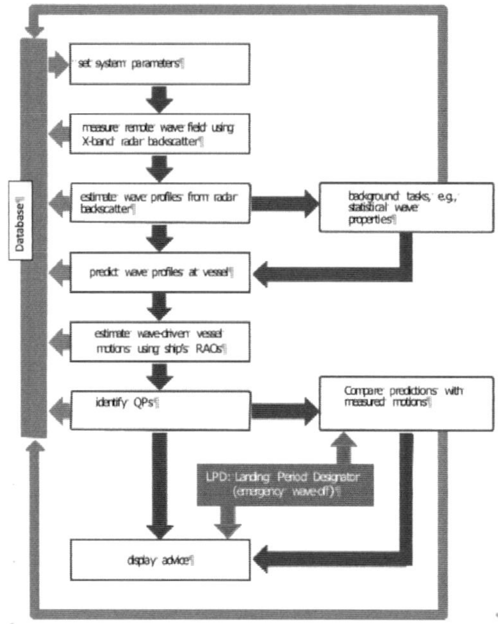

Figure 4: QPP Computational Flow Diagram

In the current QPP development work a simulator was created that enabled the identification of the ranges of values of the main QPP system parameters that allow decisions to be made on the balance between cost and performance. Such data is critical for detailed design, build and test phases a future sea going QPP system (figure 5). The primary objective of the current QPP project is the integration of the various software elements created over the last three MoD sponsored QPP and LPD program phases.

Figure 5: Typical Ship's Navigator Radar (QPP)

The key feature of quiescent periods is that they exhibit wave amplitudes significantly less than the overall local mean. Such quiescent periods alternate with large wave groups where the wave amplitudes are much larger than the mean (figure 6).

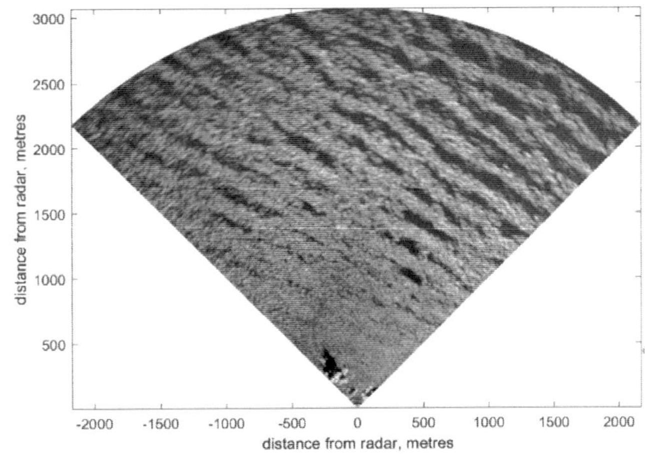

Figure 6: 2-D Backscatter Wave Plot

The ability to predict these quiescent periods means that a naval operation, which is normally only safe up to a given wave amplitude, can be safely executed in seas where the average wave height significantly exceeds this value. This safely extends the operational capability of naval assets providing both a strategic advantage and increasing their value.

The QPP software package is designed to be also capable of being integrated into a full motion simulator (in the crewed case) and a TCS (in the UAV case). This is a means to estimate system effectiveness as a function of simulated ship motion, visual environment, and synthetic operational systems, and to compare the results to related recorded at-sea data **[5]**. Such a simulator will also help ship operators gain confidence in the new technology.

A QPP system would typically also include a Landing Period Designator (LPD) sub-system to provide an emergency last minute wave-off capability.

3. QPP COMPUTATIONAL PROCESS

QPP requires first using wave profiling radar to measure the sea wave system out to approximately 2km in the region around a vessel. Second these measurements are employed in a wave propagation model to predict the actual wave forces acting on a vessel. Using the wave predictions as inputs to a vessel model makes it possible to predict the actual (deterministic as opposed to statistical) motions of a vessel.

3.1 SIGNAL PROCESSING

The method of measurement uses a typical shipboard wave RADAR and fixed time or a shipboard LIDAR. The RADAR methodology is summarized in Figure 7.

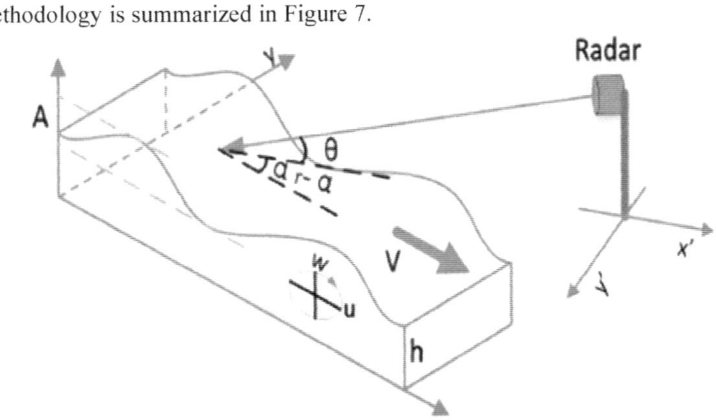

Figure 7: Radar Wave Measurement

Traditionally marine radar is used sea surface currents, and directional wave power density spectra. This only delivers statistical wave information. The novel aspect of RADAR use in QPP is to measure phase resolved directional wave spectra allowing actual wave shape to be determined. As described in reference **[6]**, ship born wave radars typically operate at acute angled incidence to achieve the desired mapping range of coverage. This introduces gaps in the data due to so called wave shadowing as illustrated in Fig 8b. The development of methods to directly invert marine radar images that can cope with such errors to return accurate sea surface elevations is the subject of the much current research.

The wave data is then re-mapped onto a uniformly distributed equivalent. Typically, the backscatter image using algorithmic filters resembles real-world wave forms as shown in figure 8 **[8]**.

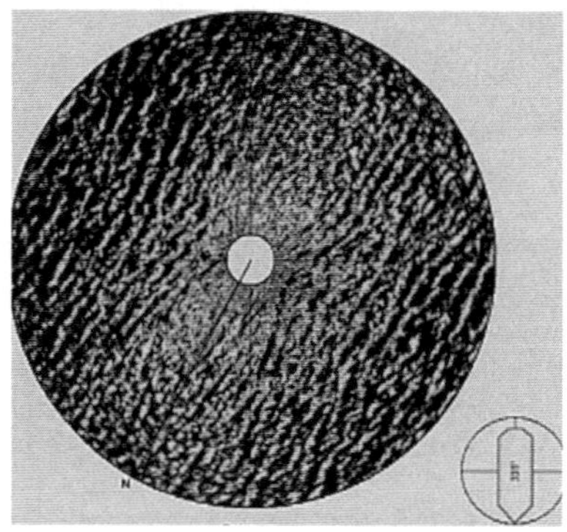

Figure 8: Backscatter Image

3.2 WAVE SHADOWING DESCRIBED

Wave shadowing shown in Fig 8a, b is acknowledged as the major source of error in the wave measurement process. Here large nearby waves, or waves with specific slope values, can block radar returns from more distant sections of the sea surface. Methods have been developed to address such errors, but they require identification of the boundaries of the regions where wave shadowing occurs.

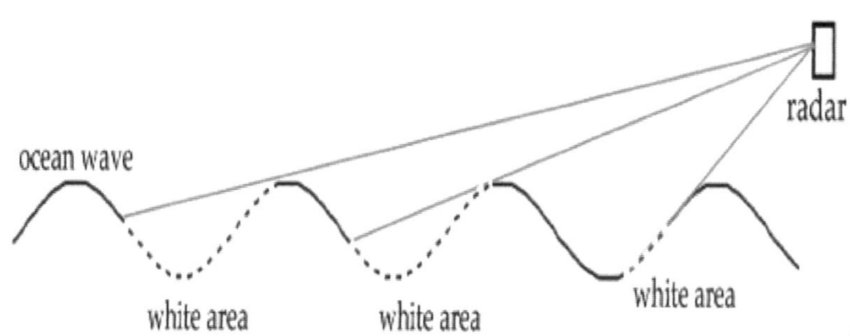

Figure 8a: Radar Wave Masking (or shadowing)

Thus, wave shadowing causes multiple regions of missing data over the region illuminated by the RADAR. Hence, the quantity described by the RADAR return signals is not a sea wave profile, but the product of a wave profile multiplied by a waveform (composed of multiple unit amplitude pulses) which is very different to the original waveform.

For small to moderate sized naval vessels various practical constraints mean that the extra RADAR antenna needed for QPP could often be positioned as low as only 15m above sea level. Figure 8 shows just how severe missing data for this case in a moderate to large sized sea and a RADAR range out to 1.2km.

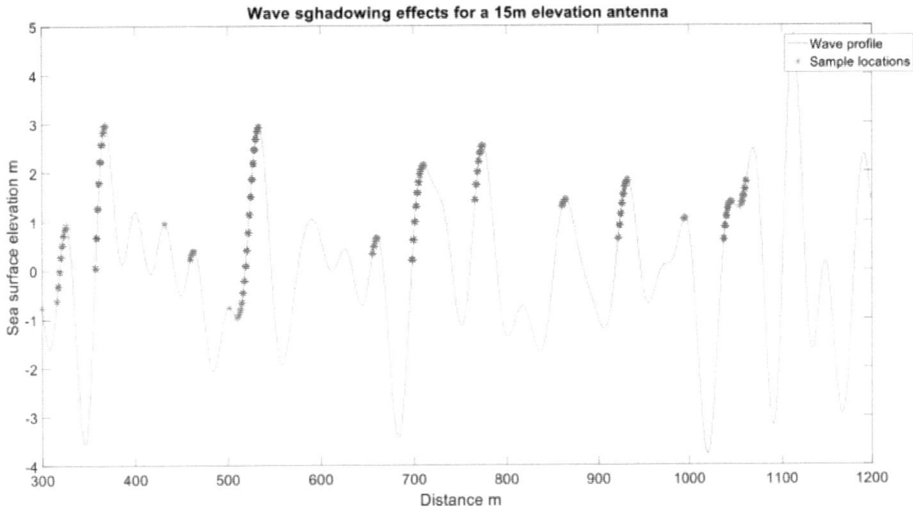

Figure 8b: Radar Wave Masking (or shadowing)

The standard approach to estimating wave profiles, **[7]** from RADAR data involves using a three-dimensional Fourier Transform (there are two wavenumber axes and one angular frequency axis). This method involves using the data from many separate RADAR scans. The wave profile evolves in shape as it propagates, hence wave shadowing is present at different places in different RADAR scans, thus by using data from many scans there is a degree of compensation of the wave shadow effects.

The effect of this phenomenon can be described as the modulation of the smooth wave profile by a sequence of pulses. This means the direction wave spectrum, $W_s(k, \varphi)$, of sea data that is significantly shadowed, is given by the convolution of the spectrum of the pulse waveform, $W_p(k, \varphi)$, with that of the true wave spectrum, $W(k, \varphi)$. Here, k is the wavenumber and φ the wave direction. Given that $W_s(k, \varphi)$ is used for prediction purposes, shadowing very significantly affects prediction accuracy. Consequently, various techniques have been developed to attempt to compensate for this effect **[9]**.

Typically, the short wavelength limit is used in such wave profiling RADAR work which neglects interference and diffraction effects. Diffraction introduces significant errors in the wave profile estimation process. To obtain experimental data on diffraction effects under highly controlled conditions, where the geometries involved are well defined, land-based experiments were conducted. In these a radar blocking fence was set up and the consequent radar returns which included the anticipated shadow region caused by the fence. The results showed that purely geometric modelling of the radar shadowing by the barrier is not adequate to describe the findings, indicating significant diffraction effects were present.

QPP radar trials were held at RNAS Culdrose, Prendannak Airfield in Cornwall (UK). A test rig composed of scaffolding was constructed upon which the QPP radar kit was installed (figure 9). The test rig featured a radar-blocking screen to reduce emissions towards the Control Tower. Cabling was strung from the test rig to the Predannak Control Tower.

Figure 9: QPP Radar Test Rig

The radar kit is vertically polarized, 8 ft, X-band radar system which has been procured for measuring the sea's surface shape around a ship in real time. The purpose of the current test is to demonstrate functionality of the ensemble of equipment with the wave defining software. To demonstrate the effects of wave shadowing; a radar reflecting fence was constructed at 100 metres from the radar installation. The second scenario was repositioning the fence another 100 metres. The third scenario was recorded without a fence. Wave shadowing could alter the wave profiling. The University of Exeter software is designed to account for the missing wave force contributions. Trials were conducted displaying the reflection from the Airfield with visual results. (Figure 10).

Figure 10: Inverse Radar Image(left) of Airfield (right)

3.3 BACKSCATTER CONVERSION

Whether X-band or S-band radar systems are used, the concept is to measure and then from such "backscatter" data estimate the wave structures. Detailed explanation on the computation of wave forms is given in reference **[10]**. The definition furnishes wave radial velocity, amplitude, direction, and spatial speed. With wave definitions various computational methodologies may be applied, including both frequency domain and deterministic analyses.

In irregular wave fronts, the radar backscatter may miss wave frequency, period, and entire waves hidden behind a closer larger wave **[11].** As might be anticipated typically, the backscatter image resembles real-world wave forms.

3.4 WAVE PREDICTOR

This part of the QPP process takes the backscatter image and converting them into measured wave components (figure 11).

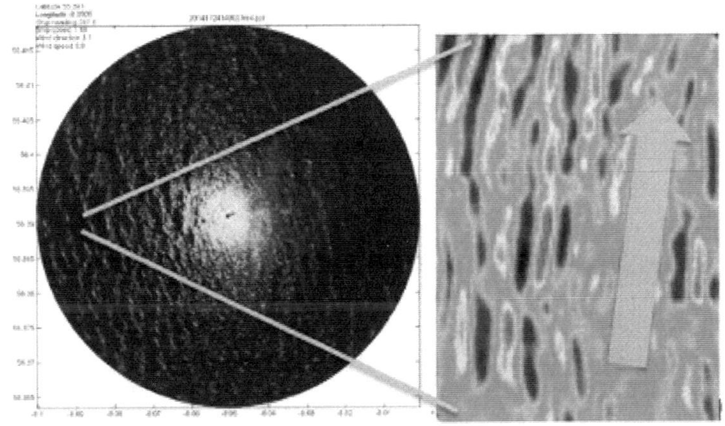

Figure 11: Backscatter converted to Wave image.

The measured waves contain most of the energies causing the ship to move. These waves are generalized into a distribution of wave frequencies at some remote location and propagates those components to the point of encounter by the ship. The ship is moving with forward speed and so the wave surface is varying both temporally and spatially. The theoretical model used to describe the propagation of the wave field is based on classical wave theory.

4. HELICOPTER – SHIP INTEFACE

4.1 SHIP VESSEL PREDICTOR

Once the measured wave elevations are known and the wave frequency model adjusted for the point of the ship encounter, a virtual motion of the ocean ship model is computed **[12].**

As shown in figure 12, the seaway frequency distribution is crossed with the ship's transfer functions (response amplitude operator) producing a distribution of harmonic components. These a summed over a short period of time to create 6 degrees of freedom time histories. The deck limits of the motion sensitive task are compared to determine if a particular heading is appropriate.

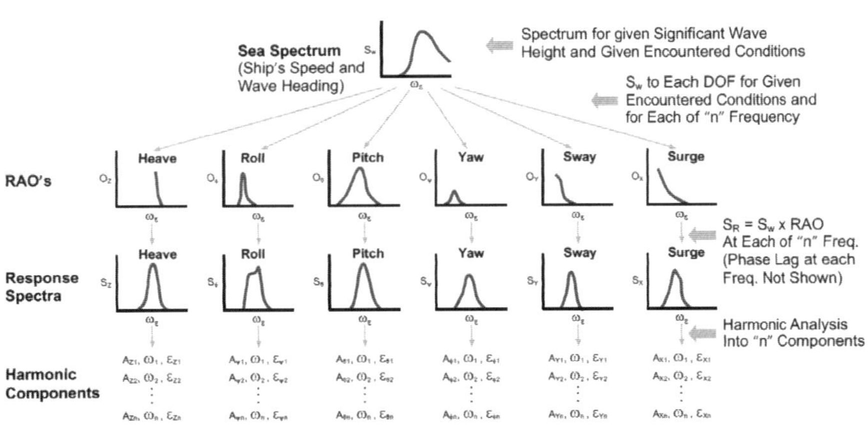

Figure 12: Conversion of Sea to ship motion

The ship response spectrum is created as the product of the RAO and the driving sea spectrum over the entire range of frequencies. In QPP the conversion to phase-preserved real-time ship responses is given by the Fourier transform in Equation 1:

$$F_i(j□□□) = W_{enc}(j□□□□V,□)\, H_i(j□□□) \quad (1)$$

where:

$F_i(j□□□)$: Fourier transform « i^{th}» mode

$H_i(j□□□)$: RAO (transfer function) for «i^{th}»

$W_{enc}(j□□□□V,□)$: Encountered Wave spectrum.

RAO; corrected ship heading □, and

Ship's speed, V

A typical time history equation is given by Equation 2:

$$A_Z = \sum_{n=1}^{k} \left(A_{Z_n} \cos\left(w_n - e_{Z_n} \right) \right) \quad (2)$$

Where:

A_Z: DOF amplitude

□: a circular frequency

e: phase angle

4.2 SHORT HORIZON MOTION ANALYSIS (LPD)

An important addition to the QPP system is the Landing Period Designator (LPD). This determines and displays real-time information about current and near- term future motion (1/4 cycle) of any vessel as a function of airframe operational limits. As it does not attempt long horizon time predictions, only rely on actual vessel motions, it can be used to provide an emergency last minute waive-off capability for QPP. Having the ability to measure real-time and near-future deck motions eliminates the need for excessively conservative roll and pitch deck limits **[13].** These limits are expressed as the ship's Energy Index, which is an empirical derivation representing ship motion as a function of aircraft limits to help guide air vehicles to safe deck landings regardless of the actual seaway.

The Energy Index formulation identifies ship's quiescent periods. These periods are interpreted in terms of aircraft and deck limit definitions. This index provides information on the motion that the ship could exhibit in the near-term future using real-time motion measurements of angular displacements and velocities and linear velocities and accelerations. It capitalizes on the rate at which a vessel can displace owing to natural hydrodynamic forces.

The energy index equation is based on an empirical derivation. It measures lateral and vertical velocities and accelerations as well as roll and pitch angular displacements and velocities weighted by dynamic coefficients. The equation is the least squares of the various parameters and terms representing real-time ship - aircraft interface motion.

Eq. (3) $EI = \sqrt{[a_1\phi^2 + a_2\phi'^2 + a_3\theta^2 + a_4\theta'^2 + a_5 z'^2 + a_6 z''^2 + a_7 y'^2 + a_8}$

where:

□□ .... Energy Index value
□ .... Roll ship angle
□ .... Pitch ship angle
□ .... Yaw ship angle
x .... Longitudinal displacement
y .... Lateral displacement
z .... Vertical displacement
a_1 to a_8 .... Weighted dynamic coefficients.

As indicated in Equation 3, the index contains acceleration, velocity, and displacement terms giving indications of the inertial content of the ship. It is an indication of the amount of motion the ship is likely to travel in the near-term future. All 8 degrees-of-freedom, their rates, and accelerations, are monitored and recorded. This includes the ship's vertical displacement, rate, and acceleration. as shown in Fig. 13.

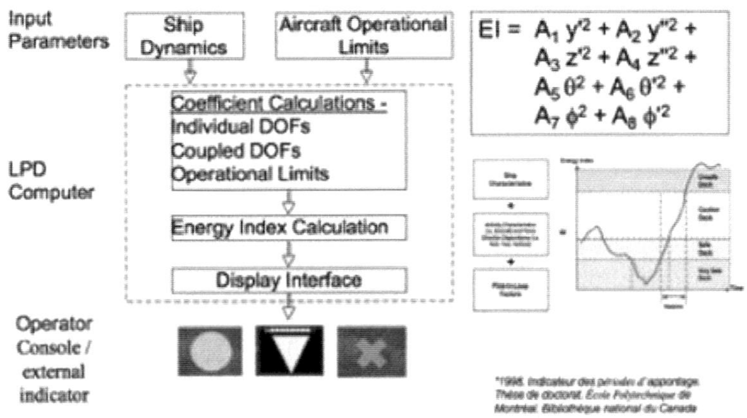

Figure 13: Deck Status and Risetime

The time required for the deck motion to rise from minimal motion (or very safe deck) to unacceptable motion is called the rise time. In terms of the EI scale, the rise time is defined as the period that is measured from the end of a green signal to the positive side of the red line. This is given as $(T_3 - T_1)$ as shown on Figure 11. The rise time is a thumb print characteristic of the ship's response and remains constant for each ship class.

The very safe deck is a special condition in which there is insufficient energy in the aircraft-ship system to raise the deck out of limit for some defined time-period or rise time, and it is this concept that was used in the simulators to indicate that the deck was safe to initiate landing [14].

Assuming the minimum average rise time for a given vessel remains constant, the aircraft descends within the rise time value and the aircraft is assured a recovery within the deck motion limits. Figure 14 displays graphically beyond static limit operations. The base envelope is taken from 10 knots of ship speed. Any points within the hour-glass structure are conditions within limits and contain no appreciable probability of out-of-limit deck motions. Outside of the structure contains motions which are considered by static reference as out-of-limit [15].

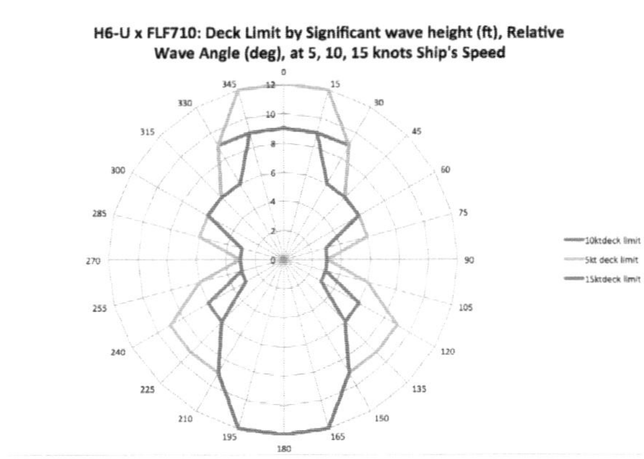

Figure 14: Safe Motion Operations

This is particularly important for robotic vehicles. A schematic representation of how the LPD is used for UAV auto recovery to signal the onset of the descent (and of deck quiescence) is displayed in Figure 15. Throughout the approach and initial hover, LPD is monitoring the deck. It is only at the low hover that the UAV autoland system would accept a Green Deck signal from the LPD. When the EI signals the onset of quiescence and through the uplink sets the air vehicle on its descent to the deck. Short horizon prediction is useful for predictions within 1/4 cycle of ship's motions. QPP is a dimension much longer in predictive horizons but uses the LPD least squares Energy Index to verify QPP's computation for launch and recovery [16].

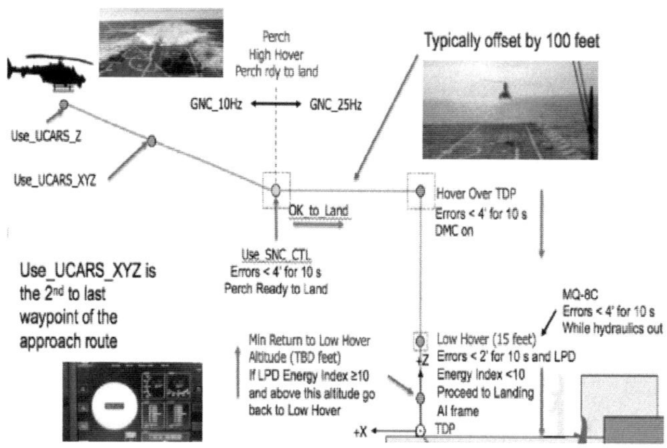

Figure 15: Notional Autoland System

In related trials it was possible to identify every period of low deck activity either by short or long horizon prediction. A simulated scenario is displayed in Figure 16. This compared to Figure 17 displaying actual launch and recovery opportunities.

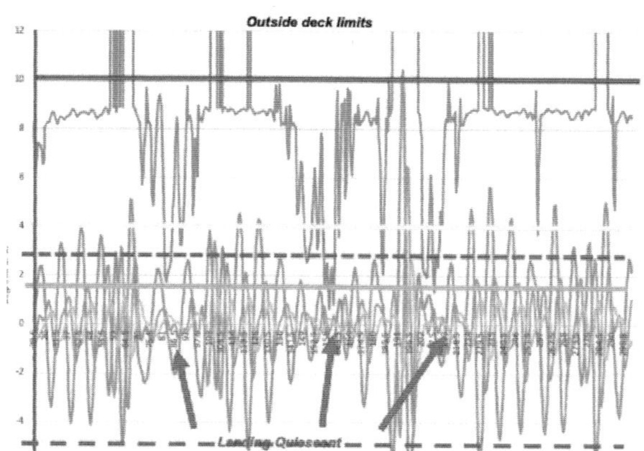

Figure 16: Landing Opportunities, simulated.

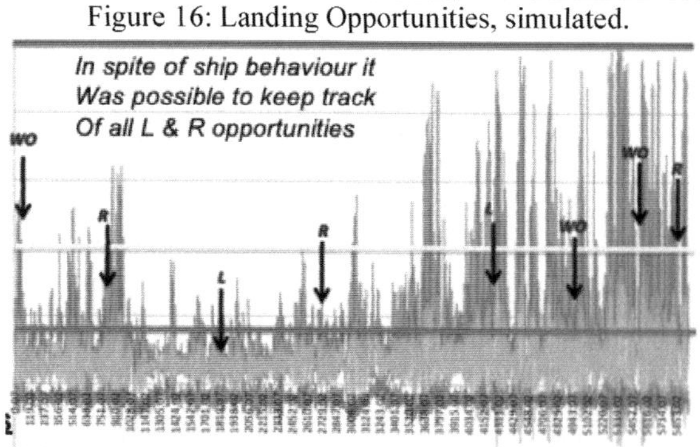

Figure 17: Landing Opportunities, recorded.

4.3 QPP BEST COURSE TO STEER

The QPP computation ultimately reduces to Command and Air Crew aides recommending appropriate heading and speed to achieve mission goals. One of the most significant mission goals is to determine the ship's heading and speed to complete motion sensitive tasks such as air vehicle launch and recovery. To best describe spatially the measured seas acceptable for flight operations, the energy index is used. It is a short horizon algorithm which is also used as verification that the capture or recovery is occurring within specified limits. As described earlier, one of the computational approaches is to use the QPP algorithms to measure in real-time the seaway around the Mother ship. The converted data are coupled with stochastic ship transfer function information to compute forecasted ship's motions. This is calculated for a given ship's heading and speed. It is estimated the speed-polar will represent the ship's motions against an operational event (e.g. air vehicle recovery) up to 2 km from the initial measurement position.

A speed-polar representation of the seaway to guide ship's command and deck hand activity, appears as a synoptic map created by the lay-over of various measured and computed parameters. Each parameter speed-polar lay-over is application specific computed as a function of the acceptability of ship's conditions to complete a specific motion sensitive task (figure 18). This is based on NAVSEA SEA 851, formally known as the Seakeeping Department, Active Operator Guidance System's post-processor [17].

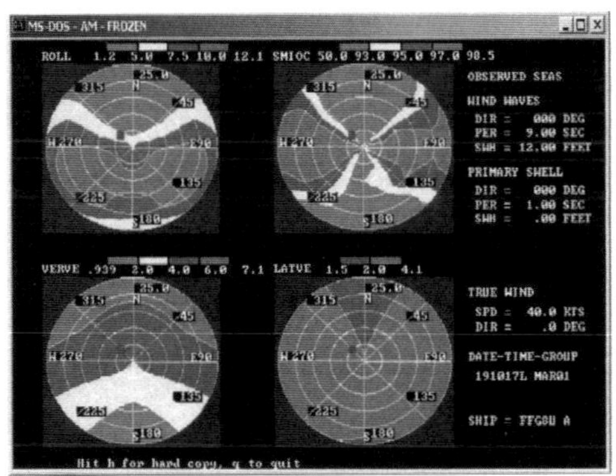

Figure 18: Parameter Map Speed Polar

A limit is defined by the impact that a certain ship motion condition may impose on the air/ship vehicle's structural integrity or that may result in undesired dynamic response. The sum of these lay overs produces an active operator's guidance map (figure 19). All the various QPP components either exist or require limited developmental modification.

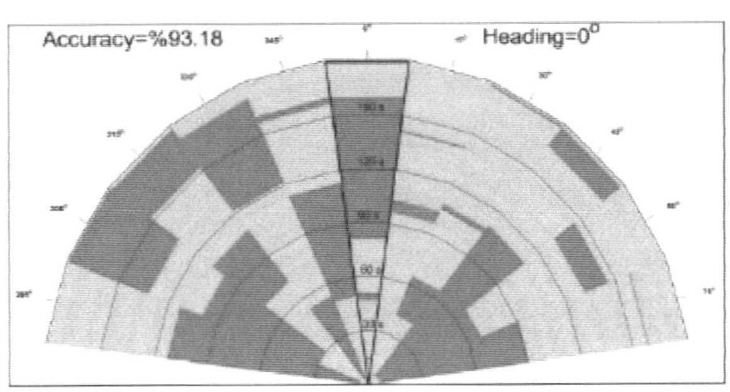

Figure 19: Best Course to Steer

By feeding the predicted vessel motions into the model of the operation being executed the QPP system can make estimates of the best time in the current timeframe in which to initiate operation. It also allows an operator to select the best course to steer. When QPP is first introduced the predictions will only serve as advice to the decision maker in charge of the operation. Thus, there is a need for a user advice interface. In the longer term when confidence has been built over time the whole process may become fully automated.

During an earlier phase of work on QPP concepts for such an interface were explored. The aim was to provide a simple to interpret display that allowed the required information to be absorbed by the operator in the shortest possible time. This is necessary because delays in decision making subtract directly from the available predict ahead time.

It was found that a polar form of visual display that showed different coloured bands advancing over time towards a central origin offered a suitable format for the user advice system. The bands were colour coded to represent the best (green) or worst times (red) to initiate the operation and their distance from the origin denoted the remaining time before the predicted vessel motion condition reached the vessel. The different polar directions indicted the different prediction histories that would occur for various ship headings adopted.

Thus, the advice system provides both advice on the best time to initiate and operation and the best course to steer. Figure 5 is a single screen shot at a given instant of time from such a user advice system. Multiple solution convergence methodologies may be offered using, for example, real-time long horizon forecasting alone or coupled with seaway real-time mapping providing stochastic (ship transfer function) based distance motion forecasts updated several times per second.

QPP can be applied to all ship motion sensitive activities. Throughout air vehicle approach and initial hover, QPP would be producing operator guidance maps allowing the ship to adopt a heading and speed acceptable for air vehicle recovery. The operator advice system would also predict the occurrence of quiescent periods and thus the best time to execute an operation. It is only during the final stages of recovery that the QPP short-horizon component would be used to verify that the deck was measured in real-time as being in a quiescent period condition.

5. CONCLUSIONS

A significant motivation for the development and deployment of VTOL UAS, is their exploitation in situations or conditions in which human exposure would by highly risky. Awareness and prediction of the ship's position and movement in the ocean is key to air operations by targeting the right time to deploy and recover. MoD and the University of Exeter's Quiescent Period Prediction (QPP) objective is to provide ship motion conditions and motion predictions with sufficient forecasted time (over a minute) to launch, recover and complete other motion sensitive tasks regardless of the seaway. The class of technology represented by the QPP approach is being researched by numerous highly competent academic groups. The fundamental concept is to measure remote sea surface profiles to predict the future wave forces acting upon a vessel. QPP's approach functions in all-weather conditions. The application of the QPP process has the potential of delivering a paradigm shift in the way vehicles and platforms approach and depart from a ship. The significant step will be the assembly of a Toughbook-based prototype in 2024. This will be followed by rigorous sea trials further defining the QPP system capabilities.

In a related but future QPP study activity, it was observed that certain ship deck condition tendencies were apparently related. Operating over deck whilst the ship is experiencing a quiescent ship motion period normally has the secondary effect of minimized air wake confusion. This is owing to fewer ship structure excursions into and out of the air flow. To better define deck airflow around the ship it is proposed to integrate a Doppler LIDAR instrumented federate. The purpose of this is to predict the future vessel air wake and look for quiescent periods in this paralleling the vessel motion QPP technique. Preliminary test results from a proof-of-concept trial, are briefly discussed. The ability to fully define a ship's motion and air wake condition as a function of a recovery device greatly enhances the overall system operational capability whilst fundamentally reducing the inherent risks associated with platform manipulation in high sea conditions.

6. ACKNOWLEDGEMENTS

The authors to thank our Team's sponsor, Commander Martin Howard, RN for his support and encouragement throughout this process. We wish to pay tribute of retired Dr. John Duncan former head of MoD's SBA, his staff and associated team members. This report is dedicated to the memory of NAWCAD's late Chief Engineer Robert Ernst, and to the late Dr. Peter J F O'Reilly and Dean Carico, fathers of the Dynamic Interface discipline.

7. REFERENCES

1. GALDORISI, G., and ROWLEY, J. (2022). "Engineering Unmanned Surface Vehicles Into an Integrated Unmanned Solution", *Naval Engineers Journal*. Vol 134, No. 1. Alexandria.

2. BELMONT, M R., CHRISTMAS, J., DANNENBERG, J., DUNCAN, J., DUNCAN, J M., and FERRIER, B., "An Examination of the Feasibility of Linear Deterministic Sea Wave Prediction in Multi-Directional Seas Using Wave Profiling Radar: Theory, Simulation and Sea Trials", *Journal of Atmospheric and Oceanic Technology*, Vol. 31, **No. 7**, pp1601-1614, 2014.

3. CROSSLAND, P., FERRIER, B., APPLEBEE, TR., "Providing Decision Making Support to Reduce Operator Workload for Shipboard Air Operations", *Royal Institute of Naval Architects*, London, 2004

4. HOENCAMP, A., "Helicopter- Ship Qualification Testing", *Ipskamp Printing*. Enschede, 2016

5. FERRIER, B., ERNST, R., "Resolving the conflict between ship design and UAV Launch and Recovery deck limits; the development of enhanced dynamic interface study tests", *American Helicopter Society Specialist Conference*, Mesa, 2019

6. STRESSER, M., et al. "On the Interpretation of Coherent Marine Radar Backscatter From Surf Zone Waves", *IEEE Transactions on Geoscience and Remote Sensing.* **Vol 60**, Toms River, 2022

7. SIMPSON, A., et al. "Wave-by-Wave Forecasting via Assimilation of Marine Radar Data", *Oregon State University Press*, Corvallis, 2019

8. NIETO, BORGE, J C., RODRIGUEZ, G R., HESSNER, K., GONZALEZ, P L., "Inversion of Marine Radar Images for Surface Wave Analysis". *J. Atmospheric and Oceanic Technology*, ***Vol. 21****, pp1291-1300*. 2004

9. BELMONT, M R., "Wave shadowing effects on phase resolved wave spectra estimated from RADAR backscatter: Part 2: The conflation technique and the deconvolution techniques for mitigation of wave shadowing effect" *Ocean Engineering.* ***Vol 267*** *113315-113330 (2023).*

10. *POULTER, E* M., SMITH, M J., MCGREGOR, J A., "S-Band FMCS Radar Measurements of Ocean Surface Dynamics". *Journal of Atmospheric and Ocean Technology.* ***Vol. 10,*** *No 4,* 2008

11. CHEN, Z., WANT, Z., ZHAO, C., XIE, F. AND HE, C., "S-Band Doppler Wave Radar System", *Remote Sensing Journal. MDPI.* ***Vol. 12****, No.1*, 2017

12. APPLEBEE, T R., BAITIS, A E., MEYERS, W G., "User's Manual for the Standard Ship Motion Program, SMP" *David W. Taylor Naval Ship R&D Centre Report DTNSRDC/SPD-0936-01*, 1981

13. BLACKWELL, J., and FEIK, R A., "A Mathematical Model of the On-Deck Helicopter/Ship Dynamic Interface (U)". *Aerodynamics Technical Memorandum 405*, Aeronautical Research Laboratory. Melbourne, 1988

14. FERRIER, B., MoD Quiescent Period Prediction LPD Software Testing Programme, UK080229-1. Bristol, 2008

15. MORRIS, E L., ZIENKIEWICZ, H K., BELMONT, M R., "Short term Forecasting of the Sea Surface Shape", *International Journal of Shipbuilding Prog.* **Vol. 45**, *No. 444*. Pp 381-400. Amsterdam, 1998

16. RINA 2022., "Ongoing relationship enhances sub rescue capability", *Warship Technology*. March 2015, London

17. FERRIER, B., Deck Motion Characterisation Report. *LPD NSRS/QPP Operations*. UKSEA130219. Bristol, 2013

8. AUTHORS BIOGRAPHIES

Michael Belmont, BSc, PhD, hold a chair at the University of Exeter UK, where he heads the Marine Dynamics group. He pioneered the new discipline of Deterministic Sea Wave Prediction which allows the short-term estimation of the details of wave driven vessel motion. This makes possible the prediction of quiescent intervals in the motion of naval platforms allowing the safe extension of the range of sea states for the launch and recovery of air, sea surface and sub-surface craft. Recently the same principals have been applied to the prediction of air-wakes over vessels. The goal of his research is the ability to allow the real time deterministic prediction of all the major environmental factors affecting maritime launch and recovery operations.

Jacqueline Christmas is an Associate Professor of Machine Learning in the Computer Science Department at the University of Exeter, UK. After many years working in industry, she obtained a PhD in the field of statistical pattern recognition in 2011. The focus of her work is the development of machine learning models for remote sensing, with an emphasis on explain ability and measures of uncertainty, especially for potentially infinite sets of real-time observations, such as are collected by long-term, continuously operating sensors. Her particular interest is in the use of inexpensive, off-the-shelf sensors, such as video, radar and still cameras, to remotely sense difficult dynamic environments in real time. Since 2012 this work has been directed towards the measurement and prediction of sea waves, based on physical oceanographic theory.

Bernard Ferrier, PhD, FRINA, FRAeS Dr. Bernard FERRIER is an Advanced Naval Systems Technical Lead at Fincantieri Marinette Marine. He is also a Systems Engineering and Technology Assistant (SETA) in NAVAIR with the primary task of integrating and deploying of UAS to sea, its technologies, and defining the operational limitations of the deck environment. Ferrier holds a Professor (hon) position at the UK's University of Exeter's College of Engineering, Mathematics and Physical Science where he is part of a development team creating a long-term ship motion and UAS launch and recovery tool for the Royal Navy. He is the Engineering Head of Syntek Technologies' Aircraft-Ship Dynamic Interface Lab in Fairfax, VA. He is a Fellow of the Royal Institution of Naval Architects and the Royal Aeronautical Society. He received his last doctorate in helicopter/ship interface engineering (Interface dynamique aéronef-navire) at the École Polytechnique de Montréal (Québec) Canada.

A STUDY INTO THE ENERGY AND PROPULSION SYSTEMS OF MODERN WARSHIPS

B A Steele BE(Mech) MIEAust CPEng MIMarEST, BAE Systems Australia - Maritime

This paper is the author's own work and the views expressed are not endorsed by BAE Systems Australia.

SUMMARY

The warships of the near-future will demand vast amounts of electrical power for supply of the new *directed energy weapons and effectors*. The *Power and Propulsion Systems* serving these ships, and delivering the large power demands, must maintain the integrity and effectiveness of present day warships to ensure the warfighting advantage can be brought to bare. Optimising the ship's average speed and range through efficiency, and demanding the best resilience to detection and classification are some factors in achieving this requirement.

The paper explores propulsion system architectures including purely mechanical, hybrid and all-electric propulsion systems, emerging technologies and alternative thinking on the efficiency, utility and survivability considerations of each. The concept of the *efficiency gap* is introduced to describe the difference in 'fuel mileage' achieved by different prime movers. An alternative hybrid-electric architecture, incorporating propulsion diesel engines, is offered to close the efficiency gap.

A secondary discussion explores the idea of the silent ship, and the design compromises necessary to achieve this – a very heavy and expensive ship. The paper introduces the concepts of *masking* or *cloaking* of the ship through acoustic signature manipulation rather than silent operation.

NOMENCLATURE

[Symbol]	[Definition] [(Unit)]
ASW	Anti-Submarine Warfare
CCG	Cross-Connect Gear
CODAD	COmbined Diesel And Diesel
CODAG	COmbined Diesel And Gas turbine
CODOG	COmbined Diesel Or Gas turbine
CODLOG-CC	Combined diesel-electric or gas turbine with cross-connect gear
CODODLOG-CC	Combined diesel or diesel-electric or gas turbine with cross-connect gear
CPP	Controllable Pitch Propeller
DDG	Destroyer, Guided Missile
DEW	Directed Energy Weapon
DG	Diesel Generator
EM	Electric Motor
EM_{max}	Maximum Ship Speed, EM propulsion
FCPP	Featherable Controllable Pitch Propeller
GT	Gas Turbine (engine)
GT_{min}	Minimum Ship Speed, GT propulsion
GT_{max}	Maximum Ship Speed, GT propulsion
GTA	Gas Turbine Alternator
IFEP	Integrated Fully Electric Propulsion
IPMS	Integrated Platform Management System
kg	kilograms (mass)
knots	nautical miles per hour (speed through water)
kW.h	kilowatt hours (electrical energy)
kWm / MWm	kilowatts / megawatts (mechanical power)
kWe / MWe	kilowatts / Megawatts (electrical power)
LPC	Loss of Power to Command
MRG	Main Reduction Gear
MSFC	Mass Specific Fuel Consumption

MW	Megawatts
nm	nautical mile
NRE	Non-Recurring Engineering (cost)
PDE	Propulsion Diesel Engine
PDE_{min}	Minimum Ship Speed, PDE mode
PDE_{max}	Maximum Ship Speed, PDE mode
PLF	Partial Electrical Failure
PM	Propulsion Motor
PSMR*	Propulsion and Steering System Redundancy in Separate Compartments (Lloyd's notation)
QPS	Quality of Power Supply
RAN	Royal Australian Navy
RPM	revolutions per minute (rotary speed)
TLF	Total Electrical Failure
UK-RN	United Kingdom – Royal Navy
URN	Underwater Radiated Noise
USN	United States Navy

1. INTRODUCTION

This paper acknowledges that near-future warships will require high capacity electrical power systems to enable the integration of emerging *directed energy weapons and effectors*. Parametric comparison and industry intelligence from public domain, and the author's own experience and analysis, are used to identify and explore different architectures and arrangements, applying technologies in different ways to describe an optimised *Power and Propulsion System* solution.

The narrative begins with a review of the basic *Power and Propulsion System* architectures of modern warships and considers the suitability of each type for meeting the electrical power need. A trade-off activity explores the options and nominates a preferred architecture on the basis of capability, including characteristics such as stealth (low noise), utility and flexibility, vulnerability and survivability.

The nominated architecture is allocated an arrangement and main machinery is selected. Analysis of propulsion system fuel consumption identifies a large efficiency problem (described as an *efficiency gap*) brought about by the nominated arrangement, and an alternative solution is offered to overcome this problem. Additional utility, providing increase in maximum ship speed, is also offered.

A secondary discussion challenges the principle that a *stealth ship* must be a 'silent ship', acknowledging that this paper has selected an architecture that permits very quiet electric propulsion at low ship speed but somewhat noisier mechanical propulsion at higher speeds. The concepts of *masking* or *cloaking* of the ship through acoustic signature manipulation are offered as an alternative to silence by design, with potential savings in complexity, weight and ultimately cost of the ship.

2. PROPULSION SYSTEM SELECTION AND PLANT ARRANGEMENT

2.1 POWER AND PROPULSION ARCHITECTURES

This section describes a number of configurations of the three propulsion architectures: All-Electric, Mechanical, and hybrid-electric. Figure 1 provides a visual aid to the following system configuration descriptions.

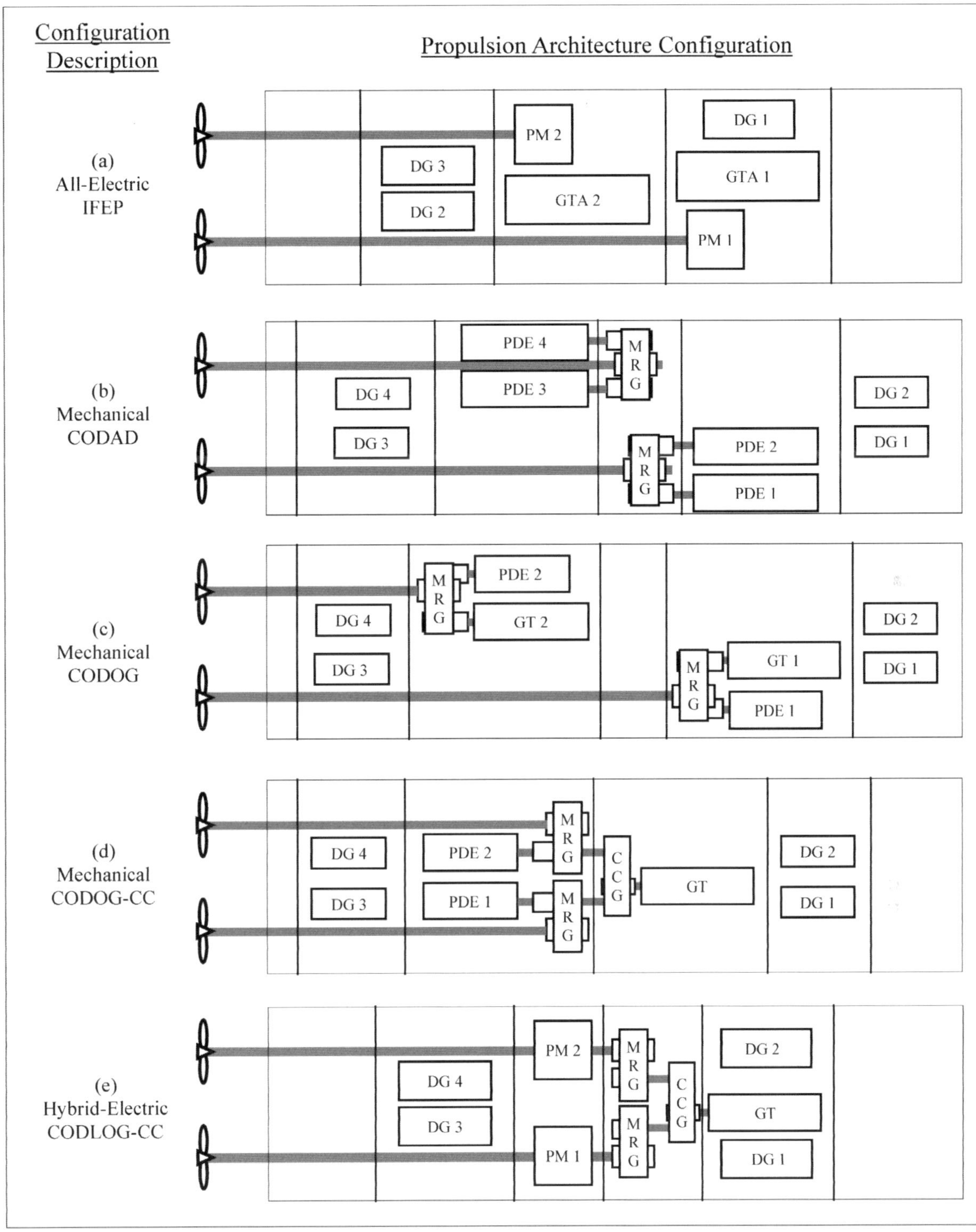

Figure 1: Plant arrangement for various power and propulsion architectures (plan view)

2.1 (a) The All-Electric Ship

The *all-electric ship* configuration optimises the power and propulsion system for high levels of flexibility, efficiency and electrical capability. Integrated Fully Electric Propulsion (IFEP) architectures place vast quantities (40 MW+) of electrical power, served from a number of *diesel generators* (DGs) and *gas turbine alternators* (GTAs), onto a central high-voltage network that can be dispatched for propulsion motors or other electrical consumers (or weapons/effectors) on command. The generating plant and propulsion motors are normally sized to the maximum speed requirement for the ship and the (much smaller) hotel services and combat/weapons loading.

IFEP systems utilise large *electric propulsion motors* (20 MW+), with one or two motors direct coupled at the forward end of each shaft. No reduction gearing is necessary, power transmission is performed in the power electronics. Electric propulsion can offer incredibly quiet ship operation, with the absence of mechanical gearing, presenting very low levels of noise and vibration.

Examples of *all-electric* ships with IFEP architectures include the UK-RN Type 45 *Daring Class* Destroyers, UK-RN *Queen Elizabeth Class* Aircraft Carriers, and the USN DDG 1000 *Zumwalt Class* Destroyers.

2.1 (b) The Mechanical Propulsion Ship

The *mechanical propulsion* ship is the traditional architecture applied by most of the world's navies. These architectures may incorporate multiple prime movers and combining gear sets, cross-connect systems and other novel configurations for high performance, high efficiency and high integrity (resilience) in propulsion. As the ship's installed power generation and distribution are segregated from the propulsion system there is no option to augment the ship's electrical power through the prime movers of the propulsion system. The propulsion plant is sized for efficient cruising and the maximum speed requirement for the ship (32 MW+), the power generation and distribution plant is sized for the hotel services and combat/weapons demands (3 MW+).

Common *mechanical propulsion* architectures in modern warships include the *combined diesel and diesel* (CODAD); *combined diesel and gas turbine* with independent shafts (CODOG), or with cross-connect gear (CODOG-CC).
Examples of each are the UK-RN Type 31 *Inspiration Class* Frigate, RAN *Hobart Class* Destroyer, and RAN *Anzac Class* Frigate, respectively.

Other less common architectures offer great power and ship speed, but low efficiency, and include the *combined gas turbine and gas turbine* (COGAG) of the USN *Arleigh Burke Class* Destroyer.

2.1 (c) The Hybrid Electric-Mechanical Ship (hybrid-electric)

The *hybrid-electric ship* was first conceived to enable ultra-quiet electric propulsion capability, for enhanced submarine hunting capability (anti-submarine warfare, or ASW), without consideration to flexibility of power network and consumers. Examples of earlier hybrid-electric ships include the UK-RN Type 23 *Duke Class* Frigates. This approach was highly successful, and despite their age, these ships remain superior ASW platforms.

Later iterations of the *hybrid-electric concept* have evolved to offer the ideal synthesis of flexibility in power allocation and ultra-quiet operation (as per *all-electric* architectures), whilst also maintaining the integrity of the *mechanical propulsion* solution. The installed electrical capacity for this type is normally much less (10 MW+), but the power can still be allocated toward propulsion motors or other electrical consumers, on command, as with IFEP systems described above.
Examples of this approach include in the German F125 *Baden-Württemberg* Frigate, Republic of Korean FFXII *Daegu Class* Frigate, and the UK-RN Type 26 *City Class* Frigate (and its variants including future RAN *Hunter Class*).

Hybrid-electric systems utilise relatively small (by output) *electric propulsion motors* (2 MW+ per shaft) for low speed loitering and patrol, to a maximum speed of around 14 knots. Whilst more powerful motors can provide higher speeds, the amount of additional electrical power required to propel at these speeds will demand an oversized power generation system – the power and efficiency arguments will be explored later in the paper.

The *electric propulsion motor* is positioned on the shaft line (either directly on, or over the shaft), or forward of the reduction gear and connected through the bullwheel by a quill shaft arrangement. In this way, the motor operates at shaft speed and no gearing is necessary as with the IFEP solution.

Power is augmented for propulsion purposes by a high-output mechanical prime mover (usually by one or more *gas turbine* engines, 32 MW+) and mechanical reduction gear transmission for traveling at high speed.

2.2 UTILITY AND FLEXIBILITY

The emergence of high power sensors, magnetic weapons and *directed energy weapons* (DEW) (microwave, laser), places additional high demands on the *energy and propulsion system* of the near-future warship. At full effector utilisation, the entire generating capacity of the ship may be consumed across the hotel services, propulsion and effector suites with no further redundancy remaining in the generating system.

Without the oversized power generation plant of an electric propulsion system, the *mechanical propulsion ship* cannot natively support the energy demands of proposed high power effectors. Options exist for these ships, such as shaft-alternators (power take-off), electrical storage, or load-shedding, but these encroach on hybrid-electric architectures and so will not be explored here. The reduced electrical flexibility, overall efficiency and electrical capacity of these systems comes with considerable advantages, where the segregation of power and propulsion improves the survivability argument. Mechanical propulsion ships can combine the power of multiple prime movers onto one shaft (*AND* logic) and achieve very high ship speeds if so designed.

A high level of utility and flexibility are delivered by the *all-electric* architecture making these ships ideally suited to the incorporation of high power sensors, effectors and weapons. Flexibility in power generation and load management combine to provide a highly fuel efficient system. Despite these advantages, the vulnerability and survivability aspects of this type demand consideration - the integrity of the power system is absolutely critical, as the loss of power (not uncommon on a complex warship) will mean total loss of propulsion and manoeuvring capability.

The *electric-hybrid ship* carries additional complexity in mechanical systems and controls, and loses many of the flexibility benefits of the electric ship – such as prime mover placement and acoustic isolation. Like the *all-electric ship* this type offers redirection of large power reserves from propulsion to other consumers, but also like the *mechanical propulsion ship* it maintains the high level of propulsion integrity in the degraded ship state. This architecture could be considered to be the 'best compromise' solution.

2.3 VULNERABILITY AND SURVIVABILITY

Ship propulsion and manoeuvring capability must be of the highest integrity, with absolute resilience assured against damage caused by impact (collision / allision), adversary action (weapon strike / shock), or mechanical failure. Redundancy is engineered into the power and propulsion systems such that the ship can continue to operate, albeit in a deteriorated capability, in such an event – refer to *class rules* such as Lloyd's Register notation PSMR* describing Propulsion and Steering System Redundancy in Separate Compartments.

Exploring the IFEP system of the *all-electric ship* it is important to understand the impact to propulsion should power be lost. In the instance that a ship of this type suffers *total electrical failure* (TLF), then propulsion and manoeuvring will be lost and the ship will experience Loss of Power to Command (LPC) [Ref 1]. Such an occurrence in a confined waterway could be disastrous, as recently experienced (March, 2024) with the cargo ship Dali at the Francis Scott Key Bridge in Baltimore, USA [Ref 2]. If this occurs in a combat situation, the ship could become defenceless, unable to fight or flee.

Electrical systems are complex on all warships; on IFEP ships, the complexity is a magnitude higher. The electrical generating and distribution system spans two (or more) islands and many distribution zones. Network integrity, the *quality of power supply* (QPS) and other parameters must be controlled such that the network remains stable and responds predictably to load changes, instability or machinery failure. The subdivision of the distribution network into multiple "islands" can instil some resilience, where any power failure is limited to only part of the overall system (*partial electrical failure, PLF)*. Island mode adds resilience but reduces the efficiency advantages of the IFEP arrangement, and therefore this protection will only be used when operational risks are high, such as confined waterways or adversarial threat. These risk mitigations reduce this risk to a partial failure, but the vulnerability remains present, particularly when the ship is at a low state of readiness when a failure event occurs.

The *mechanical propulsion ship* carries power capacity sufficient to sustain the installed ship electric loads (hotel, combat system only) with N+1 redundancy. The propulsion system is wholly independent of the power generation system and thus, any PLF or TLF event will not impede the ability of the ship to propel and manoeuvre (within limits).

In the same way as an IFEP ship, the *hybrid-electric ship* can be configured with electrical subdivisions, "islands" and zonal power management, but it also has redundancy in propulsion by means of the mechanical plant. In the event that a ship of this type suffers a TLF it is feasible that the mechanical propulsion plant can continue to operate and *ride-through* the failure, maintaining propulsion and manoeuvring despite the loss of electric power. Again the risk can be reduced to a PLF by use of segregated "island modes", however the systems must be designed to enable continuous operation in this worst case situation. Capability to *ride-through* an electrical failure is dependent on adequate gravity-fed fuel supply, low voltage power to machinery control, steering hydraulics, and engine starting (if required) - by stored compressed air.

2.4 TRADE-OFF STUDY

A trade-off study is presented at Figure 2 below, identifying that each of the three architectures offers different strengths and weaknesses and the selection must apply weightings to those characteristics more important to the ship's purpose as defined by the customer.

The engineer selecting the architecture and making machinery selection may also be influenced by the availability of suitable equipment and commercial/politically motivated procurement decisions beyond the technical considerations described below.

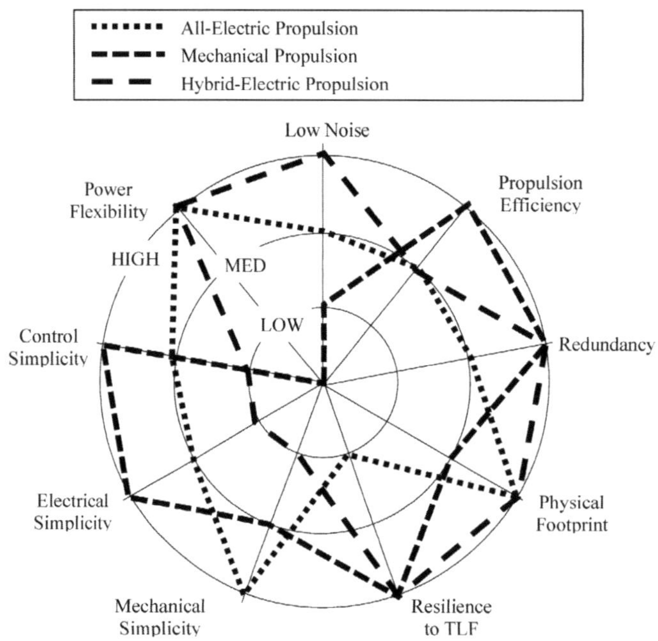

Figure 2: Trade-off analysis for propulsion architectures

2.5 PROPULSION ARCHITECTURE SELECTION

Having explored the relative complexity of the three main power and propulsion architectures, described the advantages and limitations of each type and traded-off the strengths and weaknesses of each type, it is time to select an architecture.

This paper assumes a design requirement for very low underwater radiated noise (URN) when operating at low to medium ship speeds, reducing self-noise to improve *anti-submarine warfare* (ASW) capability and avoiding detection and classification by potential threats. The combined needs of efficiency and quiet operation will tend toward an *all-electric*, or *hybrid-electric* propulsion system.

When contemplating the ship's designed mission scenarios, a customer is likely to demand a power and propulsion system with high levels of resilience to TLF, redundancy and ultimately survivability, which are hard to guarantee with an all-electric ship. Documented cases of IFEP naval ships suffering major power failures (sometimes necessitating towing or repair abroad) are not unheard of [Ref 3, Ref 4].

The author has therefore selected a *hybrid-electric* architecture for further analysis as this represents a great balance of efficiency, capability, utility and survivability. An inherent propulsion efficiency concern remains with this architecture, a study is performed later in the paper to address this short-coming. Contrary to IFEP failures, anecdotes of hybrid ships maintaining propulsion after total electrical system failures are highly encouraging.

3. PROPULSION SYSTEM DESCRIPTION

3.1 MAIN EQUIPMENT LAYOUT

The following analysis requires some definition of terms, equipment and architecture for the end-to-end *hybrid-electric* propulsion system. Figure 3 provides an elevation view of a propulsion system - the Uptakes and Downtakes are included to illustrate the need to consider the routing of these bulky systems when allocating space for propulsion machinery.

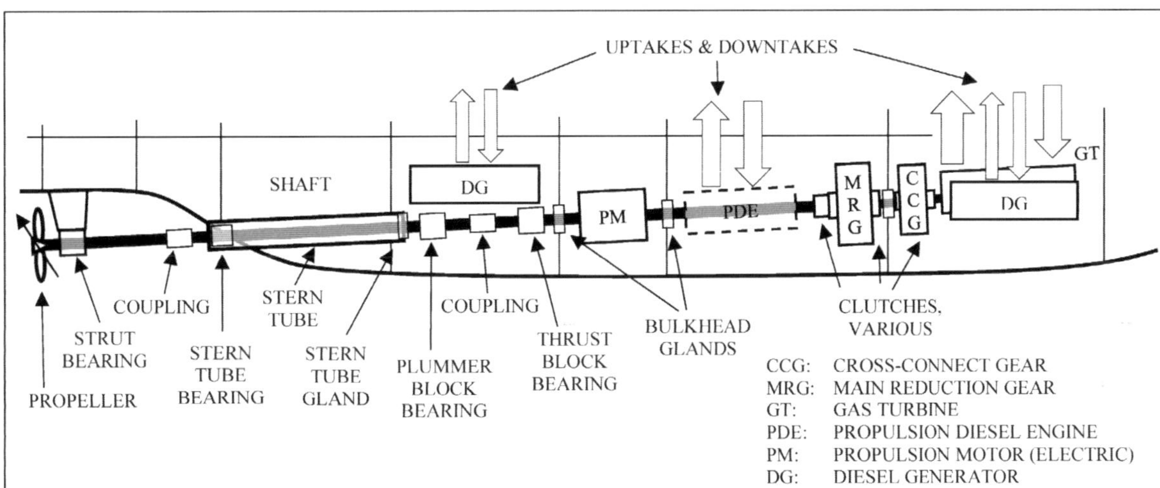

Figure 3: Plant arrangement for generic hybrid-electric propulsion system, elevation view

3.2 MAIN EQUIPMENT DEFINITION

3.2 (a) Propeller

– available as *fixed-pitch propeller* (FPP), *controllable pitch propeller* (CPP), and *featherable controllable pitch* (FCPP). Each type has its advantages, however of particular interest is the featherable CPP - by placing the blades at 90° to the shaft axis the blades will impart only a small torque on the shaft in the locked shaft condition. This capability means that the ship can steam on only one of the two shafts with no detriment, for economy or stealth.

3.2 (b) Cross-Connect Gear (CCG)

– facilitates the powering of two separate shaft lines from a single high output prime mover, such as a marine *gas turbine* (GT) engine, splitting the power, performing primary speed reduction, and reversing shaft direction as required to match propeller rotation.

3.2 (c) Main Reduction Gear (MRG)

– performs main speed reduction from the high speed output of the *propulsion diesel engine* (PDE) or from the CCG. Selection of gas turbine or diesel engine propulsion mode is made at this machine by configuring the onboard clutches as required.

3.2 (d) Gas Turbine (GT)

– provides power to the propulsion system necessary to facilitate high ship speeds. GTs provide high power in a small mass and volume, however the low-to-mid range efficiency of these machines is so poor that they are generally used as supplementary prime movers for high ship speed only.

3.2 (e) Propulsion Diesel Engine (PDE)

– provides power to the propulsion system for low speeds and cruising. PDEs provide relatively high efficiency across the power range but are limited in their maximum outputs, so a supplementary engine(s), such as a GT, is required to achieve high ship speeds. Few hybrid-electric architectures incorporate PDEs as they add considerable size, weight and complexity to the propulsion system.

3.2 (f) Propulsion Motor (PM)

– provides electric drive capability for the propulsion system. In a hybrid-electric system the PMs will provide high efficiency across the low ship speed range, operating at very low levels of noise and vibration such is as ideal for ASW.

3.2 (g) Diesel Generator (DG)

– produces electrical power for the ships network, servicing hotel loads, combat system / weapons and effectors, and propulsion motors. DGs incorporated for *hybrid-electric* systems are oversized in comparison with *mechanical propulsion* as the plant must service propulsion demands, but can be diverted as necessary to other high energy consumers such as the near-future *directed energy weapons* (DEW).

4. HYBRID-ELECTRIC PROPULSION SYSTEM OPTIMISATION FOR IMPROVED EFFICIENCY

4.1 PROPULSION ARRANGEMENT FOR STUDY

This study commenced with the knowledge that gas turbine engines are highly inefficient at partial loading, and the hypothesis that the ship would therefore suffer from high fuel consumption when operating at medium speeds through the water.

Configuration (e) at Figure 4 (copied from Figure 1) describes a Hybrid-Electric *combined diesel-electric or gas turbine with cross-connect gear* (CODLOG-CC) arrangement. This system applies *OR* logic to apply a single high-power gas turbine engine, driving a CCG, to two shafts via MRG, with shaft mounted propulsion motors.

Configuration (f) at Figure 4 introduces two propulsion diesel engines (PDE) to the Hybrid-Electric arrangement above, such that the change in fuel efficiency at medium ship speeds can be studied. The standard nomenclature describes the system as a Hybrid-Electric *combined* ***diesel or*** *diesel-electric or gas turbine with cross-connect gear* (CODODLOG-CC). A further alternative introduces *AND* logic to combine multiple prime movers for increased total shaft power and increased maximum ship speed.

Figure 4 shows the evolution from *Configuration (e)* CODLOG-CC arrangement, with the addition of two PDEs (dashed) to create *Configuration (f)*, described as CODODLOG-CC. Note the use of a cross-connect gear and a single Gas Turbine engine remains.

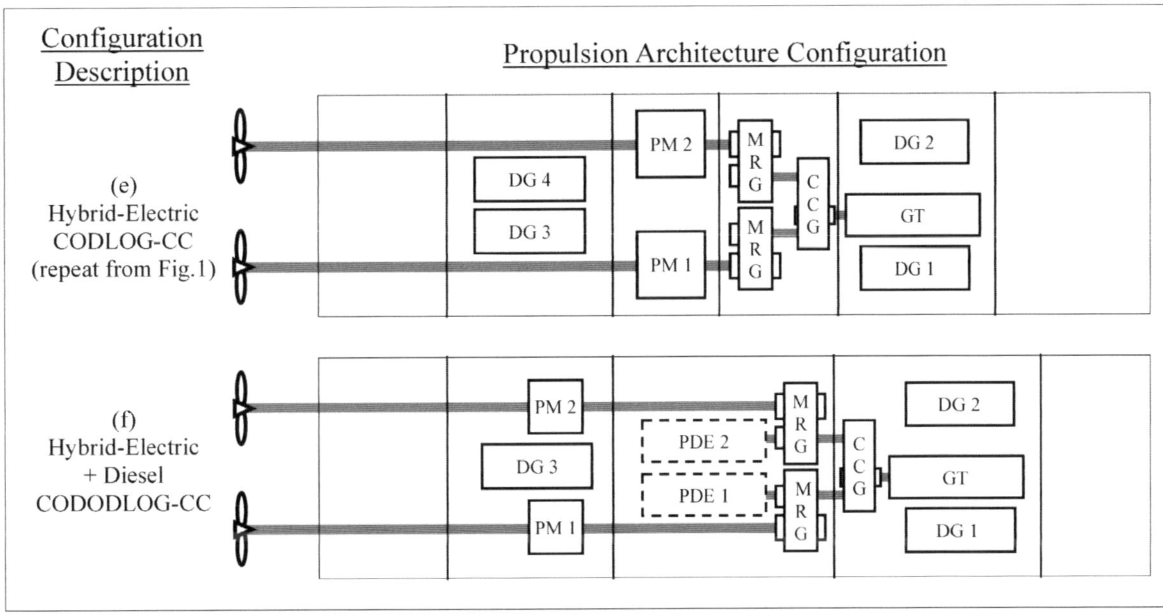

Figure 4: Plant arrangement for optimised hybrid-electric propulsion system (plan view)

4.2 PROPULSION MACHINERY ALLOCATION

The numerical model described in the next section requires the allocation of propulsion machinery for Configuration (e) and Configuration (f) as described in Figure 4. The selected machinery is detailed in Table 1.

Table 1. Plant schedule for hypothetical hybrid-electric propulsion system

Machine	Designation	Qty
Diesel Generator	MTU 20V4000M84, 2900 kWe @ 1800 rpm	3x (DG 1, DG 2, DG 3)
Propulsion Diesel Engine	MTU 20V1163M94, 7400 kW @1275-1325 rpm	2x (PDE 1, PDE 2)
Gas Turbine Engine	Rolls-Royce MT30 36MW @3600rpm	1x (GT)
Propulsion Motor	Leonardo DRS permanent magnet motor/generator 2.1 MW	2x (PM 1, PM 2)
Cross-Connect Gear	RENK-MAAG 1x 40 MW input, 2x 20 MW output with reversing idler incorporating SSS Clutches	1x (CCG)
Main Reduction Gear	RENK-MAAG 1x common input to 20 MW (CCG/PDE), 1x output (0-200 rpm)	2x mirrored to shaft axis (MRG 1, MRG 2)

5. PROPULSION SYSTEM EFFICIENCY AND RANGE MODELING

A *numerical model* was constructed in Excel to enable the following analysis. Input data was sourced from public sources. Powering curves for the ship were approximated from parametric analysis of similar ships. Fuel consumption curves for the selected diesel engines, diesel generators and gas turbine engines were defined from manufacturer data, with nominal power conversion efficiencies applied.

For hybrid-electric drive, a load factor was applied to the diesel generator (representing 'non-propulsion' loads) to define the load state of the machine for efficiency mapping, and to trigger the starting of the next machine when threshold load is reached.

5.1 POWERING REQUIREMENT

For the purposes of analysis, a powering curve was produced, shown at Figure 5. This curve is representative of a nominal load case of 7500 tonne surface combatant, and closely mimics the *cube law* for propulsion resistance where:

(eqn.1) $P \propto v^3$; or Power proportional to (speed through water)^3

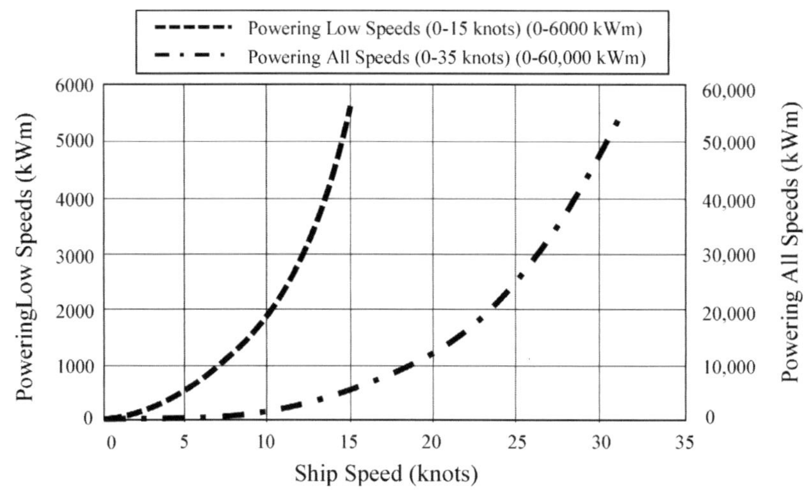

Figure 5: Powering requirement for a hypothetical 7500 tonne surface combatant.

5.2 PRIME MOVER SELECTION

Prime movers are selected to provide the following capabilities:

- Quiet operation at low speeds – optimised for reduced signature and self-noise during ASW operation
- Efficient operation at cruise speeds – optimised for efficiency and range
- High power for top speed – optimised for maximum power into the water, efficiency less important

The numerical model describes operation in pure electric drive mode (up to ~14 knots), diesel propulsion mode (up to ~21 knots) and gas turbine mode (up to maximum speed ~30 knots).

5.2 (a) Electric Propulsion Motors

Electric propulsion motors, particularly those mounted directly on the shaft line, offer incredibly quiet operation. As can be seen in the powering plots (Figure 5), a high-speed cruise of 20 knots demands about 13 MW of mechanical power, which equates to about 14.5 MWe after motor and conversion losses are considered. Whilst an electric motor can be scaled to deliver this output with little impact to its physical size and weight, considerable size and weight must be added to the electrical generation system and the distribution network to power it.

5.2 (b) Diesel Generators

Diesel generators of 2.9 MWe @ 60Hz are nominated as these machines represent the highest output for high-speed diesel generators currently available off-the-shelf. For reasons of system redundancy and capability balance, the electric propulsion train is usually limited to the power output of two such generators (i.e. about 6 MWe or 5.7 MWm) granting approximately 14 knots.

5.2 (c) Gas Turbine Engines

Gas Turbine engines deliver very high power output for (relatively) low size and weight. This *power density* comes at the cost of thermal efficiency, which is particularly evident as poor MSFC at low to medium loads.

5.2 (d) Propulsion Diesel Engines

Propulsion diesel engines (PDEs) sized for cruise speed up to 21+ knots are added to the analysis (dashed features in Figure 4, *Configuration (f)*) in an attempt to close the efficiency gap. The PDEs extend the ship's service range based on fuel efficiency and depth of fuel reserves. A speed of 21 knots corresponds to 14.8 MWm, conveniently aligning with the maximum continuous rating of the 20-cylinder variant of the MTU 1163M94 diesel engine. Above 21 knots, the efficiency gap between GT and PDE begins to close and the physical size and weight of diesel engines becomes prohibitive.

5.3 PRIME MOVER EFFICIENCY – MASS SPECIFIC FUEL CONSUMPTION

Figure 6 describes the relative thermal efficiency for each of the described prime movers described herein by the term *mass specific fuel consumption* (MSFC) in the units of kg per kilowatt-hour (kg/kW.h).

Consider the following example of the poor efficiency in the GT at low to medium loads.

At a ship speed of 15 knots, the turbine will experience a load of about 5700 kWm, and will consume fuel at the rate of about 0.35 kg/kWh. In this same scenario if electric propulsion was available, the ship would consume no worse than 0.22 kg/kWh and on propulsion engines little worse than 0.20 kg/kWh. At lower loads the GT efficiency only gets worse. This discrepancy is termed the *efficiency gap*.

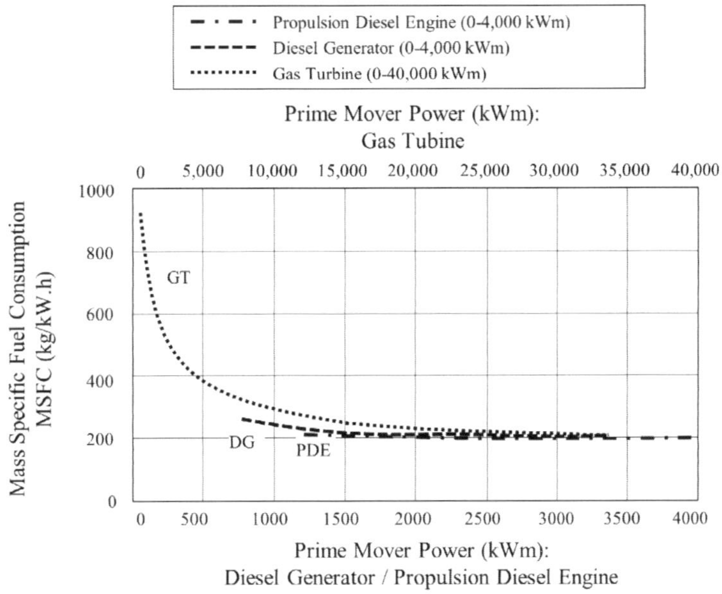

Figure 6: Mass specific fuel consumption for selected prime movers

6. DECLARING THE EFFICIENCY GAP

6.1 EFFICIENCY AS FUEL MILEAGE

In the previous sections the ship's powering curves and fuel consumption curves were introduced. The concept of the *efficiency gap* was declared to show the inherent inefficiencies in the application of a single gas turbine to low and medium speed ship's operation.

The automotive industry measures efficiency as 'fuel mileage' in the units of *litres per 100 kilometres* (L/100km). This approach allows the prediction of fuel range for a vehicle over a given set of operating conditions. The same approach can be used for a ship, where the fuel consumption in *kilograms per nautical mile* (kg/nm) can be plotted against ship speed and its range determined.

A *hybrid-electric ship* comprises multiple prime mover and transmission systems, each with efficiency and power characteristics matched to certain envelopes of ship speed. The charts at Figure 7 and Figure 8 were produced to
visualise the efficiency gaps at the main change-over points in the system, being the minimum and maximum power output for each prime mover type.

6.2 EFFICIENCY IN HYBRID-ELECTRIC ARRANGEMENT

Efficiency gaps in Configuration (e): *combined diesel-electric or gas turbine with cross-connect gear*

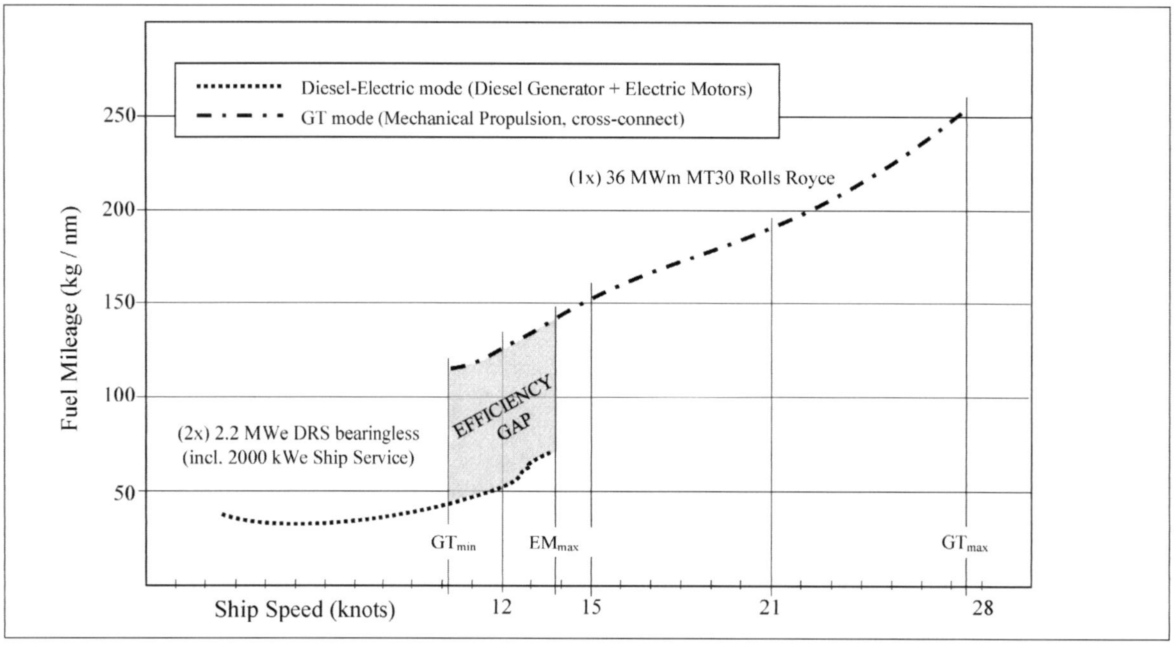

Figure 7: 'Fuel Mileage' plot for a typical hybrid-electric propulsion arrangement

First note the following:

- At speeds up to EM_{max} the ship will operate with near-silent propulsion, with much of the installed diesel generator power directed to the propulsion system.
- The GT can be used to propel the ship at all speeds, a nominal GT_{min} is used to bound the study – fuel efficiency continues to deteriorate rapidly below the nominated minimum.
- GT mode liberates the ship's diesel generators to supply other systems such as high-energy effectors instead of propulsion.

From Figure 7, consider the efficiency gap for ship speeds between GT_{min} and EM_{max}.

- At 12 knots the ship will consume just over 50 kg/nm in diesel-electric mode, or 130 kg/nm in gas turbine propulsion – this is a penalty of 80 kg/nm or 260% increase in fuel consumption to operate at the same speed!

When the ship is steaming at speeds higher than EM_{max} the GT propulsion system must be used.

- At 15 knots the fuel consumption will leap to 155 kg/nm - the fuel consumption will *triple* for the addition of only 3 knots!
- At 21 knots the fuel consumption will be 190 kg/nm, operating on the gas turbine.

6.3 EFFICIENCY IN HYBRID-ELECTRIC ARRANGEMENT WITH PROPULSION DIESEL ENGINES

Efficiency gaps in Configuration (f): *combined* ***diesel or*** *diesel-electric or gas turbine with cross-connect gear*

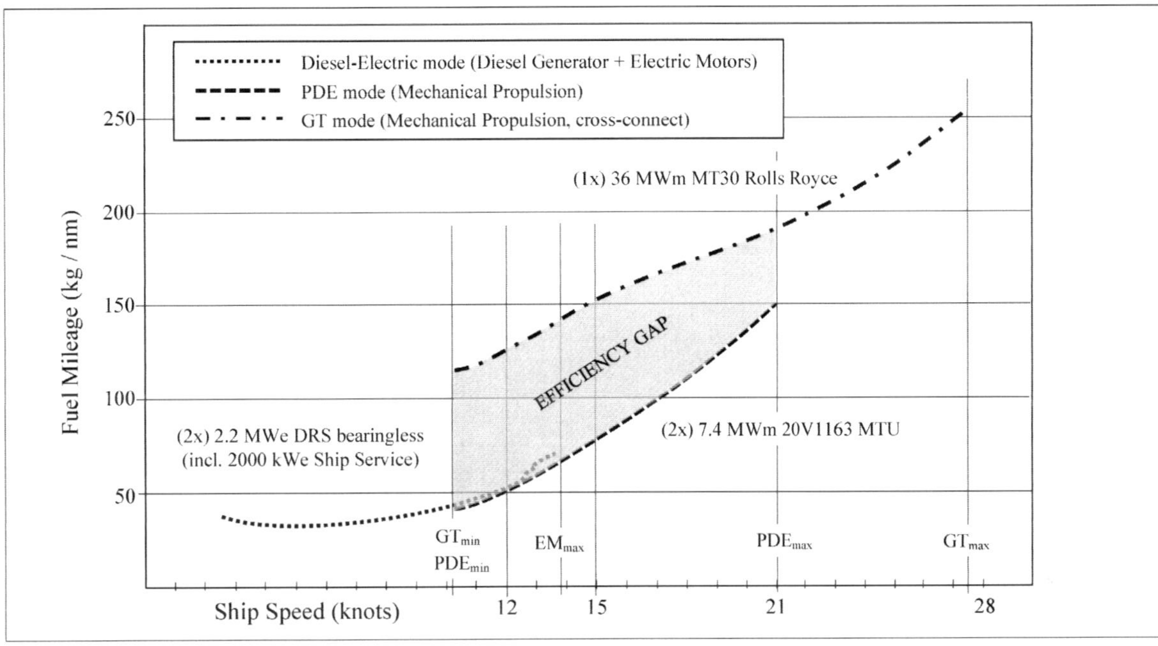

Figure 8: 'Fuel Mileage' for a hypothetical hybrid-electric propulsion arrangement including PDEs

The addition of propulsion diesel engines closes some of the gap.

From Figure 8, consider the efficiency gap for ship speeds between GT_{min} and PDE_{max} in the same way as before.

- At 12 knots the consumption remains around 55 kg/nm for diesel-electric or diesel mode.
- At 15 knots the ship will consume 80 kg/nm in diesel mode (versus 155 with no PDEs). The ship will consume nearly *double* the amount of fuel at this speed in the absence of PDEs.
- At 21 knots the ship will consume 150 kg/nm in diesel mode (versus 190 with no PDEs).

For ship speeds higher than the nominated PDE_{max} the efficiency gap continues to close, this is a consequence of GT efficiency improving with increasing load. If the reader was to extrapolate the *PDE mode* curve in Figure 8 they would see it transition into the *GT mode* curve at about 26 knots, this is the point that the GT thermal efficiency approaches that of the PDE. It is clear however that the ship would not achieve these high speeds without the high power available from the GT.

6.4 FUEL RANGE DETERMINATION

To determine the operating range of the ship it is necessary to understand both the rate of consumption (described previously as fuel mileage) and the capacity of the fuel tanks. For the purposes of this analysis a fuel bunker size of 800,000 kg has been selected.

The model calculates fuel range for the ship travelling at constant speed. The model can also predict fuel range given a specific operating scenario, where the time spent at various speeds (propulsion loads) is forecast – however this is omitted from the paper for simplicity.

Figure 9 compares Fuel Range to Continuous Ship Speed for the two configurations (e) and (f). Note that diesel-electric propulsion is available only for speeds up to 14 knots, diesel propulsion is available for speeds up to 22 knots, and gas turbine is used for all higher ship speeds.

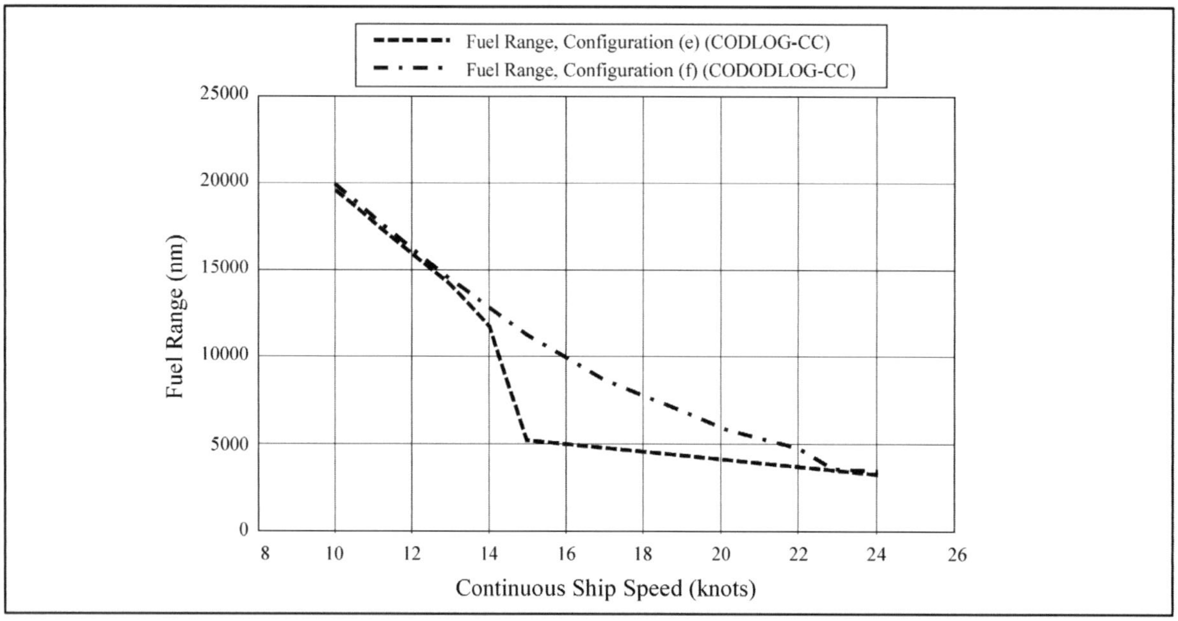

Figure 9: Predicted steaming range for each configuration variant at nominal cruise speeds

It is clear to see that the fuel range of the ship in the 14-18 knot range is improved dramatically with the addition of the propulsion diesel engine in *Configuration (f)*, and benefits continue up to 22 knots.

6.5 INCREASING MAXIMUM SHIP SPEED BY COMBINING PRIME MOVERS

The above example describes the use of the propulsion system in an *OR* logic, whereby only the electric propulsion motors, diesel engines or gas turbine are providing power to the shaft line at any given moment. This logic constrains shaft output power to the maximum rating of the specific prime mover.

The introduction of Propulsion Diesel Engines in Configuration (f) adds additional capability to the maximum speed available from the ship. Combining the output of the mechanical prime movers in an *AND* logic increases the total installed mechanical power from 36 MWm (GT) to 50.8 MWm (GT + 2x PDE) and lifts the ship's top speed by 3 to 4 knots.

The cost of enabling this additional capability is relatively low if adopted at the initial design stage, as the same machinery is already required for *OR* logic. The *AND* logic design differs only in the arrangement and control of clutches at the MRGs, and the increased torque capacity of propulsion shafts and bearings to meet the combined power and increased thrust, applying up to 25,400 kWm at each shaft.

7. PROPULSION SYSTEM DESIGN FOR SIGNATURE – SILENCE OR CAMOUFLAGE

Surface combatants of the Anti-Submarine Warfare (ASW) type are designed specifically for lowest possible underwater radiated noise (URN) levels. This approach (when taken to the extreme) results in a heavy and expensive vessel - compare the displacements of similar warships optimised for URN reduction [Ref 5] and those that are not [Ref 6]. This is attributable to the necessary high structural stiffness and damping masses required in both the ship structure and the machinery itself. The burden of acoustic isolation by resilient mounting (structural noise) and flexible hoses (fluid-borne noise) to all machinery and fluid systems adds further weight.

7.1 UNDERWATER RADIATED NOISE

Underwater Radiated Noise (URN) is managed in the ship for two reasons – reducing the noise and signature of the ship to avoid detection, classification and tracking by adversary threats (e.g. submarines), and; minimising self-noise such that the ship's own underwater sensors are able to operate most effectively to detect and track those same threats.

7.2 STRUCTURE BORNE NOISE AND VIBRATION ATTENUATION

Resonant vibration of a structure occurs when the vibration frequency of an energy source approaches the natural frequency of the structure. Structure-borne vibration can therefore be eliminated by removing the energy source (or changing its vibration frequency) or changing the natural frequency of the structure. Normal practice is to design structures with a natural frequency somewhat higher than the normally experienced resonant frequency, covering as many harmonics as reasonable – this is achieved by adding stiffness to the structure. When working with metals, a stiff structure is generally a heavy structure.

Large, heavy machinery such as main propulsion motors, main reduction gear, main engines and diesel generators can be mounted on elastic isolators to further reduce URN. The weight penalty for silencing these propulsion and power generation machines approaches approximates double the equivalent machinery and mountings of a regular ship. Isolation systems also bring a maintenance burden whereby the condition of the mounts must be carefully maintained through the ship's life.

7.3 RECIPROCATING MACHINERY VIBRATION ABATEMENT

Structural vibration from a reciprocating machine such as a diesel engine carries considerable energy and can be difficult to abate. These machines by their nature develop a vibration that corresponds with the geometry of the crankshaft and the rotating and reciprocating masses. Diesel Generators are constant-speed machines, meaning that vibration abatement solutions can be tuned to the operating speed (synchronous to the power network frequency).

One successful approach makes use of ballast (pourable resin "ballast-crete") in the subbase of the machine, such that the high frequency vibration energy from the engine is absorbed as low frequency vibration at the subbase – which is more easily attenuated to the ship's structure by rubber mounts. Whilst effective at the individual machine level, this approach adds nearly 100% additional weight to the engine assembly (i.e. the engine skid weight is doubled).

An alternative approach is to take a compartment-centric view to vibration attenuation instead of a machinery-centric view. This is achieved successfully on some classes of ships and submarines, where a *common raft* is used to support multiple machines and by its self-weight, and the accumulated weight of the machines atop it, sufficient mass exists to attenuate all high frequency, high energy vibration. This approach has the potential to reduce the additional mass to, say 10% (instead of 100%) of the equipment subject to the abatement.

7.4 SIGNATURE MASKING AND SIGNATURE MANIPULATION

7.4 (a) Hiding in Plain Sight

This paper argues that it is incredibly difficult (and expensive) to build a *silent ship*, and proposes that an alternative approach may be to build a *very quiet ship* with clever system design that presents variable noise and vibration signatures.
Consider the terminology of *cloaking* or *hiding in plain sight*. The resulting ship might be much cheaper, easier to construct and maintain, lighter (and therefore faster) and provide similar capability and survivability to the aspirational *silent* ship.

7.4 (b) Signature Manipulation by Controllable Pitch Propeller

Ships can be classified and tracked by adversaries (using passive sonar systems alone) based on the unique acoustic signature of the machinery aboard and the *turns per knot* (also known as TPK) relationship of the propulsion system. Consider the potential signature manipulation possible with a *controllable pitch propeller* (CPP).

Since power at the propeller is the multiplication of torque and rotary speed, a variation to the *combinator curve* will produce the same power and ship speed with a different TPK. Submarines tracking the ship will be unable to accurately establish ship speed and therefore accurately predict its position.

Taking this further, CPPs equipped with a *feathering mode* present the opportunity for the operator to loiter on only one shaft with the other "feathered" and locked at 90°. If operating in or around commercial shipping lanes, and operating the single shaft at the same RPM as a single-screw merchant ship (say 103 RPM) then the ship may well be masked by other maritime traffic noise and be undistinguishable. Consider revising the number of blades on the propeller – to (say) 4 blades - to improve the analogue even more.

7.4 (c) Acoustic Signature Irregularity in Ship Systems

Signature may also be attributable to non-propulsion systems on the ship. The author proposes that the use of variable frequency drives for increased efficiency of fluid and hydraulic systems will introduce further inconsistency in the ship signature and therefore inhibit classification by an adversary.

7.4 (d) Masking Technologies, Active and Passive

Surface combatants of the recent past and today apply various masking systems to lessen the impact of their otherwise noisy propulsion systems on URN:

- *Masking Air Belts* create a curtain of air bubbles beneath the main machinery, creating somewhat of an insulating layer of air. A continuous and ready supply of compressed air (such as GT bleed air) is required.

- *Prairie Masking Air* emits compressed air from the tips of the propeller blades, where unwelcome cavitation is subdued and propeller noise is reduced. Routing compressed air to the propeller blades adds considerable complexity to the shaft line, propeller hub and propeller blade manufacture.

- *Active Noise Making* systems produce deliberate noise to either cancel out (with out-of-phase noise generation) or overwhelm any machinery noise produced by the ship's on-board systems and thus mask its unique characteristics.

- Rubber *acoustic tiles* glued to the hull achieve a similar effect to masking air without the need for the compressed air supply. This solution carries a high purchase and installation cost, and high sustainment burden to maintain the tile system in good order throughout life.

8. CONCLUSIONS - PROPULSION SYSTEM SUMMARY

This paper has explored alternative propulsion system architectures; identified an arrangement that offers the best balance of capability, flexibility and integrity; closed the efficiency gap; and then considered the warfighting advantage possible through acoustic signature manipulation.

On the learnings of this study, the author sets the following challenges to the propulsion engineers and naval architects designing the next generation of warships.

1. Near future warships will require the installed power generating capacity to serve *directed energy weapons*. The Hybrid-Electric architecture provides this capability.

2. Consider flexibility in the power generation system. Hybrid-electric with propulsion diesel engines and propulsion motor regeneration (power take-off) offers further supplementary power to ship's network.

3. Consider redundancy of power and propulsion systems in the trade-off study to guarantee propulsion and steering integrity in a power failure. Hybrid-electric offers redundancy in mechanical and electric propulsion systems.

4. Consider efficiency in all aspects of system development - identify and eliminate *efficiency gaps* in the propulsion machinery arrangement. Avoid arrangements where inefficient machines are required for low load operation.

5. Optimise the ship design to be "quiet enough", achieving URN objectives at the lowest possible weight and cost. Incorporate variability in the propulsion (and other major system) to enable *cloaking* of the ship's signature.

9. ACKNOWLEDGEMENTS

Nil stated

10. REFERENCES

1. "Cure for Royal Navy destroyers engine woes in sight", 01/05/2020, https://www.navylookout.com/cure-for-royal-navy-destroyers-engine-woes-in-sight/

2. "FBI opens criminal investigation into Francis Scott Key Bridge collapse: Sources; A cargo ship lost power and slammed into the bridge last month", 15/04/2024, https://abcnews.go.com/US/fbi-opens-criminal-investigation-francis-scott-key-bridge/story?id=109235433

3. "Royal Navy Type 45 destroyer towed back to port two days after it sails", 24/11/2016, https://www.telegraph.co.uk/news/2016/11/24/royal-navy-type-45-destroyer-towed-back-port-two-days-sailed/

4. "Putting the Type 45 propulsion problems in perspective", 03/02/2016, https://www.navylookout.com/putting-the-type-45-propulsion-problems-in-perspective/

5. "Hunter-class frigate", accessed 03/05/2024, https://en.wikipedia.org/wiki/Hunter-class_frigate

6. "Hobart-class destroyer", accessed 03/05/2024, https://en.wikipedia.org/wiki/Hobart-class_destroyer

11. AUTHOR'S BIOGRAPHY

Benjamin Steele holds the current roles of Principal Maritime Engineer and Engineering Manager at BAE Systems Australia, Maritime Innovation. The Author is responsible for the development and delivery of engineering innovation projects in the maritime domain and provides engineering oversight for the Innovation Program's broader research and technology activities. The Author's prior experience includes Propulsion Engineering support across multiple naval platforms and the entire engineering lifecycle (Hunter Class Frigate, ANZAC Class Frigate, Hobart Class Destroyer, Collins Class Submarine) - and various engineering projects in reciprocating engines, including stationary power generation and rail traction.

This paper was drafted in the Author's own personal capacity and not as a representative of BAE Systems Australia. This work is not endorsed by BAE Systems Maritime Australia.

IMPACT OF HUMAN-DRIVEN MOTION CRITERIA ON VESSEL DESIGN AND THE OPPORTUNITIES CREATED BY AUTONOMOUS TECHNOLOGIES

C Borkowski, **S Smith**, **B Burgess** and **S Logan,** DSTG, Australia

SUMMARY

The design of large uncrewed vessels intended to support naval warships offers the potential to be free from human-driven motion-criteria. Such criteria include Motion Sickness Index (MSI), Motion Induced Interruptions (MII) and deck wetness limits, which are introduced to ensure the safe and efficient execution of tasks at sea. Combined with equipment-driven criteria such as ship slamming limits and vertical velocity at the flight deck, all these criteria affect the vessel's operability and, therefore, influence the design of the vessel.

With ongoing advancements in autonomous technologies and automated systems, the feasibility of a large uncrewed surface vessel to support naval operations is continuously improving. To both understand and exploit the opportunities created by removing human-driven criteria in the design of the hullform, a Design Space Exploration (DSE) activity has been conducted utilising the modelling and simulation framework InteShip, which is currently being developed by Defence Science & Technology Group (DSTG).

This paper presents a study that conducts a DSE for vessels designed with and without considering human-driven vessel design criteria for a range of selected vessel parameters. The study was conducted using InteShip to calculate vessel parameters and determine vessel performance. Study outcomes have identified the trade-offs between these selected vessel parameters that can be applied to vessel design.

1. INTRODUCTION

Autonomous technologies for ships such as autonomous launch and recovery systems (LARS) [1-4] and autonomous Helicopter Operation systems [5-8] as well as other systems [9-12] have been developed over the previous decades (see Figure 1). Due to these developments, these technologies have the potential to be installed onto autonomous naval vessels, allowing them to perform the same ocean-going tasks as a conventionally crewed naval vessel, reducing operation costs and improving vessel operability.

Depending on their mission, autonomous naval vessels can be used as part of a naval fleet, such as the US Ghost Fleet [13], or to perform lone reconnaissance and data acquisition missions [14] (see Figure 1 (f)).

In order to determine the viability of substituting a crewed vessel with an autonomous vessel, the operability of an autonomous vessel was compared to a conventionally crewed vessel for common seakeeping tasks. Vessel operability was determined for the following common seakeeping tasks;

- Operational Transit (OT)
- Replenishment at Sea (RAS)
- Helicopter landing (Helo)
- Vertical replenishment (VREP)
- Launch & Recovery (Deck) (L&R(D))
- Launch & Recovery (Ramp) (L&R(R))

Vessel operability for each seakeeping task is measured using motion-limiting criteria based on generally acknowledged standards [15-17]. Motion-limiting criteria are used to ensure that a certain seakeeping task can be completed successfully and safely. The criteria are either human-driven or equipment-driven. If humans were removed from the vessel, some restrictions could be lifted.

The hullform design of conventionally crewed vessels requires consideration of both human and equipment-driven criteria. However, due to the absence of crew members, the hull design for an autonomous vessel needs to consider only equipment-driven criteria.

Human-driven design criteria are generally considered to be;

- Motion Sickness Index (MSI)
- Motion Induced Interruption (MII)
- Deck wetness

Figure 1: (a) Fully autonomous launch & recovery ramp system recovering (Uncrewed Underwater Vehicle (UUV) conducted by Henriksen [3], (b) Designer's (Henriksen's) impression of Uncrewed Surface Vehicles (USV's) being deployed from Deck and Ramp on mother-ship equipped with autonomous LARS [3], (c) Helicopter operation conducted with Airbus VRS700 an optionally-autonomous helicopter [5], (d) Defendtex Tempest aerial drone in operation at Autonomous Warrior 2018 at HMAS Creswell, ACT [7], (e) "Snowflake" aerial delivery system developed for use by US Navy for VERTREP operation [12], (f) US Navy Autonomous surface vessel Seahawk transiting under Sydney Harbour Bridge during naval exercises with the Royal Australian Navy (RAN) [13].

Depending on the seakeeping task to be performed, specific tasks have to consider MII and deck wetness as equipment-driven criteria [15-17]. For example, an MII can also be applied to a piece of equipment, such as an aerial drone on the helipad or main deck of an autonomous vessel, and not just a crew member performing a task on a conventionally crewed surface vessel. Also, deck wetness creates a slip hazard for crew members on the main deck, as well as hitting the main deck, hull and railings with enough force to cause physical damage to the vessel.

Operability is a measure of the potential to perform a seakeeping task safely and securely on a vessel for a range of vessel speed and wave heading combinations. Motion-limiting criteria are applied to each seakeeping task to ensure safe and secure operation. However, there are no generally acknowledged motion-limiting criteria developed for autonomous vessels, only rules of thumb dependent on the experience of, and only privy to the OEM. Therefore, for this study the motion-limiting criteria are based on those outlined in STANAG4154 [15] and applied to conventionally crewed and autonomous vessels. Polar plots can be used to present operability on a scale of 0 to 1 for a range of vessel speed and wave heading combinations.

In this study, operability results for various seakeeping tasks performed on crewed and autonomous versions of the generic frigate were produced using Shipmo7, a program based on strip theory used to calculate ship motions and sea loads in various sea states. The hullform of a generic frigate was selected as a parent hullform as it suits the role of the vessel assessed in this study. The seakeeping operability of the stated seakeeping operations on the generic frigate hull were determined using InteShip [18], a modelling and simulation framework developed by DSTG that aims to holistically model vessel performance by capturing the complex interdependencies of a naval platform.

The design of conventionally crewed and autonomous vessels involves varying their local hull parameters. Average operability results for each seakeeping operation were compared against a variation of local hullform parameters to determine trends and identify parameters that maximise operability and produce an enhanced operability hullform. Polar plots of each seakeeping task were produced for comparison with the generic frigate hull and the enhanced operability hull.

2. METHODOLOGY

2.1 PARENT HULL DESIGN WITH PARAMETRIC HULL TRANSFORM MODEL

DSTG recently produced a parametric hull transform model as part of the InteShip modelling framework [19]. The model utilises the Orca 3D plugin for Rhino, which enables parametric control over global and local hull geometry.

The generic frigate hullform was generated using the parametric hullform model and is shown in Figure 2; the principal dimensions are provided in Table 1. Appendages include bilge keels, stabilising fins, propeller brackets, and rudder.

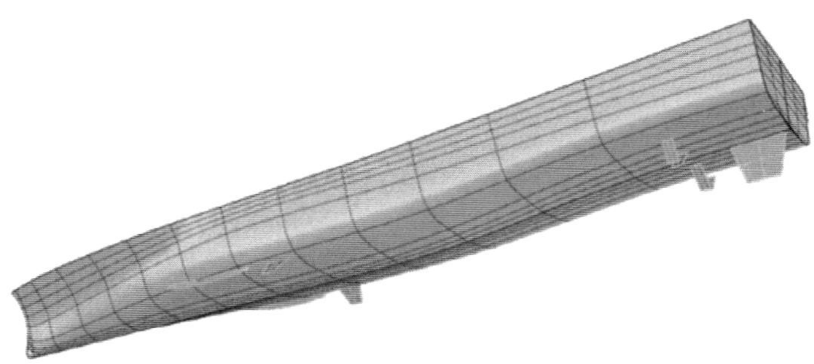

Figure 2: Generic frigate hull developed using parametric hull transform model. Appendages featuring the bilge keels, stabilising fin, propeller brackets and rudder have been added in green.

Table 1. Principal dimensions of the generic frigate hullform

Principal dimensions	Values	Units
Length between perpendiculars, L_{pp}	109.51	[m]
Beam, B	14.750	[m]
Draught, T	4.500	[m]
Displacement, Δ	3625	[t]
Wetted surface area, S_w	1667.7	[m^2]
Block coefficient, C_b	0.503	-

2.2 MODELS USED WITHIN INTESHIP FRAMEWORK

The InteShip framework features a model repository that contains a wide range of modelling capabilities that are used to predict the various performance aspects of a vessel such as stability, resistance, powering, range and seakeeping. The model workflow consists of the water properties model, hull transformation model (includes local hull parameters), equilibrium model, seakeeping RAO model and seakeeping operability models that have been designated for each task. Figure 3 presents the workflow of the models and their dependent software selected to perform the simulations in the InteShip framework. Furthermore, the seakeeping operability tasks stated previously were simulated using the generic frigate hull for both crewed and autonomous motion-limiting criteria.

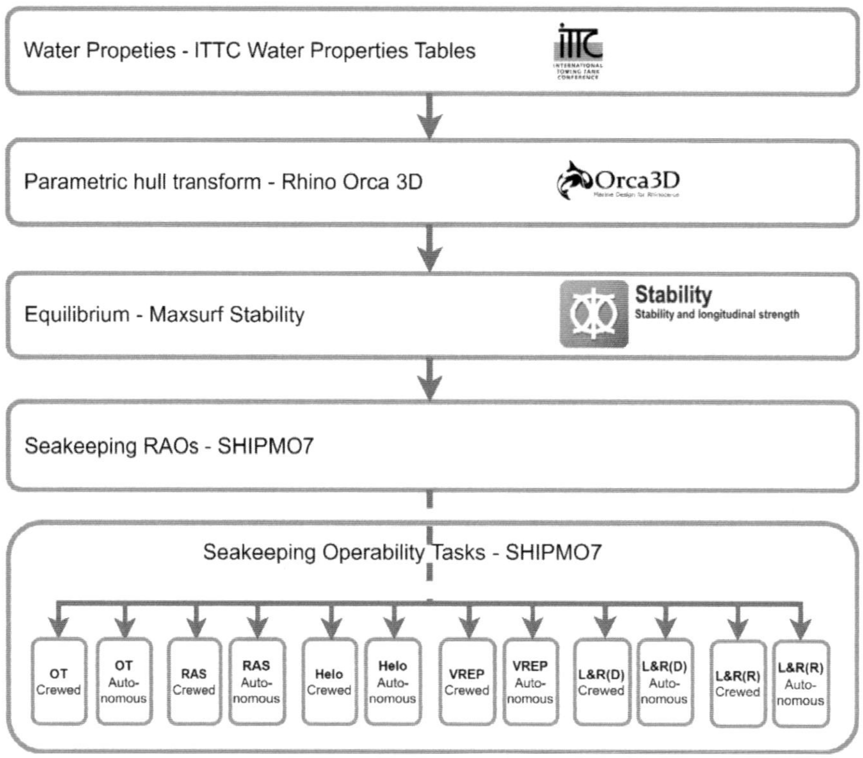

Figure 3: InteShip workflow developed to investigate the effects of local hull parameters on seakeeping operability

Within the InteShip framework, the capabilities of the parametric hull transform model will be utilised in this study to identify the optimum local hull design parameters for both crewed and autonomous motion-limiting criteria to be applied to the generic frigate hull.

3. SIMULATION SETTINGS

3.1 ESTABLISHING MOTION-LIMITING CRITERIA FOR SEAKEEPING OPERATIONS

Motion-limiting criteria for each seakeeping operation were based on NATO's STANAG4154 Edition 3 standards [15], Bureau Veritas (BV) - Rules for the Classification of Naval Ships NR483-October 2022 Part E [16], DEF(AUST)5000 standards [17] and Comstock, Bales and Gentile (1982) [20].

The human and equipment driven criteria from these standards and references were entered into the InteShip framework for conventionally crewed vessels, whilst only equipment driven criteria were entered into the framework for autonomous vessels. Table 2 presents the crewed and autonomous motion criteria limits acquired from the standards and references for each seakeeping operation used in the InteShip simulation.

For the autonomous vessel, not all human factors are omitted. Whilst MSI is strictly a human factor and always omitted, other factors, such as MII, can still be considered as equipment factors. MII_{human} and MII_{equip} criteria are each dependent on tipping coefficients in longitudinal and lateral directions. Tipping coefficients represent the potential of an object on deck experiencing a toppling or slippage event caused by wave motions. Longitudinal and lateral tipping coefficients are applied for both human and equipment-driven MII criteria (Annex H pgs. 69, 70 in DEF(AUST)5000 [17]). A lateral and longitudinal tipping coefficient of 0.7 was selected as it best represented the coefficients of equipment tipping on a surface. Tipping coefficients for MII_{human} were 0.17 for the lateral coefficient and 0.25 for the longitudinal coefficient.

Table 2. Crewed and autonomous motion-limiting criteria from the standards and references for each seakeeping operation used in InteShip simulation

	Seakeeping operations											
Motion-Limiting Criteria	OT crewed	OT auto	RAS crewed	RAS auto	Helo crewed	Helo auto	VREP crewed	VREP auto	L&R(D) crewed	L&R(D) auto	L&R(R) crewed	L&R(R) auto
deck wetness [events/hr]	30	30	0.50		0.50	0.50	0.50	0.50	0.50	0.50	0.50	
Slamming [events/hr]	20	20										
emergence [events/hr]	90	90									10	10
MII_{human} [events/min]	1		0.50		0.50		0.50		0.50		0.50	
MII_{equip} [events/min]			0.50	0.50	0.50	0.50						
MSI [% of personell in 4 hrs]	20		20		20		20		20		20	
vertical disp (rms) [m]							0.70	0.70				
lateral disp (rms) [m]											0.75	0.75
vertical velocity (rms) [m/s]					1	1	1.05	1.05				
vertical acceleration (rms) [g]									0.10	0.10	0.10	0.10
lateral acceleration (rms) [g]									0.10	0.10	0.10	0.10
longitudinal acceleration (rms) [g]												
Roll (rms) [degrees]			2.20	2.20	2.50	2.50	1.60	1.60	1.25	1.25	1.25	1.25
Pitch (rms) [degrees]			2.20	2.20	1.50	1.50	1.60	1.60	4	4	4	4

3.2 SEAKEEPING TASK LOCATIONS ON GENERIC FRIGATE

Seakeeping task locations and their location IDs are presented in Figure 4. Superstructure is added to the hull for visual purposes. Co-ordinates of these seakeeping points are presented in Table 3. The locations used in the operability calculations for each of the tasks are based on established standards [15-17].

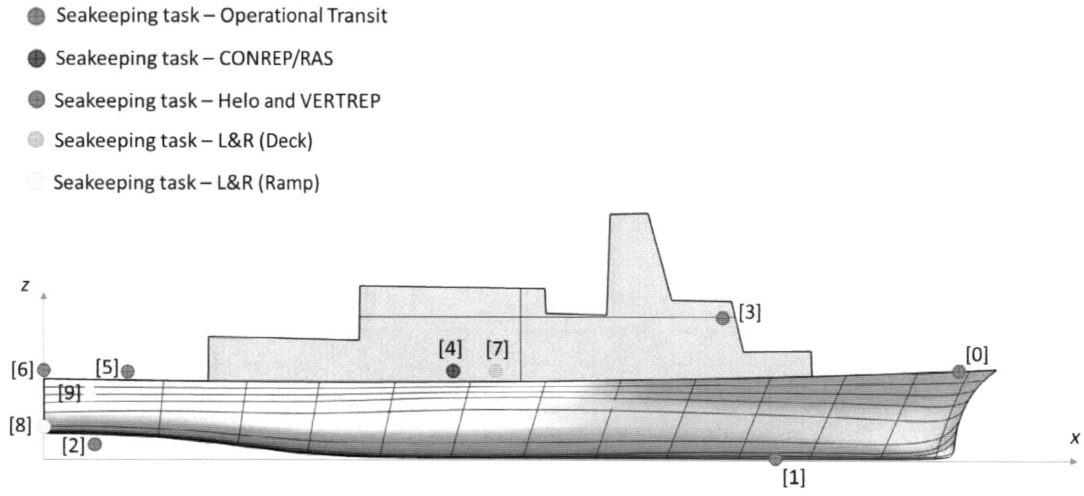

Figure 4: Seakeeping task locations considered on generic frigate

Table 3. Co-ordinates of seakeeping task locations considered in generic frigate

ID	Location	x [m] from aft	y [m]	z [m]	Deck Height [m]
0	Bow	109.515	0	10.400	10.400
1	Keel Fwd	87.612	0	0	0
2	Propeller Stbd	6.087	-3.450	1.496	1.496
3	Bridge	81.201	0	16.628	15.628
4	RAS Stbd	49	-7.400	10.300	9.300
5	Flight deck centre point	10	0	10.300	9.300
6	Flight deck corner point	0	-7.400	10.300	9.300
7	Deck, side davit	54	-6.300	10.300	9.300
8	Stern Ramp	0	0	3.550	3.550
9	Quarterdeck	5.476	-4	7.700	6.700

3.3 OPERATING CONDITIONS

The simulations were conducted at the top of Sea State 5 for all seakeeping operations with a Bretschneider wave spectrum, with wave height of 4m and average period of 9.5s [17]. Simulations were conducted for headings at intervals of 15 degrees from the head seas of 180 degrees to the following seas of 0 degrees. Ship speeds used in the simulation were at intervals of 5 knots from 5 knots to 30 knots.

4. RESULTS

4.1 OPERABILITY POLAR PLOTS – GENERIC FRIGATE HULL

The operability of a seakeeping operation is a crucial measure, quantified on a non-dimensional scale from 0 to 1, indicating the percentage of speed-heading combinations a ship can effectively perform a set or subset of tasks. At 0, or 0% operability, the ship fails to meet all motion criteria across all conditions, while at 1, or 100% operability, it fully complies with all criteria.

Tasks like transiting and helicopter operations are defined by specific seakeeping criteria, each with exceedance thresholds that the ship must not surpass to maintain operability. If these thresholds are exceeded, the ship's operability diminishes for the given operating context. To ensure a balanced comparison between ships with slight differences in seakeeping performance, gradual degradation of criteria beyond the threshold is essential. This is achieved by a linear decrease in operability from 1.0 at half the limit to 0.0 at double the limit. Weightings are assigned to each criterion relative to others, totalling 1, with each criterion specified by a location on the vessel and a performance measure. These weightings are then combined with corresponding operability values, yielding a weighted operability value for the operation.

Joint degrading polar plots of each seakeeping operation consist of an amalgamation of degrading polar plots of each motion-limiting criteria that were considered for that operation on a scale of 0 to 1. Joint degrading polar plots for each seakeeping operation from Table 2 were generated for the generic frigate hull (Figure 2) and presented in Figure 5.

Generic frigate hull: Crewed and Autonomous operation comparison

Comparing the operability plots for the crewed and autonomous versions of the generic frigate hull form reveals key differences for each of the operations. These key differences are presented below.

OT

An operability percentage of 87.65% was measured for the OT crewed operation, whilst a percentage of 96.03% was measured for the OT auto operation, resulting in an 8.38% increase in operability for the autonomous generic frigate hull.

The MSI motion-limiting criteria for the OT crewed operation for the generic frigate hull is responsible for the reduced operability when travelling between 15 knots to 30 knots for head and bow quartering seas when compared to the OT auto operation in Figure 5. The OT auto operation does experience a small reduction in operability when travelling at 30 knots at bow quartering seas to head seas ranging 150 to 180 degrees. Deck wetness and emergence are the criteria that are responsible for this reduced operability at the said speed/heading combinations.

RAS

An operability percentage of 83.97% was measured for the RAS crewed operation against a percentage of 93.33% for the RAS auto operation, resulting in a 9.36% increase in operability in favour of the autonomous version of the generic frigate.

The MSI as well as the Roll RMS motion-limiting criteria for the RAS crewed operation for the generic frigate hull are responsible for the lower operability value (light green) for all vessel speeds for beam to bow quartering seas spanning 105 to 135 degrees compared to RAS auto operation in Figure 5. Also, the combination of deck wetness, MSI and Roll RMS motion-limiting criteria for the RAS crewed operation are responsible for a further degradation in operability value (yellow) at vessel speeds of 5 to 10 knots for beam to bow quartering seas spanning 75 to 135 degrees for the generic frigate hull.

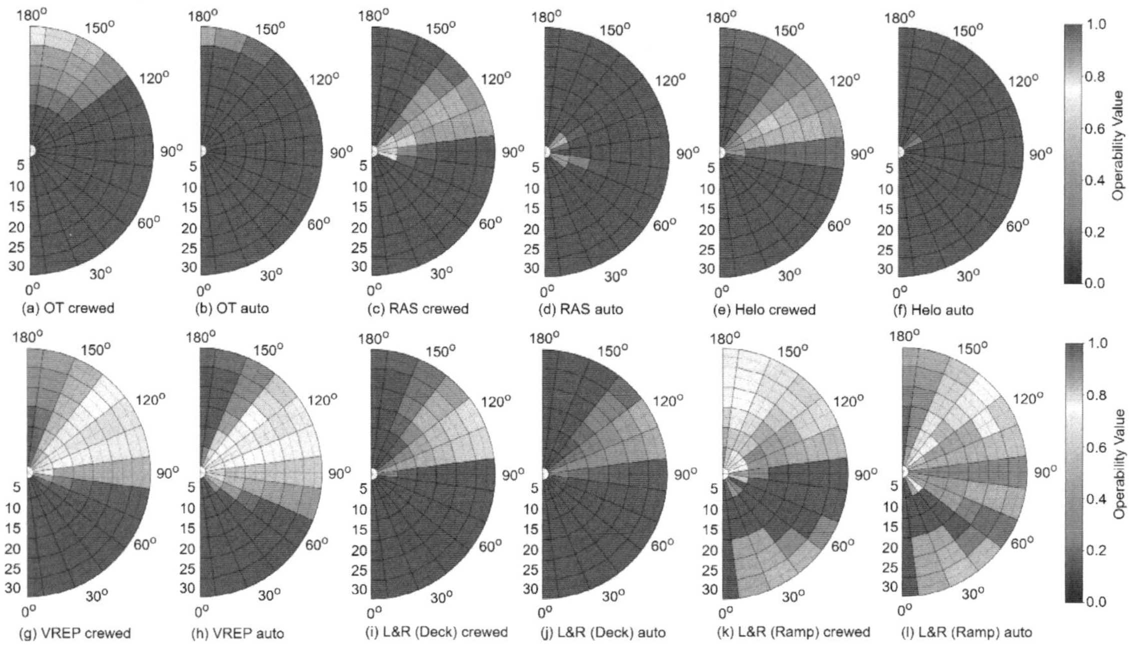

Figure 5: Joint degrading polar plots for each seakeeping operation and motion-limiting conditions (from Table 2) for the generic frigate hull (Figure 2), for both crewed and autonomous operability measures

Helo

An operability percentage of 83.25% is measured for crewed helicopter operations compared to an operability percentage of 90.76% for autonomous, resulting in a 7.51% improvement in operability. For crewed operations, the operability value shows degradation, notable at headings from 90 – 150 degrees and speeds between 5 and 30 knots. Despite the loss in operability, the analysis indicates that the vessel's performance remains favourable at lower speeds for stern seas. For autonomous systems, the operability degrades within specific parameters of headings from 105 – 150 degrees and speeds of 5 – 30 knots. It is noted that the MSI substantially affects operability for crewed operations. In addition, and in connection to MSI, excessive roll RMS motions at headings between 75 – 120 degrees across all speeds exceed twice the limiting criteria.

VREP

An operability percentage of 75.39% is recorded for crewed vessel in comparisons to 74.38% for autonomous systems. A difference of 1.01% is measured in favour of the crewed vessel. The primary driver for the difference is the consideration of more criteria, damping the impact from dynamic vessel responses such as pitch, roll, vertical acceleration and velocity. Although, the VREP crewed operation does experience reduced operability at head seas compared to the VREP auto operation. This is caused by the MSI motion-limiting criteria.

L&R (Deck)

An operability percentage of 83.97% was measured for the crewed operation compared to 84.78% for the autonomous operation, resulting in a difference of 0.81% in favour of the autonomous vessel. The polar plots for L&R (Deck) show that the reduction in operability is concentrated around headings of 105 – 135 degrees with operability reducing with increases in speed for both the crewed and autonomous vessels. The reduction in operability for both can be mostly attributed to the exceedance of the roll and vertical acceleration RMS limits in the bow-quartering seas. The difference between the two polar plots is the additional MSI criteria for the crewed vessel which leads to reduced operability for headings ranging from beam to head seas mostly confined to speeds above 10 knots.

L&R (Ramp)

An operability percentage of 69.16% was measured for the crewed operation compared to 69.49% for the autonomous operation. This has resulted to only a slight increase of 0.33% for the autonomous vessel. Operability for L&R (Ramp) for the crewed vessel is shown to be significantly reduced from beam to head seas at all speeds with the minimum operability occurring at 135 degrees and 30 knots. This is driven by multiple criteria including deck wetness, emergence, MSI, roll and vertical acceleration. Reduced operability seen in stern quartering seas is also driven by deck wetness and emergence in addition to lateral acceleration. Comparing to the autonomous vessel a significant improvement in head seas is observed which can be attributed to the omission of the deck wetness and MSI criteria.

4.2 SELECTION OF LOCAL HULL DESIGN PARAMETER FOR VARIATION

The hull geometry parameters that were varied in the simulation were confined to the forward region of the hull. This was conducted to avoid potential errors that could occur between set appendage locations and the varying hull geometry.

Local parameters that were selected to vary for the study are;

- DeadriseFwd
- SideSlopeFwd
- FlareFwd
- DeckTaperFwd (known as FullnessFwd in Orca3D)

The parameters used in the Orca3D plugin are dimensionless. These variations are made on a scale between 0 and 1. Figure 6 presents these local parameters and their variations on the generic frigate, with the scale between 0 and 1 represented by the dashed red line and red line, respectfully.

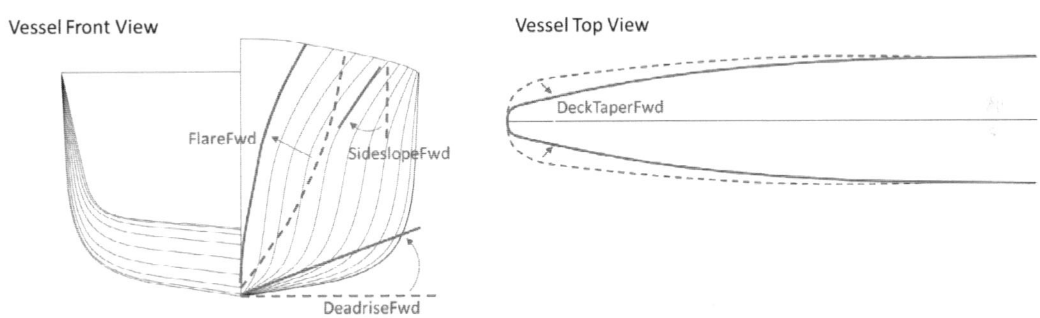

Figure 6: Local hull parameter variations on 0 to 1 scale used in Orca 3D presented on generic frigate (Figure 2).

Each local hull parameter stated previously was varied at intervals of 0.2 within the scale of 0 to 1. An interval of 0.2 was selected in order for the simulations to generate results and plots with a satisfactory level of accuracy whilst not being excessively time-consuming to generate.

4.3 INFLUENCE OF LOCAL HULL PARAMETERS ON OPERABILITY

The local hull parameters are plotted against average operability on a 2D plot for each seakeeping task to identify overall trends. Figure 7 presents the trends for average operability values of all seakeeping operations in Table 2 against the local hull parameter scale. Each plot in Figure 7 compares trends for crewed operations (crewed) against autonomous operations (auto).

The majority of plot trends in Figure 7 (the first 10 seakeeping tasks) show that as hull parameters DeckTaperFwd, SideSlopeFwd and FlareFwd approach the value of 1 on the local hull parameter scale, the average seakeeping operability value decreases. However, average operability increases as DeadriseFwd approaches the value of 1 on the parameter scale. These results suggest that a flared bow with high angle of side slope are hull characteristics that reduce operability, while a more "bloated" bow shape improves operability. DeckTaperFwd covers the greatest average operability range (1.65 to 3.22%) for the first 10 seakeeping tasks, suggesting that DecktaperFwd has the greatest influence on the operability of these seakeeping tasks. FlareFwd and DeadriseFwd cover the smallest average operability range (0.48 to 1.45% and 0.49 to 1.31%, respectively), suggesting the FlareFwd and DeadriseFwd have the least effect on the operability of these seakeeping tasks.

Results from Figure 7 suggest that the absence of human-driven motion-limiting criteria (deck wetness, MII_{human}, and MSI, see Table 2) for the L&R (Ramp) auto operation cause the average operability to increase when DeckTaperFwd, SideSlopeFwd and FlareFwd approach the value of 0 on the local hull parameter scale. Although, SideSlopeFwd appears to have a local minima at 0.6. SideSlopeFwd also performs similarly for L&R (Ramp) crewed, but its trend is mirrored horizontally about the y-axis with a local minima reduced to 0.4.

Overall, both plots for the L&R (Ramp) crewed and autonomous operations in Figure 7 show that average operability only varies over a small range (0.03 to 0.55%) for all local hull parameters. This suggests that the variation of these local hull parameters at the bow region has a more reduced effect on the average operability value of both crewed and autonomous L&R (Ramp) operations compared to the first 10 seakeeping operations.

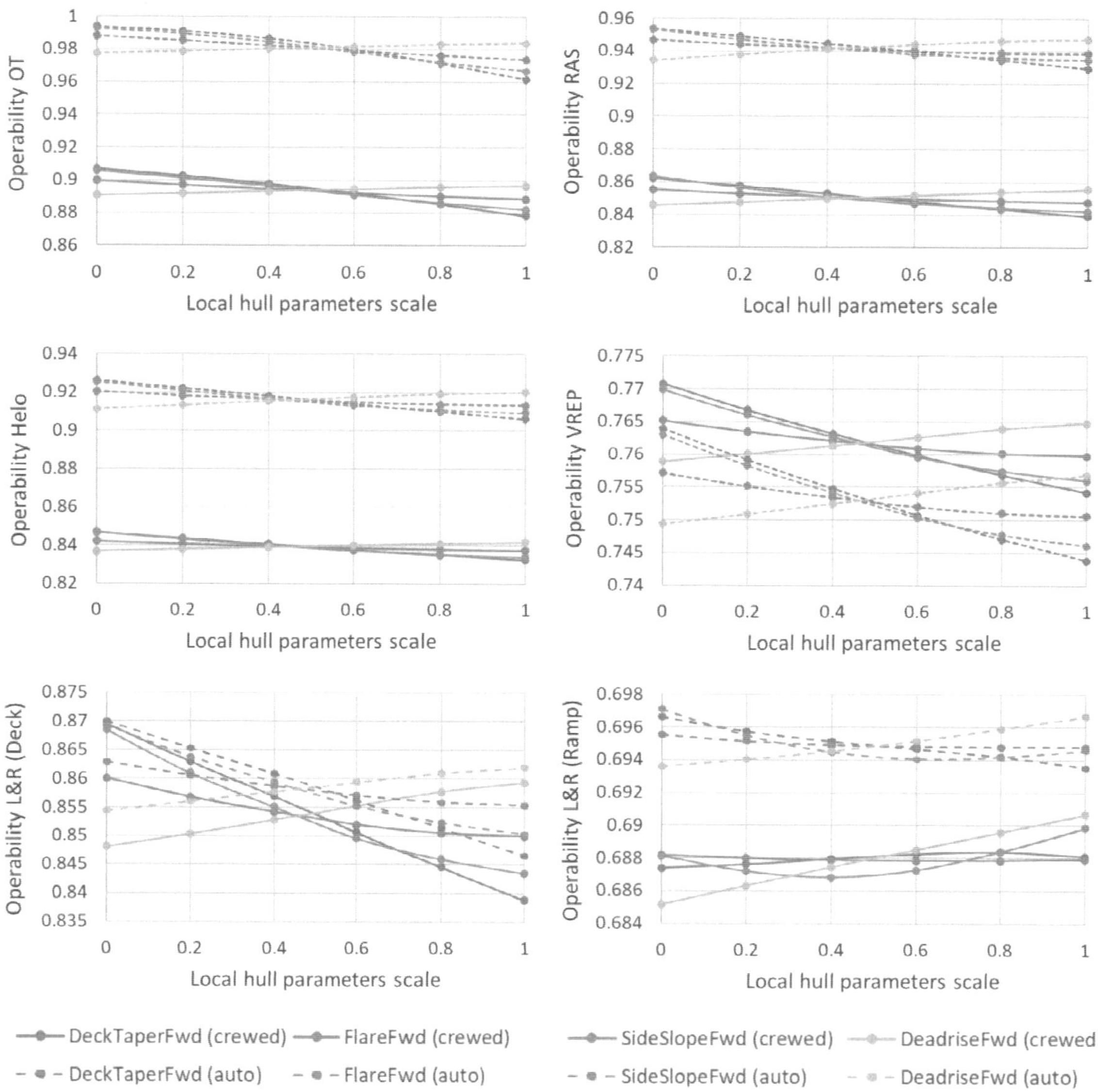

Figure 7: Comparing crewed and autonomous average operability trends against local hull parameters for each seakeeping operation.

Figure 7 shows that OT, RAS and Helo autonomous operations have a considerably greater average operability value (between 7.37 and 9.25%) than for crewed operations when comparing the average crewed and autonomous operability values at each local hull parameter interval for each local hull parameter. Average operability for autonomous L&R (Deck) and L&R (Ramp) operations is only slightly higher (between 0.04

and 0.93%) than their respective crewed operations. Conversely, the average operability for crewed VTREP operation is slightly higher (0.69 and 1.04%) than the autonomous operation.
Furthermore, greater differences are observed when comparing operability values for crewed and autonomous operations than when comparing operability values for each local hull parameter. This suggests that application of autonomous technologies to the vessel will have a greater impact on the operability of a seakeeping task than altering the vessel's local hull parameters in the bow region.

4.4 HULLFORM PARAMETERS THAT FAVOUR HIGH OPERABILITY

Figure 7 suggests that a hull design that would favour high operability for the majority of seakeeping operations would possess DeckTaperFwd, SideSlopeFwd and FlareFwd at 0, and DeadriseFwd at 1. The enhanced operability hull consists of these local hull parameters. For comparison, the generic frigate from Figure 2 used a DeckTaperFwd of 0.4 and SideSlopeFwd, FlareFwd and DeadriseFwd of 1.

4.5 OPERABILITY POLAR PLOTS – ENHANCED OPERABILITY HULL

Crewed operations: Generic frigate hull and enhanced operability hull comparison
Comparing the operability plots for the generic frigate hull and enhanced operability hull for all crewed operations shows key differences for each operation. These key differences are presented below.

OT
The OT crewed operation for the enhanced operability hull in Figure 8 experiences a higher operability percentage of 91.79% when travelling between 15 knots and 30 knots for head and bow quartering seas when compared to the OT crewed operation for the generic frigate hull in Figure 5. The remaining speed/heading combinations have high operability values (dark green) for both hulls in Figures 5 and 8. However, Figure 8 shows that the MSI motion-limiting criteria for the OT crewed operation for the enhanced operability hull does not decrease the operability values in these speed/heading combinations as for the generic frigate hull in Figure 5. This suggests that the application of the local hull parameters used to produce the enhanced operability hull is responsible for the considerable increase in operability for these speed/heading combinations.

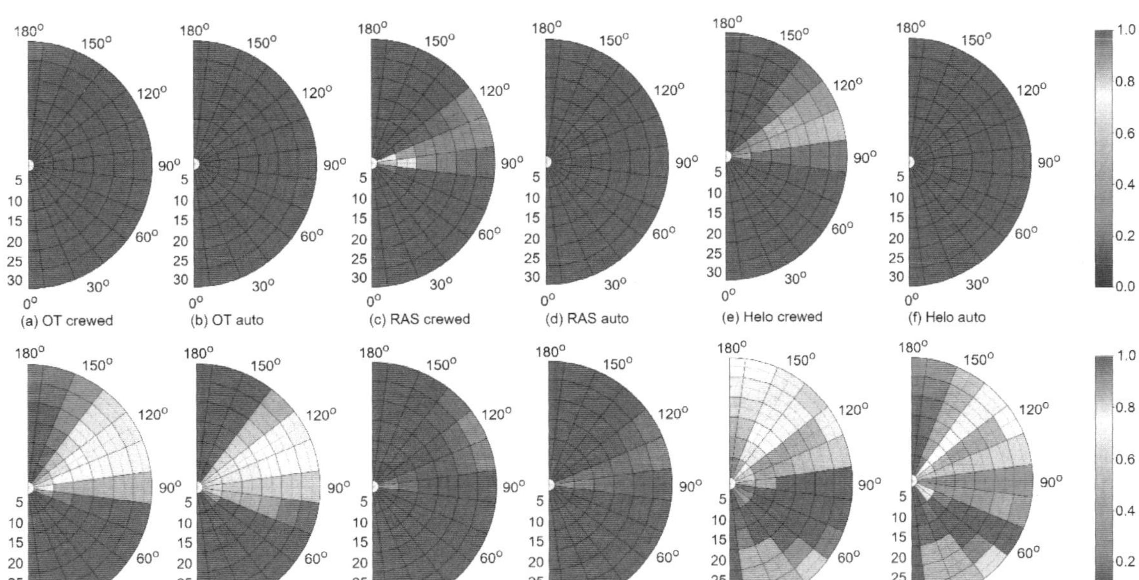

Figure 8: Joint degrading polar plots for each seakeeping operation and motion-limiting conditions (from Table 2) for enhanced operability hull, for both crewed and autonomous operability measures

RAS
The RAS crewed operation for the enhanced operability hull experiences a higher operability percentage of 89.27% for vessel speeds 15 knots to 30 knots at headings that span bow quartering bow to beam seas than the

RAS crewed operation for the generic frigate hull. The application of local hull parameters that favour high operability for the enhanced hull have considerably reduced the high MSI operability value experienced by the general frigate hull for this speed/heading combination. However, RAS crewed operation for the generic frigate hull in Figure 5 has a higher operability value for speeds between 15 knots to 30 knots at 90 degree beam seas than the enhanced operability hull in Figure 8. The primary driving factors for this trend are higher MSI and Roll values in these speed/heading combinations for the enhanced operability hull. RAS crewed operations for both hulls for vessel speeds of 5 to 10 knots at 90 degree beam seas experiences the least value of operability. The primary driving factors for this trend in both hulls are high MSI, deck wetness and Roll values in these speed/heading combinations. Interesting to note that the performance of the RAS operation is not recommended at speeds below 8 knots [21].

Helo
The Helo crewed operation for the enhanced operability hull has a higher operability percentage of 85.97% for vessel speeds 5 to 30 knots at a bow quartering to heading seas spanning between 135 to 180 degrees than the generic frigate hull. The operability values at speeds 5 to 30 knots at a heading span of 120 degree bow quartering seas to 90 degree beam seas show the lowest value for both hulls. The remaining speed/heading combinations have high operability values for both hulls.

The enhanced hull form shows a 3% enhancement to the operability score in comparison to the generic hull form. This improvement is primarily due to reduced MSI, pitch, roll, and vertical velocity recordings for all speeds and headings ranging from 120 – 180 degrees. The enhancement could be attributed to a fuller bow form, reducing the excitations from quartering seas, decreasing the overall responses of the vessel in pitch and roll.

VREP
The VREP crewed operation for the enhanced operability hull has higher operability percentage of 78.58% when travelling at 15 knots to 30 knots for bow quartering to heading seas ranging between 135 to 180 degrees than the VREP crewed operation for the generic frigate hull. The remaining speed heading combinations appear to be similar for the VREP operation for both hulls.

A 3.19% enhancement to operability is measured between the enhanced operability and generic frigate hulls. The enhancement is noted for headings of 150 – 180 degrees for all speed values for conditions of MSI, pitch and roll.

L&R (Deck)
The L&R (Deck) crewed operation for the enhanced operability hull experienced a higher operability percentage of 89.97% compared to the generic frigate hull.
Comparing the operability of L&R (Deck) for the generic frigate hull and enhanced hull shows considerable improvement which is driven by a significant reduction in vertical acceleration RMS (~40%) which also leads to the reduction in MSI.

L&R (Ramp)
The L&R (Ramp) crewed operation for the enhanced operability hull recorded a higher operability percentage of 69.43% compared to the generic frigate hull. The enhanced hull sees some improvement in terms of operability when compared to the generic frigate hull with MSI and roll actually increasing for the enhanced hull. However, this is offset by a reduction in lateral displacement and vertical acceleration RMS in stern-quartering and bow-quartering seas, respectively, in conjunction with a broadband reduction in emergence events.

Autonomous operations: Generic frigate hull and enhanced hull comparison
Comparing the operability plots for the generic frigate hull and enhanced operability hull for all autonomous operations shows key differences for each operation. These key differences are presented below.

OT
The OT auto operation for the enhanced operability hull experiences a higher operability percentage of 99.76%, but only when travelling at 30 knots for headings ranging from 150 to 180 degrees than the OT auto operation for the generic frigate hull. Deck wetness and emergence are responsible for the reduced operability values for these speed/heading combinations for the generic frigate hull. The remaining speed/heading combinations have high operability values for both hulls.

RAS

The RAS auto operation for the enhanced operability hull has a very high operability percentage of 97.11% for all speed heading combinations. The RAS auto operation for the generic frigate hull in Figure 5 has slightly lower operability value when travelling at a speed of 5 knots for bow quartering seas ranging between 120 to 135 degrees and at 5 to 10 knots for stern quartering seas ranging between 60 to 75 degrees. Roll is the primary driving factor responsible for the reduced operability value in these speed/heading combinations for the generic frigate hull.

Helo

The Helo auto operation for the generic frigate in Figure 5 shows very similar operability values as the Helo auto operation for the enhanced operability hull in Figure 8 (90.76% to 93.94%, respectively), with the exception of a lower operability when travelling at 5 knots at 120 degree bow quartering seas. Again, roll RMS is the primary driving factor responsible for the reduced operability value in these speed/heading combinations for the generic frigate hull. The remaining speed/heading combinations have high operability values for both hulls.

VREP

The VREP auto operation for the enhanced operability hull has a higher operability percentage of 78.17% when travelling at 5 to 30 knots for heading span between beam to bow quartering seas (75 to 150 degree) than the VREP auto operation for the generic frigate hull. Roll and vertical displacement are responsible for the reduced operability values in these speed/heading combinations for the generic frigate hull. The remaining speed heading combinations appear to be similar for the VREP operation for both hulls.

L&R (Deck)

The L&R (Deck) auto operation for the enhanced operability hull has a higher operability percentage of 88.93% compared to the generic frigate hull. Improvements in operability from the generic frigate hull to enhanced hull are even less than those observed for the crewed vessel with the difference predominantly due to the reduction in vertical acceleration RMS.

L&R (Ramp)

The L&R (Ramp) auto operation for the enhanced operability hull has a slightly higher operability percentage of 70.58% compared to the generic hull. As with L&R (Deck), L&R (Ramp) also sees minimal improvement in operability from the enhanced hull compared to the generic frigate hull. Increased operability is generally confined to head and stern quartering seas which is attributed to reduced vertical accelerations and lateral displacement RMS.

5. CONCLUSION

The improvement in autonomous technologies for surface vessel application in recent decades has generated the potential to design vessel hulls with the capability to perform seakeeping operations autonomously. The absence of a human crew to perform seakeeping tasks on the vessel creates some financial benefits as well as improvements in the operability of these tasks.

This study has determined that local hull parameters of DeckTaperFwd, SideSlopeFwd and FlareFwd at 0, and DeadriseFwd at 1 favour a high operability value. The application of these local hull parameters to the generic frigate hullform was performed to produce an enhanced operability hullform. Operability data from both hullforms was used to produce joint degrading polar plots, which show that the enhanced operability hullform has improved the operability values for the majority of speed/heading combinations and most seakeeping operations.

Greater values of operability for most speed/heading combinations have been observed for autonomous vessels than for crewed vessels. This is primarily due to an absence of MSI criteria in the calculation of the operability of seakeeping tasks for autonomous vessels. Furthermore, the results in Figures 5 and 8 support the use of autonomous vessels to perform all operations at sea state 5 for the majority of speed/heading combinations, with the exception of L&R (Ramp).

Deck wetness and emergence were the most influential motion-limiting criteria responsible for the reduction in operability of autonomous vessels for head to bow quartering seas for the OT operation. Roll RMS was the most influential motion-limiting criteria responsible for the reduction in operability of autonomous vessels for certain speed/heading combinations for RAS, Helo and VREP operations.

Both generic frigate and enhanced operability hulls show very high operability at all speed/heading combinations for the Helo auto operation in sea state 5, suggesting that vessels with such hulls would have the capability to conduct VTOL or drone swarm operations (see Figures 1 (c) and (d)) in rough seas without putting crew member's lives at risk.

Also, differences in operability between crewed and autonomous operations have been observed to be greater than the differences in operability created by hull modifications, the greatest differences having been generated by OT, RAS and Helo operations. The application of autonomous technologies to the vessel produces a greater improvement in operability value than altering the local hull parameters.

However, the results of this study are limited as they are based on local hull parameters from the bow region. Future work will include the further development of the parametric hull transform model to consider appendage location during hull transformation simulation, which would allow for the variation of local hull parameters from the midships and stern regions.

6. ACKNOWLEDGEMENTS

The author would like to acknowledge Mr Neil Patterson and Dr Peter Hield from the Ship Systems Analysis discipline at DSTG for their help and guidance throughout this study.

7. REFERENCES

1. NAVAL GROUP, 'World First: Naval Group successfully completes testing of its Launch and Recovery System up to and including NATO sea state 5/6 at the Maritime Research Institute Netherlands' *Wageningen, Netherlands, Naval Group*, 2019.

2. BELGIAN NAVAL and ROBOTICS, 'Belgium Naval & Robotics (BNR) solution for lateral deployment and recovery of drones successfully tested', *Belgian Naval and Robotics*, 2022.

3. HENRIKSEN, H., and EASTAUGH, P., 'Autonomous LARS for AUVs, Launch and Recovery Systems Support Minesweeping in Norway.' *Sea Technology, 22-25*, 2020.

4. VAVASSEUR, X., 'Sealartec Demonstrates Autonomous Launch & Recovery System For USV', *NavalNews, Accessed 5/2/2024. https://www.navalnews.com/naval-news/2020/06/sealartec-demonstrates-autonomous-launch-recovery-system-for-usv/*, 2020.

5. DONALD, D., 'Airbus's Optionally-Piloted Helicopter Goes to Sea', *Accessed 5/2/2024. https://www.ainonline.com/aviation-news/defense/2022-03-25/airbuss-optionally-piloted-helicopter-goes-sea*, 2022.

6. WALKER, F., 'Automating aircraft ship landings at rough seas', *TechExplore, Accessed 16 January 2024. https://techxplore.com/news/2023-10-automating-aircraft-ship-rough-seas.html*, 2023.

7. DEMPSEY, K., 'Autonomous Warrior 18', *Royal Australian Navy, Accessed 8 April 2024. https://images.navy.gov.au/Home/Search?Query=20181113ran8106604_016.jpg&Type=Filename,* 2024.

8. NAVAL NEWS STAFF, 'French Navy Tests VSR-700 VTOL UAV From Frigate At Sea', *Naval News, Accessed 9 April 2024. https://www.navalnews.com/naval-news/2023/11/french-navy-tests-vsr-700-vtol-uav-from-frigate-at-sea/*, 2023.

9. GALWAY, R. J., 'Autonomous Refuelling of Unmanned Vehicles at Sea', *Naval Surface Warfare Center Carderock Division*, 2008.

10. SCOTT, G. P., HENSHAW, G., WALKER, I. D., and WILLIMON, B., 'Autonomous Robotic Refueling of an Unmanned Surface Vehicle in Varying Sea States.' *International Conference on Intelligent Robots and Systems (IROS), 1664-71. Congress Centre Hamburg*, 2015.

11. AUSTAL AUSTRALIA, 'Austal Australia welcomes greenroom robotics to patrol boat autonomy trial', *Accessed 16 January 2024. https://www.austal.com/news/austal-australia-welcomes-greenroom-robotics-patrol-boat-autonomy-trial*, 2023.

12. HEWGLEY, C.W. and YAKIMENKO, O.A., 'Precision Guided Airdrop for Vertical Replenishment of Naval Vessels', *20th AIAA Aerodynamic Decelerator Systems Technology Conference and Seminar. Seattle, Washington, USA: American Institute of Aeronautics and Astronautics (AIAA)*, 2009.

13. MAHADZIR, D., 'U.S. Drills with Allies in the South China Sea, Ghost Fleet Ships Operate off Australia', *USNI News, Accessed 13 November 2023. https://news.usni.org/2023/10/25/u-s-drills-with-allies-in-the-south-china-sea-ghost-fleet-ships-operate-off-australia*, 2023.

14. PEDROZO, R. 'Advent of a New Era in Naval Warfare: Autonomous and Unmanned Systems.' *Autonomous Vessels in Maritime Affairs, Law and Governance Implications (Pelgrave Macmillan: Cham, Switzerland)*, 2023.

15. MILITARY AGENCY FOR STANDARDIZATION, 'Standardization Agreement (STANAG) No.4154', *Common procedures for Seakeeping in the ship design process. NATO*, 2000.

16. BUREAU VERITAS, 'Rules for the Classification of Naval Ships NR483, Part E, Additional Class Notations', *Bureau Veritas*, 2022.

17. ROYAL AUSTRALIAN NAVY, 'DEF(AUST)5000, ADF Maritime Materiel Requirement Set, Volume 3: Hull System Requirements, Part 6: Seakeeping.', *Canberra, Australia: Department of Defence*, 2003.

18. SMITH, S. and HIELD, P., 'Integrated Ship Performance Modelling (InteShip): Model Methodology', *DSTG*, 2022.

19. HOGAN, J. and SMITH, S., 'Hull form impact on whole-of-ship performance', *Indo-Pacific International Exposition, 7-9 November 2023, edited by DST Group. Sydney, Australia*, 2023.

20. COMSTOCK, E. N., BALES, S. L., and GENTILE, D. M., 'Seakeeping performance comparison of air capable ships', *Naval Engineers Journal, 94*, 1982.

21. SMITH, T., 'An Examination of Underway Replenishment Seakeeping Criteria', *National Defence Research and Development Branch, Canada*, (1992).

8. AUTHOR'S BIOGRAPHY

Chris Borkowski holds the current position of Naval Architect with Ship Systems Analysis (SSA) discipline at DSTG. The Author is responsible for vessel performance related project work, as well as the development of InteShip. The Author's previous experience includes hull design, vessel stability and ducted propeller curve research.

Dr Samuel Smith holds the current position of Naval Architect in the SSA discipline at DSTG. Sam is responsible for providing support and guidance throughout the study. Sam has extensive experience in holistically modelling the performance of naval platforms, such as the Hunter Class Frigate, as well contributing to the development of InteShip, the modelling and simulation framework being developed by SSA.

Blake Burgess started his professional career 2 years ago with the SSA Team at DSTG as a Graduate Ocean Engineer. He has spent his career contributing to the development of InteShip, a modeling and simulation framework for comprehensive performance assessments. He has conducted applied research with InteShip with a strong sense for curiosity.

Sean Logan holds the current position of research scientist in seakeeping and operability at the Australian Defence Science and Technology Group. He is responsible for examining the seakeeping-based operability of autonomous naval platforms using simulation and experimental-based approaches.

ANALYSIS OF LASER DIRECTED ENERGY WEAPONS SYSTEM INTEGRATION IMPLICATIONS FOR SURFACE NAVAL VESSELS

C Rayes, P Nguyen and **G Hale,** Defence Science and Technology Group, Australia

SUMMARY

A comprehensive understanding of how laser weapons of different output powers perform in various threat engagement scenarios is important, but how the placement and size of the required energy storage devices impact the seaworthiness and performance of the platform is also crucial for naval ship designers. Also, the dynamic behaviour of the ship can have a significant effect on a weapon's accuracy and effectiveness. This paper delves into the integration of high energy laser systems in naval anti-air capability, exploring variations in performance and their corresponding space, weight and power requirements.

Leveraging modelling and simulation tools developed by Defence Science and Technology Group, this study assesses the implications on ship stability, seakeeping, operability and range, providing valuable insights for effective implementation and operational success. The main results synthesize recommendations on the integration of laser weapons systems on generic naval platforms taking into account hypothetical limiting factors.

NOMENCLATURE

B_{WL}	Maximum breadth at waterline (m)
C_B	Block coefficient (unitless)
C_P	Prismatic coefficient (unitless)
C_M	Midship section coefficient (unitless)
DSTG	Defence Science and Technology Group
ESD	Energy Storage Device
GM_{fluid}	Metacentric height with free surface effect (m)
GT	Gas Turbine
GZ	Righting arm (m)
L_{BP}	Length between perpendiculars (m)
LCB	Longitudinal Centre of Buoyancy
L-DEW	Laser-Directed Energy Weapon
MALTESOR	Multi-Agent Laser-Threat Engagement Simulator: Operational Requirements
RAO	Response Amplitude Operator
RMS	Root Mean Square
SWaP	Space, Weight and Power
t_{dwell}	Laser dwell-time (unit not disclosed)
T_{MS}	Midship draught (m)
VCG	Vertical Centre of Gravity
Δ	Displacement (t)

1. INTRODUCTION

Designing ships is a multifaceted and intricate endeavour, shaped by numerous influences. It involves harmonising a plethora of complex subsystems to create a platform capable of meeting stringent criteria across various performance domains. The interconnected nature of these subsystems and the multi-objective performance attributes inherent in naval platforms render the design process exceptionally challenging. Altering even a single subsystem can have profound ramifications on the overall performance of the vessel, underscoring the intricacy of the task at hand.

Although decision makers may not always recognise these ripple effects, in cases where they do, they often perceive the modelling necessary to comprehend the entire scope of a design alteration or the integration of a new system as excessively costly in terms of both time and finances. Consequently, they rely on past experiences to guide their decisions regarding changes. This uninformed decision-making process has been associated with substantial cost and schedule overruns in numerous defence projects [1], as the comprehensive repercussions were not thoroughly understood at the time of the decision. To mitigate the likelihood of ineffective design changes or choices, evidence-based decision making is essential, wherein stakeholders are furnished with the requisite information to fully grasp the ramifications [2].

Directed energy technologies, specifically high energy laser systems, are continuing to be developed within industry and academia to address commercial and military needs [3]. Understanding laser-directed energy weapon (L-DEW) technologies and their performance characteristics, platform requirements, and military utility options are critical to support the integration of L-DEW onto the naval surface platforms [4]. Stronger understanding of these attributes of L-DEW will inform decision makers on decisions related to investment in technology and system development, acquisition and integration.

On account of this, an investigation into how the integration of potential L-DEW systems impact the performance of surface platforms was conducted. It focused on size, weight, power and dynamic behaviour implications for the platforms including options for L-DEW system component placement and weapon effectiveness.

2. METHODOLOGY

2.1 INTESHIP

Defence Science and Technology Group (DSTG) has developed an integrated platform modelling and analysis tool referred to as InteShip. The software leverages advanced modelling and simulation capabilities to create an integrated framework that synthesizes modelling and simulation tools, technical databases and information exchange systems. This enables InteShip to have a modular environment in which the wide range of modelling tools can be rapidly connected to create a ship performance simulator tailored to a specific design problem. The simulator is then used to explore the design space by adjusting design variables and evaluating the effect on the vessel's performance, producing technical evidence for decision makers.

In this paper, InteShip was applied to investigate the effect that the integration of L-DEW systems has on the performance of a Generic Frigate and a Generic Patrol Vessel. The analysis investigates the impact different L-DEW configurations have on ship stability, seakeeping, power and energy systems and operational range.

2.1 (a) Stability

Stability was assessed using equilibrium and large angle stability models, both of which are built upon Maxsurf Stability [5]. The equilibrium model was used to determine the attitude of the vessel, which is then used to calculate the hydrostatics and initial stability. The large angle stability model was used to produce the righting arm curve (GZ curve) and determine the large angle stability of the vessel.

2.1 (b) Resistance

Vessel resistance comprises various components that are calculated by dedicated models. Calm water resistance of the bare hull was calculated by a model built on Maxsurf Resistance [6] employing the Holtrop empirical method [7]. Appendage resistance, composed of rudder, propeller shaft, bilge keel and stabiliser fins, was calculated following the analytical methods developed by Peck [8]. The added resistance in waves model is built on Maxsurf Motions [9] which utilises strip theory to predict vessel motions with the Gerritsma & Beukelman II [10] method used to calculate added resistance.

2.1 (c) Seakeeping

The seakeeping model is based on Shipmo7 [11] software that predicts vessel motions using strip theory in the frequency domain. The model is used to obtain the vessel's Response Amplitude Operators (RAOs), which are then utilised to estimate the motions at the location where the L-DEW optical components are installed.

2.1 (d) Hydrodynamic coefficients

Wake fraction and thrust reduction coefficients were calculated via models that implement methods by Holtrop [12]. The coefficients were computed based on the vessel's principal dimensions.

2.1 (e) Propeller

The propeller model consists of a database that contains thrust coefficient, torque coefficient and efficiency curves of different series of propellers. When provided with the required vessel speed, the model uses look-up tables to determine the required propeller thrust and delivered power.

2.1 (f) Power and energy

The power and energy model is a spreadsheet-based tool which calculates the total fuel consumed during a mission for: a specified power generation system; specified hotel load; mission parameters defining propulsion power and the usage of each of the L-DEW systems during that mission. The usage of L-DEW systems was kept constant throughout the mission as well as the hotel load, which comprises the electrical loads from all other onboard systems. Fuel consumption for the Generic Patrol Vessel was calculated based on a diesel geared drive for the propulsion load complemented by an auxiliary diesel generator set supplying electrical loads. For the Generic Frigate, gas turbine (GT) mode was applied, where the propellers are driven by a gas turbine and electrical power is provided by a diesel generator.

2.2 LASER DIRECTED ENERGY WEAPON MODELS

2.2 (a) Maltesor

The Multi-Agent Laser-Threat Engagement Simulation: Operation Requirements (MALTESOR) modelling tool was developed by DSTG to simulate L-DEW target engagements that occur in surface and above-surface environments. The tool uses physics-based models to estimate L-DEW system effectiveness against targets of interest in a variety of operating scenarios.

MALTESOR characterises targets by their material type and dimensions, which are used to calculate the energy density required to defeat them. The model uses this simplified abstraction to estimate the time required for an L-DEW system to deposit sufficient energy to satisfy target defeat requirements, after accounting for L-DEW power losses due to beam divergence and atmospheric attenuation. This time-based performance indicator is referred to as dwell-time or by its symbol t_{dwell}.

It is important to acknowledge that there are several assumptions and limitations associated with MALTESOR, largely as a result of the limited amount of accessible L-DEW operational data required to properly validate the model. MALTESOR output values may change with ongoing model development, but this is not expected to affect how the model outputs are used within the analytical workflow presented here.

In this paper, MALTESOR was used to establish the relationship between L-DEW system effectiveness and the energy storage technologies required to operate them.

2.2 (b) L-DEW Energy Storage Device model

The damage inflicting component of the L-DEW modelled in MALTESOR is a diode-pumped solid-state laser. This technology relies on high-intensity and instantaneous current that can be made available through the use of an Energy Storage Device (ESD). These devices exist in various forms at different levels of technological readiness such as ultra-capacitors [13-14], mechanical flywheels [15-18] and, more commonly, electric batteries [19-20]. The L-DEW ESD model was constructed to estimate the space and weight required for the device. The model takes into consideration a range of ESD technology options and may have critical implications on weapon effectiveness. Greater power outputs naturally increase effectiveness, but are associated with heavier and bulkier ESDs.

In this paper, the L-DEW ESD model was used to estimate the size, and weight for various energy storage technologies to support L-DEW systems. The study also draws attention to the operational impacts of different ESD technologies.

2.3 Simulation workflow

MALTESOR was used to develop a dataset of L-DEW effectiveness which was fed into the L-DEW ESD model to generate space, weight and power requirements for different types of energy storage options. The resulting dataset was incorporated as a look-up table into the simulation workflow to investigate ship performance and its relationship with L-DEW effectiveness. It must be noted that MALTESOR and the L-DEW ESD model were not directly integrated into InteShip, but rather their outputs were used to generate a look-up table that is representative of various L-DEW systems and their ESDs. Figure 1 shows a simplified diagram of the simulation workflow. Models or collection of models are represented by orange boxes and the exchange or output of data is represented by the arrows.

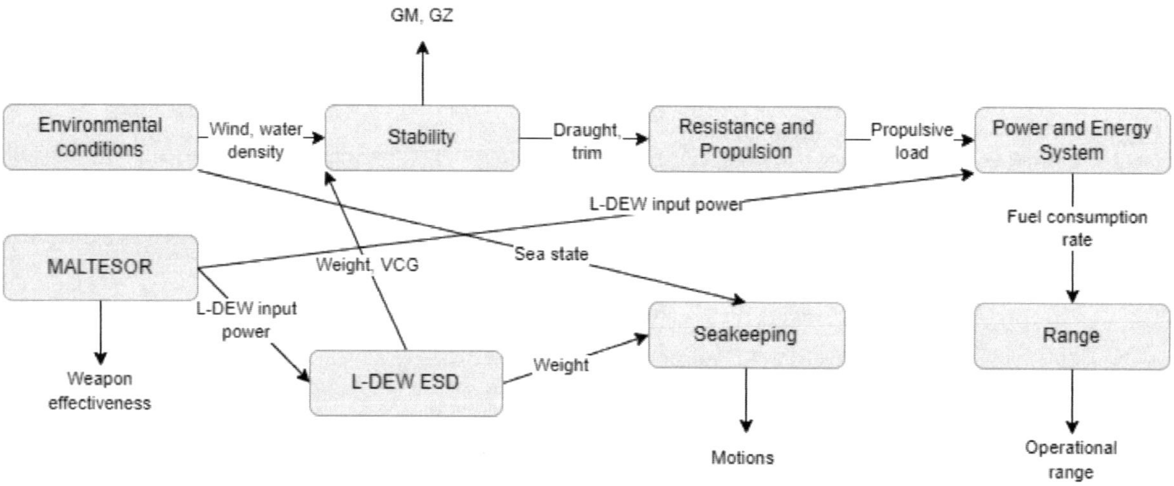

Figure 1: Simplified simulation workflow utilised in the analysis. Orange boxes represent models or collections of models and the arrows represent the exchange of data.

2.4 SIMULATION SETTINGS

This section summarises the collection of parameters upon which the simulation workflow depends to generate useful results for analyses at a conceptual level.

2.4 (a) Vessel specifications

The analysis was conducted on two hullforms: the Generic Patrol Vessel provided by the Australian Maritime College [21] and the DTMB Generic Frigate. A summary of their parameters is provided in Table 1.

Table 1. Summary of vessel parameters.

Parameter	Generic Patrol Vessel	Generic Frigate
L_{BP}	80.00 m	109.0 m
B_{WL}	12.86 m	13.97 m
T_{MS}	3.63 m	4.537 m
Δ	1621 t	3781 t
Trim	0.00 m	0.00 m
LCB	37.98 m	50.97 m
C_B	0.4477	0.5217
C_P	0.5877	0.6365
C_M	0.7465	0.8067
Propeller type	HSVA 3408/3409 [22]	B580-14 [23]
Propeller diameter	2.30 m	4.50 m
Prime mover	MTU 16V1163M74 [24]	GE LM2500 [25]
Generator set	Scania DI16 090M 640kW [26]	Scania DI16 090M 640kW
Hotel load	100 kW	200 kW

2.4 (b) MALTESOR simulation setup

MALTESOR was used to estimate dwell-times for a range of L-DEW systems. For simplicity, most L-DEW settings (wavelength, aperture size, presence of adaptive optics) were held constant across all systems with the differentiating feature being the L-DEW output power.

System effectiveness for individual L-DEW options was estimated by calculating the time required for the weapon system to defeat a single oncoming threat system. Operating conditions were chosen to be ideal for L-DEW performance. Target properties, such as velocity and altitude were held constant throughout the engagement. Simulation parameters are summarised in Table 2.

Table 2. Summary of MALTESOR simulation input parameters.

Parameter	Value(s)
L-DEW Output Power	10, 30, 50, 200 kW
L-DEW Wavelength	2 micron
L-DEW Adaptive Optics	On
L-DEW Aperture Diameter	70 cm
L-DEW Altitude	20 m above ground level
Target Initial Range	14,000 m
Target Speed	65 m/s
Target Altitude	50 m above ground level
Weather	Clear Summer
Time of Day	Day

Estimated dwell-times for different L-DEW systems were used in the L-DEW ESD model to estimate energy storage technology Space, Weight and Power (SWaP) requirements.

2.4 (c) Environmental conditions

Simulations were conducted with water density and kinematic viscosity set to salt water at 25°C. Vessel motions at L-DEW optical components location were calculated for a Bretschneider spectrum at sea state six, likely representing very rough sea conditions at which L-DEW could be operated. Added resistance in waves was calculated for the same spectrum type, but at sea state four. Sea state four is still a rough state, but a more likely state to be encountered by the vessel during its mission, therefore making it more suitable to simulate added resistance in waves. Wind resistance was calculated based on a head wind speed of 12 knots.

3. RESULTS AND DISCUSSION

3.2 L-DEW SIMULATIONS OUTPUT

3.1 (a) MALTESOR output

Results from the MALTESOR simulations are presented in Table 3. These were fed into the L-DEW ESD model to generate SWaP requirements. It is important to note that shorter t_{dwell} indicates more effective weapons.

Table 3. MALTESOR simulations output. Actual t_{dwell} units and values are classified, however they were ordered in ascending order, where $t_{dwell}(1)$ is the shortest and $t_{dwell}(4)$ is the longest.

L-DEW Output Power [kW]	Dwell-time
200	$t_{dwell}(1)$
50	$t_{dwell}(2)$
30	$t_{dwell}(3)$
10	$t_{dwell}(4)$

3.1 (b) L-DEW space, weight and power specifications

An L-DEW system is composed of several system components that can broadly be categorised as non-ESD (non-power generating) and ESD (power-generating) components. It is assumed that total L-DEW system weight and volume are the sum of the weight and volume of the non-ESD and ESD components.

Size and weight of non-ESD components were estimated as a function of L-DEW output power based on the specifications from two reference systems [28]. Reference system specifications for system components are marked with an asterisk in Table 4.

System specifications for the L-DEW systems considered in this work were estimated by assuming that size and weight components changed linearly as a function of L-DEW input power. It was also assumed that there was a 25% efficiency conversion of electrical input power to L-DEW output power, which accounts for system inefficiencies and losses due to factors such as heat generation. System specifications used in this paper are summarised in Table 4.

Table 4. Summary of size, weight and power estimates for non-ESD components for the L-DEW systems considered in this work. Reference systems specifications are marked with an asterisk (*). **All other numbers are indicative.**

L-DEW Output Power [kW]	Electrical Input Power [kW]	Total Non-ESD Weight [kg]	Total Non-ESD Volume [m^3]
10	40	260	0.48
15*	60*	510*	1.3*
30	120	1300	3
50*	200*	2300*	5.7*
200	800	9700	25

3.1 (c) L-DEW summary

This section summarises L-DEW total size, weight and power specifications for different ESD technology options. The limiting factors provided to the L-DEW ESD model allow for the calculation of the number of shots that each variation can hold. Table 5 collates the results produced by the integration of MALTESOR and the L-DEW ESD model that were looked-up during InteShip simulations. Note that lithium-ion and lead-acid batteries as well as ultra-capacitors can only provide enough energy for four shots, while nickel-zinc batteries and flywheels offer a much deeper magazine. Additionally, caution must be taken with the numbers associated with flywheels and ultra-capacitors as they have not been validated.

Table 5. L-DEW system space, weight and power estimates using different ESD technologies. **Numbers are indicative.**

L-DEW Output Power [kW]	ESD Type	Battery Mass [kg]	Number of shots	Battery Volume [m³]	Total Mass [kg]	Total Volume [m³]
10	Li-Ion	70	4	0.05	330	0.53
30		180	4	0.13	1,500	3.1
50		290	4	0.25	2,600	6.0
200		880	4	0.64	11,000	26
10	NiZn	1,100	21	0.76	1,400	1.2
30		3,300	23	2.3	4,600	5.3
50		5,300	24	3.7	7,600	9.4
200		21,000	35	15	31,000	40
10	Lead-Acid	2,000	4	1.6	2,300	2.1
30		5,200	4	4.2	6,500	7.2
50		8,200	4	6.6	11,000	12
200		23,000	4	18	33,000	43
10	Ultra-capacitor	2,400	4	2.4	2,700	2.9
30		6,500	4	6.5	7,800	9.5
50		10,000	4	10	12,000	16
200		29,000	4	28	39,000	53
10	Flywheel	14,000	50	8	14,000	8.5
30		14,000	18	8	15,000	11
50		14,000	12	8	16,000	14
200		56,000	17	32	66,000	57

3.2 INTESHIP SIMULATIONS

3.2 (a) Stability

The simulation workflow was constructed so that it allows the visualisation of design trade-offs, including the ones between weapon effectiveness and vessel stability. Figure 2 depicts one possible way of investigating such trade-off, as it relates laser dwell-time, the ships' fluid metacentric height (GM_{fluid}) and the vertical centre of gravity (VCG) of the energy storage. As discussed in 3.1, laser dwell-time is the measure of weapon effectiveness. A lower dwell time indicates a more effective weapon, which requires more energy and, therefore, a larger and heavier ESD.

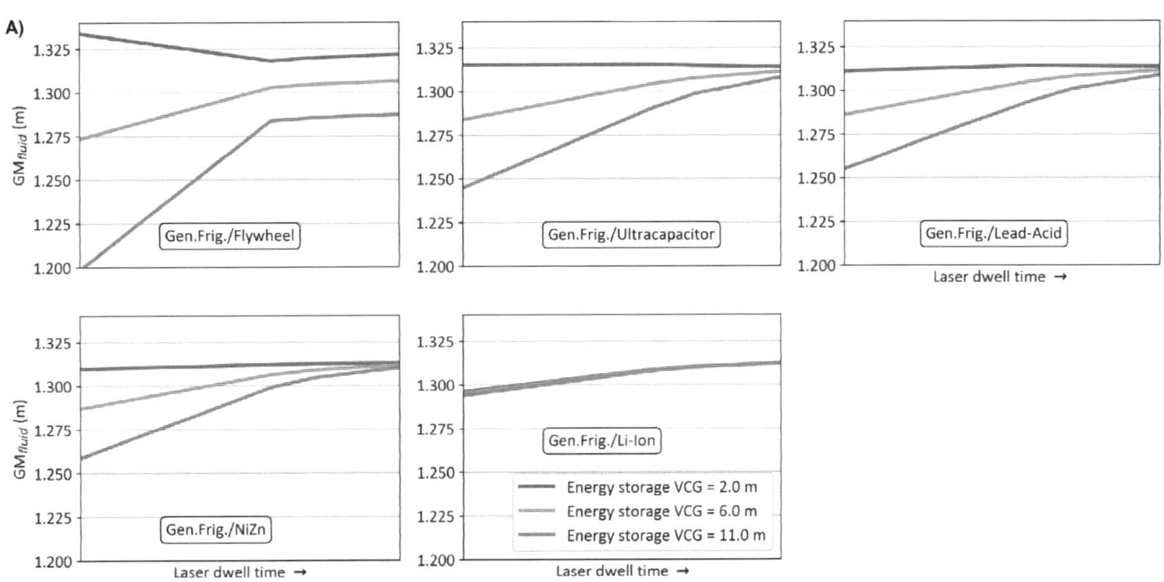

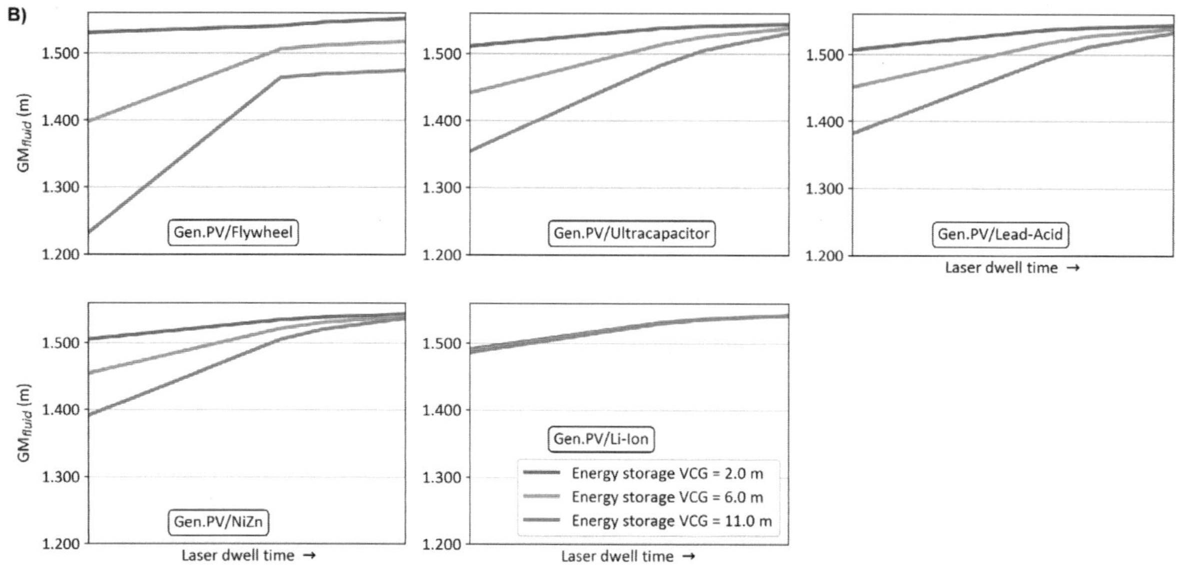

Figure 2: Plots showing the relationship between L-DEW dwell time and vessel GM_{fluid} for five different types of energy storage located at three different VCGs. A) shows the results from the simulations conducted on the Generic Frigate, while B) shows the results from the simulations conducted on the Generic Patrol Vessel. Numbers are indicative.

GM_{fluid} is a measure of initial stability of the vessel taking into account the free surface effect from its tanks. In other words, it is a measure of how stable the ship is in its upright position. A higher GM_{fluid} indicates a more stable ship that produces higher righting moments thus having a "harder" roll, while a lower GM_{fluid} indicates the opposite: a less initially stable ship with a "softer" roll about its upright position.

GM_{fluid} is highly dependent on the vessel's VCG. A higher VCG equates to a lower GM_{fluid} and vice-versa. Therefore, heavy weights placed on lower decks tend to decrease the vessel's VCG but increase its GM_{fluid}. This relationship can be observed in Figure 2.A), where energy storages with higher VCGs display lower GM_{fluid} when compared to lower VCGs for the same laser dwell time. The rate at which the trends diverge is dependent on the type of energy storages. Heavier types, such as flywheels, display a larger difference of GM_{fluid} between VCGs, while lighter ones like lithium-ion batteries, hardly show any difference between VCGs.

All curves in Figure 2.A) show similar trends but the one for a flywheel with VCG at 2.0 m. In this case, a lower laser dwell time corresponds to a higher GM_{fluid}, which is not observed in other curves. As mentioned, the indicated performance of flywheels is still pending validation, but its typical heavy weight associated with a low VCG would likely result in this phenomenon, as it lowers the overall VCG of the ship. The overall trend is notwithstanding confirmed by all other curves in Figure 2, where higher weapon effectiveness (lower dwell times) are associated with lower initial stability (lower GM_{fluid}).

Figure 2.B) shows a similar set of graphs, but these were based on the simulation results conducted on the Generic Patrol Vessel. Despite showing higher values of GM_{fluid}, the observed trends are nearly the same: a more effective weapon means a less initially stable platform. In this case, however, a trend inversion does not occur for the flywheel with VCG at 2.0 m.

GM_{fluid} is a parameter that only evaluates the initial stability of a vessel. For large angle stability assessment, one option is to analyse the ship's righting arm (GZ) curves. These can be used to compute the righting moments for a wider range of heel angles, from which several stability parameters can be derived. These parameters are usually associated with stability requirements dictated by technical standards which include maximum GZ, angle of maximum GZ, minimum area under GZ curve, among others. When overlapped with heeling moment curves, it can provide good measures of dynamic stability and how much margin the vessel holds as designed.

Figure 3 shows a set of GZ curves for various types of energy storages with their VCG at 11.0 m. The plots are zoomed in for visualisation purposes and do not show the origin, however all curves intersect it. All curves follow a similar behaviour, with an increasing GZ before reaching a peak at 45° and decreasing from there onwards. The point at which the curves cross the horizontal axis, denominated point of vanishing stability, vary between laser dwell-times and types of ESD, but they all approximate 80° of heel.

Again, due to the varying weights of ESD, the trade-offs between large angle stability and weapon effectiveness are notably different depending on their type. For heavy flywheels, the drop in GZ observed between the most effective weapon and the least effective weapon is significant, while the drop is less pronounced for lithium-ion batteries. The drop in GZ between the first, second and third most effective weapons is not distinguishable in Figure 3, indicating that, if the limiting factor was to be large angle stability, there would be no good reason to select the second or third most effective L-DEW configuration. This is an example of technical evidence that could be leveraged in decision making.

Figure 3.B) depicts the set of curves calculated for the Generic Patrol Vessel. These curves display higher maximum GZ, higher angle of maximum GZ and shoulders associated with the ship's superstructure. The Generic Patrol Vessel was modelled with its superstructure within the watertight boundary, meaning that the superstructure contributes to buoyancy. When they go underwater, the superstructure's high walls increase the vessel's second moment of area, creating the peak in GZ at higher angles of heel. Nevertheless, the trade-offs are essentially the same: higher weapon effectiveness is associated with lower righting moments. As a result, higher weapon effectiveness could decrease design margins, hindering the platform's ability to be upgraded along its lifecycle. This would also be a critical topic to be kept in mind by decision makers.

It is important to note that the results presented in Figure 3 were calculated for ESDs with a relatively high VCG of 11.0 m, so the effect on stability could be visualised more easily. Ideally, energy storages would be installed low on the vessel, mitigating their detrimental effects on stability.

It must be noted that the selection of type of energy storage depends upon more parameters than the ones presented above. Designers and decision makers need to consider how many shots the storage can hold before a recharge is needed, how long it takes for the storage to recharge itself and how safe the system is rated, to name a few important considerations.

For instance, flywheels are heavy by nature [29], possibly degrading the ship's stability and design margins as discussed above. However, flywheels offer a much deeper magazine when compared to other ESD technologies, meaning that it can hold more shots comparatively. Flywheels also have a rapid recharge rate [17], increasing the overall availability of the L-DEW system during missions. Ultra-capacitors also offer a quick recharge time [13], but cannot hold as much energy as flywheels, shortening the depth of the magazine. It must be reiterated that the performance numbers for flywheels and ultra-capacitors are indicative at this stage, requiring validation to be used in more detailed studies.

If a battery unit is chosen for the ESD, similar considerations must be made. The main drawback of having batteries compared to flywheels and ultra-capacitors is their longer recharge time and limited service life. Other notable aspects are inherent to the battery chemistry. Lead-acid batteries, for instance, are a well-known technology and relatively safe to operate, however they have greater size and weight requirements and only offer a limited number of shots. In contrast, lithium-ion batteries have much lower space and size requirements for the same number of shots, however their safe operation in naval platforms is still being studied due to the fire hazard associated with their chemical composition [30]. Nickel-zinc batteries have been proven to be a good alternative, as they have slightly less weight and size requirements than a lead-acid battery while offering a much deeper magazine [31]. It is also significantly less flammable than a lithium-ion battery.

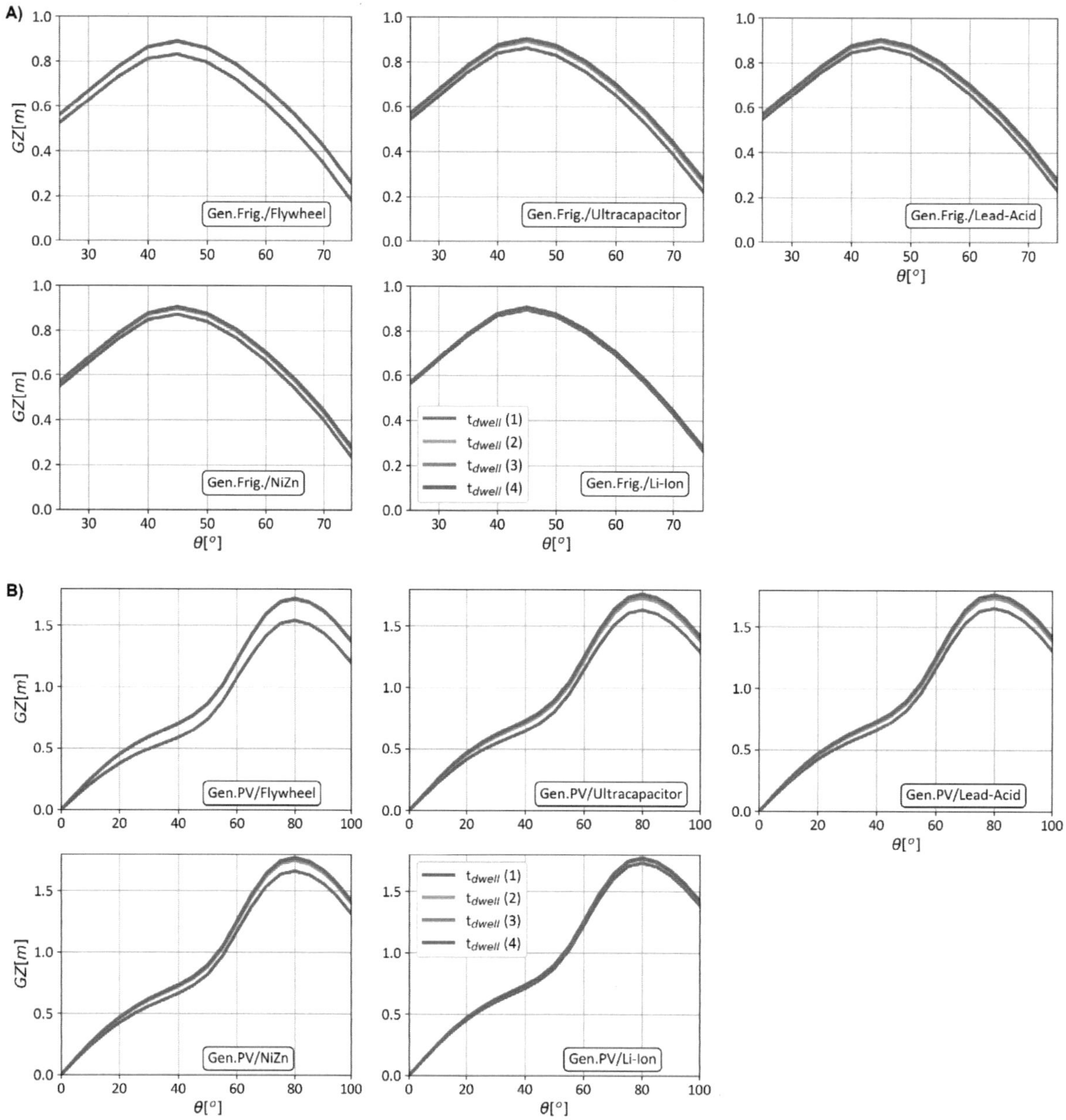

Figure 3: Plots showing the relationship between L-DEW dwell time and vessel GZ for five different types of energy storages located at VCG of 11.0 m. A) shows the results from the simulations conducted on the Generic Frigate, while B) shows the results from the simulations conducted on the Generic Patrol Vessel. Numbers are indicative.

3.2 (b) Vessel operational range

Vessel operational range calculations rely on a large set of models within the workflow. Propulsive, L-DEW and hotel loads need to be calculated or defined and passed onto the power and energy models, which then estimate the fuel consumption rate. This value, associated with the volume of burnable fuel, the vessel's service speed and several other factors, allows the simulation workflow to calculate the operational range.

Propulsive loads stem from vessel resistance and propeller performance at a given speed. Resistance is broken down into four portions: calm water, added resistance in waves, appendages and wind. Wind and appendages resistance, as the workflow is constructed, are less sensitive to vessel draught and attitude. Calm water resistance and added resistance in waves however, are more sensitive on those two parameters, as they have an impact on hulls' wetted areas and how they interact with waves. Therefore, heavy weights placed on the vessel may have a significant impact on resistance and, ultimately, operational range even if they do not consume fuel directly.

In this study, hotel load was set to a fixed value for both platforms representing all onboard systems excluding L-DEW. For the Generic Frigate, GT mode was adopted, meaning that hotel loads were applied to the diesel generator, while the gas turbine provided propulsion power. For the Generic Patrol Vessel, hotel load was applied to the vessel's generator set, while the propulsion load was fed by the propulsion engine through a geared drive. The energy required by the L-DEW system was defined by MALTESOR and ESD models' results associated with a fixed number of shots. Then, using the generator set model, the amount of fuel needed to output this required energy is calculated and the result is deducted from the total volume of burnable fuel for the supply of hotel and propulsion loads.

Figure 4 shows a subset of the results produced by the simulation workflow. In this case, the nickel-zinc energy storage was selected for further analysis and its VCG was defined at 2.0 m for both platforms. The Generic Frigate curves show a clear downwards trend in operational range with the increase in speed, which is expected at high speeds. When taking into account the accuracy of predictions offered by InteShip's range model, one can conclude that there are no significant differences in operational range between different laser dwell times for the Generic Frigate's results. The accuracy of predictions are represented in Figure 4 by the error bars.

The Generic Patrol Vessel's range curves also show a behaviour consistent with the given speed interval, initially rising before reaching a peak at a relatively lower speed and, from there onwards, adopting a downwards trend. It can be noted that the speed at which the peak is reached differs for each curve. Unlike the Generic Frigate, L-DEW loads have a more notable effect on operational range on the Generic Patrol Vessel. The lower speeds make the propulsive loads less prominent, which is observed by the larger difference in range between laser dwell time curves. However, when taking into account the accuracy of predictions for the Generic Patrol Vessel, no significant differences in operational range are observed between different dwell-times either. A larger difference can be seen when the curves from the highest and lowest dwell-times (t_{dwell} (1) and t_{dwell} (4)) subset are compared against the curves from the mid two dwell-times (t_{dwell} (2) and t_{dwell} (3)) subset, where the operational range observed in the latter is larger than in the former. This indicates a possible optimum. Nevertheless, as the error bars indicate, the difference cannot be considered significant and further studies would be required.

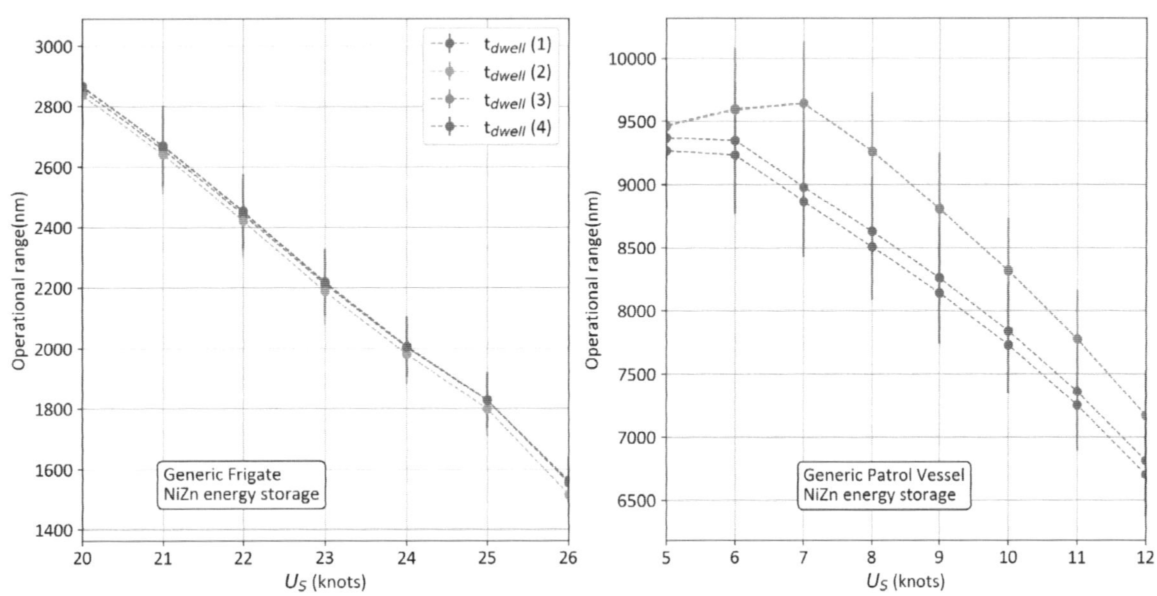

Figure 4: Plots showing the relationship between L-DEW dwell time and operational range at a range of speeds for the Generic Frigate making way at gas turbine mode and the Generic Patrol Vessel making way

with diesel propulsion engines and separate generator sets for hotel and L-DEW loads. Error bars indicate the accuracy of range predictions. **Numbers are indicative.**

Figure 4 is an example of performance prediction that could be used by decision makers when considering an L-DEW system. For the Generic Frigate, it is clear that GT mode would not predominate during a mission, but if this factor was critical for the operators and range was a limiting factor, the choice for the most effective weapon would be apparent. The same applies for the Generic Patrol Vessel, however in this case, a more detailed investigation would be required as the initial results show a possible difference in performance between different laser dwell-times.

3.2 (c) Seakeeping

The seakeeping model in the simulation workflow calculates ship motions based on the strip theory method. It leverages multiples inputs such hull geometry, mass distribution and environmental conditions. When a specific position on the vessel is specified, various parameters can be calculated to include motions in all degrees of freedom, incidences of water on deck and wave slamming. In this study, these were calculated for the location where the L-DEW optical equipment was installed.

DEF(AUST)5000 Volume 3 Part 6 [32] does not specify operational limitations for L-DEW systems, but it does furnish these for the operation of missiles. For the purposes of this study, the operational limitations of missiles were extended for L-DEW. They consist of a limit in root mean square (RMS) pitch and roll motions of 7.6°. Future studies can investigate the effect of angular motions on the ability to keep a laser beam on target, which could be useful in defining operational limitations for L-DEW.

Figure 5.A) shows polar plots of pitch and roll RMS motions calculated for the Generic Frigate making speeds from three to 27 knots and facing wave directions from zero to 180°. Due to the bilateral symmetry of the hull, only the seaways from zero to 180° need to be presented, as the other half would just be a mirrored image of its counterpart. The pitch plot on the left shows that, for sea state six, pitch RMS values are well below the limit stipulated by DEF(AUST)5000 [26], which is expected due to the typical length-to-beam ratio of a naval platform. The roll plot on the right however, shows that the roll RMS limit is marginally violated for the higher speeds and quartering seas (60° wave direction). For all other settings, vessel roll at the optical equipment location sits below the limit and would not impact L-DEW operation.

Figure 5.B) shows similar plots generated for the Generic Patrol Vessel. The speed range is reduced in this case, and it goes from three to 20 knots. The pitch RMS is slightly higher for this platform at sea state six when compared to the Generic Frigate, but the operational limits are nevertheless not violated. The calculated roll RMS motions are also higher in comparison and the region that violates the limit of 7.6° is marginally wider in terms of speed and wave direction. Nevertheless, the region is still confined to higher speeds and quartering seas.

For both platforms at sea state six, L-DEW will seldom face any situations where vessel motions would halt its operation. An operability index was calculated for both vessels in this setting, which is a measure of probability of L-DEW being operational. For both cases, L-DEW remains operational at over 95% of mission time. It is important to note that this was calculated for sea state six, which is already considered very rough seas. The choice for sea state six was made so the impact of speed and wave direction on vessel motions could be more easily visualised. For sea state four, which is still rough but a more common operational scenario, L-DEW would remain operational at virtually all times if only limitations on vessel motions are considered.

Finally, it must be noted that the damage-inflicting (non-ESD) component of L-DEW was kept at a fixed location on both vessels during seakeeping simulations. Due to the interconnected nature of L-DEW and platform, if the non-ESD component was to be moved, local motions would primarily change due to a changing relative position of the component to the vessel's centre of gravity, but also due to a shifting vessel centre of gravity itself. This dynamic effect in operability was not explored in this paper, but could act as a limiting factor in L-DEW selection.

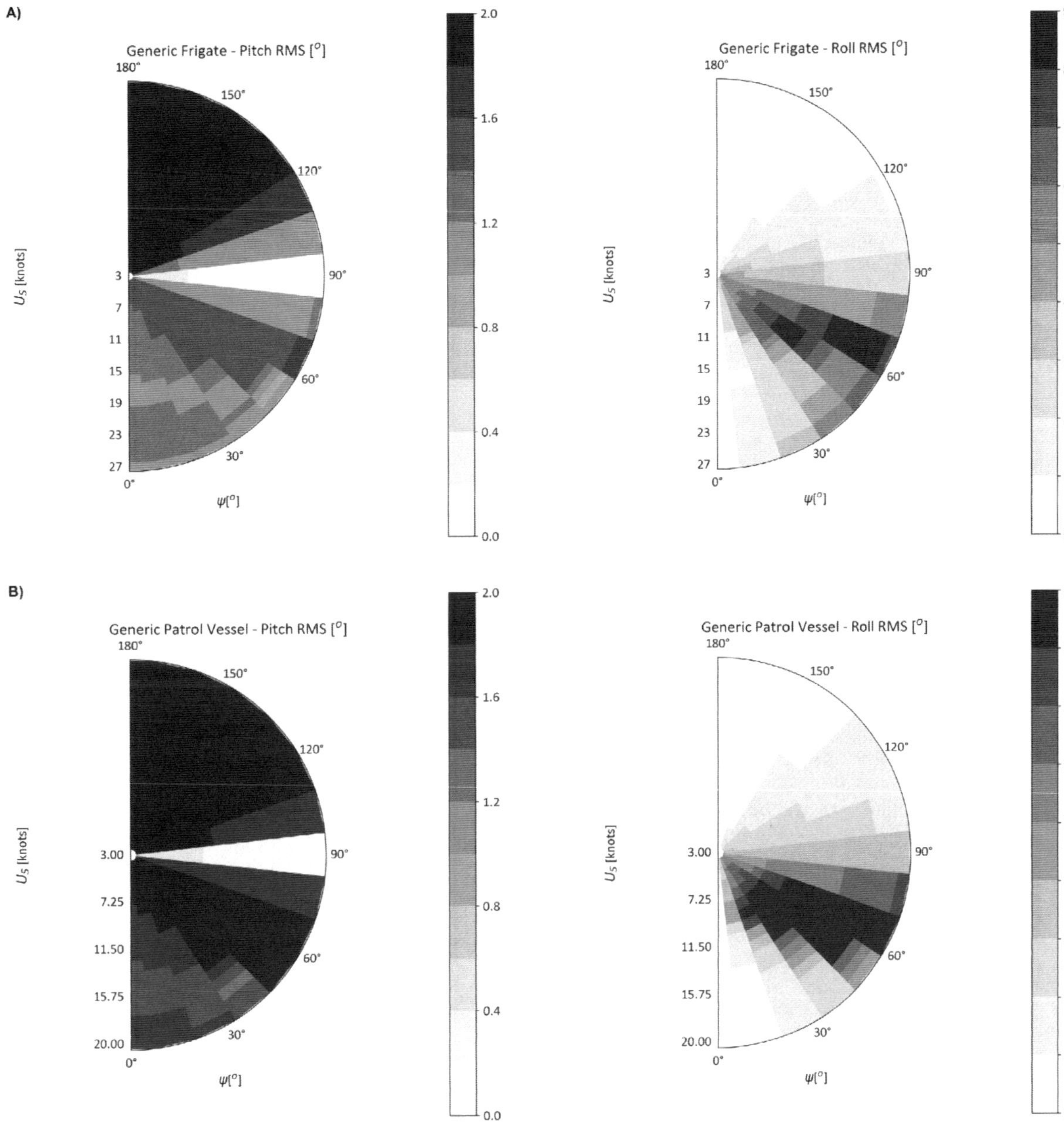

Figure 5: Polar plots showing pitch and roll RMS. A) shows the results from the simulations conducted on the Generic Frigate, while B) shows the results from the simulations conducted on the Generic Patrol Vessel. Numbers are indicative.

4. CONCLUSION

The main idea conveyed by this paper is that digital engineering tools can be used to reduce technical risks at different stages of design by generating quality evidence for a better-informed decision making process. In this case, DSTG's InteShip leveraged results from the MALTESOR and L-DEW ESD models to analyse the integration of L-DEW systems on two generic naval platforms at the conceptual level.

It was demonstrated how the simulation workflow was constructed, taking advantage of the interconnectedness of InteShip and using the outputs of some models as the inputs of others, allowing the user to investigate the effects of parameter changes on the overall performance of the platforms. The workflow is bounded by multiple conditions denominated simulation settings, which include the position of optical equipment and

ESDs on the vessel, service speed, volume of burnable fuel and environmental conditions. The limitations of the models involved were also discussed, highlighting that the analyses were conducted at a conceptual level and more detailed information on the systems and platforms would be required for more accurate results.

Simulation results were discussed with the aim of showing how the output data can be processed and presented to furnish decision makers with options and recommendations. These included advice on type of energy storage and its location, laser input power requirements and vessel service speed, indicating the limiting factors and the option to maximise capability.

The simulation framework constructed for this paper can be reutilised in a myriad of ways, catering for different types of studies with a short turnaround time. The same set of inputs provided here can be used to generate different outputs, which can be processed in a different fashion to answer other questions. The framework can also be fed more detailed data and produce results of higher fidelity that can be used in later stages of design.

5. REFERENCES

1. BROWN, A., and THOMAS, M., "Reengineering the Naval Ship Concept Design Process", *Research to Reality in Ship System Engineering Symposium* ANSE, 1998

2. BERNSTEIN, J I., "Design Methods in the Aerospace Industry: Looking for Evidence of Set-Based Practices", *Boston, USA, Massachusetts Institute of Technology*, 1998

3. COFFEY, V., "High-Energy Lasers: New Advances in Defense Applications*", Optical Society of America* , 2014

4. O'ROURKE, R., "Navy Shipboard Lasers: Background and Issues for Congress*" Congressional Research Service*, 2024

5. Bentley Systems (2020) *MAXSURF Stability User Manual*, Bentley Systems Inc.

6. Bentley Systems (2020) *MAXSURF Resistance User Manual*, Bentley Systems Inc.

7. HOLTROP, J., "A statistical re-analysis of resistance and propulsion data*", International Shipbuilding Progress* 31 (363) 1984

8. PECK, R W., "The Determination of Appendage Resistance on Ships", *Delft, The Netherlands, Technische Universiteit,* 1976

9. Bentley Systems (2020) *MAXSURF Motions User Manual*, Bentley Systems Inc.

10. GERRITSMA, J., and BEUKELMAN, W., "Analysis of the resistance increase in waves of a fast cargo ship*", International Shipbuilding Progress* 19 (217) 285-293, 1972

11. MCTAGGART, K A., "SHIPMO7: An Updated Strip Theory Program for Predicting Ship Motions and Sea Loads in Waves", *Defence Research Establishment Atlantic*, 1997

12. HOLTROP, H M G, "An approximate power prediction method", *Delft, The Netherlands, Intenational Shipbuilding Progress*, 1982

13. NAJIB, S., and ERDEM, E., "Current progress achieved in novel materials for supercapacitor electrodes: mini review", *Nanoscale Advances, Royal Society of Chemistry* 2817 – 2827, 2019

14. SIMON, P., and GOGOTSI, Y., "Capacitive Energy Storage in Nanostructured Carbon-Electrolyte Systems*", Accounts of Chemical Research* 1094–1103, 2011

15. AMIRYAR, M M., and PULLEN, K R., "A Review of Flywheel Energy Storage System Technologies and Their Applications*", Applied Sciences* , 2017

16. DAOUD, M I., et al. "DC Control of an Advance Flywheel Energy Storage Kinetic Traction System for Electrified Railway Industry", *IEEE Conference,* 2013

17. HEBNER, R., BENO, J., and WALLS, A., "Flywheel Batteries Come Around Again", *IEEE Spectrum*: p. 46 – 51, 2002

18. MOUSAVI, S M., et al., "A comprehensive Review of Flywheel Energy Storage System Technology*", Renewable and Sustainable Energy Reviews* 477 – 490, 2016

19. AGWU, D D., et al., "Review Of Comparative Battery Energy Storage Systems (Bess) For Energy Storage Applications In Tropical Environments", *IEEE 3rd International Conference on Electro-Technology for National Development (NIGERCON),* Nigeria, 2017

20. BEARD, K W., "Linden's handbook of Batteries, 5th ed", *New York, McGraw Hill*, 2019

21. DUFFY, J., and DENEHY, S., "Measurement of the free surface effect in the aft region of a generic OPV model fitted with propellers in calm waters and head seas", *Launceston, Australia, Australian Maritime College*, 2019

22. RICHTER, J., and REINHOLZ, O., "Model Tests for the SEA1905 Project", *Institution: HSVA 2020 Report Number: WP 20-018 Hamburg, Germany*, 2020

23. CARLTON, J S., "Marine Propellers and Propulsion" *Publisher: Elsevier 2007 Second edition*, 2007

25. MTU (2019) *MTU Marine & Offshore Solution Guide*

26. GE Vernova (2024) *LM2500 Gas Turbine*. https://www.gevernova.com/gas-power/products/gas-turbines/lm2500

27. Scania (2015) *Scania Marine Engines*. Edition 2 https://www.scania.com/content/dam/scanianoe/market/master/products-and-services/engines/pdf/specs/marine/marine-auxiliary/DI16090M_450-511kW.pdf

28. HUNT, E., "High-Energy Lasers", Technologies, R., Editor. *United States*, 2023

29. Amber Kinetics (2022) *The Industry's Only Long-Duration Kinetic Energy Storage System (KESS)—Enabling Highly Efficient Industrial and Commercial Applications*. Union City, California.

30. SI, R J., LIU, D Q., AND XUE, S Q., "Experimental Study on Fire and Explosion Suppression of Self-ignition of Lithium Ion Battery", *International Conference on Fire Science and Fire Protection Engineering,* Chongqing, China, 2017

31. Paay (2023) *High power NiZn Battery Initial Concept*. Osborne, South Australia,

32. Department of Defence (2003) *DEF (AUST) 5000 ADF Material Requirement Set, Volume 3 - Hull System Requirements, Part 6 - Seakeeping*. A016464. Canberra, Australia, RAN

6. AUTHORS BIOGRAPHY

Christian Rayes holds the current position of Naval Architect at Defence Science and Technology Group. The Author is responsible for the development of modelling and simulation tools for surface naval vessels. The Author's previous experience includes structural engineering projects for the offshore industry and the development of safety cases for software systems embedded in naval platforms.

Phong Nguyen holds the current position of Capability Analyst at Defence Science Technology group. The Author is responsible for evaluating emerging weapon systems to provide informed advice to Defence stakeholders and decision makers.

Garry Hale holds the current position of Capability Analyst at Defence Science and Technology Group. The Author is responsible for various aspects of the analysis of novel and emerging weapon systems. The Author's previous experience includes weapons systems propulsion development and testing, 2006 to 2022, and weapon systems operation and maintenance in the RAN, 1985 to 2005.

WEIGHT REDUCTION IN EXISTING SHIP DESIGNS

D W Whittaker, BAE Systems Maritime, Australia

SUMMARY

This paper describes a simple but systematic approach to identifying potential weight savings in existing ship designs. The proposition is that by applying these strategies for interrogating the ship designer's weight report the areas of the design with the best potential for weight saving can be identified. This is suggested as a better approach than ad hoc methods such as getting a panel of experts to brainstorm what design changes might provide a weight saving.

NOMENCLATURE

VCG	Vertical Centre of Gravity
LCG	Longitudinal Centre of Gravity
TCG	Transverse Centre of Gravity
≻	More important than
≺	Less important than
SWBS	USN Ship Work Breakdown Structure (NAVSEA 900-LP-03909010)

1. INTRODUCTION

With tightening budget constraints faced by navies, modern combatants have service lives that are measured in decades. However, they must defend their countries' maritime borders and offshore assets in a political climate that can change weekly. With an ever-evolving political climate come new potential adversaries, bringing with them new unanticipated levels of technology. As a new combatant's lead time is measured in years and its cost a major investment with limited funds, the only affordable and timely response often is to upgrade existing vessels. However, with a finite displaced volume, there is only so much capability that can be added without the need to take something else off. The challenge for designers, therefore, becomes ascertaining what to take off and from where.

2. THE TYPICAL METHOD FOR REDUCING WEIGHT

A typical response to this dilemma is to form a committee of subject matter experts to brainstorm what might be removed or replaced with something lighter, so as to make room for whatever weapon or sensor is now urgently required. While this approach will typically find some way of accommodating the required capability, it does not promise the best solution nor is it likely to keep up with an ongoing changing threat landscape that will require the process to be regularly repeated. It also has the potential of achieving the desired reduction in mass through a lot of small changes in the design while overlooking heavy, outdated equipment or details that are near due for replacement. This would achieve the same saving, or more, with less cost and effort.

Another shortcoming of this approach is that subject matter experts are usually not cheap and a group of them is even more expensive. A single naval architect studying the problem in-depth may well be cheaper. While committees are excellent for reviewing the soundness of a proposed design change and provide the opportunity to consider its impact from multiple perspectives, generating the required objective evidence to justify the final decision is not easily done by them.

The committee approach also has a bias towards single step solutions to the issue of weight. For instance, Part A can be replaced with the newer version of Part A that is lighter, or Component B could be made out of aluminium instead of steel. These sort of options seem easier for a group of people to identify than say: System C is a larger portion of lightship mass than the equivalent system on other combatants so let's see if the other combatants have a better and lighter design for System C.

3. RESOURCES AVAILABLE FOR THE WEIGHT REDUCTION EXERCISE.

It would be naive to assume that the engineer or naval architect that has been entrusted with reducing the weight of the vessel already has a comprehensive knowledge of the vessel, its configuration and systems. Yet to be effective, all parts of the ship and its systems should be candidates for reducing weight, not just those systems within the vessel the particular engineer is familiar with. Likewise, going through the drawings of unfamiliar systems will be time consuming and will not easily reveal which components are overweight, even for someone who is experienced in the systems being examined. There are however resources available that if used systematically, can identify what parts of a ship's design justify a detailed examination because they have a higher probability of providing wanted weight savings.

These resources are: the designer's weight report for the ship under consideration; the ship/work breakdown structure for the vessel under consideration; and the designers' weight reports for other classes of vessels that are of similar purpose to the vessel under consideration.

3.1 DESIGNER'S WEIGHT REPORT FOR THE SHIP UNDER CONSIDERATION

The designer's weight report for a vessel is one of those documents that always exists for a vessel but is not often used or referred to once the vessel is in service. Consequently, it can be rather hard to locate. Once located it is a daunting document because it is a tally of all of the equipment and structure of the vessel complete with their mass, centres of gravity and the calculation for the vessel's and subsystems' mass and centroids. As such, this document has the literary appeal of a telephone book.

3.2 THE SHIP/WORK BREAKDOWN STRUCTURE FOR THE VESSEL UNDER CONSIDERATION

As the weight report is such a large document it is typically broken up with subtotals that match whatever work breakdown structure was adopted during the original construction of the vessel. If you are lucky, the work breakdown structure used will be NAVSEA 0900-LP-039-9010 the USN Ship Work Breakdown Structure (SWBS). The advantage is that it has been used on multiple ship builds, both here and abroad, and as such allows for the ready comparison of the weight reports of different vessels.

If the vessel is of UK design the weight report may be broken up using Ministry of Defence Standard 02-163 Classification for Weight Groups for Surface Ships. This system has a superficial similarity with NAVSEA 0900-LP-039-9010 but deviates in detail and philosophy. It has not been as universally applied across projects as NAVSEA 0900-LP-039-9010 and so provides less opportunity for comparing the weight breakdown across classes of ships. It is also less universally applied within projects. This is because many of the non-weight report functions of a ship work breakdown structure are performed by the Fleet Area Code in the UK way of doing things.

The Fleet Area Code (FAC) is defined in Ministry of Defence Standard 02-041 Requirements for Configuration Management of Surface Ships, Annex C. The FACs break up the ship into different categories and subcategories to those used in Defence Standard 02-163 and the relationship between the two is not defined in any official standard found by this author.

If the situation is truly unfortunate, the designer will have been allowed to use their own ship work breakdown structure during the ship acquisition, in which case a specific set of strategies will need to be applied to the weight report to identify weight reduction opportunities.

3.3 WEIGHT REPORTS FOR OTHER CLASSES OF VESSELS THAT ARE OF SIMILAR PURPOSE

Having access to the weight reports of other vessels is a powerful tool if they are available. Typically, getting hold of these documents requires special permission and prolonged negotiation with their custodians. The purpose of these other weight reports for light weighting is that it allows the investigator to quickly identify what portion of their ship is heavier or lighter than the equivalent systems in other ships. For this comparison to be meaningful the other weight reports need to be of ships of similar purpose. Comparing the weight report of a landing craft with that of a tanker would be meaningless but comparing two combatants, even with very different displacements and built decades apart gives useable data.

4. STRATAGIES AVAILABLE FOR FINDING WHERE TO LOOK FOR WEIGHT SAVINGS

Each of the following strategies are designed to guide the investigator as where best to look for weight saving opportunities. They do not find the actual design change that will reduce the vessel's mass but they do direct the investigator as to where to concentrate their effort. The actual strategy(ies) used is dependent on what resources are available.

4.1 PARAMETRIC COMPARISION

This strategy consists of comparing the weight breakdown between vessels of similar purpose. The numbers that are compared are the percentage portion of lightship mass of each system. In practice, this involves overlaying graphs of the portion of lightship weight by weight group number. For example, if the vessel of interest had a higher percentage value for HVAC (512 in NAVSEA) than the three or four other designs it was being compared with, then HVAC would be a system that justified further investigation.

The first step of this deeper investigation is to inquire if the intended purpose of the vessel justifies the extra weight of the identified system. Going back to the HVAC example, if the vessel of interest is intended for service in the tropics while the comparison vessels are only intended for temperate zones, then the heavier HVAC system may well be justified.

If the identified system passes this first gate then a parametric comparison within the system is justified. Going back to the HVAC example, this would involve overlaying graphs of the relative contribution to the system's mass of components such as ducting, fans, heaters, supports etc. This level of investigation is when the results start to appear. Whatever part of the system is proportionally heavier than the other vessels needs to have its design compared with the reference vessels. In short, it is at this point of the investigation that relevant system specialists are consulted and drawings for the system are examined to identify specific items that can be made lighter.

4.2 MULTI CRITERIA SCORING

This strategy is simply an application of the weighted sum model from multiple-criteria decision analysis and has value in the following circumstances:

- Weight reports of other vessels are not available
- Weight is not the only parameter being considered
- The other weight reports available do not use the same SWBS system as the ship being investigated

The typical criteria that are considered are of course weight, but also moving the vessel's centre of gravity, the ease with which a change could be effected, and the likelihood that a deep investigation of the system will in fact identify a weight saving. Another criteria might be the likelihood of an identified weight saving actually being approved by the customer or project authority. An example of this is the combat system where experience suggests that those who want the extra capability fitted to the vessel do not want to lose any other capability to get it.

Looking in turn at the four potential criteria mentioned above.

4.2 (a) Weight

Even if there are no reference vessels to compare the weight of a system to see if it is proportionately heavy, the vessel's own weight report can still be useful in judging if an investigation into a system will be worthwhile. The logic is that achieving, say a 1% weight saving, will involve a comparable level of effort regardless of that system's weight. However, a 1% weight saving on a heavy item or system should provide a better result for that effort than the 1% weight saving for a lighter item or system. If weight is the only criteria then the weight report can be used to sort the SWBS numbers in weight order, i.e. the heaviest SWBSs first and the lighter SWBSs last. The weight investigator can then focus their attention on those Weight Codes at the top of the list, discarding any they judge are not amenable to design changes and doing a deeper investigation on those that remain. That deeper investigation again involves using the weight report and doing a weight breakdown of the items within the heavier weight codes. If there are any categories of items that contribute most of the mass for one of these heavier SWBSs then the potential weight saving has been found.

4.2 (b) Centre of Gravity

The change in vessels' centre of gravity that is usually sought is a lowering of the vessel's vertical centre of gravity (VCG). This makes sense as most new sensors need to be installed as high as possible to maximise their range and most weapons systems have to be attached to a weather deck in order for them to work. In both cases the desired increase in capability is raising the vessel's centre of gravity as well as reducing its freeboard. However, sometimes it is also necessary to correct the vessel's heel and/or trim with the intended weight reductions.

4.2 (c) Ease of Change

The ease with which the vessel's mass might be reduced will definitely be a criteria for both the customer and the project manager, even if it is not explicitly given as a requirement to the naval architect doing the study. It will often be assumed that unplugging and unbolting a piece of equipment and replacing it with a newer model that weighs less will be an easier change than, for example, redesigning a steel deckhouse in aluminium. There are multiple strategies available for reducing the weight of a system but not all of these strategies are available for every system.

A weight saving in the ship's structure (all the NAVSEA 100s Group) will require reanalysis of the structure under consideration - HARD. A weight saving in outfit items (the NAVSAE 600s Group) will often involve a change in the design of attachment details - EASIER. A weight saving in machinery (mostly the NAVSEA 500s Group) will involve straight substitution - EASIEST but may also require testing to confirm the new item meets requirements such as shock and vibration - EASY. A weight saving in the ship's propulsion system (NAVSEA Group 200) may well require a major redesign of the vessel – NOT GOING TO HAPPEN.

4.2 (d) Likelihood

The type of investigation described in this paper is a high level investigation, intended to be used to identify where effort should be focused during the next more detailed phase of looking for weight savings. As such, there is no absolute certainty that the areas identified will actually have achievable weight savings in them. There are, however, areas of greater and less likelihood of success and a judgement call can be made as to what this likelihood is.

This judgement call can be guided by a few principles. If the system has more than one strategy available for pursuing weight savings then the likelihood goes up, that is, there are more chances that at least one of the strategies will work. A piping system with lots of valves for instance may provide the opportunity of saving weight by substitution, for instance by swapping out the existing valves with ones manufactured from a stronger lighter material. The same system may also offer the opportunity of achieving the weight saving by shock testing lighter valves of the same material as the existing ones to see if the lighter version can pass the shock criteria that prompted the selection of the heavier valves. Each extra strategy that is available increases the likelihood that at least one will work.

The second principle for making an accurate judgement of likelihood is that if it can divided into a series of smaller judgements, that are combined mathematically, then the combined judgement is not only easier to make but also more likely to be accurate. The strategies for reducing the weight of a system are limited.

Basically the options are:

- Reanalyse the system in more detail,
- Make its components out of lighter materials,
- Use a different manufacturing technique that leaves less redundant material in the parts,
- Do more physical testing of the system's components to ensure only the lightest components that still pass the test are used, or
- Substitute different components that meet the required fit, form and function but are lighter.

Each system has a subset of these strategies that are applicable and their own likelihood of success for each strategy.

4.3 RELATIVE IMPORTANCE OF EACH CRITERIA

Not all of the selected criteria are of equal importance to the stakeholders. Mathematically we can deal with this using a weighting factor that reflects each criteria's importance relative to the other criteria. The more important the criteria the bigger the number. The rule here is that if the weightings for all of the criteria are totalled together then they should add up to 10. If each criteria is marked out of 10 then this ensures that the combined score for a system that takes all criteria into consideration is out of 100 and this makes the combined score easy to interpret.

So looking at criteria argued in Section 4.2 to be relevant, what is their order of importance?

4.3 (a) Weight

If a group of systems as represented by a SWBS number does not have a significant weight, then any plausible percentage reduction in its mass is not going to give much of a weight saving. Any consequent improvements in the vessel's centre of gravity will not matter and it will not matter how likely the weight saving is, nor will matter how easily the weight saving was achieved. On the basis of this logic the weight of the SWBS number is the **most important** of the criteria.

4.3 (b) Centre of Gravity

A weight reduction that increases the vessel's centre of gravity and hence reduces its stability is not likely to be appreciated. Nor will a weight reduction that introduces a heel to the vessel or an undesirable trim. If any found weight reduction opportunity is regarded as undesirable it will not matter how likely that weight change is to be achieved nor will the ease of achieving it. It will simply not be approved. On this basis, the location of the centre of gravity of the SWBS number is regarded as a **very important** criteria for considering its potential for holding desirable weight reduction opportunities.

4.3 (c) Ease of Change

The ease with which the vessel's mass might be reduced will definitely be desirable and a consideration once a potential weight saving has been found, but it is not a high motivation for picking where to look in detail to find it in the first place. On this basis the ease of achieving a weight reduction is the **least important** criteria for selecting where to look in the ship's weight report for a potential weight saving.

4.3 (d) Likelihood

A higher likelihood that a desirable weight saving can be found in a particular SWBS number will increase the motivation to look for them there. Here the question of how to judge the likelihood comes to mind and that is described in Section 4.2 (d) and 4.5 (d). Finding a big weight saving and having it move the vessel's centre of gravity in the right direction is still more important than the effort that went into finding the weight saving or getting it to work. So on this basis we will classify likelihood as an **important** criteria but not as important as weight or centre of gravity.

4.4 WEIGHTING CRITERIA

Having ordered the criteria on the basis of their relative importance, the next step is to quantify - put numbers to measure - their relative importance to us. As there are four criteria and their weightings need to add up to 10, then we could simply arbitrary apply the weightings 1, 2, 3, 4.

Table 1. Weighting of criteria.

Criteria	Importance	Weighting
The total weight that can be saved	Most important	4
Location of the weight saving	Very important	3
The likelihood of achieving the weight saving	Important	2

The ease of achieving the weight saving	Least Important	1
	Total	10

The author has found these weightings to be adequate but for those who want more rigour, there is an objective way to quantify the weightings. Setting up a table where the criteria are compared in pairs, the instances where a criteria is the preferred in a pair can be tallied. (Note this method has been adopted from a method described in Reference [5].)

Table 2. Calculation of weightings.

	Weight	**Centre of Gravity (CoG)**	**Ease**	**Likelihood**
Weight				
Centre of Gravity (CoG)	Weight ≻ CoG			
Ease	Weight ≻ Ease	CoG ≻ Ease		
Likelihood	Weight ≻ Likelihood	CoG ≻ Likelihood	Ease ≺ Likelihood	
Total Preferred[a]	3	2	0	1
Add 1[b]	4	3	1	2
Scale to add to 10[c]	4	3	1	2

Note:
a. The total of how many times the criteria is the preferred of a pair, in the table not just its column.
b. So that the subsequent calculations do not eliminate the least preferred criteria as a consideration but maintains its relative importance 1 is added to all the totals.
c. Note that if the weightings in the row above didn't add to 10 they would need to be scaled to do so and this row would therefore have different values.

Note that here the symbols ≺ and ≻ are not to be confused with < and >. The greater than > and less than < symbols are a comparison of quantities whereas the curvy symbols (≻ preferred more than and ≺ preferred less than) that admittedly look similar, are a record of preference or importance between two entities. The notation comes from the mathematics of preferences that was developed in economics.

The third last row is a total of how many times each criteria is the preferred of a pair. This leaves the least important criteria preferred zero times. So that the subsequent mathematics does not leave it with a zero weighting, and no longer a criteria, we add one to each of the counts on the second last line. The relative importance of each of the criteria has now been quantified and if they did not already add up to 10 they would need to be scaled so that they do.

4.5 SCORING AGAINST CRITERIA

The next challenge is scoring each SWBS number against each criteria. There are a lot of SWBS numbers and although no vessel uses all of them there are typically around 200 of them relevant and used for any particular design. The need is to go through the weight report quickly but accurately to identify which portions of the weight report (SWBS numbers) to concentrate the man-hours left over to find the weight savings needed. This means that the scoring of each SWBS number against each criteria cannot be a long deliberation.

A scoring system that has worked in the past with the example criteria is given in the following sections. Again, as the final combined score is out of 100 each individual criteria score needs to be out of 10.

4.5 (a) Weight

A logarithmic scale has been found to be fairly quick to apply. The idea is for those SWBS numbers that represent more of the vessels mass to get a higher score than those portions of the ship that are less significant with regards to mass. The example given in Table 3 can be scaled up or down to match the total displacement of the vessel.

Table 3. Weight scoring scale.

Mass Range	Descriptor	Score
100t to 999t	Significant	10
10t to 99t	Adequate	5
1t to 9t	Minimal	2
1kg to 999kg	Miniscule	1

4.5 (b) Centre of Gravity

The example scoring system given in Table 4 shows just how simplified the scoring system can be. Comparing the subtotal centre of gravity of a SWBS number in the weight report with that of lightship is a quick and simple thing to do.

Table 4. Centre of gravity location scoring system.

Effect of Removing Weight	Descriptor	Score
Lowers VCG, moves LCG towards amidships and TCG towards centreline	Highly desirable	10
Lowers VCG	Desirable	5
Moves LCG towards amidships and/or TCG towards centreline	Desired	2
None of above	Ambivalent	1

This example marking scale cannot be applied as blindly as the one for weight. Depending where the new capability is being added that requires the reduction in mass across the rest of the ship the impact on LCG and TCG may need to earn a higher or lower mark.

4.5 (c) Ease of Change

As stated in Section 4.2 (c) there are multiple ways of creating a weight reduction for a system but not all of them will work for every system. This example scoring assumes that there are only five ways to reduce the weight of a system or component and each has its own level of difficulty to see the weight reduction implemented. The example marking scale in Table 5 reflects a judgement of how difficult each strategy would be to implement for a completed vessel in service.

For example, if some additional analysis is required during the design, or preliminary design stage of the vessel's lifecycle then the extra effort is neither here nor there. But for a completed design that is already in service, reanalysing a system and the implementation of what could be far reaching changes is a very daunting situation. As this is interpreted as a hard strategy to implement then it gets a low score in the marking scale. At the other end of the scale a straight unbolting and replacement of an item with another of the same fit, form and function that is lighter is considered the easiest and therefore the highest scoring option.

Table 5. Ease of change scoring system.

Strategy for light weighting	Descriptor	Score
Analysis	Difficult	1
Alternate materials	Challenging	2
Alternate manufacturing Technique	Tricky	3
Testing	Involved	6
Different components	Easy	10

Note straight replacement is not always possible. Changing a structural detail for instance would not be regarded as a substitution but would need to be proceeded by some analysis and redesign. Where a system has more than one strategy available, the one with the highest mark provides the score given for that SWBS number.

4.5 (d) Likelihood

The challenge with scoring this criteria is that likelihood is a probability which means it has a value between 0 and 1 but what is required is a marking scale that can give a top score of 10. In addition, a guestimate of the probability is itself likely to be more guess than estimate. The marking scale in Table 6 can give three marks; 2, 1, and 0. Obviously this will not provide a top mark of 10.

Table 6. Likelihood scoring system.

Probability of Removing Weight	Descriptor	Score for each occurrence
60% to 100%	Likely	2
40% to 59%	Possible	1
0% to 39%	Unlikely	0
-	Unknown	0

The marking scale of Table 6 is combined with the judgement of light weighting strategies that might be available for that group of systems. This is best explained by example.

Table 7. Example of available light weighting strategies and likelihood score.

NAVSEA nnn – Dummy Weight Code						
Strategy for light weighting		Likely	Possible	Unlikely	Unknown	Total
Analysis		✓				
Alternate materials				✓		
Alt manufacturing Technique			✓			
Different components				✓		
Testing				✓		
Total ticks[a]	(a)	1	1	3	0	
Occurrence Score[b]	(b)	2	1	0	0	
Score[c]	(a) x (b)	2	1	0	0	**3[d]**

Note:
a. Total number of ticks for that columns likelihood
b. The score earned for each tick for that likelihood obtained from Table 6.
c. Obtained by multiplying the number of ticks by the occurrence score for that likelihood.
d. The sum of the values in the four cells to the left.

Each separate strategy that provides a likely or possible means of achieving a weight saving adds to the probability that at least one of the strategies will have success. If every possible strategy was judged as likely, then the maximum score of 10 would have been achieved.

4.6 A WORKED EXAMPLE

To put this multi criteria scoring system together let's use NAVSEA 0900-LP-039-9010 SWBS 111 Shell Plating. We will assume that our example vessel has a light ship displacement of 5,000 t, and that it is a combatant so its structure will be around 48.5% of lightship 2,425 t, and estimate that shell plating is around 11% of structure 267 t.

Table 8. Weight score.

Mass Range	Descriptor	Score
100t to 999t	**Significant**	**10**
10t to 99t	Adequate	5
1t to 9t	Minimal	2
1kg to 999kg	Miniscule	1

Our fictitious ship has centre of mass for the shell plating below the vessel's centre of gravity and due to a very large flight deck aft and a boxy superstructure forward the vessel's LCG is forward of amidships but the shell plating centre of gravity is close to midship. With the shell plating symmetrical port and starboard, reducing its mass is not going to move the ship's centre of gravity in any desired direction.

Table 9. Centre of gravity location score.

Effect of Removing Weight	Descriptor	Score
Lowers VCG, moves LCG towards amidships and TCG towards centreline	Highly desirable	10
Lowers VCG	Desirable	5
Moves LCG towards amidships and/or TCG towards centreline	Desired	2
None of above	**Ambivalent**	**1**

As the shell plating is part of the vessel's hull girder and change in its design will impact the ship's structural calculations necessitating redoing the required analysis. Let's assume that the original design had the shell plating made from Lloyd's Grade A steel leaving the option of changing the shell to a stronger grade of steel, so alternate materials is an option, but how practical it is depends on how much plate will be cut out and replaced. Let's assume that there is a localised thick plate due to a stress concentration. The only alternate manufacturing technique that comes to mind for shell plating is a change in the welding technology used to put it together. Any change in mass due to a different type of welding is considered near zero. The next assumption is that the shell plating was made thicker because there was a requirement to patrol the Southern Ocean so extra plate thickness was added due to concern over the steels performance in low temperature shock. This last concern might be addressed by physically testing the selected steel to confirm it's ductile to brittle transition temperature and doing some low temperature explosion bulge testing. Shell plating for ships is produced to defined standards regardless of manufacturer so replacing it with one from a different supplier will make no difference if the same grade of steel is being ordered.

Table 10. Ease of change score.

Strategy for light weighting	Descriptor	Score
Analysis	**Difficult**	**1**
Alternate materials	**Challenging**	**2**
Alternate manufacturing Technique	Tricky	3
Testing	**Involved**	**6**
Different components	Easy	10

Of the three strategies relevant to NAVSEA 111; analysis, alternate materials, and testing, the one that attracts the highest score is testing so that provides the score for this criteria.

In Table 11 the judgement call on the likelihood of each of the light weighting strategies is recorded and the score for likelihood is calculated.

Table 11. Likelihood of available light weighting strategies.

NAVSEA 111 – Shell plating						
Strategy for light weighting		**Likely**	**Possible**	**Unlikely**	**Unknown**	**Total**
Analysis			✓			
Alternate materials		✓				
Alt manufacturing Technique				✓		
Different components				✓		
Testing		✓				
Total ticks	(a)	2	1	2	0	
Occurrence score	(b)	2	1	0	0	
Score	(a) x (b)	4	1	0	0	**5**

Calculating the combined score for this imaginary example ship's shell plating consist of multiplying the scores for each criteria against the relevant weighting and then summing the results.

Table 12. Total light weighting score.

SWBS 111 Shell plating			
Criteria	**Score**	**Weighting[1]**	**Weighted score**
Weight	10	4	40
Location of CoG	1	3	3
Ease of change	6	1	6
Likelihood	5	2	10
Combined score			**59**
Notes: 1. Weightings are taken from Table 2 in Section 4.4			

This gives the shell plating (SWBS 111) of our imaginary ship a score of 59/100 for its potential and desirability as a place to look for light weighting opportunities. Chances are there will be several systems that score better than the shell plating and they will be investigated in preference to this system. However, even if the system had scored higher, the replacement of an existing ship's shell plating will never be cost effective. Obviously some common sense needs to be applied when interpreting the results of this method.

5. REVIEWING THE WHOLE OF THE VESSEL FOR WEIGHT SAVING OPPORTUNITIES

The point of this paper is to provide a tool to systematically find those weight saving opportunities that a panel of subject matter experts will overlook. The panel of experts will be limited by time available for meetings, the mix of personalities, and the range of experience available to the panel. Their biggest disadvantage though is that apart from time availability and the magnitude of the weight savings that need to be found, there is no defined start and finish to their deliberations.

The weight report on the other hand, although big, does have a first and last page. If we apply the above approach to every weight group number in whatever ship work breakdown structure was adopted for the weight report then there can be no dispute that some part of the ship was overlooked should the total of the weight savings eventually found not meet the mass reduction desired. To that end, it is recommended that the investigators final report needs to address even those weight group numbers not applicable to the vessel with a 'N/A' rather than leave them out of the report. For those weight groups that are applicable, they should be listed by their score in descending order as the priority list of most likely to least likely to provide a desirable weight saving. This will provide the project's management with a clear map of were to allocate their resources for a detailed investigation of the vessel's design.

6. FINAL STEP

Once the top NAVSEA numbers for weight reduction have been found then something as simple as a pie chart will quickly reveal where the weight saving is. Many NAVSEA numbers have a large amount of weight dominated by one type of item. To illustrate this below is a pie chart of the weight breakup typical of SWBS 512 HVAC.

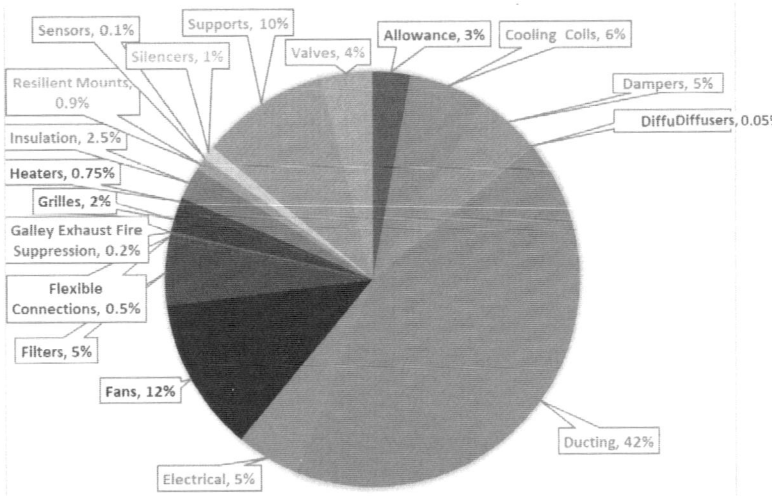

Figure 1: Typical weight breakdown within a NAVSEA SWBS number.

On the basis of this chart it would seem that the extra effort and expense of designing lighter ventilation ducting that meets the customer's and regulator's requirement may be more worthwhile than looking for lighter sensors. Within many SWBS number groups there are one or two items that contribute the majority of the mass and on most occasions these are the light weighting opportunities.

7. CONCLUSIONS

Most of this paper is devoted to having an objective method to justify where to direct effort when needing to find weight reductions in a ship. The second method of multi criteria scoring has many steps but is doable. The first method of parametric comparison of the ship's weight report with those of other vessels of similar purpose is simpler and more direct but is only possible if these other weight reports are available for comparison. In any case, the need to reduce the mass of existing vessels in order to facilitate new capability not anticipated when the vessel was designed and built will continue. The methods described in this paper have a particular advantage over ad hoc approaches when significant weight savings have to be found more than once over the life of the vessel.

8. ACKNOWLEDGEMENTS

I would like to acknowledge the supportive cooperation from my peers in reviewing this paper.

9. REFERENCES

1. NAVSEA 0900-LP-039-9010 the USN Ship Work Breakdown Structure (SWBS)

2. Ministry of Defence Standard 02-163 Classification for Weight Groups for Surface Ships

3. Ministry of Defence Standard 02-041 Requirements for Configuration Management of Surface Ships

4. KUPPURAJU, N., ITTIMAKIN, P.; and MISTREE, F., "Design through selection: a method that works", *Design Studies Volume 6, Issue 2,* April 1985, Pages 91-106

9. AUTHORS BIOGRAPHY

David Whittaker holds the current position of Principal Naval Architect at the Research and Technology division of BAE Systems Australia – Maritime. The Author is responsible for investigations to increase future growth margins of existing surface combatant designs. The Author's previous experience includes the RAN Air Warfare Destroyer Project where he reviewed design artefacts including the weight report for safety concerns and accuracy.

THE LITTORAL WARFARE CHALLENGE

E J Waters and **C M Hoey** QinetiQ, UK

SUMMARY

This paper looks at how the changing operational environment will impact the next generation of landing ships that will be required for future amphibious operations. The paper focusses on the vulnerability implications to the ship of the methods used for offload of surface connectors (such as landing craft) and the growing need for land operations to be provided with air support, especially uncrewed aircraft. It is argued that these issues will provide drivers that will change the design of landing ships when compared to contemporary and historical ships of this type.

1. INTRODUCTION

An amphibious capability allows a nation to project influence ashore, from the sea, often at significant reach from home. Over the decades, naval architects have gradually evolved a well proven formula for the design of the landing ships that make such operations possible. The authors of this paper will argue that developments in technology have radically altered the underlying assumptions that have driven those design choices. As such this implies that, to remain credible, the design of future landing ships will need to be different from the accepted legacy status quo.

This paper will discuss some of the factors that will drive these changes, and provide some thoughts on how design of these ships will have to change to accommodate them. The opinions expressed in this paper are those of the authors alone and do not necessarily represent those of any organisations to which they are affiliated.

2. LITTORAL WARFARE BACKGROUND (from the Naval Architect's Point of View)

From the ship designer's point of view, the problem can be split into two major parts, namely:

- Transporting the landing force from home base to the theatre of operations; and
- Transferring the landing force from the sea to the shore.

For blue water navies, the former requires seaworthy ships, capable of transiting long distances at sufficient speed to provide a worthwhile reaction time, these will be termed "landing ships" for the rest of this paper. The second requires specialised vehicles that can transit from the sea onto the shore. Historically there have been many ships that combined these two capabilities. The most common examples of these being large seagoing ships that could be beached to offload their payload. These vessels, such as the many tank landing ships (LSTs) used in the Second World War, present challenging design compromises in hydrodynamics, structural design and sea-keeping – these will not be considered further in this paper. As such, most current operators implement a solution whereby a specially designed vessel can transport both the landing force and the landing craft and helicopters (often termed "connectors") required to deliver that landing force to the shore. This adds a further challenge to the Naval Architect, namely that of working out how to deploy and recover the connectors from the transport vessel.

This paper will concentrate on the design issues facing the landing ship, and the features related to deploying the connectors (and potentially other deployable craft and systems). Although this paper largely ignores the challenges faced in designing the connector craft, it must be understood that the capabilities of the connectors carried will of course impact the design of the transporting vessel, and some of these impacts are discussed or implied.

3. AMPHIBIOUS SHIPPING: CURRENT STATE OF THE ART

Over recent decades, there has been a remarkable level of commonality in the overall design concepts applied to landing ships. The vessel will be large enough to carry a substantial landing force, often incorporating sophisticated on-board command and communication facilities. It will also be capable of deploying connectors in the form of landing craft and helicopters, with two different design philosophies being applied depending on the balance of surface and air connectors carried. For the purposes of this paper these are discussed as either Dock Landing Ships (LPD) or Helicopter Landing Ships (LPH).

3.1 DOCK LANDING SHIPS

The LPD's primary method of placing forces ashore is via landing craft. Most LPDs have the capability of deploying large landing craft via a well dock, and smaller landing craft via davits. The well dock being an internal space within the ship that can be flooded to allow large landing craft to float in and out of the ship and to load directly from the internal vehicle deck. The ship will incorporate the ability to operate one or two helicopters to supplement the landing craft from a flight deck, aft of a forward superstructure.

3.2 HELICOPTER LANDING SHIPS

Resembling an aircraft carrier, the LPH instead uses helicopters as the primary means of transporting the landing force ashore. Typically, the ship will embark a large number of helicopters and be capable of operating a significant number of those simultaneously to provide sufficient offload capability. An LPH is also likely to carry at least davit launched small landing craft, though larger vessels may also incorporate a well dock. The very largest of this breed, such as the Australian Canberra Class or US Navy's Wasp Class are effectively a LPH and LPD combined, these vessels are often termed LHDs signifying balanced helicopter and dock offload capabilities.

3.3 MULTI-ROLE CAPABILITY

Amphibious operations provide a challenging set of design drivers to the naval architect, but occur only relatively infrequently. As such, landing ships are likely to spend a large amount of their lives doing things other than landing forces ashore. Fortunately, the general architecture of these ships with large, easily accessed, vehicle decks and the ability to host boats and aircraft means that by their very nature that they are suited to carry out a wide range of secondary roles. This is evidenced by many recent examples such as the Royal Navy's Bay Class being used as mine warfare motherships [1] and many examples where landing ships have proven ideal in disaster relief and humanitarian situations. Traditionally these secondary roles have not driven the vessel design, but "come for free" with the features required for the primary role.

4. THE CHANGING ENVIRONMENT

The LPD, with its roots in the Second World War, and the LPH, developed from the 1950s onwards, have evolved relatively little over the intervening decades. For instance, though there are many differences in design detail, the general concept of the Royal Navy's Albion Class LPDs [2] is essentially identical to that of its Fearless Class [3] predecessors designed in the 1960s. These ships are designed to offload relatively large utility landing craft from relatively close to the enemy shore due to the slow speed of the landing craft. This was considered acceptable at the time these vessels were designed with reference to the range of threats from shore, and the defensive measures that could be applied by the escorting task group and in extremis landing ship self defence weaponry. The offload capability was, likewise, designed to support the contemporary concepts of land warfare, whilst the whole package was constrained by the technology available at that time. However the last decade, and in particular the last few years, have seen dramatic changes in all of these aspects that will inevitably have to be considered extremely carefully by those tasked with designing the next generation of landing ships.

4.1 PROLIFERATION OF ANTI-ACCESS / AREA DENIAL (A2/AD) CAPABILITIES

Amphibious operations have always had to deal with an adversary's attempts to defend their shore and the littoral waters from which these operations are launched. Historical A2/AD measures such as mines, inshore attack craft and coastal artillery remain effective, but proliferation of shore launched anti-ship missiles (including hypersonic and ballistic types) and swarming drones (air and surface) have only multiplied these threats. Additionally the widespread access to surveillance technology including satellite and uncrewed aircraft as well as traditional maritime patrol aircraft can act as a force multiplier for both simple and sophisticated A2/AD systems. The concern is not only the rising capabilities of these weapons, but also their proliferation to many countries and even non-state actors [4]. Recent examples of how effective A2/AD can be in the littoral environment can easily be found, these include the Ukrainian successes against the Russian Black Sea Fleet with uncrewed surface vessels [5] and the use of anti-ship missiles by the Houthis in Yemen [6].

4.2 REVOLUTION IN LAND WARFARE

It only takes the briefest of exposure to the media at present to see graphic examples of how technology has impacted on land warfare. To be effective land forces now require access to uncrewed aircraft systems to provide situational awareness, reconnaissance, electronic warfare and even engage adversaries [7]. Some of these systems will, no doubt, be carried ashore by the landing force and even operated from connector craft. However, it is likely that larger air systems (potentially crewed and uncrewed) will be needed to provide Intelligence, Surveillance, Targeting and Reconnaissance (ISTAR) and precision strike. As such, it is suggested that aviation facilities on the next generation of landing ships will need to take this into account as well as the landing force offload that has historically been the primary driver for the flight and hangar deck design.

It also appears that there may be less emphasis on the use of heavy armour such as main battle tanks (MBT) in amphibious operations than historically, as evidenced by recent reorganisation of the US Marine Corps [8]. This reduction in heavy armour being but one result of a move towards landing forces of overall lighter and smaller scale and logistics burden. This may have significant impacts on the design of not only the connector craft (especially as regards challenges related to high speed operation), but also the systems and equipment fitted to the landing ship to deploy and recover them.

4.3 RAPID EVOLUTION OF TECHNOLOGY

We are currently in a period of extremely rapid development of defence technology when compared to the relatively linear development over the cold war period [9]. This is well evidenced by the rapid development of systems (particularly uncrewed ones), with operational half-lives often measured in months rather than years or even decades. As such the designers of the next generation of landing ships will have to consider closely the need to operate a range of deployed systems whilst also allowing for evolving mission system fits over the life of the vessel.

5. IMPACT ON AMPHIBIOUS SHIPPING REQUIREMENTS

Taking account of the changing operational environment, we have begun to identify design drivers that would apply to the next generation of landing ships.

5.1 REDUCING SUSCEPTIBILITY

Key to this will be the need to make the landing ship as difficult a target as possible for the array of A2/AD weaponry deployed by adversaries. At present, landing ships provide a relatively predictable target, both in time and space. Contributors to this include:

- The need to approach relatively close to shore to allow slow landing craft to transit ashore in a reasonable time and or due to range constraints of helicopters;
- Constrained range of landing sites (beaches) suitable for amphibious operations;
- Limited range of weather conditions under which offload can take place;
- Mobility restrictions whilst preparing to offload (time taken to ballast down for LPDs);
- Low number of landing ships providing single point of failure.

The question then becomes how can the ship designer solve these problems? Design of the connector craft will of course have a significant impact, especially as regards ability to offload further offshore, but many challenges can only be solved through innovative design of the landing ship itself.

5.1 (a) Broadening the Surface Offload Envelope

The use of a well dock provides a number of operational constraints on the landing ship that can aid an adversary to target the ship during offload and the period leading up to it. First of these is that in order to flood the dock it is necessary for the ship to take on a considerable amount of ballast, typically many thousands of tons. For example, HMAS Canberra reportedly needs to take on board more than 4,000 tonnes of seawater prior to flooding the dock [10], whilst for the USS Wasp this figure rises to some 15,000 tonnes [11]. Clearly this evolution will take considerable time, during which the vessel's mobility will be severely restricted, or even static. Such a period of ballasting may well give an intelligent adversary sufficient warning of an imminent landing to deploy A2/AD weaponry in time to thwart the landing. It is also very likely that the significant additional free surface and trim effect from the ballasting operations will result in the vessel being in a less stable condition than during transit, this increasing vulnerability.

Operation of boats and landing craft form a well dock are typically limited to Sea State 3 [12] due to wave effects caused by vessel motion. Again, an intelligent opponent can use this knowledge to judge when landings could take place, allowing them to concentrate A2/AD efforts for periods when landings are most likely. This constraint is likely to become more significant into the future due to the pressure to increase the landing ship's stand-off from the enemy shore. Such additional distance plays dividends in that it complicates targeting and may place the landing ship out of practical range of smaller drones and / or the sea conditions in which they can operate. The down side of this is however that higher sea states are more likely further off-shore compared to more sheltered conditions close to shore. Thus future amphibious operations may well encounter sea states beyond the operating limits of well docks more frequently.

In contrast, helicopter operations from large ships do not require lengthy preparation such as ballasting, and are likely to be practical at Sea State 5 or even 6 [13], and can be carried out with the vessel operating at speed (although a designated flying course may be constrained by motions and relative wind). Helicopters however are somewhat limited in payload compared to the potential payload of landing craft, especially single heavy loads such as armoured vehicles. As such it is suggested that although helicopter offload may take an increasing proportion of the landing force, that surface offload capabilities are also likely to be needed, either in the initial landing or to provide logistic support to the force once ashore.

The practical ideal would therefore appear to be for the landing ship to be able to deploy surface connectors with similar constraints to those required for helicopter operations. This will require some form of system that does not require a long period of preparation and constrained mobility, which infers no need to ballast down, and to be able to operate in a wider range of sea states than current well docks.

5.1 (b) Force Resilience

Landing ships are High Value Units (HVU) and as such severely limited in number for most nations, and each individual ship tends to provide unique capabilities to the fleet. As such, damage to or even worse loss of, a single landing ship can effectively halt operations. Even with the support of numerous and capable escorts, it will not be possible to guarantee 100% protection to the landing ship from the complete array of A2/AD threats. One solution would be to provide the landing ship with comprehensive self-defence systems backed up with extensive vulnerability reduction methods. Though these can be effective, they are also costly and exacerbate the HVU situation as fewer ships can be afforded so each becomes ever more critical to operations and a potential single point of failure.

An alternative strategy is to make the landing ships more affordable, such that they can be acquired in greater numbers, something that may be aided by a move towards smaller landing forces. Escorts would obviously be needed to provide the majority of protection to the fleet, and a minimal level of self-protection provided for the landing ships to counter the lower end A2/AD threats. In this case, each landing ship would carry duplicate capabilities such that damage to, or even loss of, a single landing ship would not necessarily lead to a cessation of operations. Multiple ships and distributed vectors of deployment would also allow evacuation of damaged vessels if the A2/AD threat proved more capable than anticipated. Although this may seem to be a somewhat uncomfortable way of thinking, it is suggested that the return of the prospect of peer-on-peer conflict over recent years compared to the post-cold war environment means that military ship designers need to return to this sort of attritional thinking that was applied in earlier years.

Key measures to ensuring affordability will be in ensuring that the vessel size is kept under control, and the application of a carefully blended mix of military and commercial standards applied to the design.

5.1 (c) Signatures & Self-Defence

There are well established measures that can be applied to reduce the susceptibility of warships to threat weapons. Use of signature reduction, electronic warfare, decoys and hard-kill systems can all protect the vessel from a wide range of potential threats. It may then be assumed that there will be a need to outfit the landing ship of the future with a comprehensive array of such systems and design features, but these would be very costly and such a comprehensively equipped landing ship would fail the affordability aims and potentially compromise force resilience. As such it may perversely reduce force capability.

It is considered that it would not be possible to design a practical landing ship such with signatures low enough to provide strategic concealment, i.e. hide it from the array of sensors available to potential adversaries. Instead, simple low cost measures should be taken to reduce the likelihood of identification of the landing ship as a HVU target, and / or providing EW and decoy systems a reasonable chance of distracting or seducing incoming anti-ship missiles. In the former case the best approach may be that of "hide in plain sight" by employing design choices that make the landing ship indistinguishable from civilian shipping to remote sensors.

The most sophisticated of the A2/AD threats, namely anti-ship missiles (including hypersonic and ballistic types) are (unsurprisingly) the most costly to counter by hard-kill. Added to which the systems required to counter them have a profound impact on the design of any vessel mounting them which would undoubtedly undermine the amphibious capability of any landing ship unless it was very large. As such, it is suggested that defence against such threats is best left to the dedicated escort vessels that are specifically designed to deal with them.

This leaves an array of less sophisticated A2/AD threats that it may not be possible for escort vessels to deal with at reach. Shore artillery and anti-tank guided weapons should be eliminated as threats to the landing ship (though not the connectors) by the greater stand-off distances from the shore. This leaves mines (discussed separately below), inshore attack craft (IAC) and drones as the main residual threats to the landing ship. Rapid fire gun systems allied to simple air search radar and electro-optical targeting probably represent the best compromise to counter the IAC and drone threat.

5.1(d) Mine Warfare

Mines are perhaps the oldest of all A2/AD weapons, but remain highly effective at denying use of areas of sea, especially shallow waters near to the coast, which by definition are those of most interest in amphibious operations. Clearly mines are most effective when they can be concentrated on a few likely landing sites, which may be mitigated to some degree if the surface connectors are capable of landing at a wide variety of sites.

As such, mines are a threat to both the landing ship and the connectors, and the traditional approach has been to use specialist mine warfare vessels to clear a path to the landing site. This is both risky for the mine warfare vessels and time consuming. It also provides the opponent with a clear signal that a landing is coming, providing time to prepare other A2/AD assets to oppose it.

Mine warfare is however undergoing something of a revolution with the widespread adoption of remote vehicles to neutralise minefields [14]. Due to their design, landing ships can provide an almost ideal mothership to host these surface and sub-surface systems. If the landing ship is designed to accommodate these systems in addition to their offload capability, this opens the door to the amphibious force having a truly organic mine clearance capability.

5.2 PROVIDING AIR SUPPORT TO THE LANDING FORCE

As well as providing the means to place the landing force ashore, the landing ship can also provide a base from which it can be supported. Traditionally landing ships have provided command and control facilities for the landing force and used the connectors to provide logistics support. As discussed above, modern land warfare requires that ground forces are supported with ISTAR and strike capabilities to carry out their tasks. It is considered vital that the landing ship of the future should provide some of this support to the landing force if it is to be successful in its tasking.

Deep strike by large fixed wing aircraft (crewed and uncrewed) can only be delivered from an aircraft carrier or shore bases and is not credible from a modestly sized landing ship. However, for landing ships of 220m or more in length there is the prospect of operating a small number of STOVL strike fighters, such as the F-35B Lightning, in addition to the landing force as seen in the Spanish Juan Carlos. This constraint doesn't mean however that an affordable landing ship cannot contribute aviation support to the landing force ashore using a mixture of uncrewed fixed wing aircraft and rotary wing aircraft with or without a crew.

Any allowances made for the landing ship to provide such air support will need to be in addition to those required to lift those portions of the landing force transferred ashore by helicopter. As such, it is suggested that future landing ships will benefit from having aviation facilities more akin to current LPH/LHDs rather than LPDs.

5.2 (a) ISTAR

The ability to provide enduring ISTAR support is likely to be best met through the use of fixed wing uncrewed aircraft. The larger the aircraft, the more capable it is likely to be both in term of the sensors carried and range / endurance it can achieve. As such, the ability to launch, recover, control and distribute data from aircraft such as the MQ-1 Grey Eagle (demonstrated from HMS Prince of Wales in 2023 [15]) could provide a force multiplier to the landing force. This requires the landing ship to have a flight deck of sufficient area and the correct arrangement to allow for short take-off and landing (STOL), internal space for mission planning / control and topside space for the antennae required for control and data transfer. Even smaller fixed wing platforms can provide valuable capabilities, such as the QinetiQ Banshee demonstrated as part of Project Vampire [16]. Again this requires the landing ship to have a flight deck flexible enough to allow launch and recovery of the air vehicle plus the other requirements for the larger aircraft such as the Grey Eagle mentioned above.

5.2 (b) Strike

Footage from recent conflicts shows that a ground force cannot hope to operate effectively without organic air support. As such, it is envisaged that the future landing ship will need to be capable of providing significant levels of aerial support, whether in the form of attack helicopters or uncrewed aircraft. These aircraft may range from relatively simple first person view (FPV) drones through to more sophisticated "suicide" one way strike aircraft through to large, persistent, fixed wing types capable of providing close air support (CAS).

5.3 MISSION FLEXIBILITY

As stated above, landing ships tend to be inherently flexible in use due to the nature of the design. This flexibility applies both within the range of amphibious operations and extends to a wide range of other roles. As such, the future landing ship will have to embrace the ability to flex roles to match each specific mission. This will potentially require the ability to alter a variety of aspects including the balance between air and surface offload, an ability to accommodate a wide variety of connectors / daughter craft, and to bring aboard modular / containerised mission fits. Not only will such an approach result in a vessel that can tailor its capabilities to a variety of mission environments, but should also ensure that adoption of new, fast evolving, technology can be as painless as possible. If nothing else, available volume in the ship becomes a major enabler for including such flexibility.

6. POTENTIAL DESIGN SOLUTIONS

QinetiQ has been pondering the potential impacts on vessel design arising from the issues identified above, and the following section provides some of our current thinking with regards to potential design solutions that may be applicable to the next generation of landing ships. This work is still underway, and should not be considered complete, but can be considered indicative of the kind of thought processes that the designers of these vessels will need to undertake.

6.1 THROUGH DECK

A comparison was made between two notional landing ships of equal length and waterline beam, one with a typical LPD configuration and the other with that of an LPH. These vessels both have a length of approximately 180m and a waterline beam of 30m. The comparison is quite revealing, in that the LPH configuration allowed an arrangement with some 15% greater internal deck area compared to the LPD. This was achieved due to the need to extend the LPD foc'sle deck for the full length to maintain a continuous freeboard along the flight deck, which provided more internal volume than the additional superstructure decks in the LPD. This additional volume would pay significant dividends as regards mission flexibility, or alternatively allow a smaller, cheaper vessel to provide equivalent internal volume. As may be expected, the LPH arrangement allowed for a superior aviation capability, providing a flight deck with two heavy helicopter spots and three medium helicopter spots against the single heavy or twin medium helicopter spots available on the LPD. The LPH configuration also provides a much better solution for the operation of fixed wing uncrewed aircraft than the LPD where the forward superstructure would provide a significant physical obstacle and turbulence over the flight deck.

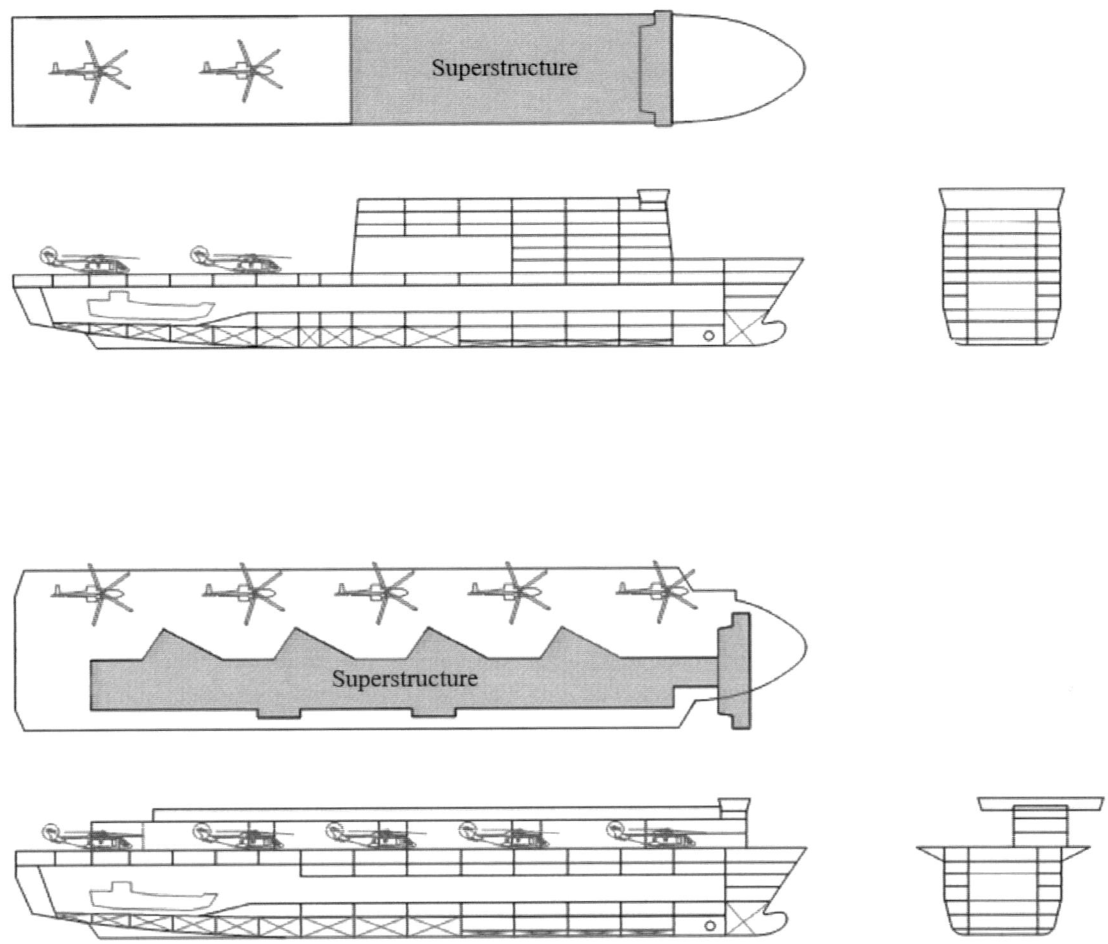

Figure 1: Comparison of LPD (top) and LPH (bottom) Configurations.

6.2 ALTERNATIVES TO THE WELL DOCK

A range of potential alternative methods to launch medium or large landing craft were considered, with the aim of improving the ability to launch and recover at significant vessel speeds, sea states and without the need to ballast down. The solution selected for further study being lifting platforms similar to those used in the "Lighter Aboard Ship" (LASH) and SeaBee [17] vessels of the 1960s and 1970s. These vessels use a stern mounted platform capable of lifting very heavy loads (over 1,000 tonnes in the case of SeaBee). Daughter vessels launch and recover from the submerged lowered platform and are lifted to / from the internal vehicle deck using a mission bay handling system.

In the QinetiQ concept, the landing ship would be provided with three platforms, one in the stern and one on each side of the vessel. The stern platform capable of lifting large landing craft (up to 30m in length) and the side platforms smaller craft of 16m in length. The platforms are secured to the vessel by guide rails, with the side platforms cantilevered from the ship side. The platforms would be lifted using chains or steel cables from the overhanging flight deck at stern and under the side sponsons of an LPH configured landing ship. As well as LASH and SeaBee vessels, such technology is very similar to that used in deck edge aircraft lifts used on aircraft carriers [18] and even to that used for a cruise ship restaurant than can descend to the waterline on the ship's side [19].

LASH and SeaBee vessels were limited in offloading to Sea State 2 due to the ability of the lifting systems being used to cope with the forces generated by ship motion. It was shown however that the ability of daughter craft to dock with the submerged platform was feasible up to Sea State 4 [20]. With one platform on each side it is expected that it will be possible to manoeuvre the landing ship to create lee to allow at least one lift to operate in high sea states. Work is ongoing to ascertain the limits of ship speed at which launch and recovery could take place and any design measures that can enhance this capability.

Given that such an approach does not require the mothership to stop to ballast down, is expected to allow launch and recovery in sea states at least as high as a well dock and at significant forward speed, further development is considered well worthwhile.

6.3 MODULARITY

QinetiQ have been heavily involved with the development of NavyPODs modular equipment fits [21]. Similar systems are very likely to be needed on the future landing ships, and their capability enhanced by careful ship design. QinetiQ's suggested approach is to ensure a backbone of essential services (power, ventilation, chilled water, information) is provided to the vehicle and flight decks such that items such as NavyPODs can be readily connected and integrated when brought on board.

In addition to large containerised equipment, it is also envisaged that there will be reconfigurable mission systems spaces on the deck below the flight deck. These will be accessible by vehicle from the flight deck and vehicle deck such that consoles, servers and related equipment can be installed / removed as necessary to flex mission system capabilities.

7. CONCLUSIONS

In order to remain effective in the face of the evolving operational environment, it will be necessary for the designers of the next generation of landing ships to think innovatively and not simply follow the formula that has been successful in the past. In short, an LPD-style configuration with a twin spot flight deck is very unlikely to provide a suitable answer to the challenge.

It appears that force resilience, in the form of a larger number of affordable ships, will be preferable to a lower number of individually more capable ships, and that a through deck design offers many advantages compared to a forward superstructure / aft flight deck approach.

Perhaps the greatest challenge will be that relating to surface offload of medium and large daughter craft. Whilst proven, the well dock has many shortcomings and may not be suitable for future needs. Submerged lifting platforms have the potential to provide a more capable alternative and are worthy of further development.

8. ACKNOWLEDGEMENTS

Not at this time.

9. REFERENCES

1. ROYAL NAVY, "Bay class act as RFA support ship completes mammoth Gulf mission", *Royal Navy Website*, 11th July 2016. Available from: https://www.royalnavy.mod.uk/news-and-latest-activity/news/2016/july/11/160709-rfa-cardigan-bay-returns-home [Accessed May 2024].

2. ROYAL NAVY, 'Equipment / Ships Albion Class', *Royal Navy Website,* Undated, Available from: https://www.royalnavy.mod.uk/equipment/ships/albion-class [Accessed May 2024].

3. McCLEARN, S., "Fearless Class (LPD) Assault Ships", *Haze Gray Website,* 2003, Available from: https://www.hazegray.org/navhist/rn/assault/fearless/ [Accessed May 2024].

4. SCHERRER, Lt. Col. A., "Force Projection in the Modern A2/AD Environment", *Marine Corps Gazette*, May 2021. Available from: https://www.mca-marines.org/wp-content/uploads/Force-Projection-in-the-Modern-A2AD-Environment.pdf [Accessed May 2021].

5. WILLS, S., "Sea Drone Swarms – Can NATO's Navies Avoid Russia's Fate?", Center for European Policy Analysis, 6th March 2024. Available from: https://cepa.org/article/sea-drone-swarms-can-nato-navies-avoid-russias-fate/ [Accessed May 2024].

6. GAMBRELL, J., "Missile attack by Yemen's Houthi rebels damages a ship in the Red Sea", *The Associated Press*, 30th April 2024. Available from: https://apnews.com/article/yemen-houthi-red-sea-ship-attacks-israel-hamas-gaza-cf858f30b4ba19731f6dc7719e9fa892 [Accessed May 2024].

7. ZAFRA, M., HUNDER, M., RAO, A., and KIYADA, S., "How Drone Combat in Ukraine is Changing Warfare", *Reuters,* 26th March 2024. Available from: https://www.reuters.com/graphics/UKRAINE-CRISIS/DRONES/dwpkeyjwkpm/ [Accessed May 2024].

8. MIZOKAMI, K., "Why the Marines Are Ditching Tanks and Howitzers to Prepare for America's Next Big War", *Popular Mechanics',* 6th July 2023. Available from: https://www.popularmechanics.com/military/weapons/a44448847/marine-corps-force-design-2030/ [Accessed May 2024].

9. MILLS, C., "Emerging and Disruptive Defence Technologies", *House of Commons Library,* 13th November 2023. Available from: https://commonslibrary.parliament.uk/research-briefings/cbp-9184/#:~:text=Artificial%20intelligence%20is%20a%20key,the%20Defence%20and%20Security%20Accelerator. [Accessed May 2024].

10. NAVY RECOGNITION, 2015,' HMAS Canberra Well Dock Receives an LCM-E Landing Craft for the First Time', *Navy Recognition Website,* 11th February 2015. Available from: https://navyrecognition.com/index.php/naval-news/naval-news-archive/year-2015-news/february-2015-navy-naval-forces-defense-industry-technology-maritime-security-global-news/2409-hmas-canberra-well-dock-receives-an-lcm-e-landing-craft-for-the-first-time.html [Accessed May 2024].

11. PIKE, J., "LHD-1 Wasp Class", *Military Analysis Network,* 9th May 2000. Available from: https://man.fas.org/dod-101/sys/ship/lhd-1.htm [Accessed May 2024].

12. PIKE, J., "Landing Ship, Dock (LSD) Amphibious Transport, Dock (LPD)", *Global Security Website*, 28th March 2019. Available from: https://www.globalsecurity.org/military/systems/ship/ld.htm [Accessed May 2024].

13. LEONARDO, 2017,' Helicopters for Naval Missions', *Leonardo Website,* 7th September 2017. Available from: https://www.leonardo.com/en/news-and-stories-detail/-/detail/helicopters-for-naval-missions [Accessed May 2024].

14. MIGAULT, P., "Mine countermeasures: the Future Belongs to the Drones", *Thales Group Website,* 26th October 2016. Available from: https://thalesgroup.medium.com/mine-countermeasures-the-future-belongs-to-the-drones-9047315c80fe [Accessed May 2024].

15. GENERAL ATOMICS AERONAUTICAL,' GA-ASI Demonstrates Short Takeoff/Landing of UAS on UK Carrier', *General Atomics Aeronautical Website*, 17th November 2023. Available from: https://www.ga-asi.com/ga-asi-demonstrates-short-takeoff-landing-of-uas-on-uk-carrier [Accessed May 2024].

16. ROYAL NAVY, 2021, 'Drones Launched from HMS Prince of Wales During Landmark Demonstration', *Royal Navy Website,* 29th September 2021. Available from: https://www.royalnavy.mod.uk/news-and-latest-activity/news/2021/september/29/210929-prince-of-wales-drones [Accessed May 2024].

17. WIKIPEDIA, 2024,' Lighter Aboard Ship', *Wikipedia Website,* 1st April 2024. Available from: https://en.wikipedia.org/wiki/Lighter_aboard_ship [Accessed May 2024].

18. US NAVY, 1997,' Naval Ships' Technical Manual Chapter 588 Aircraft Elevators', *Military Analysis Network,* 31st October 1997. Available from: https://man.fas.org/dod-101/sys/ship/nstm/588v0r1.pdf [Accessed May 2024].

19. CELEBRITY CRUISES, 2024,'Magic Carpet',*Celebrity Cruises Website,* 2024. Available from: https://www.celebritycruises.com/gb/cruise-ships/celebrity-edge/onboard-experience/magic-carpet [Accessed May 2024].

20. US NAVY, 1998,' Joint Pub 4-01.6 Joint Tactics, Techniques,and Procedures for Joint Logistics Over-the-Shore (JLOTS)', *DTIC Website,* 12th November 1998. Available from: https://webharvest.gov/peth04/20041028155630/http:/www.dtic.mil/doctrine/jel/new_pubs/jp4_01_6.pdf [Accessed May 2024].

21. ROYAL NAVY, 2022,' Maritime Modularity Concept', *Government Publishing Service*, 2022. Available from: https://assets.publishing.service.gov.uk/media/63dcece38fa8f57fc8061089/Maritime_Modularity_Concept.pdf [Accessed May 2024].

9. AUTHORS BIOGRAPHY

Caitlin Hoey holds the current position of Senior Naval Architect and Seakeeping & Stability Team Lead at QinetiQ. She is responsible for delivery of Naval Architecture consultancy and business development, and has 10 years of experience in stability, structures and solution development.

Ed Waters holds the current position of Senior Principal Naval Architect at QinetiQ where he is responsible for technical assurance of a wide variety of activities. Ed has been involved with warship design, including in technical leadership roles, for the last quarter of a century with a particular expertise in wholeship concept design. He has been involved in the design of a wide variety of ship types from submarines, aircraft carriers, frigates, auxiliaries and mine warfare vessels including internationally as well as at home in the UK.

HYDRODYNAMIC AND MECHANICAL INTERACTIONS DURING SEA BOAT LAUNCH AND RECOVERY- VALIDATION OF TIME DOMAIN PANEL METHOD PREDICTIONS

F van Walree, Maritime Research Institute Netherlands (MARIN), the Netherlands, **D Sgarioto, J Lee,** and **J Mathew,** Defence Science and Technology Group (DSTG), Australia

SUMMARY

This paper describes validation of time domain panel method predictions of multi-vessel interactions during alongside sea boat launch and recovery evolutions from an offshore patrol vessel by means of a crane. An overview of the simulation approach is provided. Validation was based on scale model test data for the alongside towing and lifting and lowering phases of a typical alongside crane-based launch and recovery scenario. Several aspects were considered in the validation, including sea boat motions and acceleration levels as well as tension in the painter-line and lifting/lowering line. Model test results showed stern-quartering waves to be preferable to bow quartering waves. They also showed that use of a pre-tension painter line and ensuring the sea boat is directly below the crane at the commencement of a hoist as important factors for increased operability. All seakeeping characteristics that influence alongside launch and recovery operability were found to be predicted with acceptable accuracy.

1. INTRODUCTION

Launch and Recovery (L&R) operations in harsh seaway environments present considerable risks to materiel and personnel, largely due to the hydrodynamic and mechanical interaction forces that develop between dissimilar-sized vessels when operating at speed in close proximity. The development and validation of computationally efficient numerical tools, capable of predicting both hydrodynamic and mechanical-based interactions during L&R, offers the potential to greatly enhance capability for these operations and the safety of personnel and materiel involved. Such tools can be used to inform requirements setting in the acquisition of new naval platforms, to assess the L&R operability of candidate designs and to develop operational guidance materials for specific combinations of parent vessel and deployed system of interest.

Limited validation studies have been published on the numerical predictions of multi-vessel interactions during alongside sea boat (SB) L&R evolutions. Carette [1] applied the time-domain code Fredyn to one aspect of the alongside L&R evolution, where a sea boat is trying to maintain station alongside a parent vessel. The author demonstrated a simplified, fast-time approach to numerical modelling of this part of the L&R without directly simulating the hydrodynamic interactions between the vessels. Van Walree *et al* [2] validated PanShip, for the case of SB motions while towed alongside a patrol boat. The PanShip code is able to model the seakeeping behavior of multiple vessels operating in a seaway, including hydrodynamic interactions between the vessels.

The present paper extends the validation of PanShip to SB L&R from an offshore patrol boat by means of a crane and considers not only the alongside towing but also the lifting and lowering phases of the evolution. It details a higher fidelity simulation approach for the SB L&R problem such that the SB is no longer assumed to remain at a constant relative positon with respect to the parent vessel. Using model-scale data from tests performed in MARIN's Seakeeping and Manoeuvring Basin, PanShip simulations were performed to validate several aspects of a typical alongside crane-based L&R scenario, including SB motions and acceleration levels; and tension forces in both the painter and lifting/lowering lines. Model test results showed stern-quartering waves to be preferable to bow quartering waves. They also showed that use of a pre-tension painter line and ensuring the sea boat is directly below the crane at the commencement of a hoist as important factors for increased operability. The paper shows detailed comparisons between simulation tool predictions and model test data. All relevant seakeeping characteristics that influence alongside L&R operability were found to be predicted with acceptable accuracy. A discussion on possible reasons for deviations between simulations and measurements is included along with recommendations for future work.

2. EXPERIMENTAL ARRANGEMENT

The main particulars of the OPV and the RHIB-type SB are shown in Table 1. The OPV was equipped with a twin propeller-rudder arrangement and four stabiliser fins. A painter line is attached to a boom at the bow of the OPV to tow the SB when in the water and at the same time keeping it at some transverse distance from the

OPV. Figure 1 shows the model arrangement for the OPV with the SB in its hoisted position. The painter line was attached to the boom by means of a spring providing the required stiffness at model scale. The painter line could be adjusted to exert a certain pre-tension on the SB during hoisting and lowering.

A generic crane with a software controlled winch was fitted on the portside of the OPV. The SB was connected to the crane by means of a single hoisting line attached above its centre of gravity. The SB was equipped with a stern drive arrangement. However, no SB propulsion was used during the model tests. The stern drive tail acted as a passive rudder. The model scale was 9, yielding an OPV model with a length of 7.8 m and a SB with a length of 0.75 m.

Table 1. Main particulars OPV and SB.

Designation	Symbol	Magnitude OPV	Magnitude SB	Unit
Length between perpendiculars	L_{PP}	70.26	6.76	m
Breadth moulded on WL	B_{WL}	12.86	2.19	m
Draught moulded on FP	T_F	3.37	0.50	m
Draught moulded on AP	T_A	2.92	0.63	m
Displacement mass in seawater	Δ_1	1661	4.46	t
Wetted surface area bare hull	S	1013	15.51	m^2
Transverse metacentric height	GM	2.30	0.15	m
Natural roll period	$T\varphi$	7.30	3.67	s

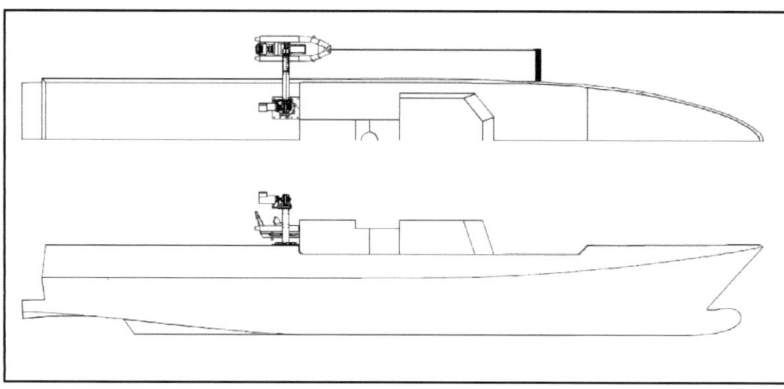

Figure 1: Model arrangement

During the tests the OPV model was free running under auto-pilot course control. The SB was towed and hoisted on the leeward side of the OPV. The winch was controlled to perform a series of consecutive hoisting and lowering cycles during each test run. The transition between the different phases of the operations was software controlled. A cycle has four stages:

- SB lowered in water: winch applies a constant tension on the hoisting line,
- SB hoisting at constant speed,
- SB hoisted with fixed position,
- SB lowering at constant speed.

When the SB is lifted by a wave crest during hoisting, the hoisting line gets slack when the lift velocity exceeds the maximum winch speed and high snap loads can occur when the line is tensioned again. In order to prevent this, hoisting started when the SB just passed a wave crest, i.e. when the winch was rolling-out more than a given threshold. The SB was lowered when the OPV heeled towards the lowering side, to avoid a drop on the deck. When the tension in the line dropped under a fraction of the SB weight the lowering stopped and the controller switched to constant tension mode.

Measurement signals included amongst others six degree of freedom motions and accelerations of the OPV and SB, relative wave heights at several locations on the OPV and incident wave heights in front of and next to the OPV model. The winch speed and operation mode, painter and hoisting line tensions and rudder and stabilization fin angles were also recorded during the tests.

Tests were performed at forward speeds of 6 and 10 knots in moderate to high steepness top SS3 and SS4 irregular sea states from stern- and bow-quartering directions. Several hoisting strategies and painter line tensions were tested. The duration of the tests for each condition was sufficient for at least 10 hoisting/lowering evolutions.
The test results showed that:

- During hoisting and lowering large pitch and yaw motions were observed when there is no pre-tension on the painter line. Through the use of a pre-tensioned painter line the motion amplitudes can be considerably reduced.
- Performing the hoisting and lowering operations of the SB in stern-quartering waves resulted in significantly lower motion amplitudes, accelerations and line loads than performing these in bow-quartering seas. Stern-quartering conditions resulted in a low wave encounter period for which the frequency of motions is sufficiently away from the natural (pendulum) frequency of the crane-SB system. The best wave direction tested was waves coming from 20 degrees off the stern. The worst wave direction tested was 135 degrees, i.e. waves coming from 45 degrees off the bow.
- The effect of a milder sea state (SS3 instead of SS4) on motions and accelerations was not very significant for stern-quartering wave directions while it was significant for bow-quartering wave directions.

3. PANSHIP SIMULATION METHOD

PanShip uses the transient Green function to obtain free-surface effects in a 3D time domain panel method. Forward speed effects are exact under the assumption of a constant speed and heading, whereas the radiation and diffraction problems are linear in combination with body-exact wave excitation and restoring forces. An impulse response function (IRF) database for radiation and diffraction forces can be used for CPU time efficient simulations for multi-body problems. Viscous reaction forces include a cross-flow drag method for manoeuvring and empirical roll damping methods. The viscous resistance is obtained from the actual wetted surface and a frictional resistance coefficient. In combination with pressure integration on the hull mesh the total resistance is obtained. Propulsion and steering is included using propeller open water characteristics, semi-empirical lifting-surface characteristics and propeller-rudder interaction coefficients. A semi-empirical waterjet propulsion and steering method is available as well. More detailed information on the method is provided by Van Walree *et al* [3, 4].

Within the assumptions of the panel method, the multi-body problem is more a book-keeping problem than a hydrodynamic problem. It essentially requires accounting for which hull panel belongs to which vessel and equations of motion for each individual vessel. For multi-body problems, impulse response functions can be determined for the mean relative position of the two vessels in the horizontal plane. The hydrodynamic interaction is then strictly speaking valid only for this mean relative position.

When the relative position between the two vessels in the horizontal plane is significantly varying an alternative approach should be taken. In this approach it is assumed that the interaction is one way, i.e. the effect of the SB on the OPV is negligible. A 3D grid covering the position variations of the SB next to the OPV is constructed for which impulse response functions for disturbed wave height and flow velocities are determined before the time domain simulation starts. In the time domain simulation the disturbed wave height and flow velocities are based on interpolation for the instantaneous SB position in the grid. Interpolation is done for each surface panel of the SB individually. This approach neglects the reflection of waves generated by the SB against the OPV side. Although a comparison between the two approaches showed that the differences are small for the current configuration, the 3D grid option was used despite its higher computational costs.

Another aspect is the hoisting and lowering of the SB which leads to large submerged volume variations. For such conditions a full body-exact approach for all force components would be best. However, simulations with such a method for multi-body applications and a large number of conditions is not practical. Therefore a simplified approach has been taken. Wave excitation and restoring forces and moments are generally the dominating force components and are actually determined in a body-exact (non-linear) manner in PanShip, i.e. using the instantaneous submerged underwater hull form including effects of motions, incident and disturbed waves. The radiation and diffraction forces acting on the SB when in contact with water are reduced with the ratio between the actual submerged volume and the submerged volume at rest. This was deemed a reasonable approach since the hoisting and lowering velocities are relatively high which means that the large submerged volume variations last typically only one second. A comparison between the full and partially body-exact methods showed that the agreement is acceptable for all excitation forces and the pitch moment but not for the roll and yaw moments for which the partially body-exact method is conservative, i.e. it predicts too high excitation moments.

The painter line was modelled as a linear spring that yields zero force when slack. Hoisting and lowering of the SB was modelled by including an additional equation of motion in the motion solver. The motion equation was simply the constant hoisting/lowering line velocity as the first derivative of the line length with time. The force exerted on the SB then follows from the actual length, orientation and stiffness of the hoisting wire. An additional damping constant was used to insert some damping in the otherwise virtually un-damped numerical system. This damping was tuned on the model test result at 10% of the critical damping. When the SB is in the water in-between hoisting/lowering sequences the line length is set to follow the SB and OPV motions while exerting the required constant tension force.

During the model tests the SB was semi-clamped when at the crane top position: the SB was pulled against the crane top against a piece of rubber that acted as a fender. This was mimicked in the simulations by means of a spring-damper model with a critical damping for the three rotational modes of SB motions.

In addition a speed-dependent linear damping term for surge has been applied when the SB was in the water to account for the fact that the impulse response approach assumes a linear dependence between resistance and forward speed while in fact it is a quadratic or cubic relationship.

In order to have a close correspondence between model tests and simulations the experimental configuration, procedures and conditions were duplicated as best as possible in the simulations. The duration of the simulations was 500 s, sufficient for at least 10 hoisting/lowering evolutions per condition. The OPV was free running under autopilot course control with active stabilisation fins. The stiffness of painter and hoisting lines was that of the model tests scaled to prototype values.

4. VALIDATION

4.1 TIME SERIES

A number of typical time series of the motions and line tensions as obtained from the model tests and simulations are shown first. The condition corresponds to top SS4, 10 kt speed, stern-quartering wave direction and medium pre-tension in the painter line. Since the wave phasing was different between model test and simulation the time traces cannot be compared one to one, however the character of the signals can be compared.

The encountered wave elevation at the centre of gravity position of the SB is shown in the Figure 2. The jumps in the model test signal are due to the start of a new run while the simulation covers the complete test duration in a single run.

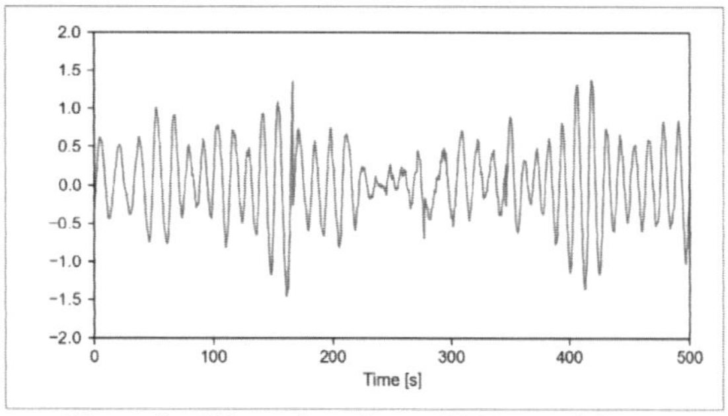

Figure 2a: Wave elevation at OPV - model test

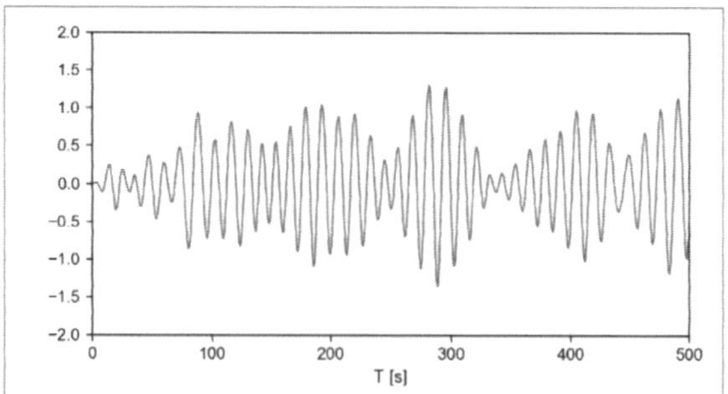

Figure 2b: Wave elevation at OPV - simulation

The heave signals in Figure 3 are similar, albeit that in the model test there is more time in between the hoists. This is due to the wave heights around 200 and 300 s in the model test which are too low to trigger a new hoist cycle. Note that there is also a difference in the reference heave position: for the model tests it is zero while in the simulations it is the height of the centre of gravity at speed above the calm water surface (approximately 0.5 m).

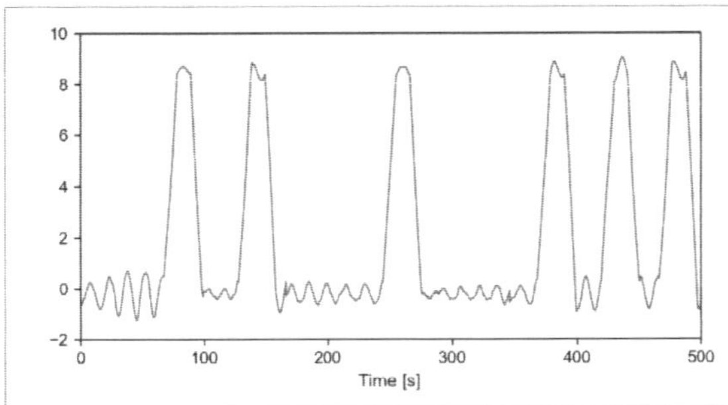

Figure 3a: Heave motion SB - model test

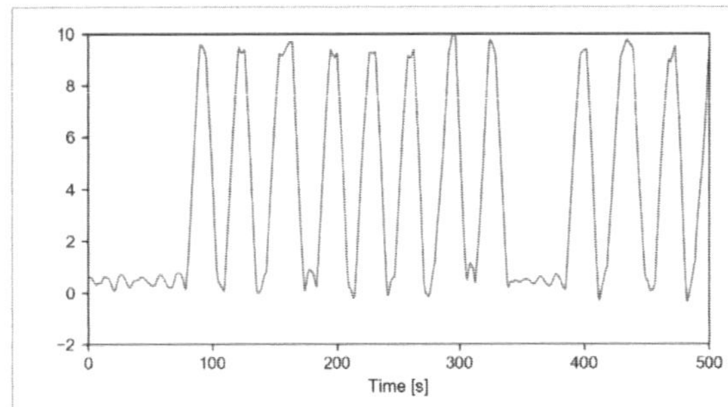

Figure 3b: Heave motion SB - simulation

The calm water pitch angle of -5 degrees (bow-up) is present in the model test signal in Figure 4a but not in the simulation signal in Figure 4b. This is due the difference in reference pitch angle: zero speed pitch for the model test and pitch at speed for the simulation. Otherwise the character of the pitch signals is similar.

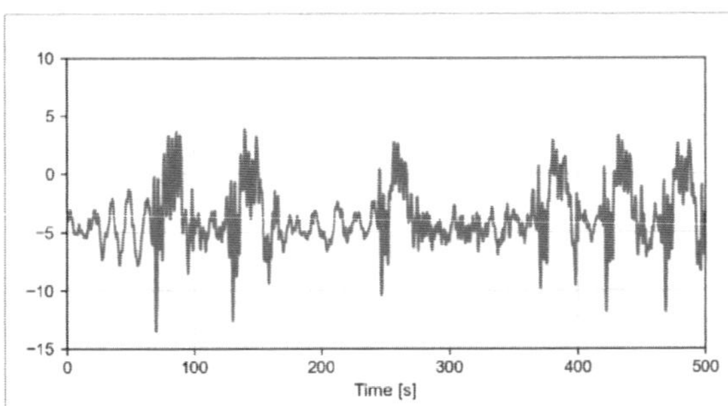

Figure 4a: Pitch motion SB - model test

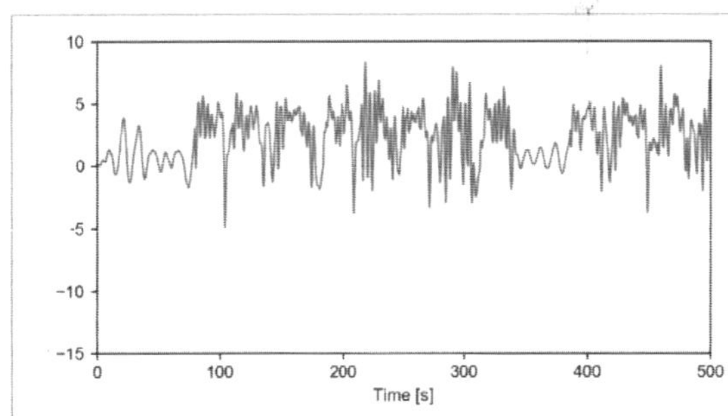

Figure 4b: Pitch motion SB - simulation

The roll amplitudes in Figures 5 are highest during hoisting and lowering for both the model test and simulation time traces.

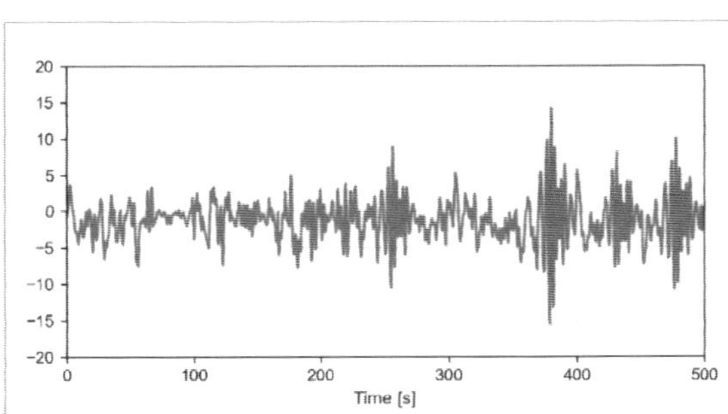

Figure 5a: Roll motion SB - model test

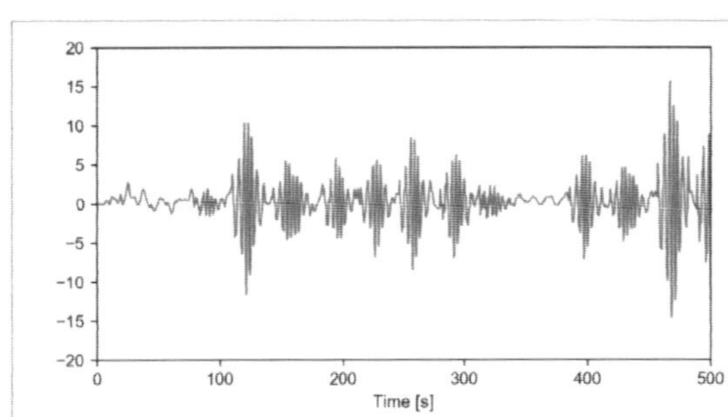

Figure 5b: Roll motion SB - simulation

The model test and simulated motion time traces have in general a similar character for the yaw motion, although the frequency of the yaw motion seems to be somewhat higher during the hoist for the model test.

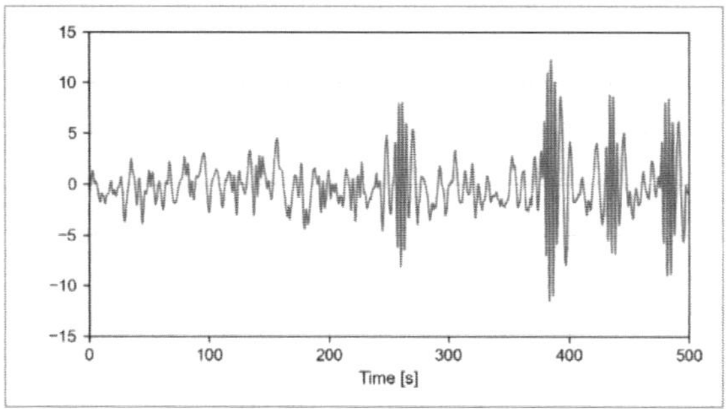

Figure 6a: Yaw motion SB - model test

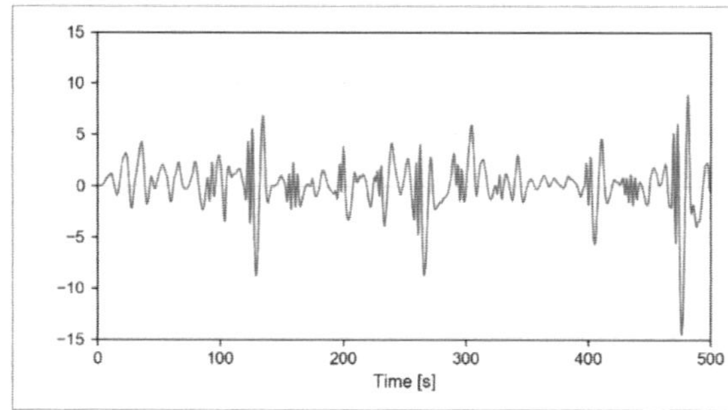

Figure 6b: Yaw motion SB - simulation

The above figures show that the motion amplitudes may be quite high and appear to be somewhat chaotic. Furthermore, a detailed inspection of experimental and simulation results shows that the position of the SB below the crane at the instant of hoisting is important. If the SB hoisting frame is not directly below the crane it starts to move like a pendulum once above water with large motion amplitudes.

4.2 MOTION AMPLITUDES

In this section comparisons are made for motion amplitudes of the SB. In view of the large motion amplitudes it is best to compare 2A1/10 motion amplitudes rather than standard deviations. These are defined as the mean of the 10% highest consecutive peak to trough amplitudes of motions and accelerations. Figure 7 shows the 2A1/10 amplitudes for roll, pitch and yaw for the tests performed at a high hoisting and lowering speed with variation of wave direction (Head), speed (Spd) and painter line pre-tension (PreTn). The data are normalised with respect to the highest experimental value.

The sea state is top SS4 with moderate steepness. PreTn 0 denotes no pre-tension, PreTn 1 denotes a medium pre-tension while PreTn 2 denotes a high-pretension in the painter line. Uncertainty margins for a 95% confidence interval are included based on procedures outlined in [5] for 10 repeat runs for two conditions with PreTn 0 and PreTn 1. During model testing no runs were performed with a variation of wave seeds so no variability can be determined for the model tests.

The figures show that the simulations correctly capture the trends with speed, wave direction and painter line pre-tension. The simulation amplitudes with uncertainty margin are also typically close to the model test values. The 20 degrees heading (stern-quartering seas) combined with a medium pre-tensioned painter line results in the lowest motion amplitudes whereas the 160 degrees heading (bow-quartering seas) is the least favourable.

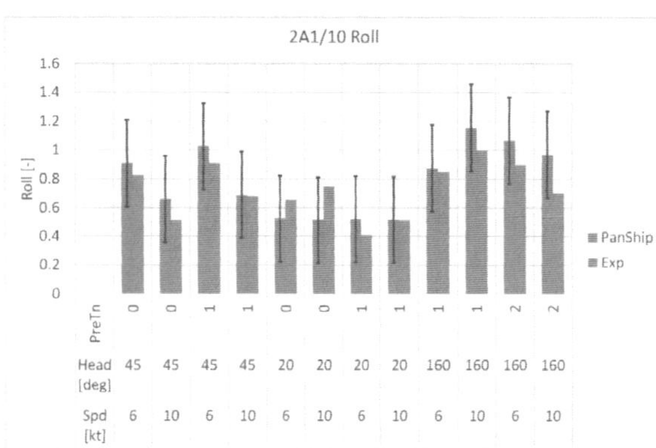

Figure 7a: 2A1/10 Roll amplitudes

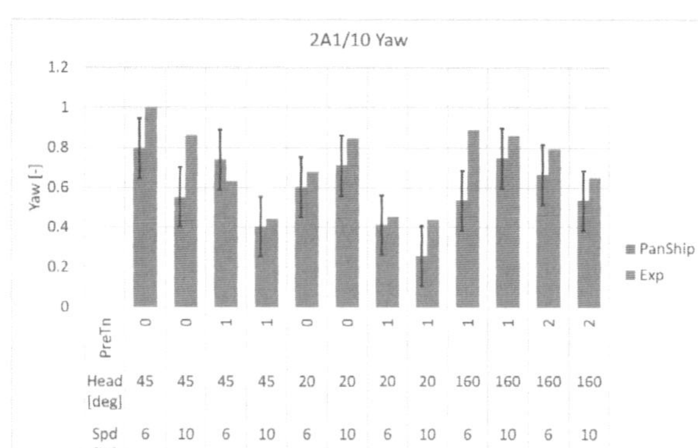

Figure 7b: 2A1/10 Pitch amplitudes

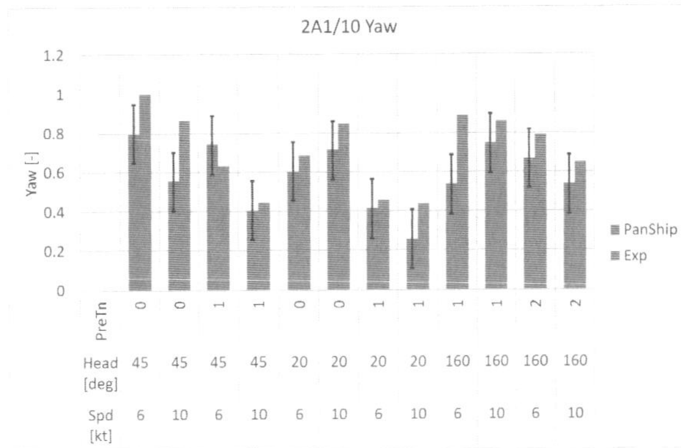

Figure 7c: 2A1/10 Yaw amplitudes

4.3 ACCELERATION AMPLITUDES

Figure 8 shows 2A1/10 amplitudes of the three acceleration components at the SB console position. The conditions are the same as for the motions in the previous section.

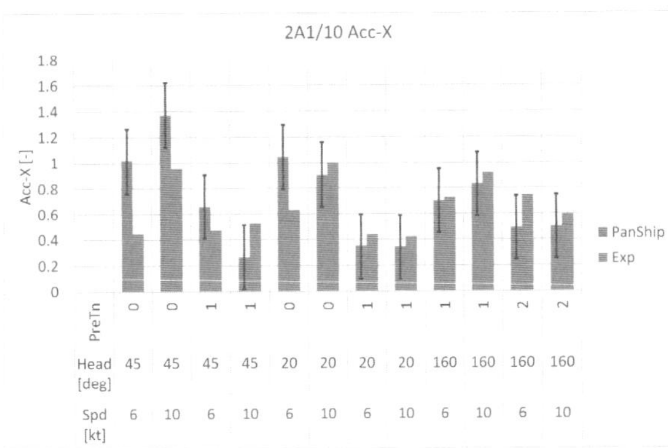

Figure 8a: 2A1/10 Longitudinal acceleration

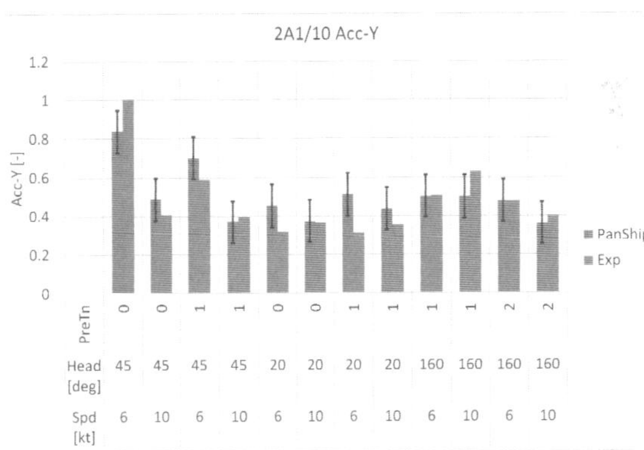

Figure 8b: 2A1/10 Transverse acceleration

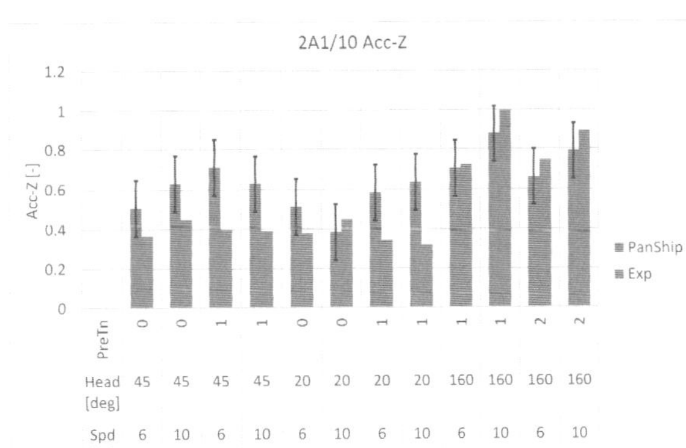

Figure 8c: 2A1/10 Vertical acceleration

The figures show that the x-component tends to be over-predicted by PanShip for the conditions without pre-tension in the painter line. The y-component matches the experimental ones reasonably well. The z-component tends to be over-predicted for all stern-quartering seas conditions. Possible reasons for the differences are:

- The experimental accelerations of the SB were not actually recorded by means of an accelerometer but derived from the motions, which is less accurate,
- The exact control of the hoisting winch was not modelled in the simulations which results in different peak accelerations at the start and end of hoisting,
- During the model tests, the SB was semi-clamped when at the crane top. This was mimicked in the simulations by means of a spring-damper model which is not an exact representation of the model test configuration. This likely leads to different acceleration peak values.

4.4 LINE LOADS

Figure 9 shows a comparison between model test and simulation for the 2A1/10 painter line and hoisting line loads. The normalisation and conditions are the same as for the motions in the above figures.

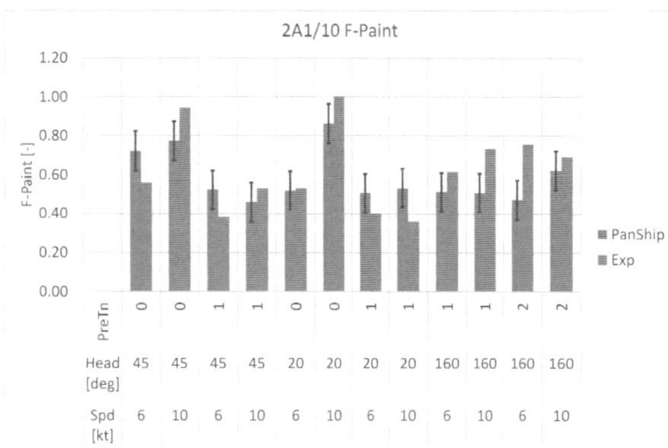

Figure 9a: Painter line loads

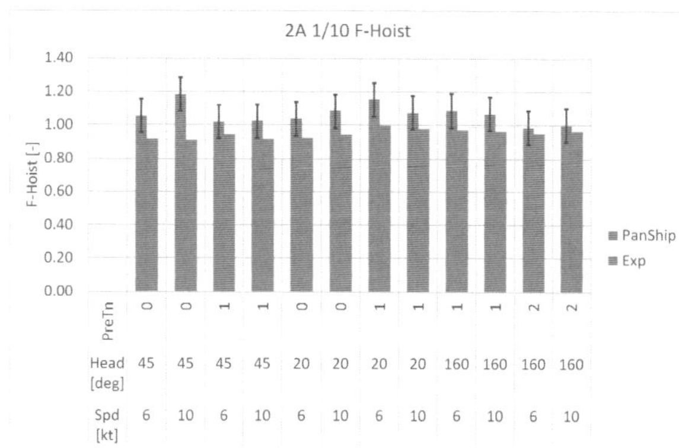

Figure 9b: Hoisting line loads

The trends in the painter line loads are reasonably well predicted by the simulations. Amplitude values are also reasonably close. The hoisting line loads show little variation over the test conditions and tend to be somewhat over-predicted in the simulations. This is likely due to an approximate modelling of the winch control in the simulations.

4.5 PROBABILITY OF EXCEEDANCE

Probability of exceedance plots for roll are shown in Figure 10 for a variation in painter line pre-tension. The conditions are the same as before while PreTn 0 again denotes no pre-tension, PreTn 1 denotes a medium pre-tension while PreTn 2 denotes a high pre-tension The plots contain simulation results for a variation of wave seeds. That means that the wave spectrum is the same but the wave phasing is different. The black line and symbols indicate the experimental results.

Some guidance on the operability of side launch and recovery operations in included in the figures by means of the 5% probability of exceedance limit for roll amplitudes of 15 degrees. A 5% probability exceedance means that in 5 out of 100 amplitudes the limit is exceeded.

The plots show that the variation in probability of exceedance with wave seeds reduces when the pre-tension in the painter line increases. Furthermore, as expected, the operability limit is more often exceeded for the low pre-tension case than for cases with a higher pre-tension. Finally, the experimental probabilities of exceedance more often lie outside than inside the spread in simulation results for the cases with a pre-tensioned painter line. In other words the simulation results tend to be more conservative than the model tests for those cases.

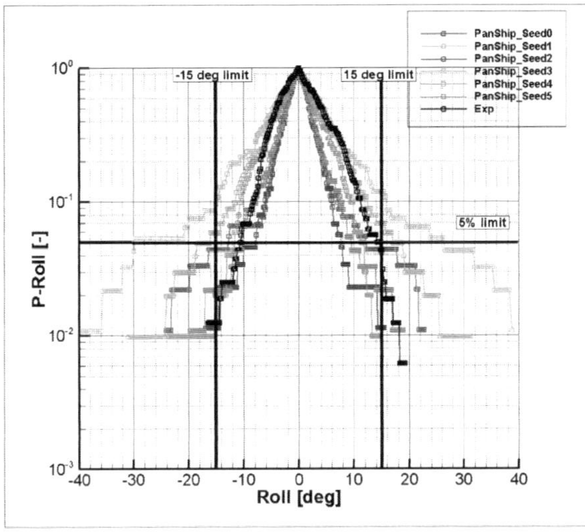

Figure 10a: Probability of exceedance PreTn 0

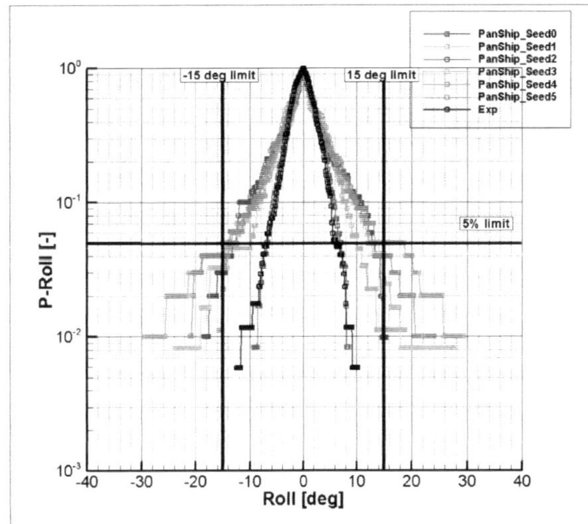

Figure 10b: Probability of exceedance PreTn 1

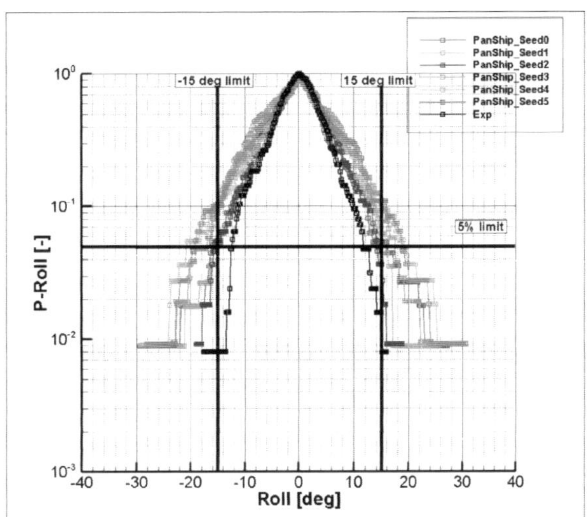

Figure 10c: Probability of exceedance PreTn 2

5. CONCLUSIONS

This paper describes validation of the time domain panel method PanShip for the prediction of several aspects of a typical scenario for alongside crane-based SB L&R from an OPV. Model tests in irregular waves were used to generate a database of vessel motions and accelerations (OPV and SB), painter line loads and lifting line loads. Findings from the model tests include the importance of pre-tension in the painter line, which can significantly reduce SB motion amplitudes during hoisting and lowering. The position of the SB relative to the crane at the commencement of the hoist was also found to be important; if the SB is not directly below the crane it starts to move like a pendulum once above water with large motion amplitudes. In general, stern quartering waves were found to be more favourable than bow quartering waves. This is partially explained by the encounter period of the waves being further away from the natural pendulum period for the crane-SB system in stern quartering waves.

The simulations were conducted with a newly developed PanShip approach to multi-vessel simulations which expands its applicability beyond cases where the relative position of the vessels is assumed to remain constant. This enhanced simulation capability will be a critical enabler for accurately determining operability for a wide-array of L&R applications, including L&R at the stern and L&R of uninhabited and autonomous surface vessels. Little difference was found in the comparison of the two simulation approaches for the given SB L&R scenario indicating that the lower fidelity approach would be sufficient in this case.

Comparison between the PanShip predictions and the model tests show that PanShip captures trends with vessel speed, wave direction and painter-line pre-tension. The 10% highest double amplitude was used as the basis for comparison and the model test and simulation amplitudes were found to be typically close. SB accelerations tended to be over predicted in the longitudinal direction for conditions without painter line pre-tension. Lateral SB accelerations were relatively well predicted, while vertical SB accelerations were over predicted for all stern quartering conditions. Possible reasons for these discrepancies included that the model test SB accelerations were indirectly derived from measured motions rather than directly measured with accelerometers. Difficulty in replicating the exact model test winch setup in the simulations may also be contributing to these discrepancy. Maximum load amplitudes in the hoisting line tend to be somewhat over predicted while painter line load amplitude predictions follow the trends and magnitudes shown by the model tests. A more detailed consideration of the influence of SB pre-tension revealed that the variation in simulation predictions for SB roll probability of exceedance with different wave seed numbers was reduced with increased pre-tension. Additionally the simulation predictions for this important operability criteria tend to be conservative with respect to the model test results when pre-tension is applied.

Given the rapidly increasing shift towards USV's as a means of supplementing/enhancing traditional naval surface fleets, it is recommended as future work to pursue validation of this higher fidelity PanShip approach for the alongside USV L&R scenario.

6. REFERENCES

1. CARETTE, N., 'Model tests and time domain simulations for launch and recovery systems with manned and unmanned vehicle', *Launch and Recovery Symposium*, November 2024, Linthicum, Maryland, USA.

2. WALREE, F V., SGARIOTO, D., TURNER, T., and CARETTE, N., 'Validation of a Time Domain Panel Code for Predicting the Behaviour of a RHIB Towed Alongside a Patrol Boat', *Proceedings of the 13th International Conference on the Stability of Ships and Ocean Vehicles*, September 2018, Kobe, Japan.

3. WALREE, F V., SGARIOTO, D., and TURNER, T., 'Validation of a Time Domain Panel Code for Prediction of Impulsive Loads on High Speed Ships', *Proceedings of 31st Symposium on Naval Hydrodynamics*, SNH'2016, September 2016, Monterey, USA.

4. WALREE, F V., MATHEW, J., SGARIOTO, D., LIN, F., DUFFY, J., and DENEHEY, S., 'Validation of a Frequency Domain and a Time Domain Panel Method for the Prediction of Hydrodynamic Interactions During Replenishment at Sea', *34th Symposium on Naval Hydrodynamics*, June 2022, Washington, D.C., USA.

5. KIM, Y., and HERMANSKY, G., 'Uncertainties in seakeeping analysis and related loads and response procedures', *Ocean Engineering*, Vol. 86, 2014, pp 68–81.

9. AUTHORS BIOGRAPHY

Frans van Walree holds the current position of senior project manager in the defence department of MARIN. He is responsible for seakeeping model test campaigns and simulation tool development for fast and advanced ships.

Daniel Sgarioto holds the current position of Discipline Lead for seakeeping and operability at the Australian Defence Science and Technology Group. He is responsible for leading research into multi-vessel interactions, seakeeping performance and operability of naval vessels.

Junghoon Lee holds the current position of research scientist for seakeeping and operability at the Australian Defence Science and Technology Group. He is responsible for designing research programs for investigating naval vessel hydrodynamic characteristics directly related to seakeeping performance and operability and analysing relevant scientific outcomes.

Jenny Mathew holds the current position of senior research scientist in the seakeeping and operability discipline at the Australian Defence Science and Technology Group. She is responsible for undertaking research in areas including multi-vessel interactions, seakeeping performance and operability of naval vessels.

THE MATE SHIP: RAPID ENHANCEMENT OF SURFACE FLEET LETHALITY

G Arnold, Naval Construction Branch, Australia

SUMMARY

The current RAN surface fleet is well-equipped in sensor and combat management systems but lacks the ability to deploy sufficient effectors. A class of vessels could be constructed based on existing high-speed vessel designs to act as additional missile magazines for existing fleet units. These vessels would provide enhanced lethality to the RAN in a short timeframe, with small crew requirements and could be built by Australian and/or allied shipbuilders based on existing designs.

DISCLAIMER

The views expressed in this paper are the author's own and do not necessarily reflect the views of Naval Construction Branch, The Department of Defence or other associated entities.

1. INTRODUCTION

The 2023 Defence Strategic Review calls for an "Enhanced lethality surface fleet…" with "…enhanced long range strike (maritime and land) and air defence capabilities." It also calls for acquisitions "…based on minimum viable capability in the shortest possible time". [1] This paper is a response to the implied problem: "How do we best enhance the lethality of the surface fleet in the shortest possible time?"

2. DEFINING THE PROBLEM

Lethality is a commonly-used but poorly defined term in military circles. Put simply it is "the capacity to cause death or serious harm or damage" [2]. In simplistic terms, the lethality of a surface fleet can be described as a combination of the capability of its sensors, combat management system and effectors. The capability of its effectors, in turn, is a function of the capability of an individual effector and the number of effectors that the fleet can deploy. Note that effectors in this instance is only considering physical effectors - the ability for non-kinetic capabilities to degrade enemy capabilities is not considered. Therefore the lethality of a fleet can be increased by improving its sensors and combat management, by increasing the quality of its effectors, or by increasing the number of effectors it can deploy.

The current RAN surface fleet has two classes of major surface combatant (MSC), the Hobart class destroyer and the ANZAC class frigate. Other classes of RAN vessel are not considered in this paper as they are not front-line surface combat vessels. The ANZAC-class frigate mounts a multispectral AESA radar system as their primary sensor system coupled with the 9LV combat system, a loadout suitable for a ship conducting area air defence. However the length of the ANZAC class vertical launching system (VLS) limits it to carrying the Evolved Sea Sparrow Missile (ESSM), only suitable for ship self-defence. Its strike capabilities are similarly minimal, consisting of 8 anti-ship missiles (Harpoon or NSM). The Hobart-class destroyer mounts the AN/SPY-1D(V) PESA radar as its primary sensor, feeding an Aegis combat management system routed through the Australian Tactical interface, a combination suitable for its role of area air defence. Its maritime strike capability is the same as the ANZAC class, but it carries a 48-cell "strike length" VLS, allowing it to carry many more missiles, including area air defence missiles, anti-ballistic missiles and land-attack missiles. This capacity is significantly smaller than the US Arleigh Burke frigate, which sports a very comparable combat system and sensor loadout.

The Hobart class' greater capacity is limited for certain missions. Project SEA 1300 involves the acquisition of Tomahawk cruise missiles for RAN warships, initially the Hobart class. This would grant the ability to conduct land attack. But would it grant the ability to conduct meaningful land attack? The USN used 59 of these same missiles in the 2017 strike at the Shayrat airbase in Syria [3]. In such a use case - an attack on a defended military installation that may possess a capability to shoot down a number of missiles - such a strike would currently require the use of 2/3rds of the RAN destroyer fleet. Moreover the ships' capability to perform their primary mission, area air defence, severely compromised, as 59 of the available 96 VLS cells would be devoted to Tomahawks.

The Hunter class frigate, currently in development, continues the pattern of current RAN combatants. Its sensor suite is designed around the CEAFAR2 radar system, along similar lines to the ANZAC class but more capable. This feeds a similar Aegis/Australian Tactical Interface combat system to that aboard the Hobarts. 8 anti-ship missiles and a 32-cell VLS provide the Hunter's effectors, a small payload given the vessels' 10,000 tonne displacement.

While the numbers of effectors that the RAN can deploy is limited, the quality of the effectors is suitable. The RAN employs modern variants of the Standard missile family and the ESSM. Tomahawks are being acquired as are the Naval Strike Missile. Upgrading to improved effectors, such as hypersonics, would provide additional lethality, but the developmental nature of these weapons mean that they do not provide a viable short-term improvement and hypersonic developments are already in-train.

We can reasonably conclude that while the RAN's sensor and combat management systems are suitable and the nature of the effectors available is appropriate, the RAN needs more strike and air-warfare effectors. Therefore the initial problem becomes "How do we best increase the surface fleet's strike and air warfare effectors in the shortest possible time?"

3. POSSIBLE SOLUTIONS

Three obvious solutions to providing the fleet with more effectors are to increase the effectors carried by each RAN MSC, to acquire additional MSCs, or to convert non-combat platforms to carry effectors.

To increase the number of air warfare and strike effectors, the most efficient means would be to increase the number of vertical launch system (VLS) cells carried by the platform, as the VLS can fit a broad range of effectors. Given the modular nature of a VLS, this approach initially appears simple. Indeed VLS cells were added to the Adelaide class frigates as part of an upgrade, demonstrating that the task can be performed. However, the availability of sufficient space and weight for a VLS aboard a MSC would be very difficult to find. The ANZAC class, for example, would have no capacity for an increase to VLS count without significant hull form change due to having no available weight margin. Any modifications that involved the modification of ship's hulls would have massive implications for almost every aspect of the vessel.

If additional MSCs were acquired to increase effector count, the Hobart class would be the clear choice as it is the RAN combatant that carries the largest number of effectors. However, MSCs require several years to build, even for designs that have been previously constructed, like the Hobart class. MSC builds also commonly run behind schedule [4]. Partial or complete construction of MSCs offshore could overcome some of these issues. However, the RAN would then have to man such vessels. With a Hobart class requiring a crew of 186 plus 16 for air crew and support, this is a significant burden for a Navy that already experiences many manpower issues.

Procurement of new "Military-off the shelf" (MOTS) corvettes or frigates would also involve issues similar to those of the acquisition of new Hobart class vessels. However existing designs cannot be expected to mount the same sensors, combat systems and effectors as use by the RAN. To change the systems to systems in-use by the RAN would involve developmental work. To not change them would involve introducing new systems into RAN service - also developmental. Thus the acquisition of "off the shelf" vessels would not be the quick fix that it initially appears.

Another option is adding missile capacity to other RAN vessels to supplement the existing MSCs. Arguments have already been raised to arm the currently in-build Arafura class Offshore Patrol Vessels with anti-ship missiles and possibly additional missile weapons. The Arafura design is intended for constabulary duties and thus has lower shock hardening and damage control. The existing radar is unsuited for higher-end warfighting and the class has no radar illuminators to guide the RAN's in-service self-defence missile, the ESSM. These issues would require massive redesign and modification to address, but if not addressed would create a platform with limited offensive capability in higher end warfighting, but huge vulnerabilities.

4. THINKING OUTSIDE THE BOX

Providing the RAN with more missiles aboard ships is not the only means by which the capability can be achieved. Other systems can either deliver missiles, or achieve the same anti-air, anti-ship and land attack outcomes as the missiles delivered from MSCs.

Ordnance delivered from aircraft, or land-based missiles could achieve similar A2AD effects to MSC-launched missiles. Relative to MSCs, aircraft lack in both range (unless supported by in-flight refuelling) and in particular in loiter capability - aircraft time on station is measured at best in hours, whereas MSCs can stay on-station for months. Land-based missiles are restricted by having to be launched from an area controlled by Australian forces, which limits where they can be employed.

The RAN could rely on the capabilities of its allies, most notably the USN, to deliver the capabilities that would be provided by additional missile capability. The USN does maintain both a significant number of carrier strike groups as well as a large number of Arleigh Burke class destroyers that carry a large number of VLS cells. Outsourcing matters of national defence does come with significant foreign policy implications as well as good deal of military risk.

Modern warfare involves an increasing use of unmanned vehicles of all varieties - UUVs, UAVs, ULVs etc. Realistically, the only unmanned systems capable of delivering a comparable effect to a MSC-launched missile would be unmanned aerial systems (UAVs) as these can strike distant land and sea targets as well as conceivably manoeuvring to intercept aerial targets. However UAVs are still constrained by the same physical laws as missiles and so would require transport to a combat area, targeting and possibly control. Thus ultimately they simply act as another missile, which existing surface combatants cannot carry enough of due to space and weight requirements.

Other weapons under development include directed energy weapons and railguns. These weapons could deliver some of the effects of VLS-launched missiles and would offer some improvements, for example, laser weapons can shoot down missiles at a lower cost-per-shot than defensive missiles. However the power requirements of such weapons are extensive, and the effort involved in refitting existing MSCs to employ them would make these MSCs unavailable for years. As developmental technologies, these options do not offer a viable option for improving RAN striking power in the short-medium term.

Thus the problem remains - how do we equip the RAN's surface fleet with more missiles?

5. THE ARSENAL SHIP

Solutions to overcome the restriction posed by limited VLS cells have been proposed previously. Most notably, the USN proposed the Arsenal Ship concept in 1996 [5]. The Arsenal ship concept calls for a class of vessels with a small crew, few sensors and minimal independent capabilities, carrying a very large number of VLS cells (around 500 cells per ship). An arsenal ship would act as an "extra magazine" for MSCs, which would launch the weapons it its cells using Cooperative Engagement Capability (CEC) or similar techniques. This would allow the MSCs to have a much larger effective VLS cell count.

Making use of an Arsenal ship would require that the RAN's MSCs be capable of conducting networked warfare - using sensors on one platform to employ effectors on another. The Hobart-class are capable of performing CEC, demonstrated in testing by the HMAS Brisbane in 2019. The Hunter-class are also designed to use Aegis, thus would reasonably be expected to be capable of performing CEC. This only leaves the question of the ANZAC-class.

ANZACs utilise the SAAB 9LV combat system, rather than Aegis, and have not demonstrated a CEC-like capability to the author's knowledge. They are however equipped with the Link-16 datalink via HF Broad and Narrowband comms, which was proposed as the secondary datalink system on the Arsenal ship [6]. The CEAFAR RADAR can also be configured to act as a datalink and may be able to perform the capability better due to the inherent flexibility of an AESA [7], though this may require software and hardware changes to achieve.

The Arsenal ship concept is not entirely appropriate to the RAN's needs - placing 500 missiles aboard a single vessel would represent an extreme case of "all eggs in one basket". However, the notion of decoupling the sensors and effectors by providing a separate "magazine" of VLS cells aboard a ship is valid. The 500-cell capacity of the Arsenal ship was defined by the USN's needs at the time - why not simply carry fewer?

6. THE MATE SHIP

If the Arsenal ship is considered as a base concept, a review of the issues that dog other solutions provides a set of criteria that would describe a solution to the question "How do we best increase the surface fleet's strike and air warfare effectors in the shortest possible time?"

- The vessel would need speed, range and endurance comparable with RAN MSCs so that it could operate in concert with them.
    - The speed and range requirements are then defined by those of the ANZAC class- the RAN's most numerous MSC, at 11000 km range and 27 knots or greater maximum speed, respectively.
- The vessel would need to be able to carry a large weight of "cargo" composed of VLS cells
    - The number of cells carried by the vessel is determined as a tradeoff of cost vs capability, as the mk.41 can be configured to have an effectively arbitrary number of cells. To add the maximum lethality to the fleet, a larger number of cells would be better. However, the cost of the system rises as the number of cells increases, particularly if the payloads are considered. . Larger numbers of cells also risk rapid depletion of limited stocks of missiles if the vessel is damaged or destroyed. A compromise measure of 64 cells was reached for the purposes of this study. This equates to a 16m x 4m x 8m, 300t cargo.

In addition a number of measures of effectiveness of solution can be identified:

- Minimise crew-per cell, thus gaining maximum lethality per person for a personnel-limited RAN
- Minimise cost-per cell, similarly maximising additional lethality for what remaining budget can be found in an environment with a large number of acquisition programs
- Minimising time-to-service. This implies building as close to an existing design as possible.

Due to the intended role of accompanying major surface combatants, names such as the "Partner Ship" or "Ship Mate" were considered before the concept was christened "Mate Ship" as a play on the traditional Australian value of mateship.

7. HULLS

The speed, range and transport requirements constrain the potential vessels considerably. However, one type of vessel that approaches the requirements listed are high speed craft (HSC) multihull aluminium vessels. These craft have been employed in a number of commercial and military applications, such as ferry operations, troop transport and as one of the hull designs of the USN Littoral Combat Ship. Fortuitously, Australia is home to two companies experienced in the construction of such HSCs: Austal, based in Henderson, WA and Incat, based in Hobart, Tasmania. Both companies have provided transport vessels for both the USN and RAN. Two of these designs, the Spearhead class Expeditionary Fast Transport from Austal and the HSV-2 Swift from Incat, were identified as potential candidates.

The first Spearhead-class was constructed in 2010 by Austal USA for the USN. The vessel was designed to ferry troops and supplies quickly. The Spearhead class has been noted as being limited in speed at higher sea states, which would limit the operations of a task group to which it had been assigned in blue-water operations. This would need to be considered if the design was intended to be used as a basis for the Mate Ship.

Figure 1: USNS Spearhead (US Navy Sealift Command, 2012)

Table 1: Spearhead class particulars

Length	103m
Beam	28.5m
Tonnage	1515t full load
Hull form	SWATH

The High Speed Vessel (HSV) Swift was constructed by Incat for the USN as part of the experimental program that resulted in the Spearhead class. The ship saw service with the USN for more than 10 years and was later damaged in a missile attack. The HSV Swift's service period involved operations worldwide which indicates that the class would be able to operate within the RAN's expected area of operations.

Figure 2: HSV-2 Swift (US Navy Sealift Command, 2004)

Table 2: HSV-2 Swift particulars

Length	98m
Beam	27m
Tonnage	955t standard 1695t full load
Hull form	Wave-piercing catamaran

The characteristics of both vessels compared with the Mate Ship requirements:

Table 3: Base vessel characteristics vs requirement set

Requirement	**Target**	**Spearhead class**	**HSV-2 Swift**
Range	11,000km	5600nm (10371km)	4000nm (7400km)
Max speed	27+ knots	40 knots	45 knots
Capacity	16m x 4m x 8m, 300t	1,900m^2 mission bay, 544t	2670m2 cargo deck, 615t

Both vessels can achieve the speeds required, HSV-2 Swift is limited in range vs the original requirement. This would therefore require modification via the installation of additional fuel tanks. Both vessels include facilities for replenishment at sea (RAS) which would somewhat alleviate the range issue.

When evaluated against the crew-per-cell measure of effectiveness, both options for the Mate Ship rate considerably better than existing RAN MSCs, as well as a common world MSC. This is to be expected as the MSCs have significantly greater capabilities outside of VLS-carriage. The Arsenal ship remains considerably more effective by this measure due to its much greater VLS count. Nonetheless this suggests that both hull options represent viable Mate Ship candidates.

Table 4: Crew-per-cell comparisons

Class	**ANZAC**	**Hobart**	**Hunter**	**Arleigh Burke (IIA)**	**Arsenal Ship**	**Spearhead**	**Swift**
VLS Cells	16	48	32	90	500	64	64
Crew	163	186	180	323	44	41	35
Crew/cell	10.19	3.88	5.63	3.59	0.09	0.64	0.55

Rather than use an existing military design as a basis, a Mate Ship could be derived from a commercial ferry design. The resultant Mate Ship would have even less military capability than an existing military transport design, unless substantial design work was undertaken, increasing cost and delaying construction. As existing designs built for the USN, the HSV-2 Swift and Spearhead would already have some level of commonality with the RAN as some RAN processes and standards are derived from USN equivalents. The design IP would be held by the USN and Austal or Incat and would need to be acquired by Australian Defence, however the strategic partnership between Australia and the US should facilitate this process.

8. COMBAT SYSTEMS

Given that the Mate Ship's central role is the provision of additional VLS cells to grant striking power to MSCs, it is clear that the ship would embark the Mk. 41 VLS. 64 cells was also chosen as it could be procured by extending the government furnished equipment (GFE) program of the Hunter-class frigates, which use a 32-cell mk. 41 VLS. As the Hunter build has already been delayed by 18 months and more delays could reasonably be expected (as almost all MSC builds deliver late), VLS systems could continue to be ordered on the original production schedule (or the buy even accelerated) and the additional systems incorporated into Mate Ships.

The primary means of defending a Mate Ship would be through MSCs directing missiles launched from their own VLS, or the VLS of the Mate Ship, however some self-defence could be included. To maintain the intent of a lower-cost, quickly built capability, these defences should be:

- Already in use by the RAN
- Require minimal additional sensors and combat systems
- Require minimal changes to ship structure and power requirements

The most obvious inclusion would be to install one or more Phalanx close-in weapons systems onto the Mate Ship. These weapons have in-built fire-control sensors, are already installed on several RAN classes and can be installed without deck penetration - though they would require strengthening of the deck as well as changes to the Mate Ship's power and cooling systems. Operating Phalanx systems 'independently' - i.e. without direction from an integrated combat system - would make them less effective, but would still provide the Mate Ship with a measure of protection against swarm threats (both surface and air) and limited last-ditch protection against anti-ship missiles. If the same approach to GFE were taken for Phalanx systems as was used for the VLS, 4 systems would be available for each Mate Ship. This large number of mounts would provide clear arcs of fire with minimal modification to the Mate Ship and would limit the area that each system had to cover. Multiple Phalanx mounts is also one of the most cost-effective means of improving defence against swarm attacks. [8]

Soft-kill protection measures, such as the inclusion of the Nulka decoy system, could also be included to provide some measure of self-protection to the Mate Ship without requiring an extensive combat system.

The Mate Ship would require a fire control system in order to launch the payloads from the VLS cells. This fire control system would need to receive track and launch information that would normally be provided by a combat ship's own sensors. The virtualised Aegis combat system demonstrated by the US Project Overlord [9] would perform this role. This would necessitate the Mate Ship carrying appropriate datalink systems, including CEC and Link 16 to receive track data and launch commands from MSCs. These communication and fire control systems would allow the Mate Ship to launch missiles at the command of MSCs, but not have an independent combat capability. This minimises the monetary and schedule cost of the Mate Ship as well as minimising the risk to other platforms if the lightly-crewed Mate Ship were to be captured.

A very basic layout of the Mate Ship concept is shown here. Due to not having access to the general arrangements of HSV-2 Swift or the Spearhead class, the Incat-built ferry Saint John Paul II is used as a basis as this hull design and layout is indicative of the vessels considered, particularly the HSV-2 Swift.

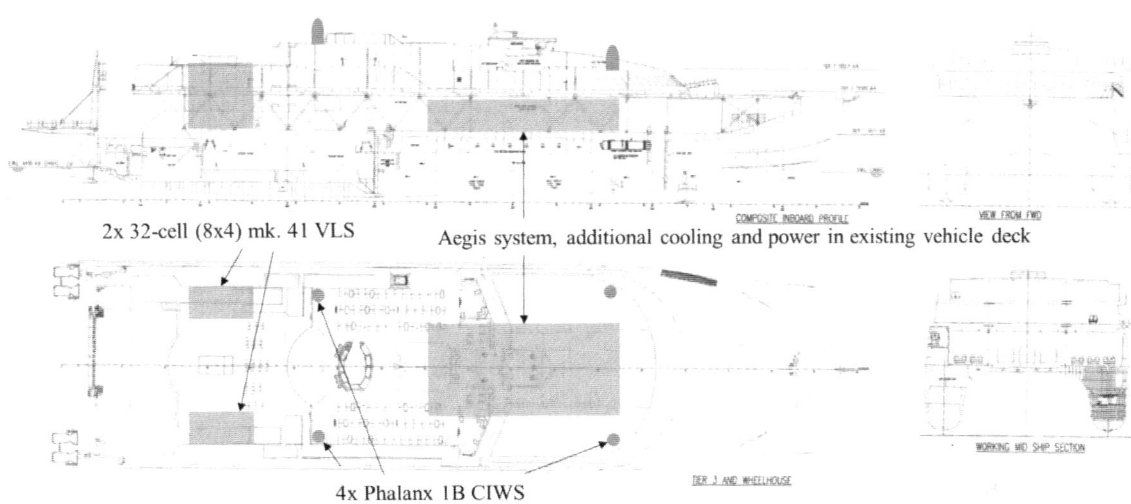

Figure 3: Basic layout of systems aboard JP2-based Mate Ship

9. SURVIVABILITY

In considering the concept of the Mate Ship, survivability is an obvious concern. The Mate Ship carries significant payload, but has little to no organic self-defence capability. So who would sign off on the acquisition of such a floating target, let alone serve aboard one?

To answer this question, we need to recall proposals to increase the capabilities of the Arafura-class OPV - by fitting a larger main gun and anti-ship missiles. This increase to capability makes an Arafura-class vessel a greater target, yet provides almost no improvement to the ability of the ship to defend itself, even if the design's existing combat system could effectively employ the its gun in an anti-missile role.

The Arafura-class carries at least 40 crew and has minimal damage control capability compared to RAN MSCs. Therefore it seems that the RAN would at least consider creating a vessel with offensive power but little survivability that carries 40 or more sailors into battle. If this is the case, why not a vessel that brings more capability for that risk?

If the Mate Ship were equipped with Phalanx and/or Nulka, then it would possess some self-defence capability, which improves survivability by virtue of lessening the chance of being hit. However, if a Mate Ship was to take enemy fire, its survivability would be very limited, as:

- As civilian, or military transport vessels for the design basis for the Mate Ship, it would not possess comparable damage control to MSCs
- The small crew of a Mate Ship would make effective damage control difficult
- The aluminium structure of a Mate Ship is more vulnerable to fire damage than the steel structure of RAN MSCs.

It is notable that the HSV-2 Swift survived enemy action in 2016 when she was hit by either an anti-tank or anti-ship missile. The ship remained afloat, though did suffer significant damage. This implies that a Mate Ship could have some capability to survive relevant weapon strikes.

The lower inherent survivability of Mate Ship would require a different approach to damage control than that employed on RAN MSCs. Current RAN damage control is intended to allow MSCs to survive and endure - albeit with possible loss of function - following damage from enemy action. A Mate Ship would likely have to use an approach more similar to civilian vessels, where the intent of damage control is to allow the crew sufficient time to abandon ship. This 'DC to abandon' approach would reflect the fact that the value of a Mate Ship is primarily found in its payload. It would require, however, that anything of value within the ship, such as whatever combat system is required to enable VLS launch, could be easily disabled before the crew abandoned the vessel.

10. SHIPBUILDING CONSIDERATIONS

As previously noted, the Incat and Austal shipbuilders are both designers and builders of HSC vessels. If the Mate Ship concept were pursued, these shipbuilders would be the logical choices for building the vessel. In-country builds would be preferred for sovereignty reasons but overseas builds would be considered if they significantly improved acquisition speed.

The Spearhead class is constructed by Austal USA in Mobile Alabama, however a derivative class has been built at Austal's Henderson facility. This suggests the possibility that Austal Henderson could produce Spearhead-based Mate Ships. However, the recent increase in ship builds planned for Henderson suggests that there may not be the capacity in the near-term. Alternatively the vessels could be constructed as part of the existing deliveries of Spearhead class vessels from Austal USA. These vessels are being delivered at a rate of approximately 1 per year, with around 2 years per build. [10] It is also possible that existing Spearhead class vessels that are currently inactive could be purchased from the USN and repurposed into Mate Ships.

The HSV-2 Swift is not currently in build and would therefore a Mate Ship based on the design would require some redesign to address obsolescence. However the vessels themselves could be constructed in Hobart. This would avoid the issues faced in Henderson with other naval builds. HSV-2 Swift was delivered 10 months after contract award, indicating that Incat may be able to produce similar vessels quickly. [11]

As Incat and Austal Australia do not have much experience in combat systems, a potential acquisition approach would involve the shipbuilder building a vessel fitted "for but not with" VLS, Phalanx and a minimal combat system and delivering such a vessel to another for the fitting of the relevant systems, handled as a "major upgrade".

While HSV-2 Swift's 10 months from contract to delivery would be excellent, it would be reasonable to think that the delivery of the Mate Ship platform would be more closely approximated by the 2-year production cycle of the Spearheads. Incorporating time for the establishment of the project as well as the installation and testing of the combat system, even a fast-tracked Mate Ship would take 5 years to service at very best. 5 years is, however, very fast for the acquisition of a naval combat capability and would increase the RAN's surface fleet lethality until other capabilities came into service.

The cost of acquisition for a Mate Ship would vary considerably depending on acquisition path and so will not be considered in detail for this paper. As a rough estimate: the Spearhead class cost approximately $180 million US ($270 million AUD) [10] per unit and a 64-cell VLS approximately $110 million US ($166 million AUD) [12]. The Aegis combat system element can be estimated at $172 million US ($260 million AUD) [13]. This equates to a cost of approximately $700 million, however this is likely a significant underestimation. However this compares favourably with the cost of MSCs.

11. MAINTENANCE AND CREWING

The Mate Ship's role would be focused around high-end warfighting as it would provide very little capability for peacekeeping and constabulary roles. As such a Mate Ship would only leave port for exercises or if it was expected that a battle group would need high-end capabilities in its operations. This would bring benefits in more opportunities for maintenance as well as for crew shore leave

Alternatively, crewing could be provided form the patrol boat crews. The Mate Ship's basic crewing requirement is similar to that of the RAN's Arafura class patrol vessels, thus the Mate Ship crews could be formed out of Arafura crews on a rotational basis to conduct exercises etc. during peacetime and, in the event of war, the Mate Ships would be deployed crewed by all Arafura crews. As patrol vessels have little to no value in a high-end conflict, this would be a net increase in the RAN's capability in warfighting with minimal crew increases. However this would require a significant amount of cross-training from the Arafura crews and would increase their sea time.

Regarding maintenance a Mate Ship would cost significantly less to replace than an MSC, as it is essentially a modified ferry, rather than a complicated high-end warship. Thus maintenance by replacement becomes a viable option, with new hulls being built and existing VLS etc. being cannibalised and installed in the new hulls. This in turn would assist in furthering Continuous naval Shipbuilding, as well as allowing opportunities for continuous improvement - Upgrade by Replacement.

12. MATE SHIP AND THE LUSV

The Surface Fleet Review announced the acquisition of 6 Large Optionally Manned Surface Vessels by the RAN, to be constructed in Henderson [14]. The vessels, based on the developments of the USN's Project Overlord [15], are intended to carry a VLS payload (nominally 32 cells) and be able to operate in an autonomous manner, though to date the RAN has stated that they will be "minimally manned". They are intended to act as additional VLS magazines for MSCs. Given that these vessels have a very similar role to the Mate Ship and are already included in Defence planning, does a role for the Mate Ship still exist?

The LOSV design is not specified, but is based upon the Project Overlord LUSV, thus this is used to assess the suitability of the LOSV.

The LUSV as based on current development appears to have range capabilities in line with Mate Ship requirements. It would also likely have sufficient endurance due to lacking crew. The hulls used for LUSV testing have as yet not included high-speed designs, so unless different hullforms are chosen it would not meet the speed requirement. The LUSV carries 32 VLS cells to the Mate Ship's 64, thus adding less lethality per vessel. However these numbers are largely arbitrary due to the VLS's modular nature. The LUSV optimises the crew-per-cell metric and also overcomes the survivability concerns around the Mate Ship, so long as the LOSV developed is run unmanned.

Crucially however, the LUSV does not provide the quick addition of lethality that the Mate Ship could provide. Currently Defence states that the LUSVs will be delivered "In the 2030s". Considering the developmental nature of autonomous vessels and regulations [16], the tendency for naval shipbuilds to run over schedule [4] and the current planning to build the LUSVs at Henderson along with four other classes of naval vessel, it can be reasonably assessed that "In the 2030s" would mean "by 2039". Thus only building the LUSVs would result in a 15-year capability gap.

This capability gap could be addressed by building Mate Ships in the interim. The RAN could then use the platform as a test bed to determine if to integrate such a capability into fleet doctrine and whether aspects of the Mate Ship concept, such as higher speed and organic defences, should be included in the LUSV designs. Furthermore, Austal USA are involved in LUSV development and are incorporating autonomous aspects into later Spearhead class hulls [17]. Thus the Mate Ship and LUSV concepts could converge as an "auto-mate ship".

13. FURTHER DEVELOPMENT

Placing only VLS cells and a minimal combat system aboard platforms designed to move hundreds of personnel leaves significant space aboard the Mate Ship. This provides opportunities for later development of the capability.

One option would be to transport not only missiles, but also unmanned vehicles into a battlespace, for the use of the escorting MSCs. This would involve significant developmental work and should thus not be considered until the original capability was achieved, but would help MSCs keep improving their capabilities whilst not adding further tophamper to their already packed hulls. Both HSV-2 Swift and Spearhead class vessels have embarked helicopters, and an aviation capability would thus also be a developmental opportunity.

The development of hypersonic weapons by Australia and its allies may produce weapons too large to fit in the launch cells of existing VLS, as achieving hypersonic speeds is a fuel-intensive exercise. The relative lack of clutter on a Mate Ship compared to MSCs would allow for the installation of launch systems for these large weapons.

14. CONCLUSIONS

To increase the lethality of the RAN's surface fleet in line with the Defence Strategic Review's recommendations, the simplest method is to increase the number of effectors. Acquisition of additional existing fleet units or modification of these vessels is impractical due to time and crewing constraints. To achieve "minimum viable capability" in short timeframes, vessels providing new capability with a combination of existing technologies are required.

By utilising existing HSC vessels loaded with minimal combat systems and multiple VLS cells, additional effectors can be carried to a battlespace and used by existing units. Two potential base vessels were identified: the HSV-2 Swift and the Spearhead class. Potential acquisition, crewing and development paths were considered for these vessels.

15. REFERENCES

1. AUSTRALIAN GOVERNMENT, "National Defence: Defence Strategic Review," *Government of Australia, Canberra*, 2023.

2. GARARD, O., "Lethality: An Enquiry," 1 November 2018. [Online]. Available: *https://thestrategybridge.org/the-bridge/2018/11/1/lethality-an-inquiry*. [Accessed 05 April 2024].

3. KOLATA, G., "Evolution of the Aerial Defence of Air Bases. Concepts and Lessons Learned," *War Studies University Scientific Quarterly, vol. 1, no. 114, pp. 53-74*, 2019.

4. UNITED STATES GOVERNMENT ACCOUNTABILITY OFFICE, "Navy Shipbuilding: Past Performance Provides Valuable Lessons for Future Investments," U.S. GOVERNMENT ACCOUNTABILITY OFFICE, Washington, 2018.

5. NINER, J., "Arsenal Ship: The Operational Commander's Perspective," *Naval War College, Newport*, 1996.

6. BAUMANN, G., ET AL., "An Arsenal Ship Design," *Naval Postgraduate School, Monteray*, 1996.

7. M. BRANDFASS, T. MOELLER AND K. WEIDMANN, "Multifunctional Radar and Data Link Functions," in IEEE International Radar Conference (RADAR), Sydney, 2023.

8. BALBUENA, D., ET AL, "UAV Swarm Attack: Protection System Alternatives for Destroyers," *Naval Postgraduate School, Monterey*, 2012.

9. NAVAL TECHNOLOGY, "Ghost Fleet Overlord Unmanned Surface Vessels, USA," 23 March 2023. [Online]. Available: https://www.naval-technology.com/projects/ghost-fleet-overlord-unmanned-surface-vessels-usa/?cf-view. [Accessed 02 May 2024].

10. GOVERNMENT ACCOUNTABILITY OFFICE, "Defense Acquisitions: Assessments of Selected Weapon Programs," *Government Accountability Office, Washington*, 2010.

11. INCAT, "Military Dossier," 16 October 2012. [Online]. Available: https://www.incat.com.au/wp-content/uploads/2018/02/military_dossier.pdf. [Accessed 01 May 2024].

12. DEFENCE SECURITY COOPERATION AGENCY, "THE NETHERLANDS – MK 41 VERTICAL LAUNCHING SYSTEM (VLS)," 2 June 2023. [Online]. Available: https://www.dsca.mil/press-media/major-arms-sales/netherlands-mk-41-vertical-launching-system-vls#:~:text=The%20total%20estimated%20program%20cost%20is%20%24110%20million.. [Accessed 01 May 2024].

13. DEFENCE SECURITY COOPERATION AGENCY, "SPAIN – AEGIS COMBAT SYSTEM," 26 June 2016. [Online]. Available: https://www.dsca.mil/press-media/major-arms-sales/spain-aegis-combat-system. [Accessed 01 May 2024].

14. AUSTRALIAN GOVERNMENT, "Enhanced Lethality Surface Combatant Fleet: Independent Analysis of Navy's Surface Combatant Fleet," Commonwealth of Australia, Canberra, 2024.

15. O'ROURKE, R., "Navy Large Unmanned Surface and Undersea Vehicles: Background and Issues for Congress," *Congressional Research Service, Washington*, 2019.

16. HORNE, R., DEANE, F., JOINER, K., and TRANTER, K., "Navigating to smoother regulatory waters for Australian commercial vessels capable of remote or autonomous operation: a systematic quantitative literature review," *Australian Journal of Maritime & Ocean Affairs, vol. 15, no. 4, pp. 496-517*, 2023.

17. WHITE, R., "A Big Step for US Navy Unmanned Programs: Unmanned EPF Apalachicola," 8 June 2021. [Online]. Available: https://navalpost.com/unmanned-expeditionary-fast-transport-epf-vessel-13-apalachicola/. [Accessed 02 May 2024].

16. AUTHOR BIOGRAPHY

Gareth Arnold holds the current position of Auxiliaries Group Lead (WA) at Naval Construction Branch. The Author is responsible for build quality assurance activities on auxiliary systems for naval builds in Western Australia. The Author's previous experience includes the Air Warfare Destroyer build program.

ACHIEVING MORE WITH LESS – AN ENGINE BUILDER'S PERSPECTIVE

G Wilson, Rolls-Royce Australia Services Pty Ltd, Australia
R Partridge, Rolls-Royce PLC, UK

SUMMARY

This presentation will examine the challenges and opportunities around electrification, environmental sustainability and digitalization in future warships, especially in power and propulsion systems, and consider how these both require and assist shipbuilders and navies to achieve more with fewer resources.

1. INTRODUCTION – THE AUSTRALIAN CONTEXT

The Defence Strategic Review placed greater emphasis on submarines and long-range fires for defending Australia, and the subsequent Surface Fleet Review detailed how this results in a changed structure for the Royal Australian Navy (RAN). The structure will include two tiers of warships to try to balance capability and cost. These changes come in a time when naval crew and defence industry labour challenges have also limited the RAN fleet. Further challenges come from the distance between Australia and its allies, which could increase supply chain cost and risk in times of conflict. This is particularly true for fuel because Australia exports crude oil and imports refined products while having only limited fuel storage capacity. In the longer term, the drive for decarbonization may further complicate the availability of established fuels and prompt a search for alternatives, but adopting alternative fuels also brings challenges.

2. FUEL RESILIENCE

While moving to a more environmentally sustainable capability by reducing the use of fossil fuels and simultaneously improving fuel resilience in defence are imperative, the transition to an alternative and/or broader range of fuels is not straightforward.

Alternative fuels can be divided into those produced from crops or biowaste and those formed from purely synthetic processes such as carbon capture, electrolysis and the Fischer-Tropsch process. These fuels differ in properties such as energy density, flashpoint, and toxicity, which may pose technical challenges to compatibility with existing marine products and technologies, and hence their ease of adoption. Even though internal combustion engines can and are being adapted and re-certified by many suppliers for some of the alternative fuels, issues with safety and logistics imply that many of the barriers are beyond the engine room and must be carefully considered by naval architects and operators/owners. They will impose additional constraints on platform design and operation, requiring some difficult choices and compromises. Due to limited infrastructure and refuelling assets, there is also the need for interoperability with allies, meaning that resolution of this complex problem cannot be done in isolation and must be entirely collaborative.

Alternative fuels may well be significantly more expensive than those derived from crude oil, and with less global security of supply.

3. ELECTRIFICATION

Electrification can have benefits in warship design such as the flexibility in layout and more options for reducing noise during missions such as anti-submarine warfare. It can also contribute to doing more with less by enhancing fuel efficiency and enabling emerging capabilities like directed energy weapons. An integrated electrical system with multiple generators can be controlled such that the prime movers are operated close to their most fuel-efficient point. This matching of supply and demand could be further enhanced by adding energy storage or by using power converters to allow variable generating frequencies.

The fuel savings electrification could bring must be weighed against the electrical conversion and distribution losses. Power quality is also a consideration because propulsion and emerging directed energy weapons can introduce large load transients that could cause excessive frequency and voltage deviations on the electrical system. Other considerations include the size and weight of the electrical equipment, the additional training required to operate it and the challenges of fault tracing and repair. However, if these are overcome, including using modelling and simulation and land-based trials during warship design, the benefits described here can be realized.

4. DIGITALIZATION

Digitalization is a broad term that loosely describes the application of technologies that use data to gain insight and enhance operation of complex systems. Increased data storage and processing capability at reduced costs has supported the development of a range of services. In ship power and propulsion systems it can be applied both at the equipment and the system level. Moreover, the propulsion system arrangement and equipment are increasingly being selected with more consideration of reliability and on-board planned maintenance burden with an aim to minimize crewing especially in the Engineering Department. For engines, digitalization can be used to monitor degradation, so maintenance is done at optimal times. At the system level, it can be used to minimize fuel consumption and engine running hours in each operating mode, or to help with fault tracing. It may also enable offline modelling and simulation that can be used for purposes such as assessing the effects of control changes, recreating faults observed at sea and for crew training.

While commercial ships can now use satellite communications to transmit live data about their engineering systems to shore for storage and processing, which can help to apply big data methods across fleets or for all installations of one model of equipment, this has challenges in warships. Warship satellite communications have limited bandwidth that is shared across all aspects of the ship operation, which will limit bandwidth for transmitting engineering data. However, increased development of edge computing allows more processing to be done in situ and so there may be a reduced need to transmit it. Furthermore, for digitalization to best help achieve more with less, it should include a human-machine interface that enables onboard crew to have real-time awareness of the status of the power and propulsion system to help prioritize tasks and ensure maintenance with a lean engineering crew.

5. CONCLUSION

In the face of budget, personnel, economic and logistical challenges, the trends of alternative fuels, electrification and digitalization can help warship designers and operators to get the most out of their available resources. In electrification and digitalization, these are continuing trends that evolve with technology and operational experience. Fuel transitions have not yet begun in navies, but they are anticipated in the coming decades, and reducing risk and achieving benefits such as resilience will require careful consideration in their choice and adoption.

6. ACKNOWLEDGEMENTS

The authors wish to thank Klaus Schmidt, Director Naval Application Engineering at Rolls-Royce Solutions, for his support with preparing this presentation.

7. AUTHORS BIOGRAPHY

Gary Wilson is the Research and Technology Manager - Defence at Rolls-Royce Australia Services Pty Ltd. He is responsible for developing relationships, assessing opportunities for technical collaboration, and managing early-stage technology development in Australia and Japan. He has worked in a broad range of roles in shipbuilding, ship classification, marine equipment and systems engineering and research and technology development in multiple countries. He studied electronics and electrical engineering and has worked in multidisciplinary teams in the commercial and naval domains.

Richard Partridge is the Chief of Naval Systems at Rolls-Royce PLC and has worked for the company since 1999. He has participated in a variety of recent naval projects including the Royal Navy's Queen Elizabeth Class Aircraft Carrier, the US Navy's Littoral Combat Ship (Monohull) and the Republic of Korea Navy FFX-II frigate. He worked for the Energy sector of Alstom Gas Turbines between 1996 and 1999, and prior to that spent six and a half years in the Marine Engineering branch in the Royal Navy, including three years serving on an Anti-Submarine Warfare Frigate.

DESIGN FOR UPGRADE – MANAGING THE PLATFORM TO COMBAT SYSTEM INTERFACE

J Cole, A Smith and **G Barden,** Department of Defence, Australia

SUMMARY

Combat systems and platforms have very different refresh cycles. This paper explores the use of design modularity and spatial arrangement to manage the combat to platform system interface and reduce the burden of combat system refresh. The work focuses on surface combatants and seeks to balance the capability of the integrated ship system in aspects such as topside design, survivability, ship performance and growth margins. The retrofit of new combat systems to existing ship designs creates many design challenges and necessary compromises. This work identifies design options available in a new platform design to accommodate current and future combat system needs. The aim is to reduce the cost and time associated with mid-life upgrades and increase the likelihood of a surface combatant design being compatible with long multi-flight build programs. The work considers examples from past programs and literature as well as presenting an exemplar surface combatant spatial design and design features.

1. INTRODUCTION

Controlling the cost and time associated with warship design, construction and modification is a core concern of all involved. Warships constitute major national investments and decision-makers are obligated to demonstrate value for money. A common factor in this argument is the warship's intended service life over which a large upfront investment is spread. Another important factor is that a warship's value is judged against its adversary in a constantly evolving technology and threat context. These two factors generate tensions between suitability and longevity. Ultimately, a balance is needed that delivers on today's capability commitments, while accommodating future capability needs.

The search for engineering solutions to relieve this tension has drawn significant attention and effort over the years. The prominent treatments available to modern warship designers are standardisation of architectures and interfaces to reduce the overhead of upgrades, as well as modularity to decouple the more rapidly changing capability elements from those that change more slowly. However, as the history of several ambitious projects and programs have shown, the practical consequences of these treatments have been mixed, and the fundamental challenge remains relevant today.

The engineering systems and disciplines of warships are often delineated between combat and platforms systems. This separation has become entrenched in their design, build, test, and even operation. While this way of thinking might be chalked up to different technologies that require unique disciplines and specialisations, there remains a critical need to ensure an optimum balance between the float, move and fight functions, which embody the collective purpose of a warship as a whole. The fundamental problem is that it is neither viable to produce a new platform for every new combat system iteration, nor retain the same combat system throughout the natural life of a platform. Whereas platform systems are generally able to operate with basic maintenance for 20 years or more, combat systems are constantly updated and upgraded to retain capability relevance. Since the vast majority of drivers for capability upgrade occur within the fight functions, tensions around return on investment and capability refresh tend to manifest at the boundaries of platform and combat systems.

Given that warships will continue to operate for decades, it is clear that they will be subject to upgrades throughout their life. Therefore, there is strong incentive for flexible and adaptable warships, which incorporate design features that reduce the cost and time associated with upgrades. This paper considers a variety of approaches to managing change within the life of a warship. The scope of change expected of future warships is also examined, with a focus on core combat system electronics spaces (communications, combat management, etc) and radar upgrades as prominent above-deck systems which have proved challenging to refit on modern warships. A novel approach is proposed that leverages past concepts and practical experiences. Many of the challenges and opportunities described are relevant to new programs based on existing designs and to the evolution of designs within a continuous build program. This paper also seeks to identify the relevant trade-offs and limitations of the approach, recognising that it is not always practicable to respond to all future threats or technologies through upgrade of an existing warship.

2. APPROACHES TO ENABLING UPGRADE DESIGN

2.1 MARGINS

A customary and often-cited measure of a ship's ability to cater for change is its 'margins'. Margins can be allocated and consumed in design, build and operation of a vessel, as well as upgrade. They are a measure of the remainder between the current vessel status and a limiting critical constraint. For example, a weight margin represents the difference between the current displacement of the ship compared to the maximum displacement possible within stability or structural limits. A large weight margin implies that the platform could withstand an upgrade which imposes additional net weight. Similarly, margins can be defined for other key design parameters such as the centre of gravity (with respect to stability limits) or electrical load (with respect to rating, quantity or configuration of generators). Unallocated vacant space can also be reserved as marginal capacity for growth (e.g. spare electronics racks or oversized compartments), but practical spatial margins at larger scales are very difficult to achieve because even small spatial rearrangements tend to cause cascading impacts which are difficult to contain.

However, the problem of integrating future combat systems is not resolved just by reserving weight, space and power. There are also complex multi-factor constraints including power distribution (frequency, quality, cabling, etc), electromagnetic interference, data transmission and storage, physical and cyber security, operator and maintainer interfaces, survivability implications, and much more. As a result of such a multitude of factors and their inter-dependencies, the fragmented management of separate margins quickly becomes onerous and does not guarantee compatibility with potential future changes. This reality limits the utility of margins to iterative (like-for-like) updates and rarely enables larger scale design changes.

Reported global margins on paper often appear larger than what is actually available in the necessary zone. A design zone budget approach has become common within modern warship design, and seeks to improve the functional allocation of margins and allow more explicit trade-offs to be made within a particular part of the ship. However, while valuable during the design phase, once a ship is built these budgets become embedded within specific zones of the ship, and are difficult to transfer to another. Beyond the initial design, there is little difference between margins and zonal budgets.

2.2 MODULARITY AND STANDARDISATION

In addition to its benefits in mission flexibility, modularity is also presented as a solution to the 'upgrade' problem and can undoubtedly offer many benefits. It sets clear managed interfaces between the platform system, combat system, and other modular elements and allows the design and development of these systems to proceed independently.

Land-based testing is a concept which leverages modularity to enable assembly and testing of equipment independent from the ship itself. Subsequent installation to the ship at a high-level of assembly minimises technical, cost and schedule risks. Although typically incorporated during new-build construction, land-based testing can also be applied to mid-life upgrades, particularly if the platform is designed to provide sufficient access (removal routes) and standardised interfaces.

Unfortunately, the benefits of modularity do not come without some significant costs. Its rigid interface constraints often force sub-optimal system designs or create incompatibility with future generations of technology. There is space and weight lost to the structural 'packaging' of modules that could otherwise be utilised by equipment. Modularity adds complexity to the sizing of platform services such as power, cooling or data connections as conservative 'worst-case' consumers define these. Incorrect assumptions in these early decisions manifest as either inadequate capacity for future demands or costly and unnecessary excess capacity.

2.3 ADDITIONAL DESIGN CHANGE TO ENABLE UPGRADE

Where margins or modularity are insufficient, upgrades can sometimes be implemented through a larger scope of design change. This larger scope is typically highly intrusive to the vessels structure or other capabilities. Changes may include lengthening hulls, enclosing open decks, adding ballast, upgrading major equipment. Offsetting design changes, such as removing obsolete equipment or less critical capabilities, provides capacity for the desired capability change. Enabling change of this type is fraught with technical risk, and carries significant schedule and cost impacts to the program. Without the ability to rebalance the overall design these approaches typically result in trade-offs in other areas, such as vessel performance for speed or rang. Design change is also the dominant method in a batch build program where changes that would not otherwise be practicable as in-service upgrades may be incorporated into ships still under construction.

3. HISTORICAL UPGRADE EXPERIENCE

3.1 MARGINS

Surface combatants very rarely have sufficient margins to support the full ambition for capability upgrade. All margins but in particular Weight, VCG, Electrical, HVAC, Chilled Water, Space, and Cable Transit margins come under pressure and limit Upgrade. [1]

For the ANZAC Class capacity within cable runs and penetrations has proven important for controlling the burden of upgrade. During the recent upgrades, the program ensured that new multi-cable transits had capacity in excess of the requirement to allow for future growth. Similarly, this has been a consideration with the Arafura Class OPV which has a low utilisation rate of multi-cable transits to ensure that there is the ability to run additional cables in the future.

3.2 MODULARITY AND STANDARDISATION

3.2 (a) ANZAC Class

The German-designed Australian-built ANZAC class frigates were designed with an acknowledgement and awareness of the need for adaptability. The Blohm & Voss MEKO 200 philosophy centred on an ability to adapt a common basic design for unique customer needs, and has been tailored to the diverse needs of eight different nations. The build philosophy also incorporated innovative modular features which accelerated the shipbuilding schedule, including block construction and high-levels of equipment assembly and land-based testing prior to installation onboard.

The ANZAC build program demonstrated several benefits of modularity in functional units during the acquisition, build and sustainment of a vessel. During acquisition, the main gun was changed with little impact to the program, and before the last of the eight Australian ANZAC frigates was commissioned in 2006, upgrades for Harpoon launchers, ESSM, Nulka and an additional sonar had been implemented, with refits to earlier ships as well as new-builds ensuring commonality across the class. The design also used standard connectors for combat system containers to connect the vessel to services, and implemented shock mounting at the container-level, allowing for simpler equipment arrangements within each to maximise spatial efficiency.

Whilst the design demonstrated the benefits of modularity for flexibility and efficiency in the initial design phase, there was a lack of consideration provided for future in-service removal routes for many functional units. This necessitated extensive structural rework, and removal of additional equipment and services to enable upgrades. Despite their location on the exterior of the ship, and incorporation of bolted flanges for the main gun, VLS and masts, a lack of standard cable interfaces meant that lengthy cables still had to be arduously rerouted throughout the ship. Furthermore, because of a cement-based cable penetration system had been used, hot work was ultimately still required which multiplied the scope of work (e.g. protection of adjacent spaces, fire watches, blast and paint, insulation removal and replacement, installation and testing of new penetrations). Another example is the upgrade of a sensor module which required the replacement of 80% of the equipment within one of the combat system containers. Due to a lack of provision of removal routes, this would have required cutting the ship to remove as a single modular unit but was instead achieved by duplicating a test environment in a spare container off the ship and then moving each piece of equipment individually to the ship.

These experiences from ANZAC upgrades show that modularity does not necessarily translate to enabling upgrades. In particular, the lack of consideration for removal pathways of modular containers during design has meant that the full potential has not been realised which highlights a need for greater consideration of upgradeability in early design.

3.2 (b) Littoral Combat Ship

The US Navy's Littoral Combat Ship program is an important illustration of adaptability in-practice. The ambitious intent was to extend the concept of modularity beyond customisation to enable day-to-day reconfigurability. This would allow fast changes of mission packs to cater for unique demands of different warfare tasks (i.e. ASW, Surface Warfare and Mine Warfare). [2]

The system architecture is built around a core sea-frame and interchangeable modules. The sea-frame provides typical platform functions as well as self-defence, C4I, navigation and the launch and recovery of unmanned vehicles. The interchangeable mission modules consist of vehicles, deployable sensors and mission manning detachments, and are typically packaged inside one or more 20ft ISO containers. The result is the possibility to dramatically reconfigure the vessel alongside a wharf within a few days. This enables fewer and smaller platforms to conduct necessary missions under the restriction that only one type of warfare is able to be performed at any given time. Another identified benefit was that the interchangeability of modules provides resilience to technological development, as the mission packages can be upgraded independently from the sea-frame.

In practice, the Littoral Combat Ship program has not met all initial expectations. The development of relevant compatible mission modules and achieving the necessary concurrent combinations of crew competencies across multiple roles has proven challenging. Additional reliability issues and cost overruns have further exacerbated program concerns. [3]

3.2 (c) STANFLEX System

The Danish STANFLEX system is another prominent example of a modular payload system, having been in use since the 1980s and applied to multiple vessel classes. It was originally introduced to enable mission modularity driven by cost incentives and a reduction in fleet size. However, in practice the system has rarely been used in this way, but provides ample benefits for maintenance through a rotatable pool of equipment and platforms, as well as providing standardised interfaces to enable efficient upgrades. [4]

By decoupling their platform systems from their combat systems using the STANFLEX system, the Danish Navy has the ability to temporarily remove sensitive equipment. This provides the benefit that classification surveys and dockings can be performed more easily in commercial shipyards, greatly simplifying scope and reducing cost of maintenance.

3.2 (d) Astute Submarine Command Deck Module

The Astute class submarine program claims to have benefited from lower level modularity during build. The implementation of a pre-integrated and tested 'command deck module' reportedly reduced build schedule by as much as 6 months [5]. Given the similarities between major warship and submarine programs, this emphasises the potential schedule impacts of parallel assembly and testing achievable through modularity and standardisation.

3.3 ADDITIONAL DESIGN CHANGE TO ENABLE UPGRADE

3.3 (a) Offsets

The last 'battleship' to be built by the US Navy, USS Missouri, was commissioned in the twilight of World War II in 1942 with a combat system based on three triple 16 inch naval gun turrets and a further twenty 5 inch guns. However, the nature of warfare changed rapidly post-war and having provided only limited shore bombardment/naval gunfire support in the Korean War, the USS Missouri was decommissioned in 1953 after just 11 years' service. Thirty years later, under pressure to rapidly increase the size of the US Fleet to 600 ships, USS Missouri was brought back into service amidst heavy contest in congress between upgrading existing or building new ships. Although not designed explicitly with adaptability to future needs in mind, extensive modifications were able to be made in 1984 to deploy Harpoon and Tomahawk missiles as well as install CIWS defences. USS Missouri's upgrade was largely achieved by offsets, such as removing other equipment, and reducing crew numbers. As it was already out-of-service, the extensive refit period did not represent a 'capability gap', and the project was broadly accepted to be cost-effective in its aim to deliver an additional ship to the fleet. Although its long-range strike lethality was demonstrated in the Gulf War, many redundant design features such as a thick steel armour deck and several large guns remained after refit. As a result, USS Missouri was a less capable warship than contemporary new-build warships, and it was decommissioned again in 1992 after only another 8 years' service or a total of just 19 years' service over the 50 years since 1942. USS Missouri was not explicitly designed for adaptability, but its upgrade was achieved by force through extensive design change and offsets.

3.3 (b) Margin Improvements

The ANZAC Class has undergone two successive major midlife upgrade programs associated with combat and radar systems on a rolling basis since 2010. These upgrades greatly enhanced combat system capabilities, including a Saab 9LV combat management system and a CEA phased array radar. This represented a complex design change, which included addressing second and third order impacts such as structural work to close-in the aft quarterdeck to provide additional reserve buoyancy, the addition of lead ballast and an increase to the limiting displacement to provide the stability and weight margins sufficient to enable the upgrade.

3.3 (c) Batch Build

In contrast to the Australian approach, the Turkish Navy embraced a batch-build approach for their MEKO 200 frigates, with newer builds responding to changing capability requirements. As a result, 3 distinct subclasses emerged, with the last two frigates built 1.3 metres longer to accommodate a planned upgrade to a 16-cell VLS. The original Yavuz subclass frigates have not been upgraded substantially since delivery and are expected to be replaced by the new Istanbul class frigates. Plans were announced in 2018 for a major upgrade for the 4 newer Barbaros and Salih Reis subclass frigates, which would include new weapons and combat management system as well as an integrated mast. This extensive upgrade scope could prove similar to that of the Australian AMCAP.

The US Navy's Arleigh Burke destroyers are another example of a batch build program with iterative designs denoted as 'Flights'. The program has successfully adapted to change over several decades of technology development and continues to provide a relevant and lethal option to warfighters. The 'Flight I' DDG-51, commissioned as the first of class in 1991, was designed to complement the more expensive cruisers in the American fleet. Budget pressures saw the withdrawal of a helicopter capability requirement and a reduction in size from the vessel originally intended [6]. The 'destroyer' design resulted a larger platform with a more powerful combat system and larger crew in comparison to its contemporary frigates, like the early MEKO 200. This relatively large overall capacity, together with some of its then novel design features (e.g. phased array radar, CBRN citadel arrangements, and reversion to all-steel structure) have stood the test of time and enabled several successful upgrades in future builds. Progress on 'Flight II' began almost immediately after DDG-51's commissioning, integrating improved electronic warfare sensors and Link 16 communications. 'Flight IIA' subsequently improved the Anti-Submarine Warfare suite, adding two helicopter hangers, readjusting the radar configuration and increasing the VLS capacity by six cells [7]. Most recently, the cancellation of the CG(X) program resulted in a 'Flight III' optimised for Integrated Air and Missile Defence [8]. This new configuration incorporates a SPY-6 radar which necessitated changes to powering, cooling, hull strength and computational processing.

4. UPGRADE LESSONS

It is common for Surface Combatants to be delivered with limited margins. This compounds the return on investment problem by both limiting the potential for upgrade and potentially limiting the service life due to inadequate service life growth margins. Adequate margins are required in a range of areas, including cable runs and cable penetrations.

Modularity and standardisation have been proposed as solutions to the surface combatant value for money problem since at least the 1970s, when several papers relating to these concepts were published [9][10][11]. Yet 50 years on, these concepts and the enabling standardisation have not yet delivered all that was hoped for. NATO has taken steps on standardisation but these are yet to be widely implemented [12]. There are significant examples like the MEKO 200 and STANFLEX where modularity has proven advantageous during design, production, maintenance and/or upgrade. However, the ANZAC frigate example shows that even where modular installations exist they do not necessarily provide the best path for upgrade. Furthermore, it is broadly recognised that there will always be cases where desired changes exceed the constraints of modular interfaces.

Although it is sometimes possible to upgrade in-service vessels, even with inadequate margins, through additional enabling design change (e.g. additional buoyancy, uprated generators), there is always a point where this becomes impractical. An alternative is to enable upgrade via offsets of existing capabilities. In some cases, change may not be feasible for an in-service vessel but can be incorporated as change during build (e.g. a batch of flight approach)

It is difficult to predict the nature of upgrades required across a vessels life. Experience shows that platform systems and some weapon or sensor systems may be adequate for the entire service life, however upgrades to computing, communication, combat management, radar and EW systems are essentially guaranteed. Standardisation in some areas, such as the MK41 VLS, manages the physical interface with some systems. Even with modular provisions in the ship design and a high degree of equipment commonality across classes, class specific installation design activities are routine and are likely to remain so. There is a need to actively manage the application of requirements to the upgrade program so that whole ship performance levels are retained, particularly for survivability.

The locations of upgraded sensors and effectors are likely to be dictated by the location of the capabilities identified for offset and removal. To ensure physical security of sensitive combat systems, an arterial or combat system gallery, between the electronics room, the command centre and critical / large volume sensor / effectors should be considered. Power, cooling, size, weight and height are main drivers for radar platform integration. [13]

The authors recognise that there is such a thing as a change too far. For example, few warships have been able to accommodate the technological step-change from azimuthing to fixed radar mid-life due to the complexity and interdependencies involved. The ANZAC upgrades are almost unique in this sense, and the challenges are also evident in new-build programs based on reference ship designs such as the Hunter Class frigate based on the RN's Type 26 and the USN Constellation Class based on the FREMM. Beyond further updates to radars, the future may hold more surprises. For example, a shift from predominantly missile to directed energy warfare could alter the entire architecture of a warship. Such a change might forego storing missiles and explosives in silos and magazines in favour of larger bulk fuel capacities and up-scaled electrical generation and distribution systems. To accommodate these changes, a new build may be required for the necessary trade-offs.

5. A CONCEPT FOR A FUTURE ADAPTABLE COMBATANT

5.1 PRACTICAL CONSIDERATIONS FOR SCOPE

Considering the upgrade experience and lessons previously discussed, the proposed concept focuses on those systems which have the most rapid development cycle and which are therefore most likely to be drivers for ship upgrades - communication, combat management, radar and electronic warfare systems. This focus is informed by recent upgrade experience. Other payload systems may require upgrade through life but it is difficult to predict which ones or the nature of the upgrade.

Enduring standardisation of platform to payload system interfaces has proven more difficult to achieve than some of our forebears had hoped when presenting concepts for variable payload or modular systems in the 1970's and 1980's. Putting aside the pursuit of an ideal solution this proposal focuses on the practical, on managing the interface between platform and combat systems at the level of the physical compartments and associated services. This approach accepts that class specific installation design will continue to be necessary and that dedicated margins on space, weight, electrical power and cooling remain critical enablers to upgrade. Where modular installations provide advantage, such as to enable installation of pre-assembled and tested equipment, then their use should be at the discretion of the installation designer but they should not create a constraint upon future upgrade installation design. Further, the aim of utilising modularity in this proposal is to benefit upgrade, and is not intended to offer mission variety.

5.2 A FOCUS ON SPATIAL DESIGN

One of the challenges of managing the platform to combat system design during the early stages is the level of uncertainty associated with each aspect and the significant inter-dependence. An enterprise approach to combat systems development, as Australia currently has, provides the opportunity for a relatively high level of maturity in the initial combat system design from the early stages of the ship design. A platform could be designed to support this initial combat system design with a relatively high degree of confidence. This initial footprint, inclusive of agreed margins, would then form a constraint to future design activities, including upgrade installation design.

A review of recent warship designs indicates that around 10% of the internal volume of the ship is allocated to the electronics spaces that form the focal point of this proposal. By the same methods, this compares to approximately 15% for propulsion systems or 8% for crew cabins and ablutions.

A spatial allocation (a set of compartments) would be set aside for these electronics systems as well as dedicated cable runs and services. The high-level goal would be to concentrate as much of the equipment as possible within two separate enclaves, as illustrated in Figure 1. Only minimal amounts of the focus combat system elements would be tolerated outside of these enclaves, and only where cable lengths or other factors made it necessary. It is important to note that this approach constitutes a significant driver on the initial arrangement of the ship and is only likely to be practicable if adopted from an early design stage.

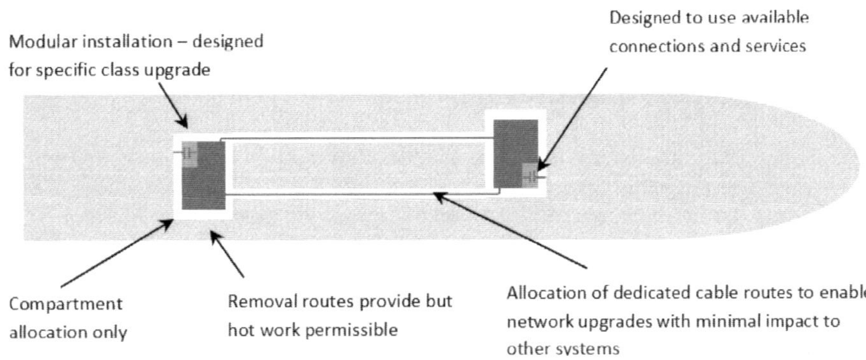

Figure 1: Conceptual diagram of spatial design features

5.3 EXAMPLE ARRANGEMENT

In an attempt to validate the proposed approach, a previous RAN surface combatant concept design was subjected to a redesign of the internal arrangement drawings. An extract of the work shown in Figure 2 shows that concentration of the identified spaces within separated fore and aft zones and the connection of these spaces by dedicated cable routes (technical galleries) was achievable without significant compromise of the overall general arrangement. This was even the case with significant competition for main deck volume created by the existence of a large mission bay. The arrangement included the routing of platform system galleries on lower decks.

Figure 2: Example Arrangement of Generic Frigate Concept with two Combat System Enclaves (Red), connecting Galleries (Pink) and Effectors (Orange)

The arrangement shows the co-located combat system spaces in red with dedicated secure combat system galleries in pink and auxiliary combat system spaces in orange. These galleries link the enclaves together as well as provide dedicated cable routes to the masts above each enclave and to the main weapon systems (VLS and main gun) forward. The separation and redundancy of multiple enclaves affords some survivability benefits whilst the galleries between enclaves are routed along different decks. The forward enclave spans two decks with the primary Combat Information Centre located above the primary equipment room. The aft enclave is secondary and hence smaller. The light blue shaded areas are machinery spaces, the green shaded areas are habitability spaces and the white areas are operational spaces or stores. Proposed combat system equipment removal routes for equipment based on 10ft and 20ft ISO container footprints are in dark blue. Note that whilst functional, this example arrangement was adapted from an existing arrangement with an intent to minimise modification. If the proposed philosophy were adopted as a central design goal at the beginning of a new design, the efficiency and functionality of the arrangement could be greatly improved.

5.4 DESIGN FEATURES FOR UPGRADEABILITY

Beyond spatial proximity, the concept is predicated on each compartment having space, weight, power and cooling allocations that are inclusive of margins which allow capacity for local upgrade. Margins would be allocated at highest-level possible, given that spatial margins at compartment levels are more versatile than embedded in individual racks. Compartments would be required to meet appropriate security requirements and the installation designs would address a consistent set of holistic requirements (e.g. shock, noise, vibration, quality of electrical power).

This illustrated concept is simplistic and focused on certain elements of the combat system. Sensor and effector changes would also need to be accounted for in the context of future upgrade. The assumption here is that upgrades for those systems would be constrained to their existing locations and could perhaps be afforded relevant local upgrade margins.

To control upgrade cost and schedule, the separation of platform and combat system designs allows them to proceed on different schedules, including the potential to incorporate an upgrade to combat system in service relatively rapidly and with minimal disruption to other ship systems. Integration between the two would only be where necessary, recognising that duplicate systems and functions may be easier and cheaper than managing complex platform / combat system interfaces. Compartments and cable runs would be arranged such that their contents could be changed out with minimal impact to other ship systems. This would include planned equipment removal routes, ideally allowing for the insertion of pre integrated and tested groups of equipment on common foundations. Equipment removal routes could be based around cutting a hole in the vessel side or deck noting that it may only be used once or twice in a vessels life, although hot work impacts should be minimised by clearly defining these locations and avoiding routing of internal structures and services or affording simple mechanisms for their temporary removal. Compartment locations would be selected with consideration of adjacency to relevant spaces such as the Ops Room, Mast, sensors and effectors.

To enable pre installation integration and testing, some form of common foundation would be important. This would allow groups of physically connected / integrated equipment to be pre-tested and installed with minimal re-work and retest. In a mission modularity context, the form of this common foundation would be prescribed but in this context the form of the common foundation would be part of the installation design, with the compartment itself defining the physical interface. This removes another standardised interface that must be managed in the context of mission modularity. It may be that foundations based around ISO 10ft or 20ft container skids or other standardised formats are selected as they aid transport between land based facilities but this would be a choice for the installation designer not a standard to conform to and it could be changed between installation versions if needed. Compartments may provide raised structure on decks, bulkheads and deckheads or other means of securing equipment installation to limit the need for hot work and the associated impacts to adjacent compartments during upgrade.

Another challenge in managing platform and combat system designs is the achievement of consistency or coherency in relation to overall design goals and holistic requirements. For example, ensuring consistent levels of zoning, separation and redundancy, between platform and combat system elements. To enable these survivability considerations to be addressed within the proposed concept, two redundant and separated combat system equipment spaces / enclaves have been included along with two separate dedicated cable routes. This allows installation designers to include separated and redundant elements as well as to concentrate dependant non-redundant equipment within one or the other of the enclaves consistent with overall survivability objectives for a combatant.

6. FURTHER WORK

It is important for project and program sponsors to understand and accommodate an appropriate level of adaptability when developing requirements. Future work would include developing practical guidance to assist this process and tailor the approach to adaptability as appropriate to the needs. The option and implications for topside design linked to this proposal warrant further consideration, in particular the option to extend overall survivability goals for separation and redundancy to key systems like a phased array radar.

As an additional area of study, the value of alternate mast arrangements for the integration of the phased array radar would be of interest. For example, the splitting of the conventional single integrated mast and separating it over two fore and aft to locations would place them directly above each of the combat system enclaves. This would provide a level of survivability in the primary sensor system that is not possible when all components are concentrated in a single mast. It also has the potential to lower the centre of gravity and minimise the wind heeling effects on the ship thereby increasing stability margins or allowing a narrower beam for increased hydrodynamic performance.

7. CONCLUSIONS

As the rules of warfare evolve, so too must warships. Sometimes these changes can be tolerated by reconfiguring an existing platform, while at other times, they may force an entirely new warship or alternative military system into play. The objective of this paper was to explore how a warship might be designed to have greater likelihood of enduring future changes and remaining relevant, in order to justify its significant capital expense predicated on a long period of service.

Underlying the proposed concept is the assumption that integration costs, including those of maintaining interfaces and disentangling combat and platform systems during design or refit are dominant cost drivers. Controlling these through the proposed architecture may drastically reduce the cost of upgrade and improve the platform design's tolerance of combat system design evolution. The approach is deliberately practical, seeking to apply the experience of past programs in a way that better enables a platform design to support an evolving enterprise combat system.

8. ACKNOWLEDGEMENTS

The authors acknowledge and value the sharing of experience and lessons learnt with colleagues across the naval enterprise, and from the ANZAC class sustainment context in particular.

9. REFERENCES

1. PEDATZUR, O., 'Weight Design Margins in Naval Ship Design', *Naval Engineers Journal*, June 2016.

2. ALKIRE, B., ET AL., 'Littoral Combat Ship: Relating Performance to Mission package Inventories, Homeports, and Installation Sites', *RAND Corporation*, 2007

3. SALISBURY, E., 'Lessons from the Littoral Combat Ship', *War on the Rocks*, November 2021.

4. SORENSEN, K., 'The Danish Experience with Flexible Warships', *ASNE Day 2015*, March 2015.

5. BOXELL, J., 'Britain's Submarines Resurface', *Financial Times*, August 2006.

6. STILLWELL, P., 'Designing the Arleigh Burke's Hull', *U.S. Naval Institute*, August 2010.

7. SAUNDERS, S., 'Fighting Ships 2015-2016', *Janes IHS*, 2015.

8. O'ROURKE, R., 'Navy DDG-51 and DDG-1000 Destroyer Programs: Background and Issues for Congress', *Congressional Research Service*, June 2010.

9. JOLLIFF, J., 'Modular Ship Design Concepts', *Naval Engineers Journal*, October 1974.

10. DREWRY, J. & JONS, O., 'Modularity: Maximising the Return of the Navy's Investment', *Naval Engineers Journal,* April 1975.

11. ABBOTT, J., 'Modular Payload Ships in the U.S. Navy', *SNAME Transactions Vol. 85,* 1977.

12. MANLEY, D., 'The NATO Drive to Mission Modularity', *Warship 2018*, September 2018.

13. WHITE, J. & BILLUPS, A., 'Navy Radar Trades at the Ship Interface', *Naval Engineers Journal*, April 2009.

10. AUTHORS BIOGRAPHY

Joe Cole is the Director of Naval Ship Concept Design within the Royal Australian Navy's Engineering Branch. Joe is responsible for leading the rebuild of the Navy Engineering capability for early stage design and trade studies in support of capability development. He has extensive experience of the RAN surface combatant program, having handed over the accountable role in Navy Engineering in late 2022. Since commencing his career with Navy Engineering in 2003, Joe has had a strong interest in early stage design and has contributed to studies and concepts associated with future surface combatants, patrol vessels and landing craft. Joe completed his Bachelor of Engineering (Naval Architecture) degree at the Australian Maritime College in 2002, graduating with Honours.

Alistair Smith is a naval architect within the Naval Ship Concept Design team, having been its inaugural member upon establishment in 2022. Alistair previously served the Arafura Class Offshore Patrol Vessel acquisition program, overseeing the development and review of naval architecture design elements between contract signature in 2018 and first-of-class launch in 2021. Alistair completed his Bachelor of Engineering (Naval Architecture) degree at the University of New South Wales in 2015, graduating with First Class Honours and the University Medal.

Gethin Barden holds the position of Mechanical Engineer in the Naval Ship Concept Design team. Gethin is responsible for supporting the RAN and other ADF maritime stakeholders in early stage concept, requirements and project support. Gethin's previous experience includes test and activation systems engineering on Hobart Class destroyers, systems engineering during early-stage acquisition phases for frigates and Army watercraft programs, ship construction assurance of Arafura Class vessels and analysis of reference ship designs to support project sponsors. Gethin has completed additional studies, achieving a Graduate Diploma in Naval Engineering.

FROM UMS TO FULL AUTONOMY; EXPERIENCE FROM A COMPLEX WARSHIP PROGRAMME

J Johnson, M Howard, Babcock International, UK

SUMMARY

The UK's Type 31 Frigate, an ARROWHEAD-140 variant, has been designed to reduce the number of personnel required in the Ship's Company, delivering the required operational capability whilst ensuring personnel safety and supporting personnel retention. The design adopts the latest high technology readiness level (TRL) solutions, such as the first use of the UMS notation in a Royal Navy Frigate and capable equipment with reduced maintenance requirements.

The paper will explore critical design features observed during the development of a lean crewing solution in a delivery programme, providing insight in future design effort and the adoption of technology solutions that enable trade-offs between cost, functionality and complexity on the path to increasing levels of platform autonomy.

NOMENCLATURE

1SL	First Sea Lord
AESA	Active Electronically Scanned Array
AH-140	ARROWHEAD-140
CIWS	Close In Weapon System
CMS	Combat Management System
DEW	Directed Energy Weapon
EO/IR	Electro Optic / Infra-Red
FOST	Fleet Operational Standards & Training
GaN	Gallium Nitride
HADR	Humanitarian Aid & Disaster Relief
LCS	Littoral Combat Ship
RAN	Royal Australian Navy
RCS	Radar Cross Section
RDN	Royal Danish Navy
RN	Royal Navy
RoE	Rules of Engagement
SCC	Ship Control Centre
TDL	Tactical Data Link
TRL	Technology Readiness Level
UMS	Unattended Machinery Space
URN	Underwater Radiated Noise
UUV	Uncrewed Underwater Vehicle
UxV	Uncrewed Vehicle
VLS	Vertical Launch Silo

1. INTRODUCTION

Many navies are experiencing difficulty in generating and retaining the required number of trained and experienced personnel to operate their warships. Data available confirms the number of personnel has reduced over recent years due to a combination of outflow from the service and a diminished recruit intake. This is the situation that both the Royal Navy (RN) and the Irish Naval Service, amongst others, are currently experiencing as shown in Figures 1 and 2. The availability of trained and experienced personnel is an issue that the Naval Service has been working to resolve for several years, as this quote from the RN First Sea Lord (1SL) in 2002 attests:

> *"A manning structure that is struggling in 2002 will certainly not meet the challenges of 2015. The status quo is therefore not an option and the need to change for the better is pressing"*

Admiral Sir Nigel Essenhigh GCB, DL, First Sea Lord; January 2002 [1]

As has been widely reported, the navies included in the Figures below have recently withdrawn platforms from service, struggled to re-generate platforms from upkeep periods or kept platforms uncrewed alongside as a result of the reduced numbers of trained personnel available. This situation is not unique to the RN and Irish Naval Service, with many others such as the Royal Australian Navy (RAN) also grappling with similar issues. With navies still required to deliver outputs and operations as directed by government policy, this situation is not sustainable in the long term.

Warships have traditionally been reliant on large numbers of personnel, with RN battleships from the Second World War requiring a war complement of almost 2,000 crew and even the ~2,500 tonne RN LEANDER Class Frigates operating in the early 1990s with ~260 personnel. Against a background of reduced numbers of personnel available, a desire to remove personnel from the hazards of combat operations where possible and in order to take advantage of reduced in-service personnel-related costs there is a concerted drive by many navies to reduce the complement required by future warships as part of the overall solution. In order to deliver a feasible solution a number of considerations should be taken into account, with a selection of these discussed within this paper.

The Type 31 Frigate is a case study into the development of a lean complement solution that remains realistic for a class of General Purpose Frigates that are currently in build for the RN, matured over a number of years. This paper includes example aspects and systems fitted in the Type 31 platform that were key in the development of this complement solution, together with a comparison to the 'parent' IVER HUITFELDT class design solution. All of the Type 31 Frigate features and attributes discussed that deliver a lean complement solution for the RN are also inherent in Babcock's ARROWHEAD-140 (AH-140) product, to support other navies experiencing similar constraints on personnel.

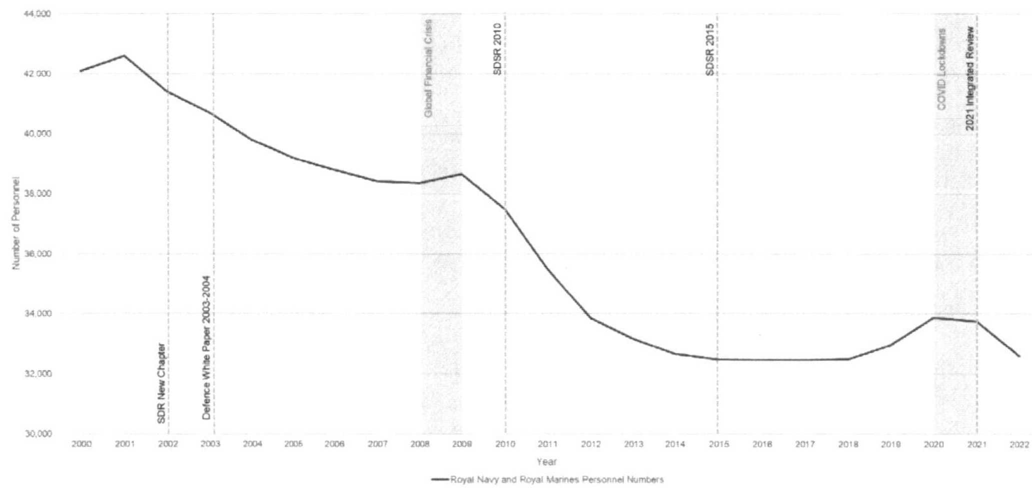

Figure 1: Royal Navy and Royal Marines Personnel Numbers 2000-2022 (data extracted from [2]).

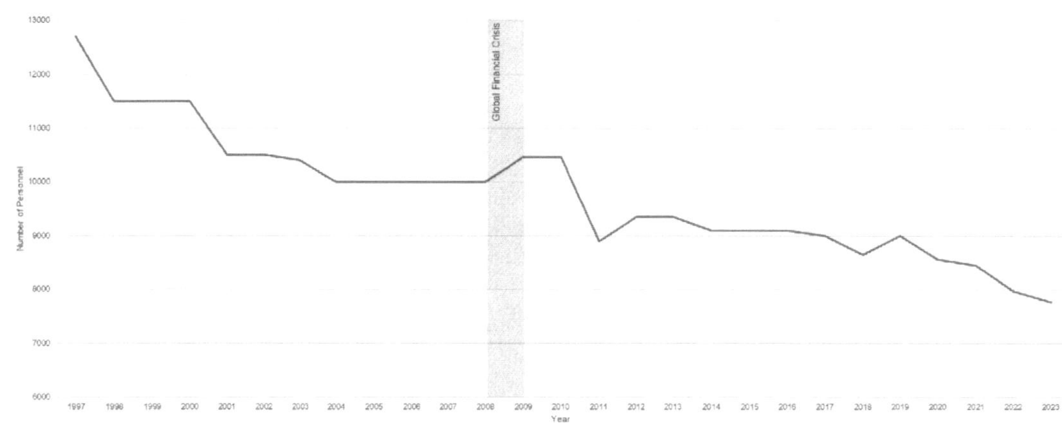

Figure 2: Irish Defence Forces Personnel Numbers 1997-2023 (data extracted from [3]).

2. PERSONNEL CONSIDERATIONS

Warships are sent to sea to deliver operational capability against a series of defence outputs that are defined by government policy. The capability of a warship itself comprises of a number of elements or 'abilities', each as critical as the others, and are currently influenced or enabled by the number of personnel onboard. Some examples of these elements, and how they relate to the number of personnel, are described in the section below.

- Susceptibility. This covers a range of aspects that include the number and configuration of hardkill effectors (missiles such as Sea Ceptor or SM-2, Close In Weapon Systems (CIWS), Turrets or Directed Energy Weapons (DEW)), softkill effectors such as decoys, and sensors such as radars and sonars. Susceptibility also covers the signatures of the platform, or reducing the ability of an enemy to detect and target the ship; this includes radar cross section (RCS) and underwater radiated noise (URN). Personnel are currently required to deliver this element of capability; as operators scrutinising the picture provided by the sensors or targeting the weapon systems, or maintaining those same weapons, sensors and the supporting infrastructure so that they are available when they are needed.

- Survivability. This covers the ability of a warship to mitigate the effects of being hit by enemy weapons, in order to carry on functioning while also preserving life onboard as far as possible. It is naive to believe that the hardkill and softkill effectors fitted will mean a platform is invulnerable to damage from enemy action, and key design effort is placed on providing aspects such as blast, shock and armour protection, and increasing the resilience of systems onboard so that they continue to function if parts of the system are destroyed.

- Recoverability. This covers the ability of the warship to react to the damage sustained and restore capability to remain in action or to counter follow-on attacks. This includes measures in the platform to mitigate the spread and effect of damage, such as fire-fighting and flood repair actions. This element of a warship's capability currently draws heavily on personnel, with these actions requiring large numbers of people to successfully prosecute.

- Lethality / Mission Capability. Covering the systems fitted to launch land strike cruise missiles, anti-ship cruise missiles or the operation of offboard vehicles as examples, this element of capability delivers the wider operational effect of the warship. Facilitated by the installation of equipment such as the Mk41 Strike Length Vertical Launch Silo (VLS) and launch & recovery systems for boarding craft this enables the warship to carry out maritime interdiction operations, combat operations at sea and influence events ashore, as examples.

A true warship requires a balance across all elements within the bounds of platform margins available; this generally favours a larger warship due to the expanded inherent margins. Unfortunately 'capability analysis' has a tendency to descend into a lone comparison of the weapon and sensors fit, to the neglect of the other critical aspects. All aspects must be considered when examining both the true capability of a warship and the number of personnel onboard in the effort to reduce this number or remove them completely.

When it comes to warship design, globally deployed navies like the RN place significant emphasis on features to support survivability and recoverability of the platform. Experience of and lessons learned from both combat in the missile era, including the Falklands Conflict (Figure 3), and peacetime incidents, such as the grounding of HMS NOTTINGHAM and flooding of HMS ENDURANCE, all drive requirements into the design. These requirements are largely met through overall design considerations and fixed systems, the nature of which means that they are configured or installed in build and not fitted whilst the platform is in-service. It is significantly easier to add sensors and effectors to a ship that is already built to be survivable and recoverable to increase lethality than it is to somehow increase survivability and recoverability in a ship already built to commercial standards that has a bristling topside full of weapons. The installation of the Mk41 Strike Length VLS in Type 31 is a good example of this type of upgrade in a warship built for survivability and recoverability, employing the dedicated margins to introduce additional lethality via a spiral development philosophy. The following sections examine personnel considerations related to some of these aspects in more detail.

Figure 3: RN Frigates and Destroyers lost during the Falklands Conflict (clockwise from top left: HMS SHEFFIELD, HMS ARDENT, HMS ANTELOPE and HMS COVENTRY). The hard-won lessons and experience from these platforms, amongst others during the conflict, spurred 40 years of UK development in survivability and recoverability aspects that are now incorporated into the latest RN warships like Type 31 and influence personnel numbers.

2.1 SUSCEPTIBILITY

In legacy platforms typically the sensors integrated into the combat system feed data to the displays of the picture compilers working in the air, sub-surface and surface domains. They interpret the information presented, including altitude, speed and Identification Friend / Foe (IFF) codes (as air-domain examples) before deciding how to classify the track. They may also contact the Bridge for the watchkeepers there to make a visual identification using binoculars before classifying the track. This compiled picture is scrutinised by Picture Supervisors, augmented by Tactical Data Link (TDL) operators feeding in and correlating the 'Link Tracks' from allied units, and overseen by the most experienced personnel. This picture is further augmented by a separate team feeding the Electronic Warfare sensor data into the 'recognised picture'. Each weapon has an operator traditionally sat behind a specialist console who will carry out targeting and engagement of threats.

A large number of highly trained and specialist personnel is therefore required in legacy ships to enable the Operations Room (or Combat Information Centre (CIC)) to deliver the required operational outputs. In enduring operations and higher threat areas this number has to be further expanded to meet the constant operational tempo; operators need frequent breaks and role rotation to ensure everyone remains alert and at the top of their game to detect and counter threats without excessive fatigue. Much operator time is spent in undertaking low-difficulty and repetitive tasks often with poorly designed user interfaces. Modern, integrated combat management systems (CMS) seek to remove this operating burden from operators through automatically undertaking the routine picture compilation and tracking functions allowing the operators to concentrate on unexpected or anomalous events. This is designed to enable a reduction in the number of personnel required on watch, without compromising the output of the Operations Room or overwhelming the operators.

2.2 RECOVERABILITY

Recoverability of a warship is a key consideration that drives the number of personnel in a platform. Efforts to contain damage from enemy action or peacetime accident draw on relatively large numbers to react to the dynamic situation, to stabilise incidents and then attempt to recover and restore equipment back to a state in which it can be used. Aspects of modern warship design have alleviated this activity; of note are fixed firefighting systems that reduce the need to send fire parties into a space to extinguish large fires, structural fire insulation that reduces the need to 'boundary cool' the external bulkheads around the fire-affected compartments to prevent the fire from spreading further through the ship, and smoke containment features that were a crucial lesson learnt by the RN from experience in HMS SHEFFIELD and other platforms.

Redundancy in system routing, such as High Pressure Sea Water (HPSW), power distribution and data networks of a modern warship reduces the need for personnel to manually re-route cables or hoses to bypass damaged sections in order to restore capability.

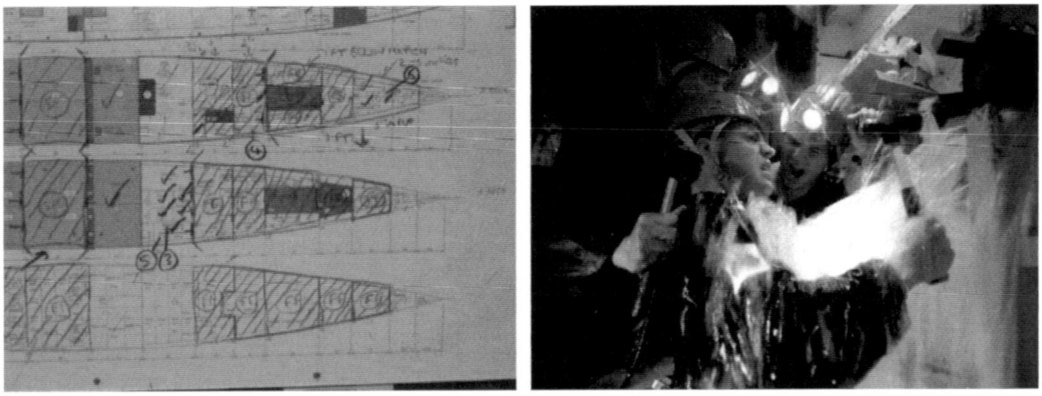

Figure 4: (image left) HMS NOTTINGHAM Incident Board in 2002, blue hatched 'FF' indicates a free-flooding section; (image right) RN Damage Repair Instructional Unit (DRIU) flood control / leak stopping training.

Beyond these features in modern warship design, damage that penetrates the hull causing extensive flooding is still difficult to counter and typically requires a large number of people to contain and manage. The incident in 2002 where HMS NOTTINGHAM suffered extensive flooding from accidental grounding required a considerable number of personnel to contain the extents of the incident in order to save the ship; an image of the incident board used during this event is included in Figure 4 above showing the breadth of this damage. This damage could occur anywhere onboard, with trained personnel able to react to the situation in order to use the correct equipment, build shoring to stem the flow (in many situations creating some inventive shoring solutions), or deploy portable pumps in the correct areas.

There has not yet been a robot demonstrated that can replicate the full flexible and adaptable capability that a trained sailor will provide in regard to damage control, able to carry out the targeted application of firefighting medium where compartment-level sprays would cause more damage than a minor fire or leak stopping in hard to reach areas, with this aspect a significant constraint to the reduction of personnel onboard.

2.3 MAINTENANCE

Without the output of the engineering and logistics departments a ship would not leave its home port. The marine environment is tough on all equipment, and the heavy use of a warship in dynamic operational scenarios exacerbates the wear on the complex systems fitted. There are two forms of maintenance that have to be accounted for; preventative and corrective. Preventative maintenance can be defined and planned in a schedule, split over activities that take place at sea, activities that take place in supported maintenance periods alongside or in a full refit. Understanding and optimising the levels of preventative maintenance of all systems is critical in the design of a warship, and this is a driving factor on the number of engineers required onboard. Corrective maintenance should be minimised, especially in new platforms, however it will inevitably grow as a burden over time as systems become worn. Allowance has to be made in the number of trained personnel onboard for this activity, enabled by automated diagnostic services, spares, tools and knowledge of the systems, in order to prevent an unmanageable workload developing and a resulting issue in retention of key engineering personnel.

Secondary duties for personnel onboard typically include the maintenance of cleanliness of spaces from bathrooms to the overheads in passageways. Dedicated effort is also required on the upperdeck to keep fixtures and fittings in working order against the harsh saltwater environment. Design features in the platform itself and examining the method of delivering this activity can alleviate the impact on personnel so that these aspects do not become unmanageable in a lean complement.

2.4 RECONFIGURABILITY

Modular and autonomous offboard systems are not miraculously maintenance free, and in a number of cases also require operators to process data and pilot vehicles remotely. Like the Flight personnel that embark with an aircraft in current platforms, these modular and autonomous systems are assumed to embark with their own teams and specific-to-capability spares in the short term at least. While these personnel are not counted as part of the core complement of a platform, they will still draw on a Ship's 'life support' including catering personnel, systems such as fresh water production and stores / victuals holdings, which would therefore have to be scaled accordingly in the design of the platform and themselves accommodated or maintained onboard. Until this maintenance and operating burden inherent in employing offboard systems can be reduced or removed, personnel will still be required at the uncrewed system's forward operating position and in turn drive personnel numbers in the core complement to support them.

2.5 DYNAMIC OPERATIONS

A warship will be required to react to unforeseen situations at short notice, in support of government direction and objectives. Following a tsunami in east Asia, a hurricane in the Caribbean or an Icelandic volcano eruption that disrupts air travel the initial government response has frequently involved deploying naval assets to carry out initial actions to assist. Beyond humanitarian scenarios, fast moving geopolitical situations also require the rapid deployment of naval assets; for example the short notice deployment of HMS DIAMOND to the Red Sea in December 2023 due to attacks on civilian shipping in the area. In order to be available to be deployed in these situations a sufficient number of personnel onboard have to be available and trained to deal with the dynamic circumstances safely.

With a reduction in the number of personnel onboard in a core Ship's Company, these kind of short-notice operations at the direction of government could be a challenge in a Ship with a lean complement. Disaster Relief operations typically rely on large numbers of personnel to distribute equipment and supplies, or to carry out initial actions to assist a civilian populace in order to provide an effective response. An option to enable the rapid augmentation of the core complement may provide a number of personnel that makes a meaningful contribution to the event. This augmentation could be supported by fly-away teams that join the Ship in order to then deploy to the scene of the incident, however the ability to hold a number of trained personnel ashore at readiness purely for contingent tasking such as this may be a challenge for a lean Navy, and instead might potentially be enabled by working with personnel from the RN Reserves or civilian organisations such as REACT [4].

In many ways this is more complicated in a short notice combat operation. A fly-away mission/combat augmentation team who are unfamiliar with the particular Ship may not be able to carry out this type of combat operation effectively or safely without significant prior continuation training on the platform and its systems and a period of integration with the core Ship's Company. This kind of 'modular' complement solution formed part of the intended philosophy behind the US Navy's Littoral Combat Ship (LCS), with the reality of this philosophy in operations now well documented [5].

3. TYPE 31 FRIGATE METHOD - ROUTE TO A LEAN COMPLEMENT

3.1 TYPE 31 INTRODUCTION

Type 31 comprises a class of five General Purpose Frigates currently under construction in Rosyth, Scotland, for the RN. Based on Babcock's ARROWHEAD-140 product line, and designed against Lloyd's Register Naval Ship Rules, ANEP-77 and a significant number of UK Defence Standards (DefStans), Type 31 possesses capabilities, levels of survivability and features not found in previous escort classes such as the Type 23 and Type 45, themselves designed against previous generations of standards. A full introduction to the Type 31 Frigate, including the novel procurement strategy and inherent capabilities, is available in a previous paper [6].

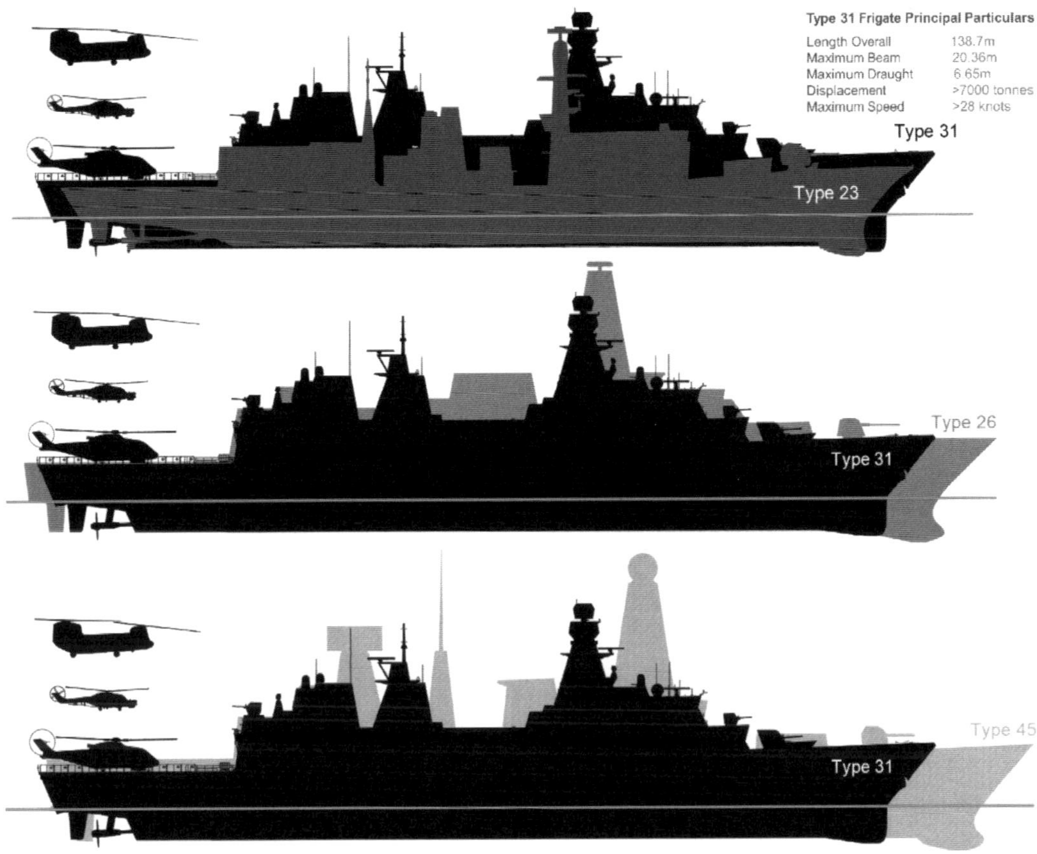

Figure 5: Type 31 Frigate size in comparison to contemporary Royal Navy escorts for context within this paper.

With the challenges of a reduced number of available trained and experienced personnel to crew RN platforms in the immediate future (as shown in Figure 1 above), against a background of contemporary platforms requiring additional personnel in order to function to their full capability, Type 31 had to be designed from the outset to be as efficient to operate as possible in terms of personnel numbers whist still able to deliver sustainable operational capability. This was a specific requirement placed on the Type 31 design by the RN to be delivered as part of the programme.

There is analysis underway by a number of organisations looking at futuristic concepts to deliver personnel reductions in future platforms, on a route to completely autonomous vessels. Type 31 required a different focus; the technology and features in the platform to deliver personnel reductions had to be available as high-TRL products and capabilities right now, in order to meet the challenging delivery schedule of the programme as the warships are currently being built in Rosyth. Overlaid with the critical safety case evidence requirements that accompany all military equipment introduced to UK service, the more futuristic ideas to reduce personnel could not be adopted. To enable some of the futuristic concepts the burden of evidence required to satisfy the mandatory safety case outside of adopting recognised standards would be prohibitive in the timelines of the Type 31 programme. The regulatory challenges encountered as part of these futuristic complement reduction concepts may benefit from prior resolution separate to the platform design phase, so the design is then able to follow a set of revised codified standards to avoid driving uncertainty, cost and delay into the warship procurement process itself.

3.2 TYPE 31 COMPLEMENT DEVELOPMENT PROCESS

The Type 31 Ship's Company is smaller than legacy RN escort platforms thanks to the latest modern technology, regulations and standards used in the design. This is shown in Figure 6 below, broken down by department. With the Type 45 Destroyers requiring a complement of 194 personnel and the Type 23 Frigates at 174 personnel [7], Type 31 represents a considerable reduction while maintaining a sustainable and effective capability.

The number of personnel in the Ship's Company for Type 31 is sufficient to operate the platform during enduring combat operations using the features of the equipment selected and fitted and the attributes of the Ship such as the Unattended Machinery Space (UMS) notation. This Ship's Company would be augmented as required by embarked forces such as those required to safely operate and maintain aircraft or offboard systems, and specialist boarding troops (Royal Marine Boarding Teams). The core complement has been deliberately configured to include the personnel needed to carry out recoverability actions in the event of damage (fires and flooding), and sufficient personnel to sustain operations without excessive fatigue through selection of latest-generation technologies.

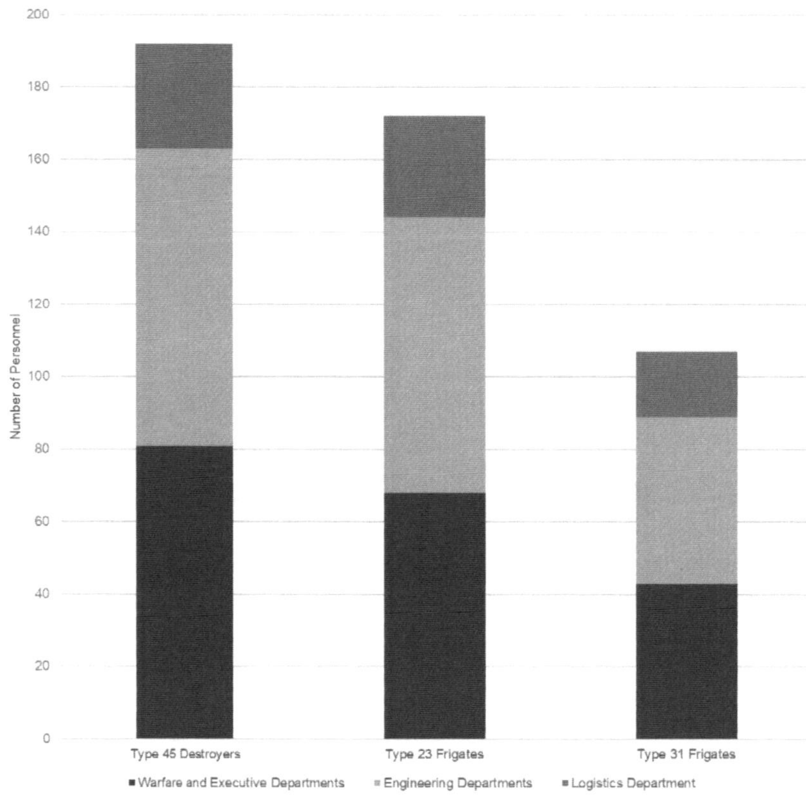

Figure 6: Comparison of the number of personnel required in the Ship's Company of contemporary Royal Navy escort platforms (data extracted from [7]).

There were no 'black box' software tools or automated databases used in the development of the Type 31 complement. A large table listed each individual by row, with states (cruising, defence watches and action) and evolutions (for example Replenishment at Sea (RAS), Boarding Operations, Humanitarian Aid & Disaster Relief (HADR) etc) listed by column. This ensured that the exact workload on individuals could be understood and tracked in each scenario, without the logic connections being lost or corrupted inside an inevitably complex software model.

This development dived down into the detail of each individual's role across the states and evolutions. For example, a junior Seaman Specialist finishing a 'Morning' watch (04:00-08:00) after weeks of an at-sea watchkeeping routine couldn't be expected to immediately become the on-call bowman for the recovery boat (called into action if a person were to fall overboard), and still be expected to remain effective at this important safety role. Another example is the load on engineering personnel; preventative maintenance tasks had to be balanced with carrying out duties as a watchkeeper for several hours every day, if this watchkeeping activity was required by the nature of the operation and the threat environment. These considerations were taken into account when generating the complement solution; statements such as 'we managed it in my day' were banished early as unhelpful contributions.

Once compiled by the Type 31 Design Authority (Babcock), this complement solution was exposed to the RN via a series of Operability Working Groups, with multiple sessions held with serving representatives from each branch (Warfare, Logistics etc) and sub-branch (Communications & Information Specialists etc). This scrutiny, applied both to the validity of the solution and the functionality of the advanced technology within Type 31 upon which the solution was based, refined the complement. For example, members of the RN Warfare branch management were exposed to the Thales TACTICOS CMS in the Operations Room simulator facility in Crawley (West Sussex) to assure the number of Operations Room personnel required in the solution.

This series of Operability Working Groups culminated in whole ship 'Fast Cruise' events, where all branches of the naval service were represented. During the 'Fast Cruises' the ship was exposed to a range of operations, incidents, emergencies and combat operations to examine if the complement solution would stand up as a coherent answer against everything that was expected of a RN Frigate; reducing the number of personnel required to operate Type 31 without placing an unmanageable workload on each person. This includes the ability for Type 31 to react to short notice events and dynamic situations such as HADR, at the direction of government policy.

This process, conducted over a number of years and through a variety of activities, culminated at a point where the first members of HMS VENTURER's (Type 31 Ship 1) Ship's Company joined in July 2023. Now that the Ship's Company have joined, their work is ongoing to further refine the complement solution to optimise the workload on individuals.

Overlaid with the work on the Type 31 complement solution itself are higher-level operating philosophy considerations, such as the 'double crewing' of platforms that are deployed as a persistent forward presence. Adopted with the deployment of Type 23 Frigates HMS MONTROSE and now HMS LANCASTER to Bahrain, a major benefit of this philosophy is that it provides more certainty to members of the Ship's Company; they are aware of which periods they will be away from home and can plan their family lives around these dates, with this certainty aimed at maintaining retention. This is allied to the operational benefits of this philosophy in avoiding long transit times to arrive at a theatre of operations, and being able to react quicker to global events in the national interest.

3.3 TYPE 31 PERSONNEL CONSIDERATIONS

3.3 (a) Technology

The next-generation technology in Type 31 plays a key part in optimising the number of operators required. As such, those developing the complement solution were trained on systems such as the TACTICOS CMS so they understood the capabilities that this system will deliver and that enable a reduction in operators whilst still enhancing overall platform capability.

For example, advanced automation in the CMS means Picture Compilers should not be required, with the system instead able to perform this role under the oversight of Picture Supervisors without these personnel becoming overwhelmed. This is further supported by the selection of the Thales NS110 Air/Surface Radar, the first 4D Dual-Axis, Multi-Beam, Active Electronically Scanned Array (AESA) Radar to be fitted to a RN Frigate. As one example against a contemporary threat; simultaneous element-level digital beamforming in both azimuth and elevation of the GaN Radar system fitted (as opposed to previous generation 3D Radars that are multi-beam capable in elevation only) enables superior detection of small drones and low-RCS missiles at considerably increased ranges, using automation to alert the operator.

The level of automation introduced in systems like TACTICOS and NS110 as key elements of the Type 31 mission system is designed to reduce the workload on the Operations Room personnel over that experienced in previous platforms. This solution alerts to the threat using equipment that does not experience the same fatigue or potential distraction as a human Picture Compiler may in enduring watchkeeping activity. Other sensors fitted such as a 360° staring Electro Optic / Infra-Red (EO/IR) surveillance system are available as an integrated resource to any operator on the CMS; with a better view than the Bridge in both day and night conditions this system enables the Operations Room to make visual classification of tracks without constant recourse to the Bridge team and expedites the process of picture validation considerably.

Other examples include the threat engagement functionality within the CMS, again employing advanced automation features to allow the platform to counter large swarms of hostile actors, using the fully integrated suite of turret and missile effectors. This system of systems delivers a capability faster than is possible with purely human operation, while simultaneously requiring a significantly reduced number of operators over legacy platforms and maintaining necessary control over weapons under the Rules of Engagement (RoE).

Beyond mission system examples, other resilient and integrated systems in Type 31 such as the centralised digital incident boards, a camera network with motion-detection capability and comprehensive sensor suites for fire, flood and equipment monitoring support recoverability actions onboard, and reduce the considerable burden of conducting damage search and system fault diagnosis activities manually.

3.3 (b) Damage Control / Recoverability

The entire Type 31 platform has been re-designed to meet Lloyd's Register Naval Ship Rules and the latest NATO / UK naval standards & regulations, which has introduced an increased capability in the redundancy and resilience of systems over legacy RN platforms including the Type 23 (the design of which pre-dates key standards such as ANEP-77) and many other contemporary Frigate designs. The impact of these standards is demonstrated in Figure 7 below, with the same location shown in the 'parent' class (designed against alternate standards and regulations) and the Type 31 detail design model; this complex system routing increases the resilience of the platform without requiring a significant increase in the number of personnel needed to carry out battle damage repair to restore systems.

The ability to carry out damage control activities required to react to both peacetime incidents and enemy action was a driving consideration. Supported by advanced monitoring systems, fixed fire suppression and containment solutions the number of personnel in the Type 31 complement was tested against various scenarios to ensure there were sufficient numbers to sustain the safety of the platform and to restore or maintain capability. This also included stress testing of the solution, in conjunction with serving members of the Fleet Operational Standards & Training (FOST) organisation and RN damage control school instructors, by removing personnel as casualties to examine whether the solution would remain effective.

Figure 7: Example of complexity introduced by alternate regulations and standards, supporting recoverability actions. (image left) 'Parent' class passageway; (image right) the equivalent location in Type 31 (extracted from the Type 31 3D detail design model).

3.3 (c) Watchkeeping

Watchkeeping is a resource-intensive activity, in previous platforms requiring trained and experienced personnel to close up in locations including the Bridge, Operations Room and Ship Control Centre (SCC) to be on hand to control the Ship and react to incidents. Sufficient numbers of personnel are required to maintain a sustainable watch cycle, driving overall numbers and imposing an extensive training burden to maintain competencies. When this sufficient number of personnel isn't available and gaps appear in the watchbill this is often cited as a contributing factor in poor retention.

Employing the functionality delivered by the UMS notation, the entire capability of the SCC and machinery monitoring service is replicated on the Bridge; a feature not found in legacy RN escorts. Therefore, when the operational tasking and threat environment allow, the SCC can be closed with the systems overseen by a single engineer in the Bridge. Removed from the obligations of maintaining watches, the remaining engineers could devote more time to the preventative maintenance schedule and work a number of hours each day that are more manageable over the span of an enduring operation. CMS access is also provided on the Bridge to enable routine picture supervision activity by a smaller number of personnel in lower threat environments, which can reduce this watchkeeping burden on the remaining Operations Room specialists.

3.3 (d) Maintenance

Due to the importance of providing a sufficient number of personnel to carry out the preventative maintenance activity of all the systems fitted to the platform, a considerable amount of analysis was carried out on the maintenance burden in the development of the Type 31 complement solution. Drawing on decades-worth of maintenance data from in-service platforms that are currently supported by Babcock, and aligned with the assured data provided by the suppliers of the next generation equipment for Type 31, a detailed analysis of this preventative maintenance workload for all systems onboard was undertaken. This preventative maintenance workload was broken down by engineering department section to assess the load required to keep the ship running.

To reduce this maintenance burden Type 31 has taken advantage of the latest technology. This modern equipment, such as the BAE Systems Bofors 40mk4 shown as an example in Figure 8, places a significantly reduced workload on maintainers over alternative systems. The maintenance required by systems was a key element considered in equipment selection for the Type 31 programme, assured through evidence provided by manufacturers to substantiate claims against real-world operation, as part of the wider effort to deliver a platform that could be operated by fewer personnel whilst still increasing capability over the previous Frigate classes.

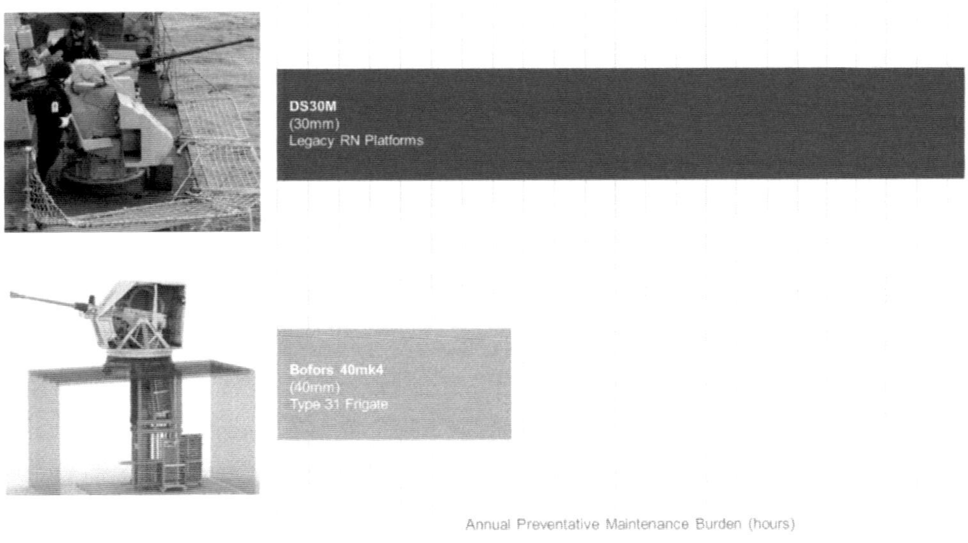

Figure 8: Type 31 annual preventative maintenance burden example (BAE Systems Bofors 40mk4), shown as a comparison to legacy RN equipment (data extracted from RN Unit Maintenance Management System (UMMS); axis scale redacted).

3.4 (e) General Activities

As an example in the area of general / whole ship activities onboard; in previous platforms the embarkation of supplies such as food and engineering parts, and discharge of waste (garbage) are handled by a 'clear lower deck' of all members of the Ship's Company passing items in a long human chain over the gangway. In a high endurance platform such as Type 31 these activities would take a significant amount of time and require a large number of people, so therefore the platform is fitted with a transverse handling system that extends over both sides from midships, that is able to pick up whole pallets of stores from the dockside and bring them into the ship, which are then processed using dedicated internal handling equipment and routes. Processed waste is offloaded using the same system. Use of this handling system minimises the burden of these traditionally personnel-intensive activities from the shipborne workload. As an additional benefit this handling system, located within the connected forward Mission bays, could also be employed in the future for the launch and recovery of uncrewed systems (UxV) such as REMUS UUVs separate to the boat handling systems.

The types of modern deck coverings selected throughout the ship can be kept clean with no polishing required, bathroom units are designed for quick cleaning and daily routines can be optimised to maintain the required levels of cleanliness onboard without imposing an excessive burden, particularly on junior members of the Ship's Company as a secondary role.

4. COMPARISON TO THE 'PARENT' DESIGN

The 'parent' design to the ARROWHEAD-140 product is the IVER HUITFELDT class in service with the Royal Danish Navy (RDN). Key amongst the inputs in the early stages of development of the Type 31 complement solution in 2018 was the Watch & Station Bill for the IVER HUITFELDT design as employed by the RDN. However the RDN operates in a different overall organisational philosophy to the RN, based more on a conscription model [8], so this could only provide guidance for the Type 31 complement rather than be used as the solution directly. The complement in Type 31 has to function within the overall philosophy of the RN personnel, branch and training structures, to fit within the overall Fleet as personnel move from ship to ship as part of the current career progression route.

This difference in organisational structure can be illustrated by looking at the Marine Engineering (ME) department of both the Type 31 and the IVER HUITFELDT class, shown in Figure 9 below. The RN use a pyramid structure where personnel are brought into the organisation at the Junior Rate level before being promoted up to Senior Rate based on experience and aptitude. The RDN structure employs a larger number of Officers, who take the leadership roles, a small number of Senior Rates who typically perform administration activities, and a large number of Junior Rates.

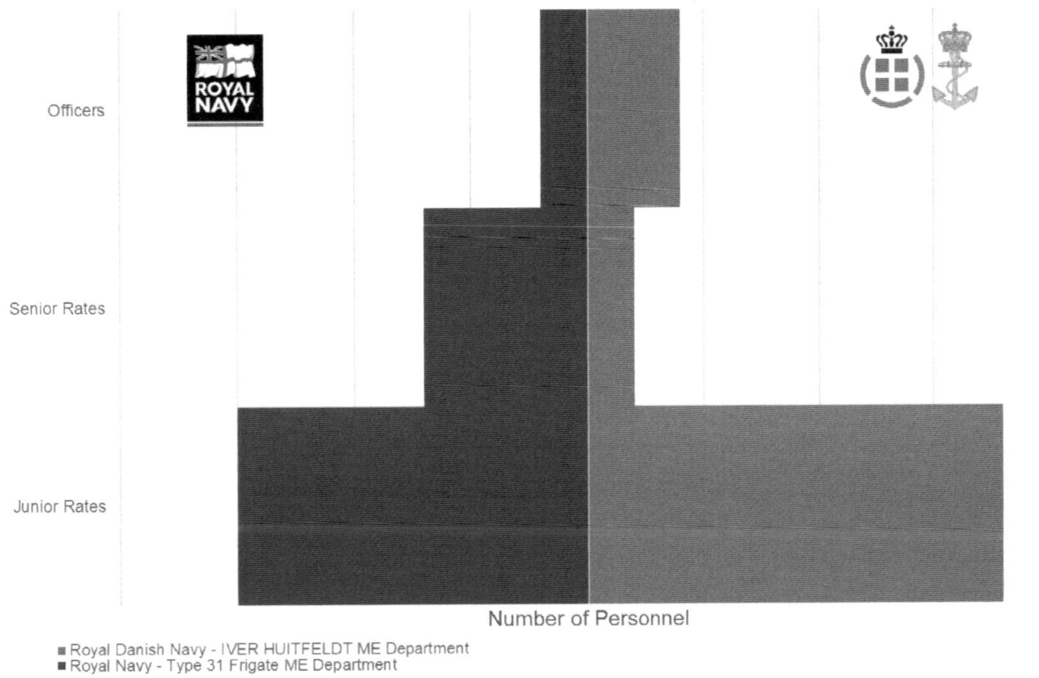

Figure 9: Comparison of the number of personnel in Marine Engineering (ME) Departments by rank / rate (axis scale redacted).

The RDN Junior Rates are given responsibility for systems and equipment that resides at the Petty Officer (Senior Rate) level in the RN. Personnel either enter the RDN for a defined period via conscription, or they can repeatedly join and leave the RDN as they see fit for their careers. These Junior Rates are assigned responsibility for a system or equipment as they join a ship, such as responsibility for maintaining the main engines or the reverse-osmosis equipment. Disadvantages of this philosophy include the reduced experience of wider warship operations such as damage control and firefighting, as this is not always a transferable skill inherent in someone joining in a 'sideways entry' from a job repairing diesel truck engines for a haulage firm (as an example) and requires additional training.

This is in contrast to the RN, where any Senior Rates leaving the Service take with them knowledge and years of experience of the systems they were previously responsible for that can only be replaced by personnel coming into the system from the Junior Rate level. A reduced number of Junior Rates entering the system would also cause a 'black hole' in the system for years to come. The advantage of the RN system is in the experience built up of wider warship operations, drilled into personnel from the most junior level at the start of their careers and built upon over time.

5. FUTURE

The Type 31 Frigate complement represents an optimum solution at a point in time. Defined by the latest high-TRL technology available that can be delivered into a class of warships in build now (Figure 10), and the various naval rules and regulations against which the ship was designed, Type 31 has the smallest effective and sustainable size of Ship's Company possible for a 7,000 tonne General Purpose Frigate that is capable of global combat operations.

Figure 10: HMS VENTURER (Type 31 Frigate Ship 1) and HMS ACTIVE (Ship 2) in build in Rosyth, Scotland (photo taken April 2024).

This complement solution and philosophy has also been built into the ARROWHEAD-140 product line, drawing on the work conducted to date over years of development activity and the scrutiny applied by the RN as part of the Type 31 delivery programme. Additional separate work within Babcock has identified potential options to further reduce the Ship's Company within the ARROWHEAD-140 design, through an investigation into alternate operating philosophies.

There are also wider studies into future measures that could reduce the number of personnel to the point where there are no human operators required onboard at all. To provide a true capability as a combatant warship however, there are a number of obstacles to overcome and considerations that will have to be made.

Significant amongst this number of obstacles is the ability to conduct damage control activity, in particular flooding following either accidental damage or enemy action. Either a new philosophy in the design and construction of the hullform will have to be developed, one that can tolerate extensive flooding without intervention while retaining the speed, manoeuvrability and internal functionality of compartments expected of a Frigate or Destroyer, or the entire philosophy on the attritional nature of the platform will have to be considered.

The requirements on the number of personnel driven by damage control actions could be removed completely if it is accepted that the warship can be lost following damage. This is a controversial decision that would have to be made at senior navy and political levels; a warship represents an expensive investment of taxpayer's money, and the optics and fallout of a total loss due to the platform accidentally hitting a semi-submerged shipping container one night may be unpalatable. If a minimal crew is retained onboard for other purposes such as weapon control to satisfy Rules of Engagement, however is not sized to deal with damage control, then consideration to their safety must be made in both peacetime and in the presence of the enemy as essential moral and ethical concerns and against duty of care legislation. Where commercial ships tend to follow established shipping routes and are therefore not far from another vessel that can effect rescue, warships are deployed widely as required by the operational tasking and may therefore be extremely isolated; the incident in 2008 involving HMS ENDURANCE is an example [9]. If the crew's only available action on taking accidental or hostile damage is to abandon to the liferafts and watch as their ship goes down, many lives may then be lost to exposure or capture before they can be rescued; this philosophy also poses a possible deterrent to both recruitment and retention of personnel.

Maintenance is another consideration. As outlined above the marine environment is tough on systems, and equipment can suffer unforeseen damage particularly in heavy sea states. The ability to keep a highly complex uncrewed platform functioning at sea on enduring operations, to deliver an equivalent capability to current Frigates and Destroyers, will rely on resolving the question of how this preventative maintenance activity on its systems will be achieved.

These are two elements amongst many that have to be resolved, either through technology, policy development or an overall change in philosophy before the crew of a warship can be reduced to a point where a completely uncrewed warship is operationally feasible.

6. CONCLUSIONS

The number of trained and experienced personnel available to several Navies is reducing. Through a combination of recruitment and retention challenges, this reduced number is a significant concern against an uncertain global situation and the enduring requirement to provide safe platforms at sea in order to deliver government policy. However, as stated in JDP 0-10 a key attribute of maritime forces is their versatility to react rapidly to events in the national interest. Efforts to reduce the number of personnel required onboard warships, driven by recruitment and retention challenges, should still remain cognisant of this fundamental attribute and not degrade a Navy's overall ability to exercise maritime power.

> *"Maritime forces are uniquely versatile, easily changing their military posture, while undertaking several tasks concurrently and remaining available for rapid re-tasking. Deploying with inherently high levels of readiness, a warship can transition from a peacetime state to a combat ready one in a matter of hours."*

Joint Doctrine Publication 0-10; UK Maritime Power [10]

The process followed for Type 31 that is outlined in this paper generated a complement solution that was considered the minimum viable size for the versatile combat operations of a globally deployable General Purpose Frigate that is in build today. Taking all aspects into account this solution did not seek solely to reduce numbers at all costs; instead significant focus was also devoted to the workload on personnel to ensure the scope of tasks placed on individuals was realistic, in an effort to support retention of experienced personnel. Note that it has only been possible to provide a high-level summary of some of the considerations and solutions for Type 31 in this paper, with other critical aspects such as alongside harbour watches not discussed and comparison charts provided with redacted axes; the full capability of Type 31 is not exposed in open-source.

Type 31 represents an optimum solution that is deliverable today for a globally deployable warship. There are studies both within Babcock and in other organisations looking to reduce this number still further, ultimately seeking a viable solution to deliver a fully uncrewed warship.

7. REFERENCES

1. CHADWICK, R R., 'CNEO Symposium and Conference; The Changing Environment', *Journal of Naval Engineering*, June 2002.

2. UK PARLIAMENT, 'To ask the Secretary of State for Defence, how many Royal Navy personnel there have been on average in each year since 2000', *Written Question UIN 10474,* answered January 2024.

3. MACROTRENDS, 'Ireland Military Size 1985-2024', *www.macrotrends.net [online]*, date accessed February 2024.

4. REACT DISASTER RESPONSE, 'REACT; About', *https://www.re-act.org.uk/about/ [online],* date accessed April 2024.

5. SAPIEN, J., 'The Inside Story of How the Navy Spent Billions on the "Little Crappy Ship"', *ProPublica [online] https://www.propublica.org/article/how-navy-spent-billions-littoral-combat-ship*, September 2023.

6. JOHNSON, J., HOWARD, M., 'Type 31 Frigate; Complex Warship Design for a Dynamic Operational Environment', *Conference proceedings of INEC 2022 [online] https://library.imarest.org/record/10665?v=pdf*, November 2022.

7. UK PARLIAMENT, 'To ask the Secretary of State for Defence, what the maximum crew complement is of each of the Royal Navy's (a) Type 45 Destroyers and (b) Type 23 Frigates', *Written Question UIN 24292*, answered February 2016.

8. ASSOCIATED PRESS, 'Denmark plans to expand military draft to women for the first time and extend service terms', *AP News [online] https://apnews.com/article/denmark-conscription-men-women-defense-military-service-182028ad1d0dafa0dd17f5ca0eaebff1,* March 2024

9. SHARPE, T., 'Mayday in Magellan; Leadership Lessons on flooded HMS ENDURANCE', *Wavell Room [online] https://wavellroom.com/2018/12/16/mayday-in-magellan-leadership-lessons-on-flooded-hms-endurance/*, December 2018.

10. UK MINISTRY OF DEFENCE, 'UK Maritime Power; Fifth Edition', *Joint Doctrine Publication 0-10 (JDP 0-10)*, October 2017.

8. AUTHOR'S BIOGRAPHIES

J Johnson. A delegated Design Authority for the Type 31 Frigate, and the Transversals Engineering Manager. Responsible as Design Authority for the operability, complement and future adaptability solutions of the platform, alongside managing transversals engineering aspects including security, standards, noise & vibration, human factors, materials compliance, operating environment, Bridge / navigation solution and Design Authority detail design liaison. He was previously a Warfare Officer in the Royal Navy.

M Howard. ARROWHEAD-140 Technical Director responsible for the export of this proven design, which has been selected for both the Indonesia and Poland (MIECZNIK) programmes (at the time of writing). His previous role was the Type 31 Frigate Chief Engineer, 2017 to April 2022, which included Design Authority responsibility for the platform and gun systems. His tenure as T31 Chief Engineer included the Competitive, Basic, Functional and Detailed Design phases.

COMPOSITE REPAIRS: PRACTICAL IDEAS

FOR AN INCREASE IN CAPABILITY AND OPERABILITY WITH LESSONS FROM THE OIL AND GAS INDUSTRY

J LeCoultre and **P Butt,** Fuze, Australia, **MT Williamson,** Floating Solutions Consulting, Australia

SUMMARY

Composite repairs in a marine environment have been a practical reality for the past 30 years and have been applied in a wide range of commercial and naval applications. Despite the potential to increase capability and availability, there are barriers to implementing structural composite repairs. This paper does not seek to address the advantages or disadvantages of composites; however, it does aim to identify barriers to adoption and a route to implementation where the evidence is that they are the optimum solution. This will be done by synthesising past experience, reviewing organisational best practice, providing illustrative case studies, as well as addressing the barriers to adoption in both naval and civilian contexts.

1. INTRODUCTION

Following significant research conducted both before and during the 1990s and early 2000s [1], [2], [3] and [4], composite repairs have been trialled on a number of marine assets in a number of locations. Even with the volume of research and supporting evidence available, the take-up has not reflected the promise of the early work. Despite this, there have been several instances of successful composite repairs both in oil and gas and naval vessels [5], [6]. This paper seeks to outline the reasons behind the current low adoption of composite repairs, the barriers to adoption, some best practice in getting acceptance and approval and highlight some lessons from the oil and gas industry that have wider benefits to the capability and operability of assets in both civilian and naval contexts.

NOMENCLATURE

ASME	American Society of Mechanical Engineers
CFRP	Carbon Fibre Reinforced Polymer
CALM	Catenary Anchor Leg Mooring
CAPEX	Capital Expenditure
Class	Classification Society (ABS, DNV, BV, Lloyds Register, etc.)
FEA	Finite Element Analysis
FRP	Fibre Reinforced Polymer
FSO / FPSO	Floating Storage and Offtake / Floating Production, Storage and Offtake
FVF	Fibre Volume Fraction
IMO	International Maritime Organisation
ISO	International Organization for Standardization
JIP	Joint Industry Project
MOC	Management of Change [process]
PMC	Polymer Matrix Composites
RAN	Royal Australian Navy
SDIP	Strategic Defence Industry Priority
UK HSE	United Kingdom Health and Safety Executive
VaRTM	Vacuum assisted Resin Transfer Moulding

1. HISTORY AND DEVELOPMENT OF THE INDUSTRY

Within the marine and naval context, this paper considers structural composite repairs made from Fibre Reinforced Polymers (FRP) applied to any marine structure. In general, composites are made from two or more different materials with different physical or chemical properties that, when combined, are stronger than those individual materials by themselves [7]. Structural composites are a subset of Polymer Matrix Composites (PMC) where the mechanical performance is repeatable, with high strength and stiffness, often superior to metals [8]. To achieve this high performance and reliability the structural composite combines the properties of a pre-determined orientation, stack of continuous reinforcing fibres such as; glass, carbon, aramid, and boron, embedded within a matrix which spreads the loads between the fibres and adhered substrate (marine structure). In the case of structural composites an epoxy-based system is commonly used for the highest mechanical performance from the resin choices available [8], [9]. A structural composite repair ultimately looks at the performance of the combination of the fibres, resins, and metallic substrate as a composite itself [10].

Figure 1: Structural composite repair on ship bottom plate using Carbon Fibre Reinforced Polymer (CFRP) [11]

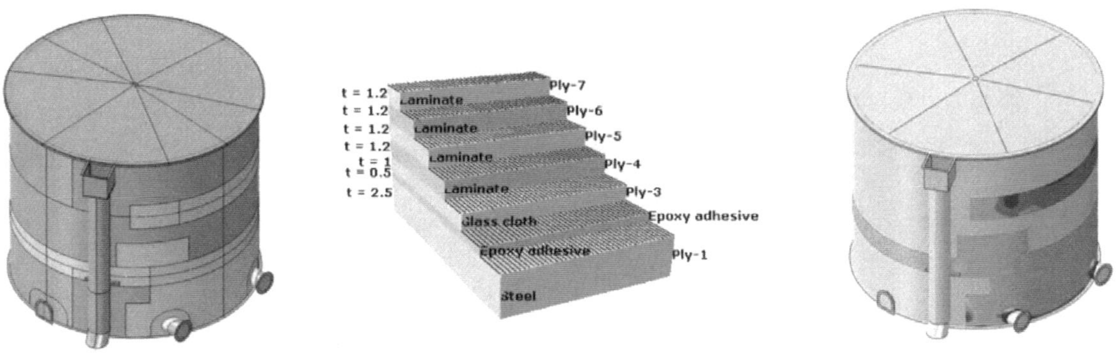

Figure 2: Structural composite repair FEA adhesion review, example [12]

1.1 CONSTRUCTION INDUSTRY

Traditional Wet Lay-Up for structural composite repairs to infrastructure has inherent problems and limitations [8], [9]. However, the system allows for widespread application by low skilled workforces. Infrastructure applications have looked to improve performance through process change [13], [14]. In the 1980s carbon fibre laminate strips were manufactured in a workshop and then secondarily bonded to bridge undersides for strengthening. This improvement from Wet Lay-Up manufacturing meant that the fibre performance was made consistent and reliable, but its adhesive performance was poor. This process allowed for concrete skilled workers to achieve repeatable applications (although not maximising the structural composite's performance). Other occurrences of this approach to structural composite repair involve combined Wet Lay-Up and pre-cured strips [15], [16] applied to concrete and steel infrastructure, including examples such as pre-cured carbon fibre strips applied to London Underground steel structures for strengthening.

The lessons from construction and particular bridge strengthening using composites are;

- Open sharing of information on research and applications allowed for rapid and low risk applications of composites to be delivered into the field.

- Establishing bridge composite strengthening codes was a faster process because of the open sharing and significant cost savings offered.
- With codes in place technology development continued. Improvements and industry cost savings were therefore realised quicker.

1.2 OIL & GAS

In the oil & gas industry the use of composites was initiated with the use of fibre glass pipes, introduced in 1948 [7] and has grown on the use of composite products in CAPEX projects. These are produced in workshop environments and are typically manufactured using pultrusion and filament winding methods enabling high volume manufacture.

For in-field manufacture and repairs to metallic components using composites, the uptake has been strong globally because of the significant commercial benefits, specifically allowing for repairs to occur under operating conditions. This in-field application under live conditions is leveraged in ships to prolong time at sea and reduce or remove the need for dry-docking. In these scenarios the cost of the repair is insignificant compared to the costs associated with time not at sea or lost production. The oil & gas industry has seen similar drivers to the applications of composites on bridges, where a large workforce unskilled in composites is required to undertake repairs. To do this, basic wet-layup systems have been deployed globally offering low cost but low performance repairs. Codes by ISO [17] and ASME [18] for the application of composites to pipelines, pipework, tanks and vessels were put in place in the early 2000s to provide structure and risk management to the application of these repairs. The first iterations of these codes considered basic wet-lay-up processing as this allowed rapid deployment of composites across the industry. Later iterations of the ISO and ASME codes have increased the requirements on installer skills, quality controls and long-term testing of the composite systems. Also included were discussions on consolidation and elevated temperature control as established processing controls for improved composite performance [19].

For oil & gas industry applications on floating assets there have been recent developments of vacuum forming to improve the reliability and performance of composite repairs. DNV (2015) [20], [21] and ABS (2021) [22] have issued guidance notes for structural repair using these systems. Although piping and tank repairs are structural in many ways, the ISO and ASME codes are not generic structural codes for the use of structural composites. Existing codes such as Class rules, do allow for alternative materials if their performance is understood. However, the lack of definition as to how the alternative materials can be proven to be acceptable is a barrier to their adoption.

1.3 NAVAL

Naval applications of composites have taken a different path to the oil & gas industry. The ability to access and control more easily the specialised trade skills required for Structural Composites saw vacuum consolidated and temperature controlled composite repair applications in the late 1980s and early 1990s [1], [23] Applications of composite patches for structural repair, specifically for the prevention of crack growth in underlying metal, has been previously used on warships [6], [24]. The prediction and modelling of this type of repair is well understood with the US Navy Ship Structure Committee [6] stating that "little difference between FEA and experimental results, implying that the finite element modelling is effective, and the results are reliable." The basic composite processing controls learned from the Marine and Aerospace Industries were picked up early for naval applications, compared to the decades later introduction in the oil & gas Industry.

1.4 DEVELOPMENT TIMELINE

A timeline of the key developments and in-field applications for naval and oil & gas applications are highlighted below.

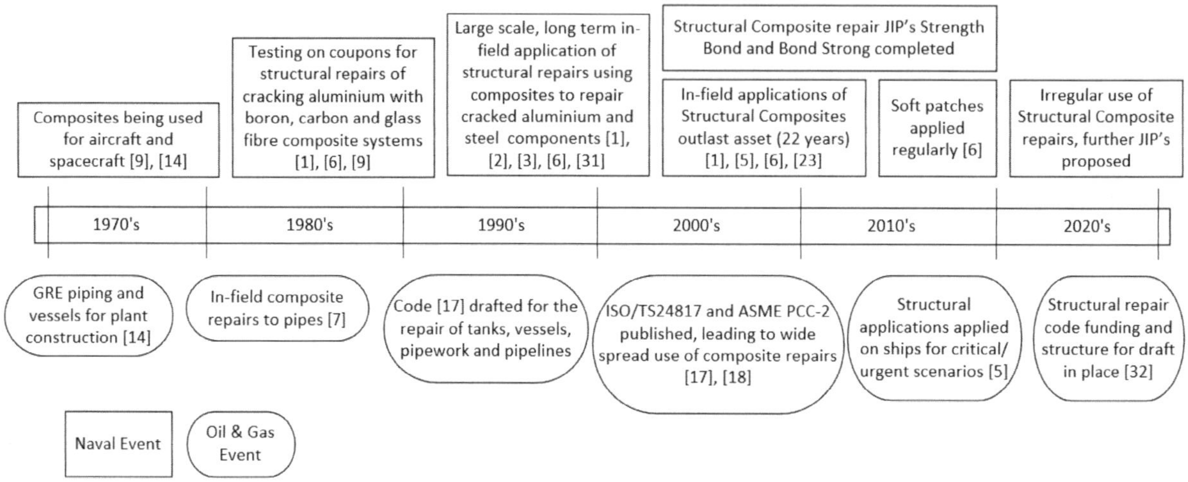

Figure 3: Structural composites development timeline

2. CURRENT INDUSTRY GAPS (NAVAL AND OIL & GAS)

2.1 STATE OF THE INDUSTRY

In general, understanding of the gaps in structural composite repairs across naval and oil & gas industries is muted due to all parties; operators, class societies, and vendors not sharing information about their composite repairs. The UK HSE review of composite repair on offshore assets [19] found that the records of composite repairs were poor, noting that although a detailed repair register is a requirement of ISO and ASME codes, these records are not well kept or maintained in the oil & gas industry.

Although there is a significant body of academic data on composite repairs spanning more than 50 years [1], [6], [14]. The information on structural composite repairs in the field is available but hard to find. This makes the education process of supply chains slow and costly. In turn this means that it is harder to provide evidence to demonstrate that currently novel and contentious applications for structural composite repairs do in fact show reduced risk during engineering, installation and over the life of the repair [25], [26].

From industry and academic data reviewed as part of this paper the technology supporting structural composites is available for industry use. The industry gaps appear to be in education, understanding, procurement and engineering approval processes (including code compliance and understanding of the approval system) [10, [14], [25].

2.2 BARRIERS TO ADOPTION

Despite a wealth of supporting research and isolated individual applications, there is not a widespread adoption of composites to repair naval structures [26]. The barriers can be grouped in various areas as illustrated in tables 1 - 4.

The tables compare two types of composite repair and a traditional "like for like" steel repair. The difference between the two types of composite repair is that an ISO 24817 compliant repair is mainly for pipework or for instances where a defect may result in a loss of containment (watertight integrity) but not affect the structural integrity of the asset. A "structural composite application" is one where the defect repaired using composites can be considered to take all loads (local and global) that the original structure is designed for.

The colouring of the tables gives an indication of the areas where the most significant barriers to adoption are experienced.

Table 1. Technical barriers to adoption

	ISO 24817 Compliant applications, O&G Industry	Structural Composite Applications, Naval or O&G	Traditional steel or similar, Naval or O&G
Capacity for structural integrity	> 30 years of in-field evidence and a structure for new product to be tested and engineered	Although not always accepted, a comprehensive body of academic and in-field evidence is available	Capacity is available and well understood.
Materials	Existing test standards and program to classify material performance	Available products are continuously evolving and installation performance varies.	Globally available, easy to obtain and comparatively cheap with material properties well defined
Life span	Defined life to 20 years, with inspection, maintenance, and re-validation processes	Difficult to demonstrate in a repair context or get accepted despite known materials	Readily accepted as "permanent"

Table 2. Production and design barriers to adoption

	ISO 24817 Compliant applications, O&G Industry	Structural Composite Applications, Naval or O&G	Traditional steel or similar, Naval or O&G
Level of engineering skill required	Low - Engineering skill directed to following ISO 24817 [17] or ASME PCC-2 [18] direction.	High level of engineering skill needed and understanding of material properties, failure modes, rules and regulations.	Low level required, with limited or no calculation or sophisticated engineering required.
Design workforce	With the code direction operational teams can conduct their own reviews.	Limited in availability and generally not knowledgeable of both composites and structures.	Widely available in most ports and jurisdictions worldwide.
Production and installation workforce	There are clear installation experience and skill requirements in the codes.	Not widely available. Training is not generally available.	Widely available in most jurisdictions worldwide.
Industry knowledge understanding	Level of familiarity has contributed to unsophisticated approach where only low risk applications are utilised.	Lack of embedded corporate experience, knowledge or understanding. Lack of willingness to share or publicise applications.	Embedded corporate experience knowledge and understanding within many organisations. Success stories are widely shared as part of general publications.

Table 3. Regulatory barriers to adoption

	ISO 24817 Compliant applications, O&G Industry	Structural Composite Applications, Naval or O&G	Traditional steel or similar, Naval or O&G
Approvals (regulatory and other)	Operational and third-party approvals are well established.	The approval process is not clear and generally by "special consideration" even for situations that aren't novel or contentious.	Clear routes to approval with established and understood timelines. Special consideration only required for novel and contentious situations.
Guidelines	Established and well used, even if incorrectly.	Limited or non-existent Complicated	Widely available or codified (IACS 47) [27] Simple needing low level of technical ability to understand and implement"

Table 4. Project barriers to adoption

	ISO 24817 Compliant applications, O&G Industry	Structural Composite Applications, Naval or O&G	Traditional steel or similar, Naval or O&G
Impact on operations	Small and limited with the avoidance of hot work. Can be compatible with not decommissioning critical systems or spaces.	Small and limited with the avoidance of hot work. Can be compatible with not decommissioning critical systems or spaces.	Can be considerable particularly given the need for hot work and the impact on potentially "clean" spaces.
Schedule	Well understood and easy to define.	Difficult to predict. Lack of generalised embedded industry knowledge and understanding makes forecasting of time and resource to achieve approval difficult and unreliable	Well understood and easy to define.
Project risk	Low	High	Low
Appetite for use	High Often used in applications incorrectly because the appetite is so high.	Low Effort required to adopt and overcome barriers is high.	High Default position. Normally unquestioned.

2.3 BARRIERS DISCUSSION

A key learning from the oil & gas industry is to utilise composites regularly in areas where the barriers are low. By doing this the operations, engineering and approvals teams increase their understanding and capability to manage the risks associated with composite repairs. Practically this means, repairs to pipework following the established ISO 24817 repair code and existing Classification Societies' product approvals before approaching scopes for repair of vessel structures [22]. A visual representation of this is seen in tables 1 - 4, where the application of composites for pipework, even critical pipework, can be relatively barrier free.

The biggest barrier to implementing composite repairs in an operational context tends to be the approvals and justification of the design approach used for the composite repair [25]. The challenge is multifaceted and difficult to address in the normal time constrained and risk averse context of a project involving a repair.

The following is an expanded list of the regulatory or approvals barriers:

- Not all Class Societies have developed guidance [10].
- Those guidelines that do exist may not have been widely adopted within industry with well publicised and numerous examples of implementation.
- Within the approval organisation, acceptance may need to be done by a specialist section that may not be clearly identified or local to the place where the repair is being completed.
- Many guidelines place restrictions on the use of composite repairs. The justification for these is not necessarily apparent or in line with published experience or the capabilities of composites (e.g. more than 6mm of remaining steel required in the context of corrosion, or composites may not be used in areas where fatigue is suspected, or they may require demonstration that the repair is "non-structural" for approval).
- There are limited research funds and actors to develop structural design guidelines.

2.4 CONSEQUENCES OF BARRIERS

The consequence of the barriers to adoption is a self-perpetuating cycle. The lack of industry knowledge and understanding without common embedded experience of where composite repairs can be successful perpetuates a perception of high-risk unproven products and techniques. This means that composite repairs are not adopted and hence the cycle continues.

3. PROCEDURE TO HELP ENSURE A SUCCESSFUL REPAIR

3.1 CHOICE OF CODE

The choice of code will depend on the nature of the repair. There are many established codes for the repair of pressure vessels that are frequently used to justify the design of structural repairs that are not considered structurally effective for global loads. For long life (>5 years) repairs of structure that have to be effective against such loads, Class rules need to be considered. Given that not all Class Societies have current guidelines, it may be that guidelines or rules from a Society that is not Classing the asset in question may be acceptable.

Given the previously discussed barriers to adoption, notably the difficulty in being able to define a timeline for a repair to be accepted, early engagement of the acceptance authorities is key.

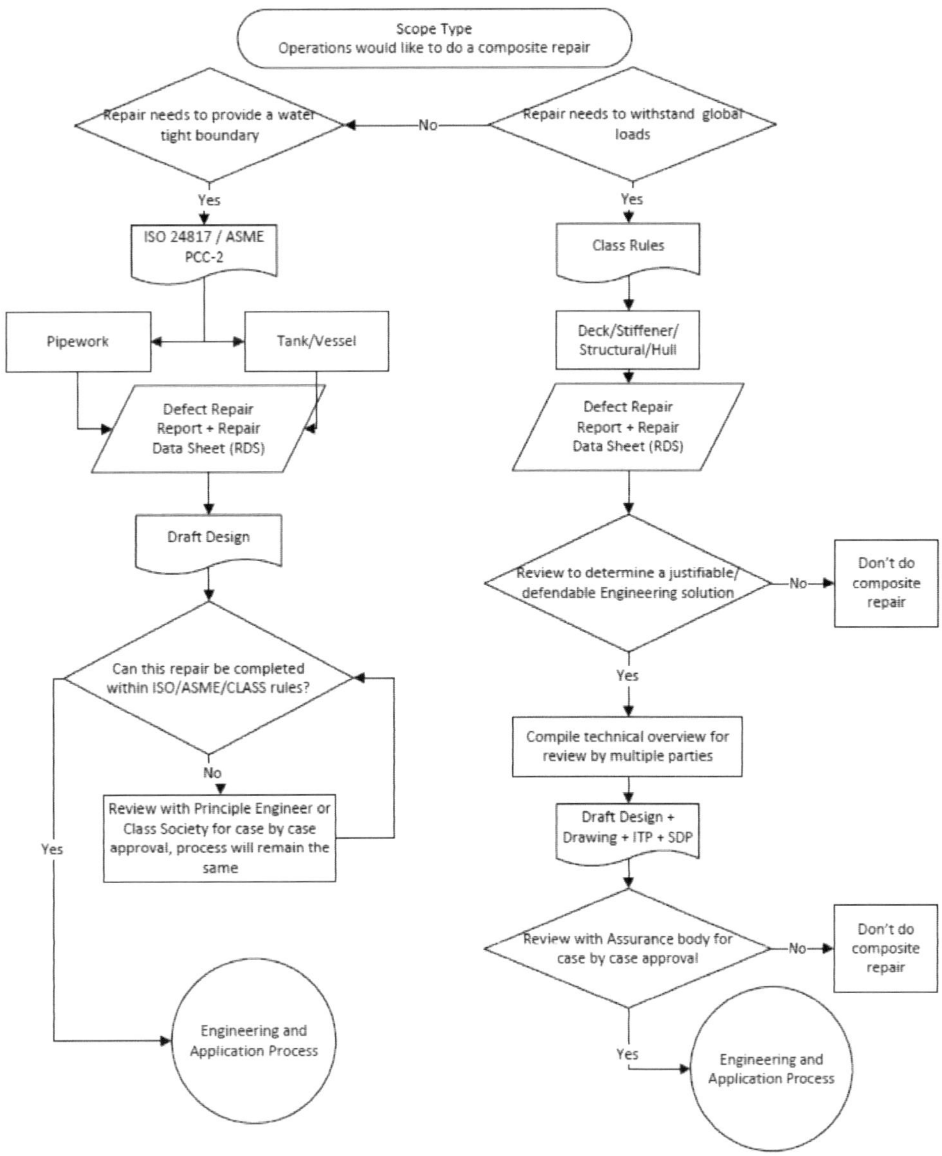

Figure 4: Typical composite repair selection and execution process.

3.2 USE OF FINITE ELEMENT ANALYSIS (FEA)

For simpler repairs that are not subject to global loads, Finite Element Analysis (FEA) is not normally required, with simple code checks defined in the ISO/ASME rules. However, care must be taken with the definitions of the inputs as some interpretation is required as codes developed for pressure vessels do not necessarily describe vessel structures clearly and do not address failure mechanism review [10].

For repairs that have to be demonstrated to withstand global loads, the use of FEA is recommended [28], [29], [30]. Ideally a pre-existing FEA model of the asset exists. Preferably this will have been previously accepted by the Class Society for the vessel in question. A local model of the composite repair should be built that can demonstrate that stresses and strains are below acceptable limits for the required load cases and combinations. Although for steel, there are established guidelines for developing and approving FEA models [28], the local model of a composite repair will need close liaison between the Class Society and the composite designer to ensure it is acceptable.

4. CASE STUDIES

It should be noted that both navy case studies presented here involve ships that were not Classed and where the Design Authority was "in house" with the Naval Authority supportive of the proposed composite repairs.

4.1 TYPE 42 COMPOSITE REPAIRS

On the now decommissioned UK Royal Navy Type 42 destroyers, from 1997 to 2003, composite patches were applied in 35 locations following a standard repair that was developed for where the two fitted lift trunks intersected with 1 and 01 deck (Officers and Senior Rates Food Lifts). Fatigue cracking would often be observed in this location. To produce an effective steel repair, a major redesign of the food lift trunk structure would be required, whilst like for like repair involved intrusive hot work on multiple compartments.

As an alternative, a reinforcement by Carbon Reinforced Plastic was developed. This was possible given the then current knowledge of composite repairs within the UKMOD and the fact the structure was not considered critical and caveated that any cracks had not propagated into primary structure [2], [31].

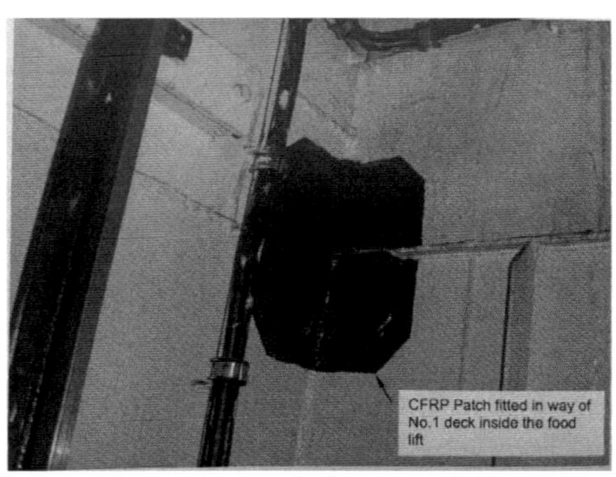

Figure 5: Carbon fibre repair patch fitted to a UK T42 Destroyer [31]

The learnings from this case study are:

- Although traditional steel repair or redesign was possible, a composite repair was a more optimised solution.
- A composite repair was feasible because of the knowledge of the Design Authority and the Naval Authority meant that it was not onerous from an approvals point of view.
- Application of composites to fatigue cracking is a successful way to prevent crack growth.

4.2 HMAS SYDNEY

HMAS Sydney was the third Adelaide Class guided missile frigate built at Todd Pacific Shipyards (launched 1980). The first four ships of the class experienced some significant fatigue cracking in the aluminium superstructure. For the follow-on ships, lessons learnt meant that improved production and design details addressed the cracking issues from the earlier vessels. For HMAS Sydney, in 1993, two carbon fibre “strip” reinforcements were bonded to 02 deck on the port and starboard sides. Subsequent work and investigation showed that the strips were effective against fatigue. Only minor repair and maintenance ended up being required between 1993 and 2000. The strips were still in place when the vessel was paid off in 2015. Both, FEA modelling and physical strain measurements showed that the critical stress concentrations had been reduced [1], [6]. This was confirmed by the absence of any deck cracking occurring within the boundaries of the composite patches.

This is an example of “permanent” repair using composites that lasted for the life of the asset. Although like for like costings are not readily available, the remediation for the design and build issues experienced for the initial vessels of this class came at a considerable cost. The cost of the carbon strips was much less and significantly less intrusive.

Figure 6: HMAS Sydney [24].

The learnings from this case study are:

- At the time, the academic teams were the only composite trade experienced personnel available to complete the installation. Finding verified competent structural composite teams is critical to the success of composite repairs [1].
- To achieve the performance structural composites can provide requires controlled consolidation (vacuum bagging) and temperature control [1].
- Structural health monitoring utilising strain gauges and ultrasonic inspection are a valid long term and reliable for inspections [1], [20], [21], [25], [32].
- Maintenance and repair of structural composite repairs can occur successfully over the life of a repair such that the composite outlasts the underlying substrate [1], [5], [6], [25], [32].

4.3 CALM BUOY AUSTRALIAN OIL AND GAS CASE STUDY

A Catenary Anchor Leg Mooring (CALM) buoy in Australian waters had developed holes through corrosion. The repair needed to have the following characteristics:

- Class approval
- Small footprint
- Be applied under the water line and on a live leak
- No lifting of equipment or materials
- No hot work.

The treatment of the CALM buoy hull as a tank for the purposes for ISO 24817 provided a route to approval, with justification that the structure of the hull was fit for service once the repair was complete.

The installation of the composite repair under live conditions and under the water line required a novel installation method, in part a VaRTM system [8], [9]. This method was not part of the initial ISO 24817 specification and structural composite code compliant performance. However, with close collaboration between the operator, Class and vendor the material performance testing guidance provided by ISO 24817 was used to create a short-term (<4weeks) test program to demonstrate the novel installation would produce adhesion performance equal to or greater than that required by ISO 24817 and Class approval requirements.

Multiple patch repairs were installed under live leak situations and the composite repairs remained in place until the asset was decommissioned. This repair was completed before 2015 but the details of the repair are lost due to the industry wide poor information sharing problem.

The learnings from this case study are:

- Although the use of composite repair for a hull is not compliant with a pipework or pressure containing code, the usefulness of this repair code and the defined test programs to confirm adhesive and structural composite performance provides the basics from which to design naval or other structural repairs.
- The importance of close working relationships between composite vendors, the execution team (with composite trade qualifications), Class engineers & surveyors, and the operator's engineering team are crucial.
- A rapid verification test program is an effective way to build confidence and knowledge in the performance of structural composites.

5. HOW COMPOSITE REPAIRS CAN CONTRIBUTE TO AUSTRALIAN CONTINUOUS NAVAL SUSTAINMENT

Within Australia a government policy for continuous naval sustainment has been clearly articulated and published [34]. It is suggested that facilitating structural composite repairs can help deliver several aspects of this policy. This is demonstrated in Table 5 where compliance is mapped to some of the requirements. Providing the naval sustainment environment with the direction to deliver the capability to complete composite repairs is a suggested route to delivering on the stated naval sustainment policy.

When comparing to the civilian context, it is believed that a "top down" approach to addressing the barriers to adoption detailed in Tables 1 – 4 is possible. This is because within the naval sustainment environment, the reduced total number of assets, all regulated by a naval authority (as compared to a civilian regulator), coupled with a singular maritime doctrine [35] means that a clearer sense of purpose can be fostered when compared to the civilian fleet. This means that all the key stakeholders that can influence the barriers to adoption can deliver an environment where these are reduced or eliminated. i.e. the naval regulators, capability managers, suppliers and end users can all respond to a direction to ensure that composite repairs are an available capability.

Table 5. Composite repairs and compliance with Australian continuous naval sustainment.

Requirement	Source	Route to closure	Actions required
"Certification that enables seaworthy operation"	SDIP 2 page 3 [36]	Accept existing academic and other research as well as documented field trials thus building on existing standards and codes, develop naval standards or guidelines that provide a route to certification.	The requirement to be able to conduct composite repairs on naval ships be declared as a key capability
"Enhance performance and reduce costs"	SDIP 2 page 3 [36]	Life of structure can be extended. Complete work scopes faster, while out of dry-dock, with a smaller labour footprint and reduced impact on the asset systems and personnel.	Ensure that the ability to complete composite repairs is no more onerous, from an approvals perspective, than a steel repair.
"Support industry-led collaborations between industry, researchers and end-users"	Defence Industry Development Strategy Fig 3 page 18 [34]	Through sharing of knowledge and experience of actual repairs	Sponsor a JIP to facilitate and publish information sharing on composite repairs in naval and commercial assets.
"Achieve availability targets while maximising the speed of upgrade"	Defence Industry Development Strategy Fig 3 page 18 [34]	Provide projects supporting the sustainment of naval assets the ability to choose composite repairs if it is the best technical solution for their asset.	ANP 4412-4315-4 [37] and other documents to be updated to include composite repairs.

6. RECOMMENDATIONS

6.1 INDUSTRY WIDE

The following four recommendations are believed to be key to addressing the barriers to adoption detailed in Tables 1 - 4. It is thought implementing these will break the current self-perpetuating cycle preventing the adoption of composite repairs. Once the cycle has been disrupted, other barriers such as perceived project risk and lack of appetite for use will fall away.

- Create a database of current composite repairs that is available to industry and contains details such as: date of repair, vessel IMO number, Class Society, temporary or permanent repair, fire rating (A0, A60 etc.), material used, structure being repaired (watertight bulkhead, main deck, parent plate, transverse frame, longitudinal stiffener etc.). This would be ideally suited to a Joint Industry Project (JIP).
- Within guidelines remove any unnecessary restrictions on the use of composites or provide a clear route to provide justification on a case-by-case basis.
- Form a JIP with the objective of developing and industry wide standard for implementation of repairs of structural steel using structural composites in a marine field. This would be an equivalent to IACS 47 [27].
- When communicating strategic requirements to the education sector as to what the marine industry requires, ensure that the need for education and practical training for composites in a marine context is included.

6.2 IN AN AUSTRALIAN NAVAL CONTEXT

Three clear recommendations are made in order to facilitate the ability for composite repairs to be made available to RAN sustainment organisations:

- At a senior level within defence, it is recommended that the ability to conduct composite repairs of naval assets be explicitly declared as a key capability. In an Australian context the justification for such a statement would be to provide a route to deliver on the Australian Defence Industry Development Strategy. In particular, 'SDIP 2 Continuous naval shipbuilding and sustainment' [36].
- Ensure that any structure repaired using composites is considered as having the same baseline as a traditional like for like steel repair (no Baseline Configuration change).
- Clarify and explicitly define the route to certification. In particular ensure that approval of standard composite repairs need not require a higher level of approval than a steel equivalent.

7. CONCLUSIONS

Given the parallels between oil & gas and naval assets (and organisations), there are similar advantages to adopting composite repairs to ensure ongoing commercial viability or capability and availability of the assets. Noting this, it can be seen that the barriers to the wider adoption of composite repairs (in both industries) are not necessarily related to their technical properties or general performance. By adopting some of the recommendations in this paper, more options will become available to deliver an optimum repair, improving capability and operability. Furthermore, these will also support the delivery of a number of the requirements of the Strategic Defence Industry Priorities.

8. REFERENCES

1. GRABOVAC, I., et al. "Are Composites Suitable for Reinforcement of Ship Structures," *12th International Conference on Composite Materials*, 1999.

2. TURTONA, T J., et al. "Oil platforms, destroyers and frigates—case studies of QinetiQ's marine composite patch repairs," *Elsevier*, 2004.

3. "Repairing steel ships with composite patches," Shiprepair and Conversion Technology 4th Quarter, 2003.

4. MCGEORGE, D., et al. "Repair of floating offshore units using bonded fibre composite materials," *Elsevier*, 2009.

5. BELEKAR, Y D., et al. "A Review On-Study of Use of Bonded Fibre Composite Materials for Repairs," *International Journal of of Engineering Research, vol. 5, no. Special 1, pp. 179-183*, 2016.

6. SHIP STRUCTURE COMMITTEE, "STRENGTH AND FATIGUE TESTING OF COMPOSITE PATCHES FOR SHIP PLATING FRACTURE REPAIR," www.shipstructure.org, 2015.

7. Composites Lab, "History of Composites," 2024. [Online]. Available: https://compositeslab.com/composites-101/history-of-composites/index.html. [Accessed March 2024].

8. Gurit, "Guide to Composites GTC-6-0417-1," 2022. [Online]. Available: https://www.gurit.com/wp-content/uploads/2022/12/guide-to-composites-1.pdf. [Accessed 2024].

9. HAYWOOD, M A., "The investigation of a vacuum consolidated and heat curing process for applying advanced fibre reinforced polymers to rehabilitate infrastructure," *Perth*, 2008.

10. GRAVE J H., et al. "Bonded patch repairs for metallic structures - A new recommended practice," *Journal of Reinforced Plastics and Composites*, 2014.

11. MODEC, "MODEC obtains Approval by ABS for New Offshore Repair Method for Hull Structures of Floating Production Facilities," 2020. [Online]. Available: https://www.modec.com/news/2020/20201223_pr_CFRP.html. [Accessed April 2024].

12. FUZE, Structural Composite Repair, Perth: Fuze Group Pty Ltd, 2024.

13. KARBHARI, V M., et al. "Interlaminar and intralaminar durability characterization of wet layup carbon/epoxy used in external strengthening," *Elsevier, Composites Part B: Engineering*, 2006.

14. BHATT, A T, et al. "Primary Manufacturing Processes for Fiber Reinforced Composites: History, Development & Future Research Trends," *IOP Conf. Series: Materials Science and Engineering 330*, 2017.

15. SIWOWSKI. T., et al. "Strengthening Bridges with Prestressed CFRP Strips," *SSP - JOURNAL OF CIVIL ENGINEERING*, 2012.

16. HU W., et al. "Review of Experimental Studies on Application of FRP for Strengthening of Bridge Structures," *Advances in Materials Science and Engineering*, 2020.

17. ISO, 24817 Petroleum, petrochemical and natural gas industries — Composite repairs for pipework — Qualification and design, installation, testing and inspection, 2018.

18. ASME, PCC-2 Part 4 "Non metallic and Bonded Repairs", 2022.

19. UK HSE, HSE Good Practice Guide – Management of Engineered Composite Repairs, 2021.

20. DNV GL AS, Design, Fabrication, Operation and Qualification of Bonded Repair of Steel Structures, DNV RP-C301, 2015.

21. DNV, DNV-ST-C503 Offshore Standard – Composite Components, DNV GL AS, 2022.

22. ABS, "Guidance Notes on Composite Repairs of Steel Structures and Piping," 2021.

23. SAAB, "Visby Class Corvette," [Online]. Available: https://www.saab.com/products/visby-class-corvette. [Accessed March 2024].

24. US NAVY, "HMAS Sydney (FFG 03) underway during Pacific Bond 2013," [Online]. Available: https://commons.wikimedia.org/w/index.php?curid=29958155. [Accessed March 2024].

25. SOURISSEAU, Q, et al. "A new adhesively bonded composite repair for offshore," *CICE2023-11th International Conference on Fiber Reinforced Polymer (FRP) Composites, Rio deJaneiro, Brazil*, 2023.

26. SOURISSEAU, Q., et al. "Adhesively Bonded FRP Reinforcement of Steel Structures: Surface Preparation Analysis and Influence of the Primer," *ASME2022-41st International Conference on Ocean, Offshore and Arctic Engineering, Hamburg, Germany*, 2022.

27. IACS, "Recommendation Number 47 Shipbuilding and Repair Quality Standard," September 2021.

28. WILLIAMSON, M., et al. "Lessons learnt from oil and gas: Practical opportunities to increase operability and capability through optimising inspection, engineering and repairs," *IMC, Sydney*, 2022.

29. MADELPECH, P., et al. "Bonded composite patch to repair metallic structures: fatigue behaviour of a disbond", *ICAF 2009, Bridging the Gap between Theory and Operational Practice, Rotterdam, The Netherlands,* 2009.

30. BARKER, A., et al. "A procedure to assess disbond growth and determine fatigue life of bonded joints and patch repairs for primary airframe structures," *Elsevier, International Journal of Fatigue*, 2020.

31. UK MOD Defence Equipment and Support, Sea Systems Directorate, "Type 42 Destroyers Survey and Repair Guide," 2008.

32. ALEXANDER, C., et al. "Fatigue Growth Behaviour of Cracks in Pipelines Reinforced by Carbon Composite Wraps," *Pipeline Pigging and Integrity Management Conference, Houston*, 2022.

33. TAVAKKOLIZADEH, M., et al. "Galvanic corrosion of carbon and steel in aggressive environments," *Journal of Composites for Construction*, 2001.

34. Australian Government, "Defence Industry Development Strategy," 2024.

35. Royal Australian Navy, Australian Maritime Doctrine: RAN Doctrine 1, Canberra, 2010.

36. Australian Government, SDIP 2 Continuous naval shipbuilding and sustainment, 2024.

9. AUTHORS BIOGRAPHY

Matthew Williamson is a Principal Naval Architect with Floating Solutions Consulting in Perth Australia where is role is as Lead – Client Delivery. Matthew is a Chartered Naval Architect with a career that has spanned Oil and Gas, Defence, and Commercial Shipbuilding.

As a Professional Engineer with a proven record in technical delivery and project management in the Marine Field. Experience includes the full range of projects from conception and bid through contract negotiation, detail design, build and in service. This has been whilst working for suppliers, prime contractors, regulators and clients internationally and within Australia. This understanding of the full lifecycle of marine projects is now being leveraged with Floating Solutions Consulting to deliver optimised inspection and repair scopes for marine assets.

Jason LeCoultre is the founder and current Managing Director of FUZE Solutions, a composite technology focussed business that provides solutions to the Oil & Gas, Mining and Defence industries.

Mr LeCoultre has two decades of experience in engineering company management in the Oil & Gas Sector utilizing composite materials.

Peter Butt is the Chief Technology Officer at FUZE Solutions. Mr Butt is responsible for the technical delivery of products and services at FUZE. He runs the R&D programs at FUZE that develop, trial and implement new technology for the company.

Mr Butt has experience developing and promoting technologies across the oil and gas sector from exploration to decommissioning, prototyping and field testing to asset sustainment and life extension.

SURVIVE THE SWARM! WARSHIP VULNERABILITY TO SWARMING UNCREWED AERIAL VEHICLE ATTACKS

J W Neill, QinetiQ Ltd, United Kingdom

SUMMARY

Methods of vulnerability reduction for surface vessels against a swarming UAV attack are investigated by modelling a salvo of fictitious bombs against a fictitious frigate using QinetiQ's Survive software capabilities. The nature of such a threat is also explored. The number of bombs incident on the ship is varied, up to a maximum of 15 concurrent strikes. The best and worst case damage cases for the 15 bomb scenario are interrogated. Implications for damage control following a swarming UAV attack are also discussed.

NOMENCLATURE

AAW	Anti-Air Warfare
ASROC	Anti-submarine Rockets
ASuW	Anti-Surface Ship Warfare
ASW	Anti-Submarine Warfare
CIWS	Close-In Weapon System
CODLAD	Combined diesel-electric and diesel
DCFF	Damage Control and Firefighting
FREM	Fast Running Engineering Model
FRPP	Fire Repair Party Post
N	Number of bombs that strike the frigate
UAV	Uncrewed Aerial Vehicle

1. INTRODUCTION

As the 21st century progresses, Uncrewed Aerial Vehicles (UAVs) continue to gain popularity for navies modernising their forces and this trend is likely to continue. While these new technologies aim to enhance the capabilities of naval platforms, the ability to endure the threat posed by attack from such craft is also being considered. This paper aims to analyse likely outcomes of a swarming UAV attack against a surface warship, with support from the Survive software, to identify lessons in survivability for warship design.

Naval platforms today are typically armed with either intercept missiles or Close in Weapon Systems (CIWS) to deal with incoming air threats such as missiles and aircraft. Crew may be less experienced in use of the combat system and air defence systems to track and address larger numbers of smaller air threats simultaneously than is conventional. If a swarming UAV attack was to occur, it is plausible that the air defence capabilities of a vessel becomes overwhelmed, due to the inability to employ sufficient countermeasures for all avenues of attack, leading to multiple successful weapon strikes being delivered to a single target.

Survivability principles of redundancy and separation, as well as concentration and protection are typically considered for damage from a single threat weapon. How these principles fare in the context of multiple weapon strikes is explored through modelling the vulnerability of a fictitious frigate against a salvo of small guided bombs using the Survive software.

This is a novel type of vulnerability assessment whereby a large number of threat weapons are modelled as striking the same target vessel as part of a single damage scenario. Usually when carrying out typical vulnerability assessments, only a limited area of the vessel is damaged and vulnerability reduction design philosophies are typically developed with this in mind. Investigation of a scenario whereby damage is inflicted to several locations on the vessel will therefore give a fresh view as to what vulnerability reduction strategies need to be considered for the modern threat of a swarming UAV attack.

2. SURVIVE CAPABILITY

It is important to understand the Survive software capabilities in order to frame the results and conclusions of the study.

2.1 WHAT IS SURVIVE?

Survive is a Fast Running Engineering Model (FREM) primarily used for the vulnerability assessment of surface ships.
Being a FREM, it is best used for comparative studies to determine the performance of different design options, however predictive analysis can also be carried out with appropriate caveats.

Survive operates as a FREM by using empirical algorithms underpinned by over fifty years of trials data, real world incident data and detailed modelling. Its powerful visual representation of results allows for design optimisation, real world event audits and in service alteration analyses (including role change).

2.2 HISTORY

Development on the Survive software capability first started in 1994, QinetiQ used the first version internally in 1998 and it was first released commercially in 2007.

2.3 PREVIOUS USES

Survive capabilities have supported delivery to a wide range of customers in defence including the UK MOD, BAE Systems, Babcock, Thales, BMT, UK MOD Naval Design Partnership as well as merchant shipping organisations such as Oil Companies International Marine Forum.

Survive capabilities have supported programmes such as UK Royal Navy frigates (Type 26, Type 23), destroyers (Type 45), Queen Elizabeth Class Carrier and Royal Fleet Auxiliaries.

3. SCENARIO CONFIGURATION

3.1 SHIP MODEL

The Survive model used for this study is a fictitious general purpose frigate with combined diesel-electric and diesel (CODLAD) propulsion. In order to conduct a vulnerability assessment of this vessel, the survivability of its main mission capabilities has been examined. These are: Anti-Air Warfare (AAW) using a vertical launch missile system used for self-defence and protecting allied assets; Anti-Submarine Warfare (ASW) using either a helicopter equipped with torpedoes or an Anti-Submarine Rockets (ASROC) system and a 5" gun for horizon ranged Anti-Ship Warfare (ASuW) or naval gunfire support. A Close-In Weapon System is also used as an additional option for self-defence AAW.

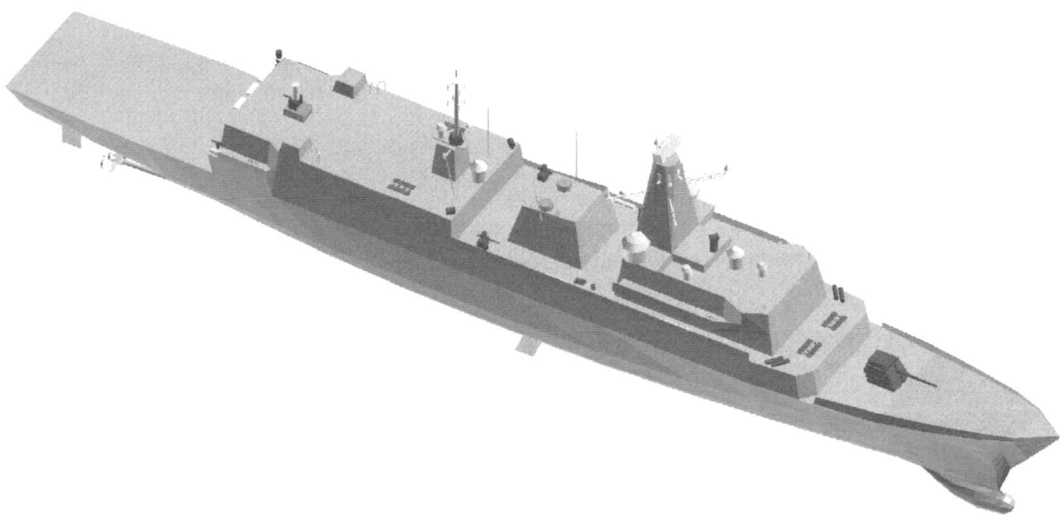

Figure 1: The fictitious, general purpose frigate used for all analyses in this study.

The mobility of the frigate is also an important capability to be aware of. The ability to move at full speed and reduced speed are both of interest and are represented separately within the ship model to gain a better picture of how vulnerable it will be post attack and whether it will have any ability to retreat. The move capabilities are termed as Max Propulsion and Min Propulsion. For Max Propulsion both drive chains, both diesel engines, both electric motors and both rudders are required. For Min Propulsion one drive chain powered by a single diesel or electric motor and both rudders are required or both drive chains are required with either a single diesel engine or an electrical motor available for each (no rudders required).

3.2 SWARM ATTACK REPRESENTATION

This study has analysed likely damage outcomes of a fictitious frigate struck by several bombs at different angles in a short space of time, delivered by a swarm of UAVs. This has been done for a varying number of bomb strikes. A set of possible strike trajectories that the UAV delivered bombs can take has been chosen to give a spread of damage to the exposed parts of the ship above the waterline. A total of 596 possible strike trajectories have been chosen, coming directly from starboard and port, from the bow and from the stern. This is shown in Figure 2.

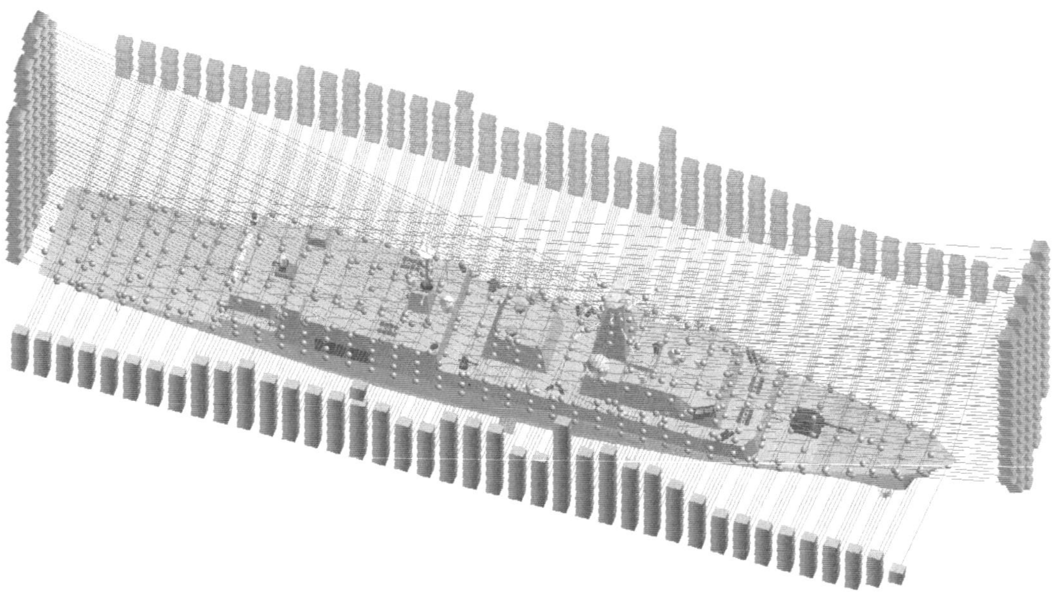

Figure 2: All possible strike locations on the fictitious frigate. Cubes represent the trajectory start location and spheres indicate where the bomb detonates. Lines are drawn to show the bomb trajectory.

From all of the possible trajectories, N are chosen at random, with replacement, where N is the number of bomb strikes to be modelled for the given scenario. This is repeated 1000 times such that there are 1000 damage scenarios that are analysed for each choice of N bomb trajectories.

3.3 WEAPON MODEL

A fictitious small guided bomb has been used to cause damage in this study, consisting of small warhead with medium weight fragments in a pattern emanating in all direction (360 degrees azimuthal and elevation). A contact fuze is used which results in external detonations.

3.4 SCENARIO CONFIGURATION DISCUSSION

During a swarm attack, it is possible that not all drones would be operating fully and errors or issues may occur. Bomb guidance system malfunctions could also occur. A lack of experience in the use of drones in a swarm may also have an effect on the number of successful bomb strikes that occur during a swarming UAV attack.

Attempted hardkill and/or softkill countermeasures will also have an effect on the number of successful bomb strikes that occurring during a swarming UAV attack.

With the above factors in mind, it is therefore difficult to estimate how many bombs might be successfully delivered onto the ship target even if the swarm size and payload per drone was known. For the purposes of this study, a range of different bomb strike numbers were analysed to account for this uncertainty. The maximum number of bombs analysed was chosen by observing the vulnerability of the ship's main capabilities as the number was increased.

The method used to determine all available strike locations that can be chosen from assumes that each one has an equal chance. In practice this is unlikely to be the case due to hotspots in the signature of the ship. It is possible that multiple strikes will be drawn towards the same location and this will be dependent on the homing mechanisms.

For example, the first weapon may be laser or TV guided to strike a specific part of the ship, then multiple infrared weapons may be used in quick succession by the swarm to further increase damage to the same area. If such a strategy is used, the effectiveness of the overall attack on the vessel may have somewhat of a dependency on the success of the initial strike(s). If multiple waves of weapons are directed towards the same area of the vessel then it is possible that damage to the structure from initial strikes opens up pathways for subsequent strikes, resulting in damage deeper in the vessel.

However, there are several other factors that might alter the strike location of each bomb despite the use of a guidance system, many of which interact, including but not limited to:

- Weather
- Sea conditions and resulting motions of the target vessel
- Target vessel's velocity
- Method and accuracy of the weapon's guidance system
- Any reactive movements of the target vessel initiated in advance of the incoming attacks
- Attempted hardkill or softkill countermeasures that alter the trajectory of the bomb or UAV but do not prevent a strike to the vessel

Many of the variables listed are hard to quantify which makes it difficult to determine the likely strike areas for any given attack. With all of these factors in mind, removing susceptibility from the equation and using a random strike location for each bomb is still a reasonable approach.

3.5 ASSUMPTIONS

When Survive models multiple strikes in a single damage scenario, it is assumed that pressure in any compartments affected by blast from a weapon returns to ambient before the next set of damage is evaluated. The modelling of multiple sources of blast in a short enough timescale to cause interaction is a more complex physics problem and is beyond the scope of Survive as a FREM.

The bomb strike locations are chosen randomly with replacement such that the same strike trajectory can be chosen more than once. When this occurs, any damage to structure from a previous strike will affect the path and fuzing of subsequent strikes.

All possible trajectories are direct strikes to the ship. No near miss bomb detonations resulting in underwater explosions have been considered.

The delivery vehicle is assumed to employ a method of loft bombing to deliver its payload, staying at a sufficient height and range to skirt the short-range air defence installed on the frigate. The guidance method and flight control surfaces on the bomb would be utilised to allow it to glide the required range while maintaining accuracy. With these factors taken into account, a shallow dive angle for the bomb of 20 degrees to the horizontal has been chosen. This can be seen in Figure 2.

No flooding has been modelled for near waterline bomb strikes, no magazine consequence reactions have been modelled and residual strength calculations have not been included in the analysis, however each of these are discussed.

4. RESULTS

4.1 NUMBER OF BOMBS SENSITIVITY

Studies were carried out on scenarios where swarming UAVs struck the target ship with 1, 2, 3, 5, 10 and 15 bombs. The vulnerability of each of the ship's main capabilities is shown in Figure 3. Minimum Propulsion is not plotted on the graph as this capability had 0% vulnerability for all analyses. The vulnerability of all capabilities except Max and Min Propulsion exceeded 90% in the 15 bomb analysis so no quantity of bomb strikes beyond this were considered.

The 15 bomb analysis is used henceforth as a focus for discussion. Worst and best case scenarios have been determined with regards to capability vulnerabilities. Results are presented for the best and worst case damage scenarios of the 1000 permutations of 15 bomb strikes analysed.

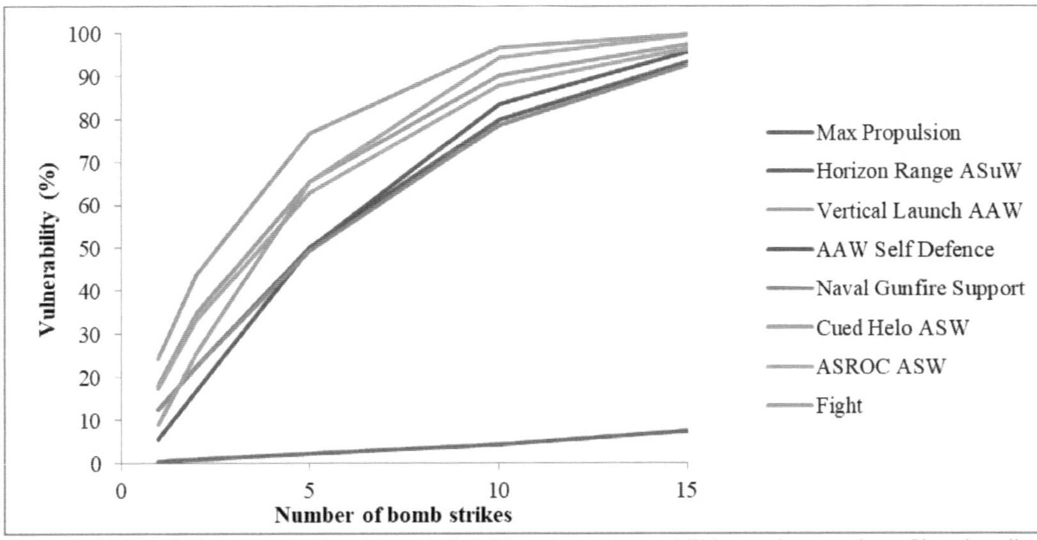

Figure 3: Vulnerability statistics for the modelled frigate's main capabilities against number of bomb strikes.

4.2 WORST CASE DAMAGE SCENARIOS

All of the worst case damage scenarios in the 15 bomb analysis resulted in 100% vulnerability for all of the fictitious frigate's main capabilities except for Max Propulsion and Minimum Propulsion. Six damage scenarios had this outcome. The location of the 15 bomb strikes and the damage caused to the outer shell from one of the worst case scenarios is shown in Figure 4. The damage caused to internal equipment from that same scenario is shown in Figure 5.

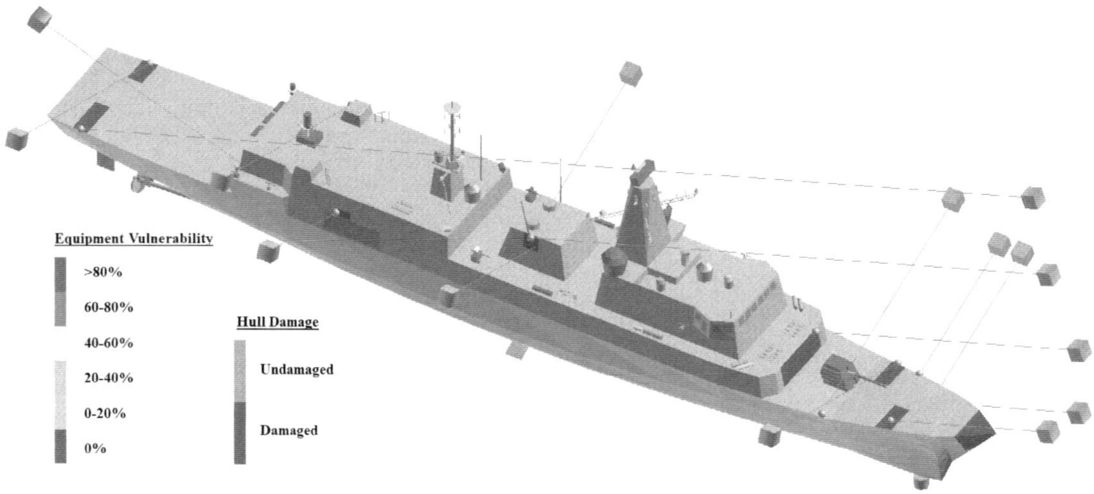

Figure 4: Bomb strike trajectories and hull damage for one of the worst case, 15 bomb scenarios.

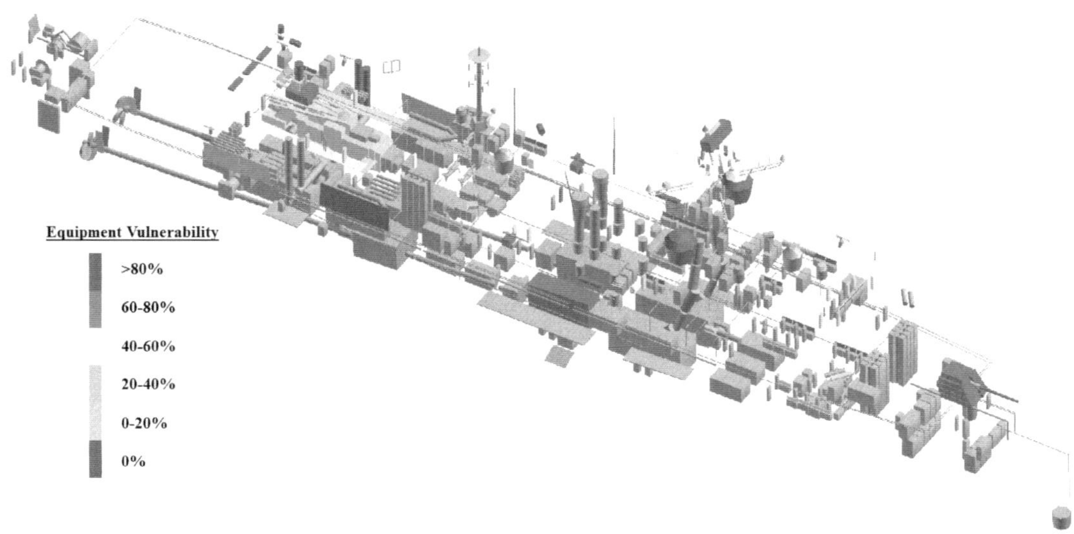

Figure 5: Internal equipment damage from one of the worst case, 15 bomb damage scenarios.

The reasons that most of the main capabilities had 100% vulnerability in this scenario are as follow:

- One diesel engine was destroyed by fragmentation making it impossible to travel at full speed
- The gun and radar of the Close-In Weapon System were destroyed by fragmentation, affecting self defence AAW
- The long range radar was destroyed by fragmentation, making Vertical Launch AAW inoperable
- The helicopter and landing aids were destroyed by fragmentation making the ASW systems inoperable
- The 5" gun was destroyed by fragmentation

Upon investigation of the other five worst case scenarios, some additional themes explaining the loss of the main capabilities include:

- One of the cables runs from the aft switchboard providing electrical supply forward passes somewhat close by the forward switchboard so enough damage in this area (2 or more bombs) defeats the redundancy of electrical supply to some of the forward Electrical Distribution cabinets. This results in loss of electrical supply to elements such as the vertical launch missile uplinks, communications room equipment and the bow sonar.

- Switchboards are placed close to the hull, so the swarm attack overcomes separation and redundancy by causing damage at the hull from multiple angles.
- The ultra/very high frequency antenna are often taken out by fragmentation resulting in loss of communications for Naval Gunfire Support
- In rare cases, bomb strikes low down on the side of the hull resulted in either diesel engine or electric motor kills

4.3 BEST CASE DAMAGE SCENARIOS

The vulnerability results of the three best case scenarios are presented in Table1. As can be seen, some capabilities are still likely to be destroyed; however, some capability is still maintained.

Table 1. Best case scenario vulnerability statistics.

Capability	Vulnerability (%)		
	1st Best Case	2nd Best Case	3rd Best Case
Min Propulsion	0	0	0
Max Propulsion	0	0	0
Horizon Range ASuW	68.8	100	100
Vertical Launch AAW	100	100	100
AAW Self Defence	100	0	7.1
Naval Gunfire Support	0	100	100
Cued Helo ASW	11.8	0	4.2
ASROC ASW	2.3	0	4.2

As can be seen in Table 1, even in the best case scenario, there is the loss of some of the ship's main capabilities. This is due to the damage done to critical external equipment such as radars and the CIWS. It is also observed from the best case scenarios that the bomb strikes were more focused on certain parts of the ship which left some capabilities with little to no damage. This can be seen in Figure 6 and 7 where most strikes are focused on the central two thirds of the ship where there is superstructure. In all cases, the ability to move was unaffected, maintaining retreat as a possibility following attack.

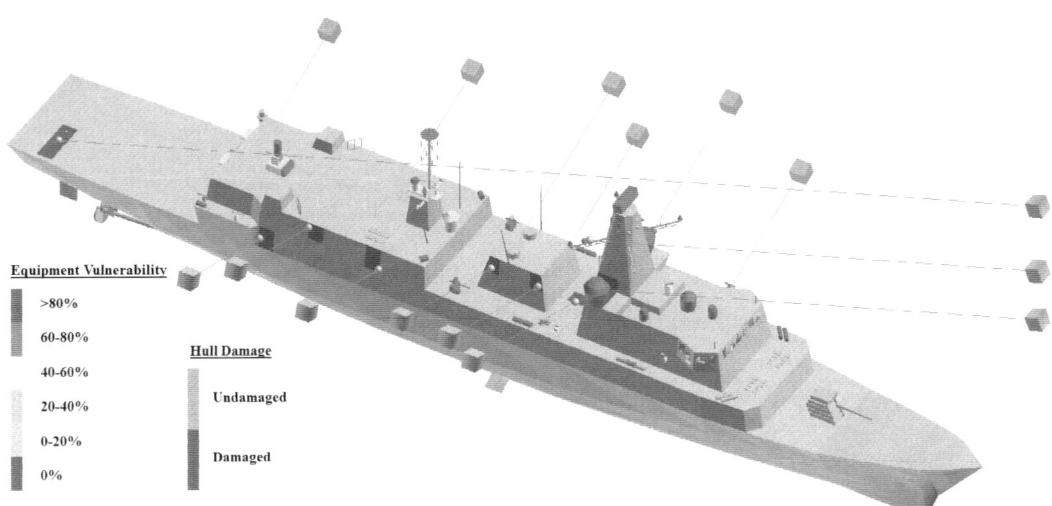

Figure 6: Bomb strike trajectories and hull damage for the best case, 15 bomb scenario.

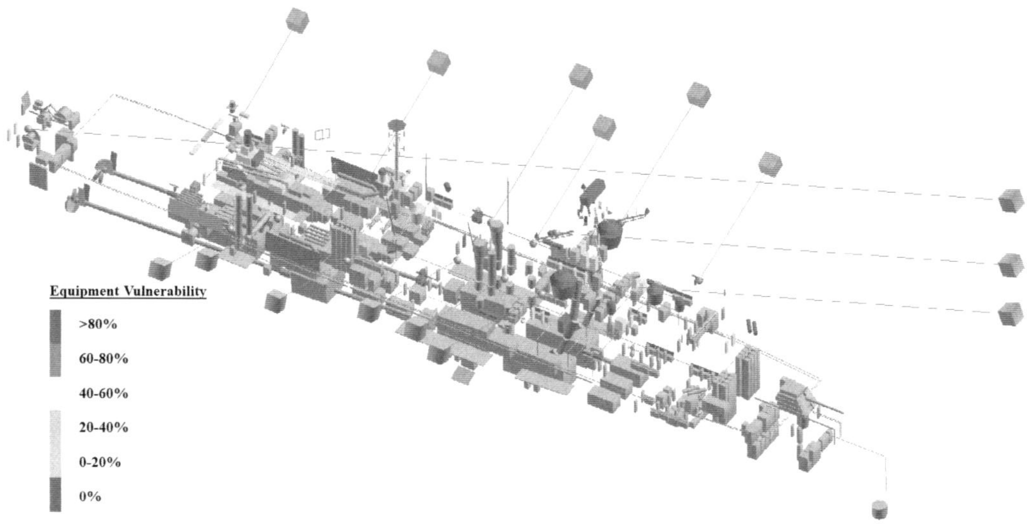

Figure 7: Internal equipment damage from the best case, 15 bomb damage scenario.

5. DISCUSSION

5.1 KEY THEMES

Based on the worst case scenarios identified, the following overarching themes have been highlighted:

- An important factor to consider is that most of the ship's main capabilities have critical elements situated externally. Due to the nature of the externally detonating warhead, any external equipment that is close enough and is in line of sight of the detonation will experience highly energetic fragments and this is likely to result in the equipment becoming disabled or destroyed. Thus, any systems with critical components situated externally above the waterline are likely to be vulnerable to the style of attack modelled.
- Redundancy and separation has reduced effectiveness when considering multiple sources of damage dispersed over the ship. This is especially the case for systems that contain less resilient equipment.
- Internal equipment that is close to the hull of the vessel above the waterline is also vulnerable to the fragmentation of the warhead since only the hull needs to be penetrated and the fragments will still have sufficient energy to cause damage.
- The propulsion systems have a much lower vulnerability than the mission capabilities due to all of equipment required for their operation being deep within the ship and thus protected by the internal bulkheads and standoff that this provides with respect to the above water bomb threat.
- Due to the likelihood of damage to the main capabilities of the fictitious frigate, damage prevention through signature management, hardkill and softkill will play an important role in maintaining capability during a swarm attack.

5.2 DAMAGE MITIGATION FROM SWARM ATTACKS

For one of the key observations made, that most of the ship's capabilities rely on critical components that are placed externally, there is no immediately obvious was of protecting all of these from widespread external fragmentation. The medium calibre gun, radars, communications antennae and CIWS are all vulnerable to this type of threat and simply having redundancy for these components might not be enough.

This issue aside, damage mitigation methods recommended to reduce the vulnerability of the fictitious frigate against the nature of swarm attack modelled are as follows:

- Redundancy and separation is still valuable despite having reduced effectiveness when facing multiple threats simultaneously. Redundant and separated elements of a system could also be armoured where appropriate. In the case of the ship modelled, this would involve ruggedising or armouring around switchboards and electrical distribution cabinets. In particular, a modern, lightweight armouring system used to catch fragments, placed around critical internal components could work well without being invasive to the overall design.

- In addition to the point above, reinforced electrical cabling would further add to the resilience of widespread fragment damage.
- The hangar could be armoured to reduce damage sustained by the helicopter when it is embarked.
- Internal, critical equipment above the waterline could be placed in a more centralised location within the hull, to increase standoff from any potential externally detonating weapon and to provide extra protection from the internal bulkheads and other equipment.

Making some of these alterations may improve the vessel's performance against an externally detonating, fragmentation warhead however this may lead to increased vulnerability from other threats. It is therefore important to consider and analyse the full threat basket to make balanced design choices.

5.3 CASUALTIES AND DCFF

The 15 bomb vulnerability analysis shows that the chance of losing at least one of the two Headquarters positions is 26% and the chance of losing at least one of three Fire Repair Party Posts (FRPPs) is 23.6%. The worst case scenario in terms of casualties and deaths from the 15 bomb analysis shows 66 of 104 crew remain uninjured and in all 15 bomb scenarios there are some casualties and deaths. There are mostly deaths with much fewer casualties across the board which are mostly attributed to fragmentation due to the nature of the weapon. A small blast component to the warhead with a contact fuze means only some crew near the hull will have been affected by blast. The fragmentation from the warhead, on the other hand, is energetic enough to penetrate the outer hull and cause fatal injuries to the crew it can reach at or near the hull.

Despite these statistics, it is likely that some Damage Control and Firefighting (DCFF) activity would still be possible following the swarm attack modelled. However, the extent of damage across several areas of the ship would be an unusual and difficult challenge with regards to the planning and carrying out of DCFF activity. If a swarm attack started multiple fires in different areas of the ship, this has the chance to overwhelm firefighting effort, leading to loss of the vessel. Fixed firefighting systems would be valuable in such a scenario.

It is still plausible that all FRPPs are destroyed and a different swarm threat than the one analysed may have more chance of doing so. In such a scenario, DCFF efforts would be extremely hampered and additional support may be required to deal with such widespread damage.

Recoverability analysis is within QinetiQ's modelling capabilities; however, this was beyond the scope of this study.

5.4 RESIDUAL STRENGTH

For this study, the residual longitudinal strength was not modelled as a result of the damage done by swarm attacks.

If multiple smaller warheads where spread along the whole length of the ship then it is unlikely that this would lead to a catastrophic loss of longitudinal strength.

If several smaller warheads were detonated in the area defined by the damage extent of an equivalent, larger warhead then it can be assumed that the smaller warheads do no more damage than the larger one. However, if the smaller warhead detonations were distributed across the span of a transverse section and caused damage to a larger number of stiffeners in that section than the single warhead then this could start to present problems to the longitudinal strength of the vessel by creating a weakened section.

5.5 SECONDARY MAGAZINE EVENTS

Secondary magazine events were not modelled as part of this study. Multiple bombs could initiate such a response, potentially with a difference in likelihood compared with single, conventional strikes and this is worth further investigation to determine appropriate armour solutions.

Magazine consequence analysis is within QinetiQ's modelling capabilities; however, this was beyond the scope of this particular study.

6. CONCLUSIONS

The modelling carried out in this study has highlighted specific changes that could be made to the fictitious frigate design to improve its survivability. In addition, general learning points for mitigating damage from a swarming UAV attack delivering multiple, externally detonating, fragmentation warheads have been identified. An understanding of the nature of such attacks has also been highlighted.

A swarming UAV attack is now an additional threat to be considered for any navy when designing next generation warships and for protecting its existing fleet. It is important to consider and analyse UAVs as part of the full threat basket that a vessel could face to make effective and balanced design choices.

The main capabilities of the fictitious frigate were most vulnerable when damage was dispersed along the full length of the ship.

Equipment close to the hull is the most vulnerable to this type of attack and this should be borne in mind when employing other vulnerability reduction measures.

Carrying out traditional damage control methods following a swarming UAV attack may be challenging due to the possibility of a large number of casualties, the widespread damage caused and the potential for several fires being started.

Traditional vulnerability reduction measures such as separation and redundancy, and concentration and protection should continue to be employed. In the case of separation and redundancy, armouring can also be employed to protect against a widespread fragmentation threat affecting a large portion of the vessel. This could take the form of modern, lightweight protection, placed around critical internal components to catch fragments. This would minimise any impact on the overall design.

One unsolved issue is that external components critical to the ship's main capabilities (such as radars, guns and antennae) are also highly vulnerable to widespread external fragmentation and it is challenging to determine a solution to protect all of them. Redundancy and separation of such equipment is less effective against this nature of swarm attack.

It follows that the likelihood of damage to the main capabilities of the fictitious frigate is quite high from a swarm attack of this nature and damage prevention through signature management and hardkill and softkill countermeasures will play an important role in maintaining capability.

7. AUTHORS BIOGRAPHY

Josh Neill holds the current position of Senior Survivability Engineer at QinetiQ Ltd. The Author is responsible for leading on survivability assessments of maritime vessels and design and testing of new algorithms for Survive as part of QinetiQ's Maritime Survivability Team. Josh is also the Product Owner for the Survive software, governing software development priorities to meet QinetiQ and customer needs. The Author's previous experience includes almost ten years of working in the maritime survivability sphere including almost daily use of the Survive software suite. Josh has lead on several MOD funded procurement and research survivability projects.

SCALING UP MARITIME AUTONOMY – LARGE UNCREWED SURFACE VESSEL (LUSV) VISION

J Rigby, BMT

SUMMARY

The naval landscape is evolving and so must our operating strategies. Whilst the concept of a Large Uncrewed Surface Vessel (LUSV) is not new, many navies debating their future fleet composition have yet to formally consider the use of these types of high endurance vessels to enhance its surface fleet. By making the vessel large you unlock greater capacity and capability as well as ensuring it can be a truly deep water vessel operating all year round in the highest sea states. In order to fuel the debate and discussion BMT has outlined a vision for an autonomous future with the potential of adding mass to an existing or future fleet. When considered within a mixed fleet approach, the LUSV demonstrates compelling technical feasibility. However, as we chart this course towards the future of naval capabilities, it will be imperative that we further explore the economic viability to ensure a balanced and sustainable naval strategy.

This paper outlines a vision for an adaptable, affordable and autonomous vessel of the future for discussion. Focusing on the six key challenges of command & control, sensor data management, low carbon engineering systems, vessel & mission resilience, maintenance & logistics and modularity; this paper outlines the potential benefits and creates a technology roadmap of the developments required to turn the vision into reality.

1. INTRODUCTION

The concept of a Large Uncrewed Surface Vessel (LUSV) is not new; LUSVs have been seen as the key component of the United States (US) Navy's ASW Continuous Trail Unmanned Surface Vessel programme since 2010. More recently, the US Navy has also looked at LUSVs to address the projected reduction in vertical launch strike missile capacity as their Ticonderoga Class cruisers are retired. The key element which makes these systems "large" is their requirement to operate on open ocean and higher sea states as well as the scale/size required to host the modular capability and support it for long durations of time. While the Royal Navy is keen to integrate and exploit uncrewed systems and is making significant progress in areas such as mine warfare and maritime air power, alongside many navies it has yet to formally consider the use of LUSVs to enhance its surface fleet.

BMT has previously conducted studies into Highly Autonomous Warship Technologies (HAWT), creating a vision of a safe and secure lean crewed warship. The vision had the aim of reducing crew numbers to a minimum by embracing onboard automation whilst still maintaining the flexibility of the vessel and evolving its use into the battlespace of the future. However, despite the potential effectiveness of lean crewed warships, they are not the full solution on their own. A future fleet of lean crewed warships enabled with high levels of automation systems could be more expensive than standard crewed ships and could lead to a loss of mass/hull numbers for a fixed fleet unit production cost. From our work in the HAWT project we identified that the Future Battlespace will be Continuous, Global, Uncertain and Collaborative. Ship numbers are needed to provide continuous presence in a large number of locations. Intensity of operations will vary between long periods of no activity to short periods of intense operation. This profile cannot be achieved using expensive and intelligent crews alone, and cannot be provided by small offboard units.

The core challenge then becomes how can we effectively provide additional mass to increase the operational capability of the fleet in a cost-effective manner. A mixed fleet solution must be sought, to allow different vessels with different roles and different capability levels to work together as a flexible protean force. Autonomy is a great enabler, but without changing the operational model the cost benefits cannot be claimed i.e. without shifting to introducing fully uncrewed vessels for parts of the fleet you cannot see the full cost savings. Fully autonomous platforms will not require hotel services, nor will they necessarily be built to a warship standard to protect the crew and loss of life. A force-mix of crewed systems operating as the mission commander alongside uncrewed protean platforms operating up-threat could transform the battlespace. This led to the creation of the LUSV vision – a simplified supporting vessel that can utilise the full benefits of autonomy to help reduce costs.

2. CONTEXT

In the 2022 Maritime Operating Concept (MoD, 2022) the Royal Navy stated their intention to pivot to a systems of systems architecture distributed protean force. This pivot is required to meet the demands of a continuous, global, uncertain and collaborative future battlespace; with a limited ability to predict where and what threats will emerge this flexibility and availability of assets is key to success.

To achieve this pivot there are three underlying challenges:

i. Ship Numbers

A common problem for many modern navies is a lack of hull numbers, the increased complexity of modern warships and hence increase costs and production times prevent large numbers being produced:

- **Lack of combat mass** – The US Navy recently confirmed that Chinese shipyards have a building capacity 232 times greater than that of the USA; the Chinese Peoples Liberation Army Navy already had 60 more 'battle force ships' than the US Navy in 2020 (Trevithick, 2023). While numerical superiority is not the only measure of combat effectiveness, the ability to be able to absorb losses is still important and clearly a concern. These additional hull numbers will be needed in the future battlespace to provide a persistent global presence operating against uncertain threats. A great example of such a rapidly evolving threat environment is seen in Ukraine's made effective use of small maritime surface drones to challenge Russian naval superiority in the Black Sea, conducting damaging strikes on capital ships and military tankers. It is threats like this that will need greater numbers of flexible protean assets to counter.
- **Reduced flexibility / Limited Surge Capacity** – Unexpected events such as non-combatant evacuation operations that require warships to be sent at short notice produce a ripple effect through the surface fleet's programme. All other armed forces, the Army and the Air Force hold vehicles in a state of readiness and peak performance in preparation for surge requirements. It is only naval forces that constantly exercise and utilise all their assets, this is mostly due to cost of the assets, but it could be argued that this is part due to the personnel requirement as even alongside personal are needed to maintain the vessel.

ii. Personnel and Training

Another common issue is availability and training of personnel. With growing pay expectations and opportunities in wider economy on the rise, it can be extremely difficult to recruit and retain the numbers of personal required to crew the existing fleets. Autonomy can break the link between fleet mass and number of trained personnel. Additionally, by providing options where crew can work from shore or have a more effective work-life balance some of these issues can be overcome. These are both solutions that could be realised through the introduction of greater autonomy on board vessels.

iii. Long Duration Tasking

The modern naval battlespace is no longer black and white between conflict and peacetime, but vessels need to battle through the sub threshold and maintain persistent engagement and maintain a continuous situational awareness picture. To support this grey space operation, vessels need to provide a long duration watchkeeping capability and provide area denial.

Can Autonomy Provide a Solution? In common with other uncrewed systems, LUSVs are ideal for 6D (Dull, Dangerous, Dirty, Dear, Deep and long Duration) mission sets and an essential component of the described navy pivot. They will use fewer personnel than those required for current / near-future warships and will free up these valuable assets for other tasking. The clear global interest in these vessels from other navies also represents a useful export opportunity that could be used to drive down unit cost.

Figure 1: The Large Uncrewed Surface Vessel Vision

These core challenges have led to creation of a LUSV vision (Figure 1) with the following key features:

- **Large** – Roughly 80m in Length - To be able to survive high sea states in the North Atlantic, provide a stable modular platform and carry the required cargo and fuel
- **Connected** - With sensor and communication separation- to reduce vulnerability from attack
- **Low Signature** – Focusing on underwater radiated noise and radar cross section
- **Low Resistance** – For Long duration missions with minimal fuel usage
- **Long Endurance** – Capable of operating for 30 days or more without physical human intervention
- **Adaptable** – Working Deck with TEU Storage – To provide a range of modular capability options.
- **Affordable** – £50m to £100m per vessel Unit Production Cost for a core navigable vessel – To provide added mass within current budgets
- **Autonomous** – Autonomy could open the opportunity for a Distributed Protean force that Persistently Engages with State and non-state actors

3. OPERATING SCENARIOS

3.1 ANTI SUBMARINE WARFARE

Anti Submarine Warfare (ASW) is a resource intensive endeavour. In order to counter a threat from entering a sensitive area, such as the Greenland Iceland UK (GIUK) Gap, a mixed fleet approach will often be used deploying a combination of submarines, Maritime Patrol Aircraft and expensive, high-performance front-line warships (UK Defence Journal, 2023). A search requires personnel to work in Defence Watches (6 hours on, 6 hours off) often in heavy seas while maintaining strict adherence to 'Quiet States' to reduce the chances of counter detection. These activations can last from several weeks to a few months. The work is tiring, repetitive and there is an increased risk of damage to equipment or injury to personnel. As the timing of these activations cannot be planned, there are adverse consequences for these front-line warships subsequent availability as maintenance, training and leave plans must be rescheduled.

There are many visions of what the future of anti-submarine warfare could look like in science fiction and doctrine. For example, in the novel Ghost Fleet (Singer and Cole, 2015), uncrewed vessels are deployed ahead of the surface vessels to hunt for and intercept underwater threats and allow the main fleet to focus on the surface/air domains. A similar principal has been envisaged for the LUSV, by operating upstream of the main fleet an mixed fleet solution can be deployed and utilise the strengths of each vessel type. The benefits of LUSVs for ASW include the possibility of persistence, at a lower cost and the significant reduction of risk to human life.

LUSVs could be deployed to cover the GIUK Gap or similar maritime area on a permanent / near-permanent basis as required by intelligence-led indicators and warnings. Equipped with towed array and sonar buoys, these vessels would be capable of proceeding to a patrol station and maintaining a pre-planned patrol pattern (or respond to remote orders) to conduct either barrier operations or search functions. They would be capable of exchanging real-time data with both shore-based control centres and other assets (crewed and uncrewed) including those from other NATO countries. In this scenario, the crewed assets need only be activated once a positive detection is made. This means other more complex assets spend less time away from other, planned, commitments and the impact on personnel is also reduced.

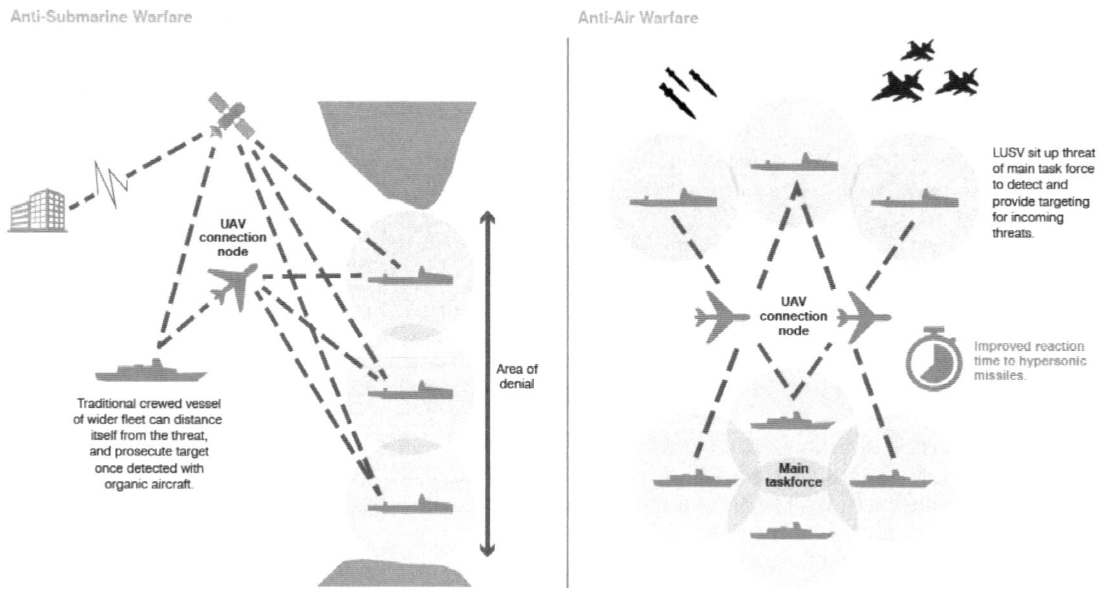

Figure 2: Operating Scenarios for the LUSV Vision

3.2 ANTI-AIR WARFARE

Just as in the previous use case the vessels are used to provide an anti-submarine warfare piquet, these lower cost vessels could also be used to provide an early warning for incoming air threats to a task force. By positing the LUSV up threat of the main task group they can provide a shield and additional detection barrier, critically increasing the time for the main force to react.

3.3 OTHER SCENARIOS

Another driving use case is the increasing threat to Critical National Infrastructure. Currently, these are primarily undersea cables carrying internet traffic (such cables carry c.$10 trillion of daily financial transactions worldwide (NY Times, 2015)) and gas and oil pipework (77% of UK Gas imports are via undersea pipelines from Norway (Commons Library, 2023)). In the future however, this will include power cables such as the Xlinks UK – Morocco Power Project (X Links, 2023) meaning that the UK's energy security is also further at threat. The Royal Navy must expand quickly to meet this threat. The modular capability of the LUSV could allow a wide range of sensors and actuators to be fitted to the vessel making it a flexible and highly capable asset.

The current vision has been predominantly designed around low resistance long endurance operations, but the collapsible bulwarks and large working deck could open up a whole range of new scenarios, currently considered operational roles could include:

- Humanitarian support
- Logistic transfer and supply
- UxV deployment
- Toop Deployment and Accommodation
- Missile Barge

It is however noted that future design iteration would be required in order to effectively incorporate these features.

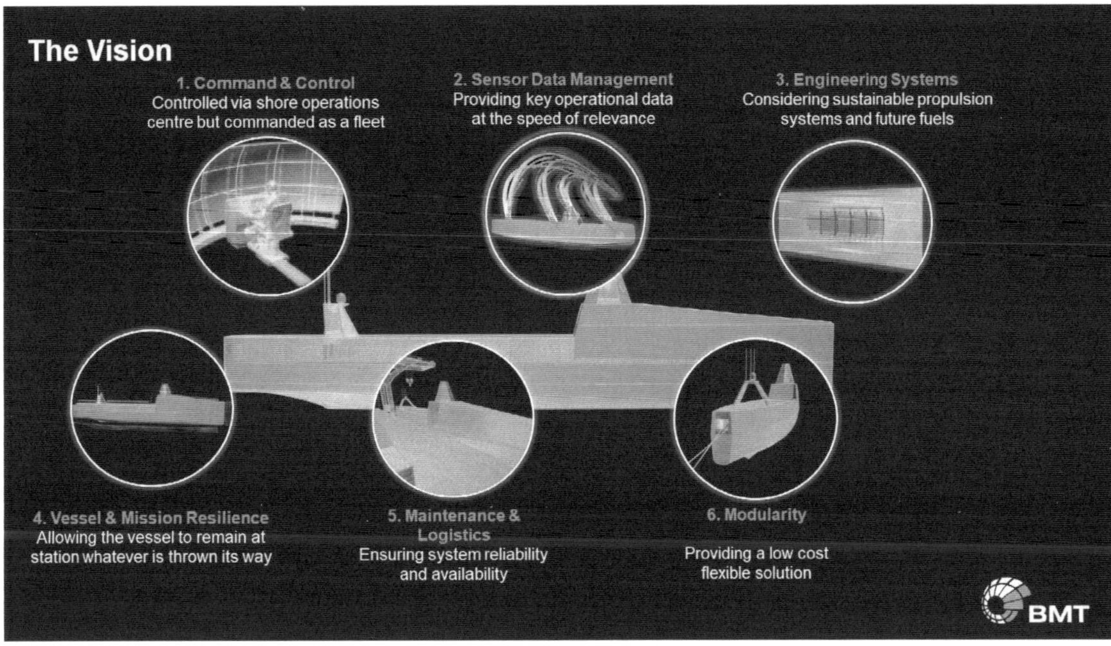

Figure 3: Six key challenges of deploying a LUSV (BMT, 2023)

4. CHALLENGES

In order for the LUSV to become a reality there are still considerable challenges to overcome. These challenges can be summarised into the following six themes:

i. Command and Control

The nature of LUSV and its operations demand high endurance, long range and robust seaworthiness. The core challenge for command and control is ensuring that the system can be reliably operated for months at a time maintaining navigational situational awareness with minimal human interaction, as well as providing a weaponised asset which adheres correctly to the rules of engagement.

ii. Sensor Data Management

Data is often stated as the world's most valuable asset (The Economist, 2017) and for the LUSV this statement is especially accurate. The clear common role of the LUSV is as a sensor package or picket capability. Due to the modular nature of the vessel a wide range of sensor data could be collected and transmitted at a range of security classifications, from simplified data logs to terabytes of complex high fidelity sonar imagery. This creates the core challenge of connectivity especially in a possible satellite denied environment.

iii. Engineering Systems

By creating a vision for the future, it is important to consider that the regulatory and operational environment of the future will likely require alternative power and energy solutions. The LUSV provides an opportunity to incorporate alternative fuel solutions, without the same safety concerns of a crewed vessel. Although removing people from the vessel allows a re-evaluation of space for machinery or fuel, the energy density of alternative fuels still causes a significant headache for Naval Architects.

iv. Vessel & Mission Resilience

The LUSV could be required to operate in some of the most challenging and dangerous environments, even as a lower cost capability the loss of any one asset would be a significant financial and operational setback. The Core Challenge is getting the right balance between keeping a modest cost versus respecting the need to minimise the risk of losing the capability when its needed.

v. Maintenance & Logistics

Possibly the greatest single challenge for completely uncrewed ships is maintenance especially when the capability is required to be at sea for long durations. All the likely LUSV scenarios require persistent (interruption free) capability over a timeframe of weeks/months. This maintenance and logistics challenge applies equally to vessel itself and its mission payload, the latter whose continuous functioning is pivotal to the mission and likely concerns complex, high-end equipment.

vi. Modularity

The use of modular systems has become a byword for the "future warship", in which it is envisaged that warships will embark capabilities according to mission or even re-configure mid deployment. As an enabler to providing increased mass, LUSV provides the opportunity of a larger hull to embrace modularity and create flexibility when needed. However, in common with other themes the use of modularity has to reflect simplicity and affordability.

These six themes outline how there is still significant work required, further papers from BMT will explore the solution space for these different areas.

5. TECHNOLOGY ROADMAP

In order for the LUSV vision to become a reality there are a few key driving technology areas that need to mature as outlined in Figure 4. There is no one single development item, but instead a stream of key development areas that come together to form a solution.

Figure 4: LUSV Technology Roadmap (BMT, 2023)

6. CONCLUSIONS

In Summary, the Large USV gives you true deployability in open ocean and enables modularity for wider roles while you are there, but the challenges of autonomy do not scale linearly with size. The LUSV is the missing link to provide high capability added mass to future blue water fleets. We have described the clear demand from current naval forces for long duration low cost added mass and discussed the feasibility and viability challenges in making this work.

Working as part of a mixed fleet solution the operational scenarios and missions of LUSVs can vary from ASW to Anti-Air Warfare to Logistic Supply. Using the flexible modular capability the embarked capability can be adapted to suit the required situation.

We understand that the LUSV vision still needs significant development especially to incorporate wider mission capabilities including organic aircraft, however it serves as a first marker in the sand to help identify the challenges and the technological solutions required.

7. ACKNOWLEDGEMENT

BMT would like to thank all those who supported the production of this report and all our customers; without their support none of this would be possible.

The authors would like to especially thank Andy Kimber, Mike Williams, Will Alexander, Aaron McMaster, Chloe Yarrien, Thomas Beard, Eshan Rajabally and Ian Savage for their contributions to the content.

8. REFERENCES

1. ANDREWS, D., 'The Multi-Role Combatant – Jack of All Trades, Master of None?, *RINA Warship 2019, The Royal Institution of Naval Architects,* 2019

2. BMT, The Large Uncrewed Surface Vessel Vision [Online]. Available at: [https://www.bmt.org/news/2023/charting-new-horizons-bmt-showcases-vision-for-lusv/] (Accessed: 20 December 2023), 2023

3. MANLEY, D., 'The NATO Drive to Mission Modularity', *Warship 2018: Procurement of Future Surface Vessel*, 11-12 September 2018, London, UK. UK Ministry of Defence, 2018

4. MINISTRY OF DEFENCE. Maritime Operating Concept: The Future Navy [online]. Available at: [https://assets.publishing.service.gov.uk/government/uploads/system/uploads/attachment_data/file/1102484/20220629-Maritime_Operating_Concept__OFFICIAL___Publication.pdf] (Accessed: 20 December 2023), 2022

5. PAWLING, R., ET AL., 'Survivability: The Human Element', *Warship 2015: Future Surface Vessels, 10-11 June 2015*, Bath, UK. The Royal Institution of Naval Architects, 2015

6. SINGER, P W., and COLE, A., 'Ghost Fleet: A novel of the next world war', *Boston: Houghton Mifflin Harcourt*, 2015

7. THE ECONOMIST. The world's most valuable resource is no longer oil, but data. [online] Available at: https://www.economist.com/leaders/2017/05/06/the-worlds-most-valuable-resource-is-no-longer-oil-but-data [Accessed 14 August 2023]. (2017)

8. TREVITHICK, J., 'Alarming Navy Intel Slide Warns Of China's '200 Times' Greater Shipbuilding Capacity", *The Drive, Available at: 1* (Accessed: 20 December 2023). (2020)

9. UK DEFENCE JOURNAL. The GIUK Gap: The chokepoint in Britain's backyard. [online] Available at: https://ukdefencejournal.org.uk/the-giuk-gap-the-chokepoint-in-britains-backyard/ [Accessed 20 December 2023]. (2023)L. T. De Luca, Propulsion physics (EDP Sciences, Les Ulis, 2009)

9. AUTHORS' BIOGRAPHIES

Jake Rigby is the Global Head of Innovation and Research at BMT, responsible for the portfolio management of internal research projects. He is a chartered engineer and Member of the Royal Institute of Naval Architects. He is also an Honorary Associate Professor at the University of Exeter.

COMBAT SAFETY - THE ROLE OF VULNERABILITY ASSESSMENT AND THE PURPLE FIRE TOOL IN MEETING THE DUTY OF CARE IN WARTIME

J S Schofield, Survivability Consulting Limited, UK, **Prof. D Manley**, University College London, UK

SUMMARY

The concept of Combat Safety addresses issues raised in UK Supreme Court rulings that there exists a legal Duty of Care to service personnel in wartime. It delivers a structured process to execute and record the survivability design process of a warship. Survivability is defined as "The capability of a weapon system to continue to carry out its designated mission(s) in a combat threat environment", and in this case it refers to the ability of the warship as a whole to protect its crew and the systems that they rely on to keep the ship fighting. Survivability is a function of susceptibility (the probability of being hit), vulnerability (the probability of being disabled when hit) and recoverability (the probability of rectifying the disability after being hit). Of the three areas, arguably vulnerability management has the greatest impact on whole-ship design, in terms of the benefits to be gained if considered at every stage of the design process. The assessment of vulnerability throughout a design is a well-established process which pre-dates the Combat Safety concept, but is given even more importance because of it. This paper explores Combat Safety and vulnerability management in detail, including the capability of the Purple Fire tool and examples of its use.

1. INTRODUCTION

Three important concepts are first introduced; the existence of a "Duty of Care", the naval interpretation of "Survivability", and "Combat Safety" which spans the two.

1.1 A DUTY OF CARE IN WARTIME

The general concept of a Duty of Care is not new, having its origins in a ruling by Lord Atkin (Donoghue v Stevenson, 1932) in which he formulated the following principle: *"You must take reasonable care to avoid acts or omissions which you can reasonably foresee would be likely to injure your neighbour. Who, then, in law is my neighbour? The answer seems to be – persons who are so closely and directly affected by my act that I ought reasonably to have them in contemplation as being so affected when I am directing my mind to the acts or omissions which are called into question"* [1].

Whether there is a Duty of Care to service personnel in wartime had been the subject of discussion over many years, but in 2013 the UK Supreme Court confirmed the existence of MOD's Duty of Care to its personnel in action under specific conditions. Supreme Court Case UKSC 41/2013 [2], Allbutt, Ellis and Smith v MOD, commonly known as the "Snatch Land Rover case", followed the deaths of service personnel in lightly armoured vehicles which were damaged by IEDs in Iraq in 2005 and 2006. In the words of the court judgement, *"Their armour was designed to provide limited protection against ballistic threats, such as those from small arms fire. It provided no protection, or no significant protection, against improvised explosive devices ("IEDs")* [2].

Figure 1: Snatch Land Rovers in Iraq (UK MOD)

An earlier case concerned Hercules XV179, shot down by insurgents in Iraq in 2005. Ground fire caused the separation of the port wing, with damage to and subsequent explosion in one of the aircraft's external fuel tanks believed to be the cause. The MOD Military Aircraft Accident Summary, from November 2005 stated that *"It was recognised that the lack of a fuel inerting system in the wing fuel tanks left the aircraft vulnerable to an explosive fuel/air mixture developing in the tanks"*. An inerting system suitable for use in the aircraft type was available and in use in other Air Forces. [3]

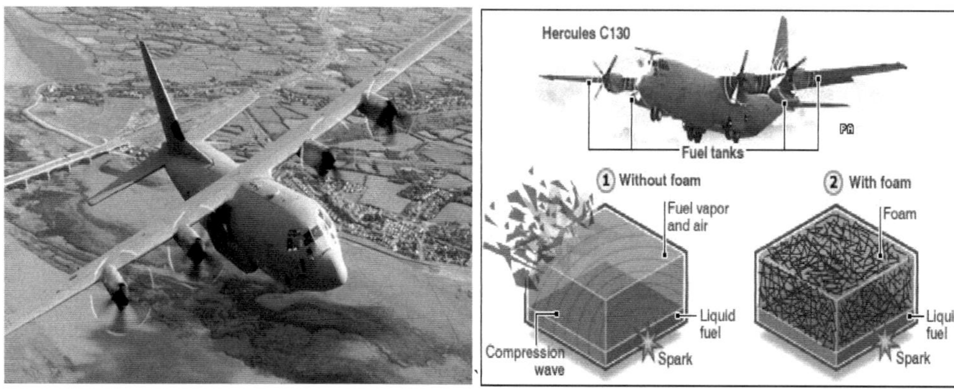

Figure 2: Hercules and the inerting principle (UK MOD)

Battle damage can, of course, lead to casualties, but if there are mitigation measures which can prevent them, is there not a legal duty to do so? Or at least to consider doing so? And if so, when? UKSC 41/2013 confirmed the need to consider the risk to personnel from enemy action when making design and deployment decisions. The judgement did not intend to impact operational aspects, rather its primary focus was on provision of suitable equipment, justification of design decisions in particular relating to protection features and ensuring awareness by the operators and those responsible for operational tasking of the capability of that equipment.

1.2 SURVIVABILITY

The Duty of Care is aimed at ensuring the provision of suitable equipment and ensuring that there is an awareness of the protection that the equipment provides, i.e. those planning operations need to be aware of the survivability aspects of the equipment. In similar timeframes to the Duty of Care being established, it was becoming evident in discussions with operators that there were gaps and misconceptions regarding the survivability and personnel protection aspects of modern warships.

Survivability, in the naval context, is *"the capability of a system to continue to carry out its designated mission(s) in a combat threat environment"* [4]. It is a function of susceptibility (the chance of being hit), vulnerability (the chance of loss of capability given a hit) and recoverability (the ability to regain lost capability after damage control and reconfiguration) [1].

- Susceptibility is concerned with being detected, targeted and hit by an incoming threat. It is commonly seen as being driven by the vessel's signatures. However, the reality is significantly more complex since, as well as signatures, it encompasses own end enemy force levels and capabilities, hard and soft kill systems on board ship and in the deployed force, rules of engagement, the environment, enemy and own crew training and a number of other aspects. The key measure of effectiveness for susceptibility is Probability of Escaping Hit (PEH).
- Vulnerability looks at the short term effects of a successful weapon engagement. Vulnerability reduction is primarily concerned with the design of the vessel, the arrangement of (and redundancy within) its systems, armour and protective measures, and system reconfiguration to restore lost capability within short timescales (up to minutes). The key measure of effectiveness for vulnerability is Probability of Escaping Critical Damage (PECD).
- Recoverability addresses the longer term spread of damage (for example, fire, smoke and flooding), damage control and firefighting and post-hit capability restoration. In extreme cases, where the vessel is not recoverable a transition is made into evacuation, escape and rescue. Hence there is a need for escape routes and equipment to themselves be effective after damage and for the vessel to be capable of acting as a safe haven for ships staff and a safe platform from which to evacuate should the need arise.

1.3 THE CONCEPT OF COMBAT SAFETY

The Snatch Land Rover ruling prompted work within Navy Command and wider MOD to determine how safety of a warship's crew exposed to the damaging effect of a hostile act could be defined and managed, throughout the platform lifecycle, and hence how MOD's duty of care could be addressed and managed. The outcome of this work was the development of the principle of Combat Safety. Now enshrined in the MOD Shipping Regulations, Combat Safety requires the Safety Case for MOD shipping to take account of safety during peacetime, tension and war. [1]

The Combat Safety principle delivers a structured process to execute and record the survivability design process of a new ship. The primary focus is on personnel protection. However, the ship as a whole contributes to that protection, including through functionality such as emergency systems to aid firefighting, damage control, lifesaving and abandonment (recoverability), propulsion to avoid staying in harm's way, along with hard and soft kill systems to avoid being hit (to reduce susceptibility). The continued availability of all of these systems is ensured by vulnerability reduction. Therefore, Combat Safety can be seen to be linked to the overall system-level survivability of a vessel. It can essentially be considered as a subset of survivability.

Combat Safety sits within the UK safety management system and is linked with the Royal Navy Survivability Strategy. Inherent within Combat Safety is the requirement to provide suitable information to operational commanders and planners on the personnel protection characteristics of their equipment so that informed decisions on the safety risks likely to be faced in combat can be made. Combat Safety management provides an auditable record of the development and maintenance of the elements of the design that impact on risk to life in combat. The close coupled nature of Combat Safety and survivability means it is sensible to manage the overall survivability of the design through a joint process, providing a complete survivability record. Ultimately, this information is conveyed through the creation and issue of a Combat Safety Summary (CSS). [1]

Combat Safety also applies for existing ships, giving a process to provide survivability information for ship's staff and operational planners, highlighting shortfalls against current Royal Navy survivability & personnel protection policy which may not have existed at the time the ship was designed.

2. UNDERSTANDING AND MANAGING COMBAT RISK

Combat Safety follows a goal based approach. The Combat Safety goal is to ensure that risk to life in action, "Combat Risk", is properly considered. The existence of that goal, enshrined in law, requires us to understand the performance of ships in wartime and the risks being taken in their use. To manage the Combat Risk through the platform lifecycle and measure that performance requires a process, threat information and assessment tools.

2.1 THE COMBAT SAFETY PROCESS

Combat Safety, alongside wider Survivability, must be considered at each phase of the lifecycle if the goal is to be reached with a rigorous set of evidence of decision-making in place as a future audit trail.

2.1 (a) Requirements

At the earliest stages it is essential that baseline levels of performance are defined. In the UK a set of requirements, the Key Threshold Requirements (KTRs), are defined based on experience of what has been achieved with survivability features on previous classes and the level of performance that these bring. Thus, the requirement starts with a set of mandated features which are known to provide the basis for a survivable design. These features can include a list of spaces which are to be protected to a given level, and are not to become part of the trade space. Beyond the mandated features, the requirement specifies a level of performance for key functions, in particular in terms PECD for vulnerability, which is to be met. Such numerical requirements cannot be measured without defining the threat (see section 2.2) which the vessel might face, and against which it is to be designed. An important consideration is which key functions will have requirements against them. It is important to avoid deciding that all capabilities which the ship might have are "key". It becomes impossible to effectively protect every capability on board, instead it is best to ensure good performance of a smaller set of functions, and then possibly at an acceptable degraded level. This allows the best possible result for those functions. By definition, this requires survivable service systems such as electrical and cooling, which will give benefits to other lower-priority functions too.

2.1 (b) Initial Design

Once design begins, there is typically a set of potential concepts, but all of these should include the mandated features (which will be dependent on the type and size of ship being designed). At all stages, a record of design decisions is essential for the audit trail. It was all too common for the reasons behind survivability design features to be lost, risking their removal later in the process. Assessment of survivability performance can and must begin as soon as a design is considered, as it is at these stages that the biggest gains can be achieved for the least effort, in particular for vulnerability. Vulnerability assessment, covered more in Section 4, should include both system-level PECD assessment and crew casualty assessments, each with consideration of improvement measures beyond the mandated features to start the process of continual monitoring and mitigation through the design.

2.1 (c) Design Options Assessments

Possibilities for mitigating Combat Risk, along with many other options for the design and capability of the platform, will become part of the cost-capability trade process, but it is essential that the documented requirements prevent the possibility of mandated features becoming part of the trade space. For a feature seen as essential for survivability, and therefore combat safety, to be removed could surely be seen as going against the Duty of Care. Even where options might be excluded from the design process based on the cost-capability trade, their performance should still be logged as they may remain potential additions to the design or future upgrades in service. Equally for options which are excluded the reasons for this must be explained and retained.

2.1 (d) Detailed Design

The assessment process continues through detailed design, with iterative checking against the survivability requirements. As the design matures it is likely that new issues will be identified through the process of detailed layout and down-selection of ship's systems, which will require identification of mitigations to keep the design compliant against the requirements. Given the design span of complex warships, the threat spectrum may also change. Mitigating new threats not part of the original requirement may not be possible without additional funding, and where issues cannot be mitigated this is again recorded.

2.1 (e) Build & Acceptance

Survivability features should be the subject of specific assurance activities, to ensure that the mitigation measures give the level of performance expected and that they are correctly installed. This will extend to testing/inspection of structures (including specific areas for blast and ballistic protection), overall systems (such as effective provision of required redundancy), equipment (such as shock trials) and equipment installation (such as the presence of the correct mounts and clearances). Documentation remains important.

2.1 (f) In Service, including Legacy Ships

The final embodiment of the initial Combat Safety process is the provision of the Combat Safety Summary to Ship's Staff and operational planners, backed up by the record of decisions taken throughout the design process, documented in a Survivability Case. Once in service though, it is critical that performance is maintained to show that the Duty of Care is being maintained. This will include checking and necessary maintenance of protection measures, assessment of any changes to the vessel which may impact upon survivability, and review of the threat which may change as weapons evolve or depending on the sorts of operation being considered for the vessel.

Here is a convenient place to discuss legacy ships. As mentioned above, the concept stage is where the biggest survivability gains are possible and it is clearly not feasible to apply the full Combat Safety process to a ship which may have been designed decades ago. However, this does not stop the owner and operator considering the Combat Risk which the crew may be exposed to, nor does it stop the performance being quantified and survivability features being documented. Thus, a Combat Safety Summary for legacy ships is perfectly possible and a good way of considering the Duty of Care. The UK has created Combat Safety Summaries for all warships, auxiliaries and many smaller boats. This process of measuring existing performance (for example in terms of functional vulnerability and casualties which may be expected against a reasonable threat set) can identify "quick wins" which may then be considered for integration and inform the level of threat which may be acceptable.

2.2 THE THREAT

The threat spectrum was mentioned above, but deserves specific consideration. It would not be reasonable to think that the crew are protected against every hostile act; there will always be the possibility of a hostile act which the crew cannot survive no matter what the design features. The Combat Safety process seeks to ensure the crew are reasonably protected against the foreseeable hostile acts, where it is practicable to do so. This implies the need to consider the threats which a vessel might reasonably be expected to face and be designed against. This could of course be different between each class of ship; an aircraft carrier might reasonably be expected to withstand larger threats than a patrol ship.

Further, different levels of threat would be expected to have different levels of impact and the requirements placed on a design should account for this. The KTR (see section 2.1 (a)) have a base set of threats designated "Design" and "Extreme" for each category of ship (a function of size and role). Against a "Design" threat a vessel would be expected to retain a level of performance, albeit to a degraded level, after being attacked. An "Extreme" may be likely to cause loss of warfighting functions and perhaps even the loss of the ship, but there is still a priority to ensure that it remains able to protect surviving crew, including moving out of harm's way and self-defence, and enable their safe abandonment if necessary.

There is a temptation to overly focus on the precise detail of threats, or include an excessive number of different threats in the requirement set. This is to be avoided. A smaller set of threats of a sufficient range of types and sizes is sufficient to identify Combat Risks and develop mitigations, which allows focus to be retained without overly burdening the process. Where vessels may require information pertinent to specific threats that they are likely to encounter in a deployment, this is provided as a mission-specific addendum to the CSS.

3. THE COMBAT SAFETY SUMMARY

The CSS (published as an Annex to the Command Safety & Environmental Summary) provides:

- The Commanding Officer (CO) knowledge of the survivability features and performance built into their platform, which may influence the approach to fighting the internal & external battles; and
- Wider command infrastructure a general understanding of the combat risks involved in operation of the platform.

A CSS is required for each class of surface ship and must be reviewed and re-issued if required with each update to the Command Safety & Environmental Summary or if there is a major change to the equipment fit or significant change to the threat which the platform might face.

The content of the CSS is defined by a specific MOD document MAP 01-474 "Combat Safety Summary Generation Guide" and is aligned with the three areas of Survivability.

3.1 CSS Contents: Susceptibility

For above water susceptibility, the CSS identifies design features present for Radar Cross Section (RCS) and infrared signatures, along with reference to the ranging (measurement) of the signatures and actions whilst might affect it. For example, RCS features might be angled sides of the hull and superstructure, reduced micro-geometry and bulwarks to screen equipment, but the performance of these might be affected by opening doors or heeling the ship relative to a threat radar. For infrared, design features might be plume coolers and special paints, but signature could be affected by prime mover operation, bridge window heaters and accumulation of soot in uptakes.

The above water susceptibility performance is then given in terms of the PEH against a variety of threats for the deployment of relevant countermeasures. For example, the PEH against homing weapons as affected by the number of threat weapons fired, various hard and soft kill anti-air warfare systems available on board, or the ability of each gun system to counter surface threats such as Fast Inshore Attack Craft and the mix of weapons which they may fire.

The content for underwater susceptibility is similar, for magnetic design features such as ship de-perming and degaussing, and acoustic design features such as mounts and low noise propellers. It also stresses the need for appropriate operation, such as mount maintenance and speed management. PEH data against underwater threats such as torpedoes is given, with an additional key result present being the safe activation depths for mines, being the minimum depth of water in which the mine will not be activated, given the signature and operation of the ship.

3.2 CSS Contents: Vulnerability

Prior to the Combat Safety process, it was becoming clear that warship designs were achieving increasing gains in vulnerability reduction, but that ship's crew did not necessarily know about the features employed in terms of their existence, reason for inclusion or capability to protect them.

Therefore, the vulnerability section first provides descriptions and location diagrams for protection measures such as:

- Blast protection in the form of specific structures designed to reduce the ability of blast to spread within the ship;
- Penetration protection in the form of armour against bullets, fragmentation or shaped charge jets.

More holistic design features for the mitigation of vulnerability are then discussed, with examples being shock protection provided for key equipment, the level of redundancy built into warfighting and lifesaving functions and coverage of any Lloyds Rules Military Notations included in the design.

Numerical vulnerability performance data is given in terms of:

- Tables of PECD for key functions against design threats;
- Keep-out distances for small arms;
- Shock Safe Damage Depths, the minimum depth of water at which the mine will activate but whose detonation will not cause damage to the ship's systems, thanks to the protection in the design;
- External blast vulnerability information.

3.3 CSS Contents: Recoverability

The recoverability section has relatively short sub-sections covering flooding limits, damaged stability, damaged structural strength and combat system recovery examples. These are designed to give Command a brief initial overview, with much more detail for incident preparation and response already in specific documents on board. The CSS is intended as a concise summary.

3. VULNERABILITY MANAGEMENT

Once we have an understanding of the risks being taken in wartime, we seek to mitigate them. Vulnerability has the greatest impact of the three aspects of survivability on whole-ship structural and systems design. Therefore, it is important that a specific vulnerability management process is in place. During the course of designing several classes of ship, an efficient and effective means of achieving low vulnerability has been established in the UK.

In order to specify that a design should be low vulnerability it is of course necessary to set numerical requirements and measure the design against those requirements, as explored in the Combat Safety process above, which includes mandated features in design. Experience from several recent classes has resulted in a particular set of features that are known to produce the basis for low vulnerability. Mandated vulnerability features can relate to:

- Strengthening of certain key bulkheads to protect other compartments or zones;
- More focussed protection of certain compartments;
- Shock mounting of key systems;
- Separation of propulsion machinery;
- Separation of power generation machinery;
- Separation of steering machinery;
- Systems design for damage tolerance, in particular zoning of key service systems such as chilled water and power distribution, including separated redundant supply routes (pipes and cables).

Measurable vulnerability requirements relate to what level of capability is expected from the design after damage from the set of threats selected for the vessel. More specifically, with what likelihood do we expect different functionality to be retained after a hit? The chance a given capability (e.g. ability to move at 5 knots) is available after a hit from a given threat (e.g. 1000lb bomb) is the PECD referred to above, which of course must be set realistically – a lower PECD requirement should be set for a 20 knot capability than say a 5 knot capability, whilst the PECD requirement against a large missile will be lower than that against a smaller missile.

Requirements are typically set for the traditional "Float", "Move" and "Fight" capabilities of a platform. The set of PECD requirements expected for a given type of ship design against a specific threat set is encoded in MOD policy.

Having started with a robust but realistic set of mandated features and numerical requirements, the vulnerability of a design must be managed from stage to stage through to build and acceptance. As each design phase introduces a greater level of detail this invariably means additional potential failure modes. The management process involves iterative checking of the design against the requirements, avoiding the introduction of vulnerabilities as the design matures and mitigation of vulnerabilities as they are identified. The UK MOD's Vulnerability Reduction Strategy [5] is a prioritized list of measures to be applied in this management process:

- Prevent catastrophic loss (magazine explosion, sinking);
- Remove single point failures;
- Concentrate remaining critical elements;
- Separate alternative sources of capability;
- Protect remaining vulnerabilities (improve build standard, use armour or shock mounting systems);
- If all else fails don't place critical items in areas most likely to be hit.

There is nothing complicated in the above common-sense steps, but they act as a simple basis to focus the analyst and designers. Sometimes critical components or locations are obvious, but with increasing warship complexity the interplay between seemingly unrelated systems becomes important. This is why it is desirable to track vulnerability during the evolution of the design and not just at the end in an audit role. This means measuring against requirements as a design aid to identify where and why vulnerability occurs to generate recommendations and test potential design changes that mitigate vulnerability.

Quantitative vulnerability assessment is therefore a key part of the design process from the earliest stages. The biggest gains are to be made by considering vulnerability in the earliest structural layouts (hence the good sense of mandated features). The later vulnerability is considered, the more limited the scope for design change and the greater the cost of the more limited changes available.

This vulnerability management approach is not new, it stems from approaches during the Type 45 design in the early 2000s and before, but Combat Safety gives a new focus and legal basis. All recent classes of surface ship (Type 26 Frigate, Tide Class tankers, Type 31 Frigate, Fleet Solid Support, RFA Proteus and RFA Stirling Castle) have had their vulnerability assessed in the Purple Fire tool.

4. PURPLE FIRE

Purple Fire is SCL's vulnerability/lethality assessment software. It performs fast-running simulations of weapons against targets, primarily in the naval domain but it can also be used for commercial maritime targets and is applicable to other domains.

It is supported under a multi-year contract to ensure that the UK MOD can direct development to give it a version of the tool continually enhanced to meet the evolving threat, in support of new designs and the in-service fleet, including to meet the requirements of the Combat Safety process.

4.1 OVERVIEW

Purple Fire has been designed to facilitate the sorts of assessment required for modern platform designs, weapon programmes and operational analysis. Notably, this includes large parameter spaces of simulations, which as we will see is necessary for supporting the Combat Safety process.

It consists of three parts:

- Data manipulation in a User Interface (UI) based on a spreadsheet-style layout (Figure 3, right) which allows the user visibility and control over many objects simultaneously. The style means that many familiar operations (e.g fill down, cell colouring and copy/paste) can be undertaken and there is easy data exchange to/from Microsoft Office tools and other simulation codes;

- 3D visualisation of targets and results within the UI (Figure 3, left);
- A numerical simulation engine calculating damage and system vulnerabilities for the required parameter space, controlled by the UI.

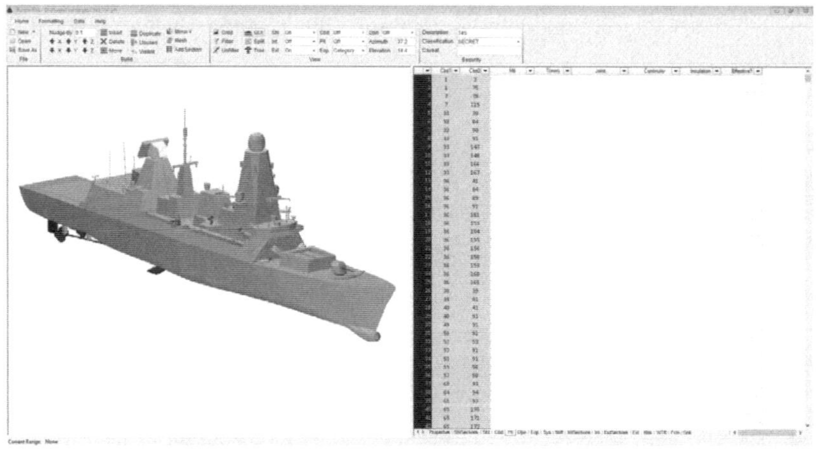

Figure 3: Purple Fire User Interface (data redacted)

The primary entities in the physical model of the ship (Figure 4) are, as might be expected, an external surface to represent the hull and superstructure, internal subdivision of decks and bulkheads along with the openings (doors, hatches, soft patches, etc.) within them, and the equipment which allow the ship to do its job.

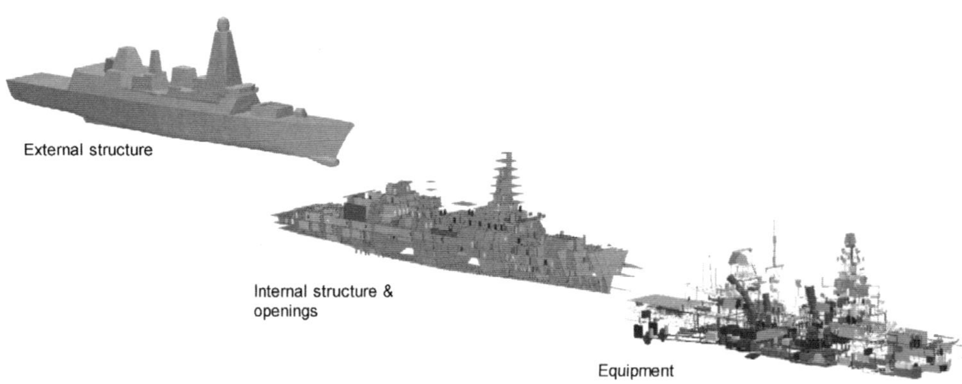

Figure 4: Main Purple Fire target model entities

The physical model is augmented by an integrated functional model to represent each capability of the ship (e.g. Anti-Air Warfare), which are represented by combinations of equipment in the form of fault trees shown as intuitive, user-friendly block diagrams (for example Figure 5, intentionally blurred). Each block represents individual equipment items or sub-systems. Simply put, in a given damage case, where a route can be traced from the top to the bottom without passing through a damaged equipment, the function survives. The actual analyses are more complex, taking a probabilistic approach, with Monte Carlo simulation used to find equipment vulnerability.

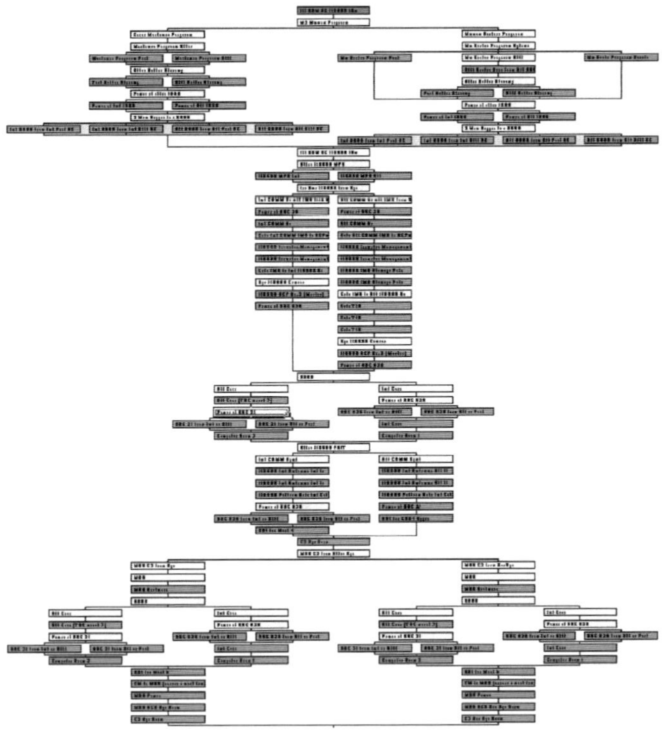

Figure 5: Purple Fire fault tree example (intentionally blurred)

A typical single vulnerability assessment proceeds as per Figure 6, with the input representations of weapon and target being combined in a defined scenario (which will define aspects such as threat direction). The assessment tool processes the primary damage (such as penetration of the threat, detonation of its warhead and subsequent blast and fragmentation spread), the secondary damage (such as flooding and fire) and potentially the reaction to damage (evolving, in the worst case, to the simulation of abandonment). Multiple threats may be combined to calculate cumulative damage, and importantly any vulnerability tool will reset and repeat to establish the potential vulnerability at all potential threat locations. Numerical results allow comparison against requirements, whilst visual results allow the analyst to identify sources of vulnerability and potential mitigations.

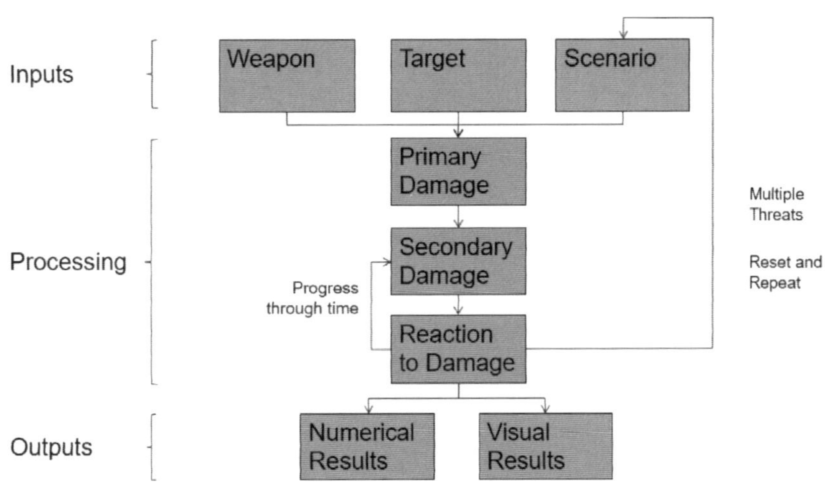

Figure 6: Simple vulnerability assessment process

This single, simple, process does not cover the whole story of any analysis however, which will always require a parameter space of multiple weapons, targets or scenarios to be considered. Until the development of Purple

Fire this could be an intensive and manual process, but Purple Fire offers a single sheet interface for populating a parameter space, where a single line of data can define a "batch" of thousands of scenarios, such as combinations of target, weapon, fusing, azimuth etc. There is no need for external pre-processors or user input to generate a separate file per scenario. This saves substantial time and improves the ability to check parameter spaces, drastically reducing user error and the need for re-runs. Another single sheet interface then automatically runs the parameter space, deploying the batch across a user-specified number of cores, which can be changed as simulations progress. Results of interest are extracted automatically from each scenario as the batch progresses. Finally, automatic analysis tables can be generated, such as showing variation of a given system vulnerability with azimuth/elevation for a given weapon. This is the process shown in Figure 7.

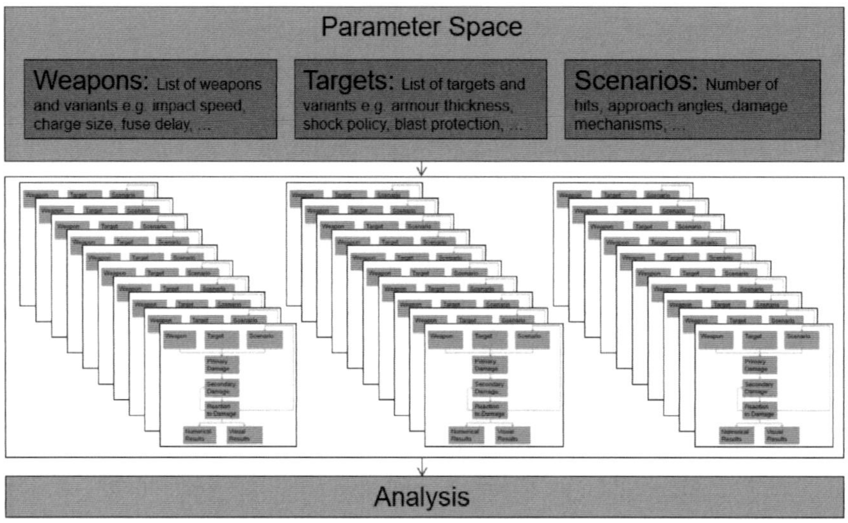

Figure 7: Purple Fire automatic parameterisation, simulation and analysis

4.2 INPUT TO COMBAT SAFETY

SCL maintains a configuration controlled repository of models for all vessels in the Royal Navy and Royal Fleet Auxiliary fleets and the full set of threat weapons defined for the Combat Safety process across all sizes and roles. One of the major challenges of maintaining such a set of models is the availability of data on the fleet which is provably up-to-date. To show it is meeting the requirements of the Combat Safety process it is incumbent on MOD to ensure that such data is available and well-controlled.

Thanks to the efficiency of Purple Fire and availability of the library it is possible to rapidly create a single, consistent set of vulnerability data across the fleet, as part of a periodic "Fleet Vulnerability Review", which is directly used for input to Combat Safety Summaries in the form of PECD tables. Associated analysis allows comparisons to be made across the fleet and between different ages of vessel to inform future policy.

A Vulnerability/Lethality Ready Reckoner spreadsheet tool has been developed to allow the interrogation of previously-produced parameter spaces of Purple Fire results. Selected results from what may be gigabytes of output files can be viewed in a readily accessible form, without the need for an analyst familiar with Purple Fire. Therefore, the ready reckoner is suitable for distribution to analysis study customers/stakeholders and operational planners, including those who have to consider the Duty of Care in their decision-making.

The ready reckoner is configuration controlled by SCL and can "make" a lookup file for its own use by extracting data from a parameter space of runs. The tool itself therefore contains no data, so it is not classified and does not require additional configuration control by the different projects on which it may be used in future.

The parameter spaces of simulations are characterised in terms of:

- A set of platforms;
- A set of weapons divided into various types;
- A set of azimuth and dive angles for each weapon against each platform (note that the angles are not relevant for all weapons, e.g. underwater mines);

- A set of systems representing the main capabilities of each platform.

The types of result shown, as in Figure 8 are:

- Vulnerability/Lethality (probability of kill) of capabilities of interest for each platform/weapon/angle combination. These are in terms of numerical results on average and for various selectable aim point methods (see Figure 9 for further detail), and visual results at each potential hit point (a so-called "damage footprint");
- Casualties for each platform/weapon/angle combination (in the same way as for Vulnerability/Lethality).

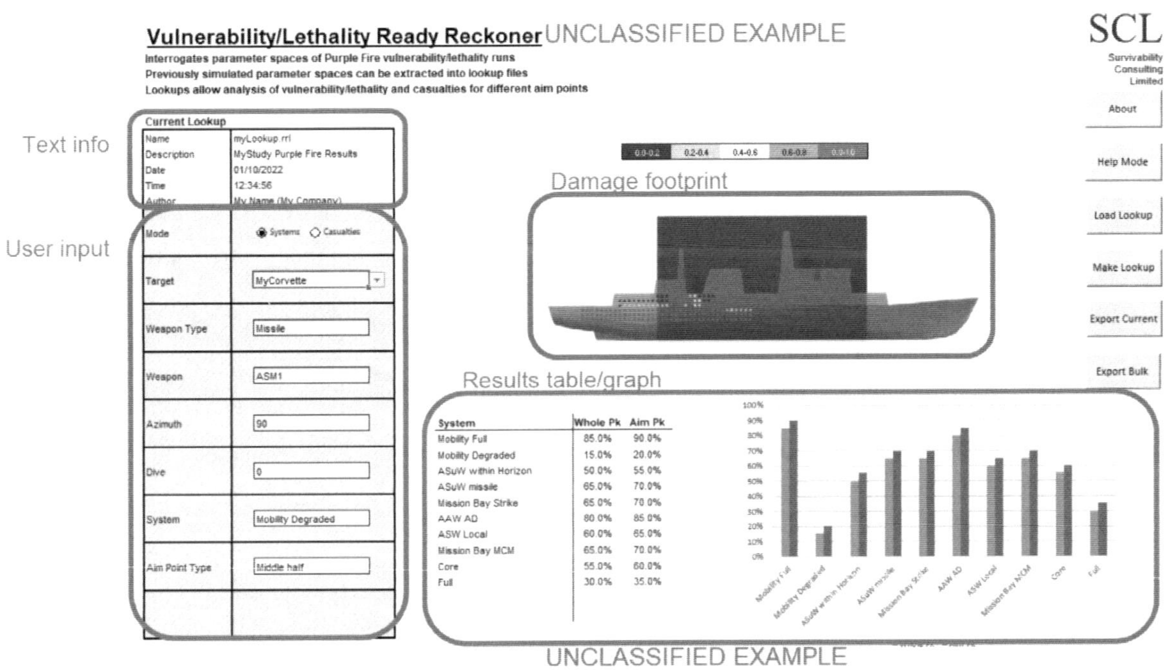

Figure 8: Vulnerability/Lethality Ready Reckoner with "Middle Half" aim point

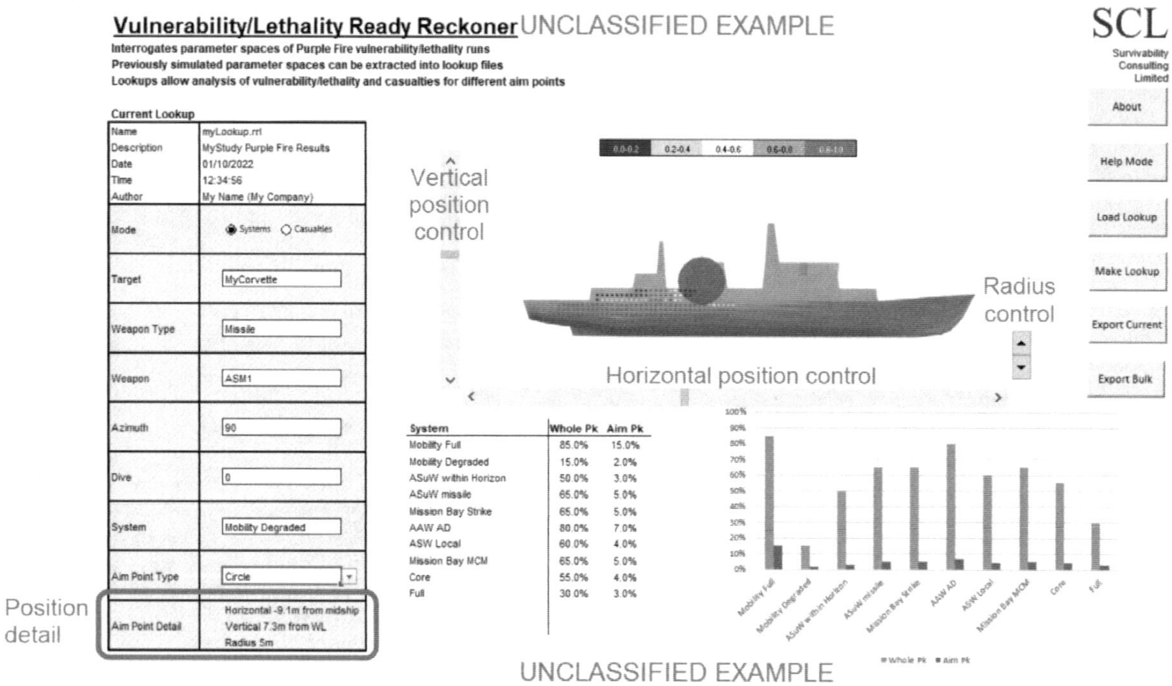

Figure 9: Vulnerability/Lethality Ready Reckoner circular aim point detail

4.3 OTHER TYPES OF ASSESSMENT

In addition to the standard types of vulnerability/lethality assessment which feed directly into vessel/weapon design analyses and the Combat Safety Summaries, various more complex Purple Fire assessments are possible, including:

- Tandem weapons – where the detonation of the main warhead occurs a given time or distance after the precursor;
- Salvoes/Swarms – where cumulative results after a given number of hits, accounting for attrition of incoming threats (such as may be caused by hard or soft kill systems) if desired, are calculated with statistical convergence on average results after that number of hits;
- Number of hits-to-kill – to find the number of hits of a given weapon needed to achieve a desired probability of kill against a given function, useful for weapons effect planning;
- Beaten zones – to represent small arms vulnerability/lethality where the firer attempts to inundate a small region of a ship with repeated fire;
- Intercept debris – in order to assess the potential damage to a ship after a successful hard kill intercept results in debris from the inbound threat weapon, even for long range intercepts;
- Magazines – in terms of the risk to stowed munitions and the consequences of their reaction, which is automatically handled to consider the potential waves of subsequent reaction which may occur after an initial accidental or hostile stimulus;
- Crew movements – using a link to the maritimeEXODUS tool from the University of Greenwich, primarily useful for abandonment assessments but also for examining internal layout optimisation for aspects such as watch changeovers and recoverability activities;
- Dynamic stability – using a link to the FREDYN tool from MARIN to assess the flooding of a Purple Fire ship in a sea state;
- Criticality – a threat-independent assessment of a target model's systems fault trees to determine the level of redundancy and quality of that redundancy, by establishing at equipment, compartment, flat and zone level what is critical, redundant or dual-redundant.

5. CONCLUSIONS

The concept of Combat Safety addresses issues raised in UK Supreme Court rulings about the Duty of Care in wartime. It delivers a structured process to execute and record the survivability design process of a ship, whose primary focus is on personnel protection, but which covers wider survivability due to the intrinsic relationship of the ship and its capabilities to protect the crew.

The end stage of the Combat Safety process is the existence of Combat Safety Summaries, delivered to the Ship's Staff and planners, which help inform them of the Combat Risk which may be inherent in the operation of the vessel in given environments. Combat Safety Summaries must be produced for each ship class, must be up to date, and are summaries of key things established in the wider Combat Safety process. More detail on decisions made through the life of a design and into service relating to Combat Safety are documented in a more substantial Survivability Case.

In the naval domain Survivability is considered as a combination of Susceptibility (the chance of being hit), Vulnerability (the chance that hit leads to loss of functionality) and Recoverability (the chance of regaining functionality). Arguably vulnerability is the biggest influence on whole-ship design. A comprehensive process can be put in place for vulnerability management, the key part of which is iterative numerical assessment against an appropriate range of threats. This has been possible for some time, but Combat Safety gives increased onus and rigour to the process.

The Purple Fire software tool is now covered by a multi-year contract from MOD to support, maintain and develop the capability and its substantial library of vessel and threat models. It offers an efficient and affordable method to track the vulnerability of a new vessel from the earliest concept stage through to service, establish the performance of legacy platforms and perform a range of other assessments related to design and operational analysis.

6. REFERENCES

1. MANLEY, D., 'Combat Safety and Survivability. Combining survivability and safety techniques to address crew safety in combat', *INEC 2018 Conference Proceedings*, October 2018.

2. UK Supreme Court, UKSC 41/2013 Judgment, Smith and others, Ellis, Allbutt and others v MOD, 19 June 2013.

3. Hercules MOD Military Aircraft Accident Summary, November 2005.

4. NATO, ANEP43 Ship Combat Survivability 2nd Edition, September 2002.

5. DE&S, MAP 01-105 The Vulnerability Reduction Compendium, Issue 2, March 2015.

9. AUTHOR BIOGRAPHIES

James Schofield holds the current position of Managing Director at Survivability Consulting Limited and is the lead developer of Purple Fire. His 23 years of experience in naval vulnerability and lethality includes vulnerability assessments of a number of developing designs and the in-service fleet, research into damage physics, fast-running algorithm development and engagement with MOD on policy and requirements definition. James is a Fellow of the Institute of Mathematics and its Applications and a Member of the Royal Institution of Naval Architects.

David Manley holds the current position of Professor of Naval Architecture at University College London lecturing on warship design, acquisition, and naval operations. His research interests include the use of wargaming for concept design analysis, seabed security and future maritime survivability, David is also the UK MOD's Head of Specialism for platform survivability and the DE&S / SDA Senior Specialist Fellow for maritime platform vulnerability reduction. David is a member of the Royal Corps of Naval Constructors and a Senior Fellow of the Royal Institution of Naval Architects.

ENHANCING FUTURE SURFACE COMBATANTS THROUGH ADVANCED SHIP DESIGN DIGITAL TWINS

L Wong, SIEMENS DI SW, Australia, **Dr. R Perez Fernandez**, SIEMENS DI SW, Spain, **M Shipley**, SIEMENS DI SW, UK

SUMMARY

This paper explores the transformative potential of ship design digital twins in optimizing the performance, resilience, and operational capabilities of future surface combatants. Digital twins, advanced simulations mirroring real-time vessel behaviour, are poised to revolutionize naval engineering by providing unparalleled insights into the intricate dynamics of maritime operations.

By integrating cutting-edge technologies such as artificial intelligence, data analytics, and sensor networks, these digital twins offer a comprehensive understanding of a ship's structural integrity, propulsion systems, and onboard functionalities. This paper elucidates how leveraging ship design digital twins can empower naval architects and operators to enhance decision-making processes, streamline maintenance procedures, and bolster overall mission effectiveness. The study underscores the imperative role of digital twins in fostering innovation and adaptability within the naval domain, heralding a new era for future surface combatants.

NOMENCLATURE

AI	Artificial Intelligence
CAD	Computer-Aided Design
CFD	Computational Fluid Dynamics
FEA	Finite Element Analysis
USV	Unmanned Surface Vessels

1. INTRODUCTION

1.1 OVERVIEW OF THE SIGNIFICANCE OF SHIP DESIGN DIGITAL TWINS IN NAVAL ENGINEERING

Naval engineering stands at the intersection of tradition and innovation, where centuries-old maritime principles meet cutting-edge technologies. In this context, ship design digital twins emerge as a revolutionary tool with profound significance for the field of naval engineering. These digital replicas of physical vessels, operating in real-time, offer a paradigm shift in how naval architects conceptualize, design, and operate surface combatants. Below are the crucial key points highlighting the significance of ship design digital twins in naval engineering:

- **Optimizing Design Processes**: Traditional ship design involves iterative cycles of modeling, testing, and refinement. However, this approach is time-consuming and costly. Ship design digital twins enable engineers to simulate and evaluate various design configurations virtually, significantly reducing the time and resources required for prototyping. This optimization of the design process leads to more efficient and effective surface combatants.
- **Enhancing Performance Prediction**: Understanding a vessel's performance characteristics is crucial for mission success and operational efficiency. Ship design digital twins allow engineers to simulate a wide range of operating conditions, including different sea states, speeds, and maneuvering scenarios. By accurately predicting performance metrics such as speed, fuel consumption, and stability, digital twins enable naval architects to fine-tune designs for optimal performance in diverse environments.
- **Improving Structural Integrity and Safety**: The structural integrity of a vessel is paramount for its safety and longevity. Ship design digital twins provide detailed insights into how various factors, such as load distribution, material properties, and environmental conditions, affect the structural behavior of a ship. By simulating stress, strain, and fatigue, engineers can identify potential weak points and design robust structures that withstand harsh operational conditions, enhancing overall safety and reliability.
- **Facilitating Lifecycle Management**: A ship´s lifecycle encompasses design, construction, operation, maintenance, and decommissioning. Ship design digital twins serve as a central repository of information throughout the vessel's lifecycle, facilitating seamless data exchange and collaboration among different stakeholders. From initial concept development to ongoing maintenance and upgrades, digital twins provide a holistic view of the vessel's history, enabling more efficient lifecycle management strategies.

- **Enabling Real-time Monitoring and Diagnostics**: In the dynamic environment of maritime operations, real-time monitoring and diagnostics are essential for maintaining operational readiness and responding to emergent situations. Ship design digital twins integrate sensor data from various onboard systems, enabling continuous monitoring of performance parameters and early detection of anomalies. By leveraging advanced analytics and predictive algorithms, engineers can identify potential issues before they escalate, enabling proactive maintenance and minimizing downtime.
- **Supporting Training and Simulation**: Ship design digital twins serve as invaluable training and simulation platforms for naval personnel. Crew members can familiarize themselves with vessel systems and procedures in a realistic virtual environment, improving readiness and proficiency. Additionally, digital twins facilitate scenario-based training exercises, allowing crews to practice response strategies for various operational scenarios, from routine maneuvers to complex combat situations.

Ship design digital twins represent a paradigm shift in naval engineering, offering unprecedented capabilities for designing, analyzing, and operating surface combatants. By leveraging advanced simulation, data analytics, and connectivity technologies, digital twins empower naval architects to optimize design processes, enhance performance prediction, improve structural integrity and safety, streamline lifecycle management, enable real-time monitoring and diagnostics, and support training and simulation activities. As the naval industry embraces digital transformation, ship design digital twins are poised to play a central role in shaping the future of naval engineering and maritime operations.

1.2 BRIEF EXPLANATION OF DIGITAL TWINS AND THEIR POTENTIAL APPLICATIONS IN SURFACE COMBATANT OPTIMIZATION

A digital twin is a virtual representation of a physical object or system that operates in real-time, capturing its behavior, performance, and characteristics through data. In the context of naval engineering and surface combatants, a ship design digital twin is a sophisticated computer model that mirrors the attributes and dynamics of a physical vessel. This virtual replica integrates data from various sources, including design specifications, sensor readings, and operational parameters, to simulate the behavior of the actual ship in different scenarios.

The potential applications of ship design digital twins in surface combatant optimization are manifold:

- **Design and Prototyping**: Digital twins enable naval architects to create virtual prototypes of surface combatants, allowing for rapid iteration and evaluation of design concepts. By simulating different configurations and performance scenarios, engineers can optimize vessel designs for efficiency, stability, and mission effectiveness before physical construction begins.
- **Performance Prediction**: Digital twins provide insights into a ship's performance characteristics under various operating conditions, such as different sea states, speeds, and maneuvering maneuvers. By accurately predicting performance metrics such as speed, fuel consumption, and maneuverability, engineers can optimize propulsion systems, hull shapes, and operational strategies to maximize efficiency and effectiveness.
- **Structural Analysis and Safety**: Ship design digital twins facilitate detailed structural analysis, simulating stress, strain, and fatigue on various components and materials. By assessing the structural integrity of the vessel under different loading conditions, engineers can identify potential weak points and design robust structures that ensure safety and resilience in challenging environments.
- **Maintenance and Lifecycle Management**: Digital twins serve as comprehensive repositories of vessel data throughout its lifecycle, enabling proactive maintenance and lifecycle management strategies. By monitoring equipment health, predicting component failures, and optimizing maintenance schedules, digital twins help minimize downtime, extend operational lifespan, and reduce lifecycle costs.
- **Real-time Monitoring and Decision Support**: Digital twins integrate sensor data from onboard systems, enabling real-time monitoring of performance parameters and operational status. By providing continuous insights into vessel health, condition, and environment, digital twins support data-driven decision-making, situational awareness, and response planning in dynamic operational scenarios.
- **Training and Simulation**: Ship design digital twins serve as training and simulation platforms for naval personnel, allowing crews to familiarize themselves with vessel systems and procedures in a realistic virtual environment. By simulating diverse operational scenarios, from routine maneuvers to complex combat situations, digital twins enable crews to practice response strategies, enhance readiness, and improve operational proficiency.

Ship design digital twins offer a versatile and powerful toolset for optimizing surface combatants throughout their lifecycle. By leveraging advanced simulation, data analytics, and connectivity technologies, digital twins empower naval engineers and operators to enhance design efficiency, performance prediction, structural integrity, maintenance management, decision support, and training effectiveness, ultimately ensuring the effectiveness and resilience of future surface combatants in diverse maritime operations.

2. UNDERSTANDING SHIP DESIGN DIGITAL TWINS

2.1 DEFINITION AND CHARACTERISTICS OF SHIP DESIGN DIGITAL TWINS

Ship design digital twins are virtual representations of physical vessels that operate in real-time, providing a detailed and dynamic simulation of the ship's behavior, performance, and characteristics. These digital twins leverage advanced modeling, simulation, and data integration techniques to replicate the intricate dynamics of maritime operations, enabling engineers and operators to analyze, optimize, and manage surface combatants throughout their lifecycle.

Ship design digital twins are sophisticated computer models that replicate the structure, systems, and behavior of a physical vessel in a virtual environment. These digital replicas integrate data from various sources, including design specifications, sensor readings, historical performance data, and operational parameters, to simulate the ship's operation and response to different environmental conditions, mission requirements, and operational scenarios. Ship design digital twins provide a holistic and dynamic representation of the vessel, enabling naval engineers and operators to assess performance, predict behavior, optimize design, and support decision-making throughout the ship's lifecycle.

- **Dynamic Simulation**: Ship design digital twins operate in real-time, continuously updating and adjusting their simulation based on inputs from various sources, including sensor data, environmental conditions, and user interactions. This dynamic simulation capability allows digital twins to accurately reflect the evolving behavior and performance of the physical vessel in different operational scenarios.
- **Multi-disciplinary Integration**: Ship design digital twins integrate data and models from multiple disciplines, including naval architecture, marine engineering, systems engineering, propulsion, hydrodynamics, structural analysis, electronics, and control systems. By combining diverse datasets and models, digital twins provide a comprehensive understanding of the complex interactions and dependencies within the ship's systems and subsystems.
- **High-Fidelity Representation**: Ship design digital twins aim to replicate the physical vessel with high fidelity, capturing detailed geometric, kinematic, dynamic, and behavioral attributes. This high-fidelity representation enables engineers and operators to analyze and assess various aspects of the vessel's design, performance, and behavior with accuracy and precision.
- **Predictive Analytics**: Ship design digital twins incorporate advanced analytics and simulation algorithms to predict the ship's performance, behavior, and response to different operational conditions. By simulating various scenarios and assessing potential outcomes, digital twins enable stakeholders to anticipate challenges, optimize strategies, and make informed decisions to enhance vessel performance and mission effectiveness.
- **Lifecycle Support**: Ship design digital twins support the entire lifecycle of the vessel, from concept development and design optimization to construction, operation, maintenance, and decommissioning. By serving as a central repository of vessel data and knowledge, digital twins facilitate collaboration, data exchange, and decision-making across different stages of the ship's lifecycle, enabling efficient and effective lifecycle management strategies.
- **Interoperability and Connectivity**: Ship design digital twins are designed to be interoperable with other digital systems, tools, and platforms used in naval engineering and maritime operations. This interoperability enables seamless integration of data, models, and analyses from various sources, allowing stakeholders to leverage existing infrastructure and capabilities while maximizing the value of digital twin technology.

Ship design digital twins are advanced virtual models that replicate the structure, systems, and behavior of physical vessels, providing a dynamic and comprehensive simulation of maritime operations. With their dynamic simulation capability, multi-disciplinary integration, high-fidelity representation, predictive analytics, lifecycle support, and interoperability, ship design digital twins empower naval engineers and operators to optimize vessel design, performance, and operation throughout the ship's lifecycle, ultimately enhancing mission effectiveness, operational efficiency, and maritime security.

2.2 EXPLANATION OF HOW DIGITAL TWINS SIMULATE REAL-TIME VESSEL BEHAVIOR

Digital twins simulate real-time vessel behavior through the integration of advanced modeling, simulation, and data analytics techniques. The process involves creating a virtual replica of the physical vessel and continuously updating and adjusting this replica based on inputs from various sources, including sensor data, environmental conditions, and user interactions.

- **Geometric Representation**: The first step in creating a digital twin of a vessel is to generate a geometric representation of its physical structure. This typically involves using *Computer-Aided Design* (**CAD**) software to create a detailed 3D model of the ship, including its hull, superstructure, compartments, systems, and components. The geometric model serves as the foundation for the digital twin, providing a visual representation of the vessel's physical characteristics.
- **Physical Dynamics Simulation**: Once the geometric model is created, engineers use physics-based simulation techniques to simulate the dynamic behavior of the vessel in real-time. This involves applying principles of mechanics, hydrodynamics, aerodynamics, and other relevant disciplines to model the motion, forces, and interactions that occur within the vessel and its surrounding environment. Advanced simulation algorithms calculate the vessel's response to external forces such as waves, wind, currents, and operational inputs such as propulsion, steering, and control.
- **Sensor Data Integration**: Digital twins incorporate data from onboard sensors and external sources to enhance the fidelity and accuracy of the simulation. Sensors installed on the physical vessel measure various parameters such as position, velocity, acceleration, orientation, temperature, pressure, and load. This sensor data is transmitted to the digital twin in real-time, allowing the virtual replica to mirror the physical vessel's state and behavior as closely as possible. Additionally, environmental data such as weather conditions, sea state, and navigational information may be integrated into the simulation to simulate realistic operating conditions.
- **Feedback Control Systems**: Digital twins may incorporate feedback control systems to regulate and adjust the vessel's behavior in response to changing conditions or user commands. These control systems use sensor data and simulation outputs to compute control actions that optimize performance, stability, safety, and efficiency. For example, automatic steering systems may adjust rudder angles to maintain course, while propulsion control systems may adjust engine thrust to achieve desired speeds or fuel efficiency.
- **Machine Learning and Predictive Analytics**: Advanced digital twins may employ machine learning and predictive analytics algorithms to enhance the accuracy and predictive capabilities of the simulation. Machine learning techniques can analyze historical data, identify patterns, and learn from past experiences to improve simulation accuracy and anticipate future behaviors. Predictive analytics algorithms can forecast potential outcomes, assess risks, and optimize strategies based on real-time data and simulation results.
- **Real-Time Visualization and Interaction**: Digital twins provide real-time visualization and interaction interfaces that allow users to monitor the simulation, analyze data, and interact with the virtual vessel. Visualization tools display the vessel's geometric model, simulation outputs, sensor data, and environmental conditions in a user-friendly format, enabling engineers and operators to gain insights into vessel behavior and performance. Interaction interfaces allow users to input commands, adjust parameters, and conduct what-if analyses to explore different scenarios and optimize vessel operation.

Digital twins simulate real-time vessel behavior by integrating advanced modeling, simulation, and data analytics techniques to create a virtual replica of the physical vessel. By continuously updating and adjusting the simulation based on sensor data, environmental conditions, and user interactions, digital twins provide a dynamic and comprehensive representation of vessel behavior, enabling engineers and operators to analyze, optimize, and manage vessel performance in real-time.

2.3 DISCUSSION ON THE INTEGRATION OF ADVANCED TECHNOLOGIES LIKE AI, DATA ANALYTICS, AND SENSOR NETWORKS

The integration of advanced technologies such as *Artificial Intelligence* (**AI**), data analytics, and sensor networks plays a pivotal role in enhancing the capabilities and effectiveness of ship design digital twins. Below is a discussion on how these technologies are integrated and their impact on surface combatant optimization:

- How the AI technology is used:
  - AI algorithms are utilized within ship design digital twins to improve decision-making processes, optimize operational strategies, and enhance predictive capabilities.
  - Machine learning algorithms analyze vast amounts of data collected from sensors, historical performance records, and simulation outputs to identify patterns, trends, and correlations that human operators may overlook.
  - AI-driven predictive analytics enable digital twins to anticipate potential failures, optimize maintenance schedules, and predict vessel behavior under different operating conditions with greater accuracy and reliability.

  - AI-based control systems can automate certain tasks and processes, such as route planning, propulsion control, and fault detection, to improve efficiency, safety, and autonomy in surface combatants.
- Data Analytics:
  - Data analytics techniques are applied within ship design digital twins to extract actionable insights from large and complex datasets, facilitating data-driven decision-making and performance optimization.
  - Descriptive analytics techniques are used to summarize and visualize data, providing engineers and operators with a comprehensive understanding of vessel behavior, performance trends, and operational patterns.
  - Predictive analytics techniques enable digital twins to forecast future outcomes, identify potential risks, and optimize strategies based on historical data, simulation results, and real-time sensor inputs.
  - Prescriptive analytics techniques recommend optimal courses of action to achieve specific objectives, such as minimizing fuel consumption, maximizing mission effectiveness, or optimizing maintenance schedules, based on data-driven models and simulations.
- Sensor Networks:
  - Sensor networks are deployed onboard surface combatants to collect real-time data on various operational parameters, environmental conditions, and system states.
  - Sensors measure physical quantities such as position, velocity, acceleration, temperature, pressure, vibration, and load, providing insights into vessel behavior, performance, and health.
  - Sensor data is transmitted to the ship's digital twin in real-time, where it is integrated into the simulation to reflect the vessel's current state and environment accurately.
  - Sensor networks enable proactive monitoring, predictive maintenance, and condition-based management of onboard systems and components, improving reliability, efficiency, and safety.

Integration of these advanced technologies within ship design digital twins enables naval engineers and operators to achieve the following benefits:
- Enhanced decision-making processes through data-driven insights and predictive capabilities.
- Improved operational efficiency, reliability, and safety through automation and optimization.
- Increased situational awareness and mission effectiveness through real-time monitoring and analysis.
- Proactive maintenance and lifecycle management strategies to minimize downtime and lifecycle costs.
- Adaptability and resilience to changing operational conditions and mission requirements.

The integration of advanced technologies like AI, data analytics, and sensor networks within ship design digital twins enhances surface combatant optimization by enabling data-driven decision-making, predictive analysis, automation, and optimization across various aspects of vessel design, operation, and maintenance.

3. OPTIMIZING PERFORMANCE AND RESILIENCE

3.1 UTILIZING DIGITAL TWINS TO ASSESS AND ENHANCE STRUCTURAL INTEGRITY

Utilizing digital twins to assess and enhance structural integrity represents a significant advancement in naval engineering, offering unparalleled capabilities for ensuring the safety, reliability, and resilience of surface combatants. Digital twins enable naval engineers to simulate and analyze the structural behavior of vessels in real-time, providing insights into potential weaknesses, vulnerabilities, and optimization opportunities. Below are key strategies for leveraging digital twins to assess and enhance structural integrity:
- **Virtual Stress Testing**: Digital twins enable engineers to subject virtual vessels to simulated stress tests under various operating conditions, including different sea states, loads, and maneuvers. By analyzing stress, strain, and deformation patterns, engineers can identify potential areas of concern and assess the structural performance of critical components such as hulls, bulkheads, decks, and superstructures.
- **Fatigue Analysis**: Digital twins facilitate fatigue analysis by simulating the cumulative effects of cyclic loading on vessel structures over time. By modeling operational scenarios and environmental conditions, engineers can predict fatigue-related failures and assess the remaining service life of structural components. This information enables proactive maintenance strategies and design optimizations to mitigate fatigue-related risks.
- **Damage Assessment and Repair Planning**: In the event of damage or structural degradation, digital twins provide valuable insights for assessing the extent of damage and planning repair strategies. By simulating damage scenarios and evaluating the structural response, engineers can identify optimal repair methods, assess repair feasibility, and estimate repair costs and timelines. This proactive approach minimizes downtime and ensures the rapid restoration of vessel operability.

- **Material Selection and Optimization**: Digital twins support material selection and optimization by simulating the behavior of different materials under varying conditions. Engineers can evaluate the performance characteristics of materials such as strength, stiffness, corrosion resistance, and weight, and identify optimal material combinations for specific structural applications. This iterative process enables the design of lightweight, durable, and cost-effective structures that meet performance requirements while minimizing material usage.
- **Real-time Monitoring and Health Management**: Digital twins integrate sensor data from onboard monitoring systems to provide real-time insights into structural health and performance. By continuously monitoring structural parameters such as strain, vibration, temperature, and corrosion, digital twins enable early detection of anomalies and degradation, allowing for proactive maintenance interventions to prevent catastrophic failures and extend structural lifespan.
- **Optimization through Iterative Design**: Digital twins facilitate iterative design optimization by enabling engineers to rapidly evaluate and refine structural configurations. By simulating alternative design concepts and assessing their performance against predefined criteria, engineers can identify optimal design solutions that balance structural integrity, performance, and operational requirements. This iterative approach accelerates the design process and ensures that structural designs meet or exceed safety and performance standards.

Utilizing digital twins to assess and enhance structural integrity offers a proactive and data-driven approach to ensuring the safety, reliability, and performance of surface combatants. By leveraging advanced simulation, analysis, and monitoring capabilities, digital twins empower naval engineers to identify potential risks, optimize designs, and implement proactive maintenance strategies, ultimately enhancing the structural resilience and operational readiness of naval vessels in diverse operating environments.

3.2 ENHANCING PROPULSION SYSTEMS EFFICIENCY THROUGH SIMULATION AND ANALYSIS

Digital twins offer a transformative approach to assessing and enhancing the structural integrity of surface combatants by providing real-time insights, predictive analytics, and simulation capabilities.

- **Real-time Monitoring**: Digital twins integrate sensor data from onboard monitoring systems to provide real-time insights into the structural health of the vessel. By continuously monitoring parameters such as stress, strain, vibration, and temperature, digital twins detect anomalies and deviations from normal operating conditions, enabling early identification of potential structural issues.
- **Simulation and Analysis**: Digital twins simulate the structural behavior of the vessel under different operating conditions, including sea states, loads, and maneuvers. Advanced *Finite Element Analysis* (**FEA**) and *Computational Fluid Dynamics* (**CFD**) techniques enable engineers to assess the structural response to various environmental factors and operational scenarios, identifying potential areas of concern and optimizing structural designs for strength, stiffness, and durability.
- **Predictive Maintenance**: Digital twins leverage historical performance data and predictive analytics algorithms to forecast potential structural issues and anticipate maintenance requirements. By analyzing trends, patterns, and degradation mechanisms, digital twins enable proactive maintenance strategies that prevent structural failures, minimize downtime, and extend the operational lifespan of surface combatants.
- **Optimization of Material Selection and Design**: Digital twins facilitate virtual testing and optimization of material properties, configurations, and designs to enhance structural integrity. By simulating different material options, structural layouts, and construction techniques, engineers can identify the most suitable solutions for maximizing strength, reducing weight, and minimizing vulnerability to fatigue, corrosion, and other degradation mechanisms.
- **Risk Assessment and Mitigation**: Digital twins enable engineers to conduct risk assessments and evaluate the potential consequences of structural failures. By simulating failure scenarios and analyzing their impact on vessel performance and safety, digital twins help prioritize mitigation measures and design modifications to reduce risk and enhance resilience against unforeseen events.
- **Lifecycle Management**: Digital twins support comprehensive lifecycle management strategies for maintaining structural integrity throughout the vessel's lifespan. By serving as a central repository of structural data and knowledge, digital twins facilitate data-driven decision-making, documentation, and communication across different stages of the vessel's lifecycle, ensuring continuity, consistency, and accountability in structural management practices.

By integrating real-time monitoring, simulation, predictive analytics, and lifecycle management capabilities, digital twins empower naval engineers to optimize structural designs, mitigate risks, and extend the operational lifespan of surface combatants in diverse maritime environments.

3.3 IMPROVING OPERATIONAL RESILIENCE BY IDENTIFYING VULNERABILITIES AND MITIGATING RISKS

Improving operational resilience through the identification of vulnerabilities and the mitigation of risks is a critical aspect of naval operations. Digital twins offer advanced capabilities to enhance operational resilience by providing real-time insights, predictive analytics, and simulation capabilities.

- **Risk Identification and Assessment**: Digital twins integrate data from various sources, including sensor networks, historical performance data, and environmental conditions, to identify potential vulnerabilities and risks. By analyzing trends, patterns, and anomalies in operational data, digital twins enable naval operators to identify weak points, failure modes, and critical dependencies within the vessel's systems and subsystems.
- **Scenario Analysis and Simulation**: Digital twins simulate a wide range of operational scenarios, including normal operations, emergencies, and combat situations, to assess the impact of potential risks on mission success and operational effectiveness. By conducting scenario-based analysis and simulation, digital twins enable naval operators to evaluate different response strategies, contingency plans, and risk mitigation measures, ensuring preparedness and resilience against unforeseen events.
- **Dynamic Risk Management**: Digital twins provide real-time monitoring and predictive analytics capabilities to dynamically assess and manage operational risks. By continuously analyzing sensor data, environmental conditions, and performance metrics, digital twins detect changes in risk levels and trigger automated responses, such as alarm notifications, system adjustments, or mission reconfiguration, to mitigate potential threats and ensure mission continuity.
- **Optimization of Resilience Strategies**: Digital twins facilitate the optimization of resilience strategies by evaluating trade-offs between risk, cost, and mission objectives. By simulating different risk mitigation measures, such as redundancy, diversification, hardening, and adaptive control, digital twins enable naval operators to identify the most effective strategies for enhancing operational resilience while optimizing resource allocation and minimizing impact on mission performance.
- **Training and Preparedness**: Digital twins serve as valuable training and preparedness platforms for naval personnel, allowing them to practice response strategies and emergency procedures in realistic virtual environments. By simulating realistic scenarios and stress-testing operational systems, digital twins enhance crew readiness, decision-making skills, and coordination abilities, ensuring effective response to emergencies and maintaining operational resilience in high-pressure situations.
- **Continuous Improvement and Learning**: Digital twins support a culture of continuous improvement and learning by providing feedback mechanisms and performance metrics for evaluating risk mitigation strategies and operational effectiveness. By analyzing simulation results, lessons learned, and operational feedback, digital twins enable naval organizations to identify areas for improvement, refine response strategies, and adapt resilience measures to evolving threats and operational requirements.

Improving operational resilience through the identification of vulnerabilities and the mitigation of risks is essential for ensuring mission success and maritime security. Digital twins offer advanced capabilities to enhance resilience by providing real-time insights, predictive analytics, and simulation capabilities for identifying vulnerabilities, assessing risks, optimizing response strategies, and enhancing preparedness across various operational scenarios. By leveraging the transformative potential of digital twins, naval organizations can enhance their ability to withstand and adapt to challenges, ensuring operational continuity and mission effectiveness in dynamic and uncertain maritime environments.

4. STREAMLINING MAINTENANCE PROCEDURES

4.1 PREDICTIVE MAINTENANCE ENABLED BY DIGITAL TWINS

Predictive maintenance enabled by digital twins revolutionizes how surface combatants are maintained, ensuring optimal operational readiness, reducing downtime, and minimizing maintenance costs.

- **Real-time Monitoring**: Digital twins continuously integrate data from onboard sensors, monitoring systems, and historical performance data to provide real-time insights into the health and condition of critical components and systems. By monitoring parameters such as temperature, vibration, pressure, and fluid levels, digital twins detect early signs of degradation, anomalies, and impending failures.
- **Condition-based Maintenance**: Based on real-time data and predictive analytics algorithms, digital twins enable condition-based maintenance strategies. Instead of adhering to fixed maintenance schedules or waiting for components to fail, condition-based maintenance focuses on monitoring the actual condition of equipment and performing maintenance only when necessary. Digital twins trigger maintenance alerts

or notifications when deviations from normal operating conditions are detected, enabling proactive interventions to prevent failures and extend the operational lifespan of components.

- **Predictive Analytics**: Digital twins leverage predictive analytics algorithms to forecast potential failures and anticipate maintenance requirements based on historical performance data, trends, and degradation patterns. By analyzing data from similar operating conditions, equipment usage, and environmental factors, digital twins predict the remaining useful life of components and systems, enabling naval operators to plan maintenance activities, allocate resources, and optimize maintenance schedules for maximum efficiency and effectiveness.
- **Failure Prediction and Diagnostics**: Digital twins simulate the behavior of components and systems under different operating conditions, enabling engineers to identify potential failure modes and diagnose underlying issues. By analyzing simulation outputs, sensor data, and performance metrics, digital twins identify trends, patterns, and anomalies indicative of impending failures, enabling early intervention and corrective actions to prevent costly downtime and mitigate risks.
- **Optimized Spare Parts Management**: Digital twins support optimized spare parts management by providing insights into the usage, performance, and availability of spare parts and inventory. By analyzing historical maintenance data and simulation results, digital twins enable naval operators to anticipate spare parts requirements, optimize inventory levels, and ensure timely availability of critical components, minimizing downtime and disruption to operations.
- **Lifecycle Cost Reduction**: By enabling proactive maintenance strategies and optimizing maintenance activities, digital twins help reduce lifecycle costs associated with maintenance, repair, and replacement of components and systems. By minimizing unplanned downtime, reducing unnecessary maintenance tasks, and extending the operational lifespan of equipment, digital twins enable naval organizations to achieve significant cost savings over the lifecycle of surface combatants.

Predictive maintenance enabled by digital twins revolutionizes maintenance practices for surface combatants, enabling proactive interventions, optimized resource allocation, and cost-effective operations. By leveraging real-time monitoring, predictive analytics, failure prediction, and optimized spare parts management capabilities, digital twins enhance operational readiness, reliability, and efficiency, ensuring surface combatants are always mission-ready and capable of meeting operational requirements in dynamic and challenging maritime environments.

4.2 REAL-TIME MONITORING AND DIAGNOSTICS OF ONBOARD SYSTEMS

Real-time monitoring and diagnostics of onboard systems are essential aspects of ensuring the reliability, safety, and performance of surface combatants. Digital twins provide advanced capabilities for real-time monitoring and diagnostics, enabling naval operators to detect anomalies, assess system health, and identify potential issues before they escalate.

- **Sensor Integration**: Digital twins integrate data from onboard sensors installed throughout the surface combatant, including machinery, propulsion systems, electrical systems, environmental controls, and mission-critical equipment. These sensors measure various parameters such as temperature, pressure, vibration, fluid levels, power consumption, and operational status, providing real-time insights into the health and condition of onboard systems.
- **Data Fusion and Aggregation**: Digital twins aggregate and process sensor data from multiple sources, integrating information from diverse systems and subsystems into a unified data model. By fusing data from different sensors and systems, digital twins provide a comprehensive view of the surface combatant's operational status, enabling operators to monitor overall system health and performance in real-time.
- **Anomaly Detection and Alarm Generation**: Digital twins analyze sensor data using advanced analytics algorithms to detect anomalies, deviations from normal operating conditions, and potential issues. When abnormalities are detected, digital twins generate alarms, alerts, or notifications to alert operators to the presence of a problem, enabling timely intervention and corrective actions to prevent failures and minimize downtime.
- **Predictive Analytics and Failure Prediction**: Digital twins leverage predictive analytics techniques to forecast potential failures and anticipate maintenance requirements based on historical performance data and degradation patterns. By analyzing trends, patterns, and anomalies in sensor data, digital twins predict the remaining useful life of components and systems, enabling operators to plan maintenance activities, allocate resources, and optimize maintenance schedules proactively.
- **Remote Monitoring and Control**: Digital twins support remote monitoring and control capabilities, enabling operators to access real-time sensor data, diagnostic information, and operational controls from anywhere, at any time. Through user-friendly interfaces and dashboards, operators can monitor onboard

systems, analyze performance trends, and initiate commands or adjustments to optimize system operation and response to changing conditions.

- **Condition-based Maintenance**: Based on real-time sensor data and diagnostic information, digital twins enable condition-based maintenance strategies, where maintenance activities are performed only when necessary, based on the actual condition of equipment and systems. By prioritizing maintenance tasks according to the health and performance of onboard systems, digital twins optimize resource allocation, minimize unnecessary maintenance activities, and reduce maintenance costs while maximizing system reliability and availability.
- **Integration with Decision Support Systems**: Digital twins integrate with decision support systems, enabling operators to make informed decisions based on real-time monitoring and diagnostic information. By combining sensor data with operational parameters, environmental conditions, and mission objectives, digital twins provide actionable insights to support decision-making processes, enhance situational awareness, and optimize operational strategies for mission success.

Real-time monitoring and diagnostics of onboard systems enabled by digital twins provide naval operators with critical insights into the health, performance, and condition of surface combatants. By integrating sensor data, analytics, and decision support capabilities, digital twins empower operators to detect anomalies, predict failures, optimize maintenance activities, and ensure operational readiness, reliability, and safety in dynamic and challenging maritime environments.

4.3 PROACTIVE IDENTIFICATION OF POTENTIAL ISSUES TO PREVENT COSTLY DOWNTIME

Proactive identification of potential issues to prevent costly downtime is a crucial aspect of maintaining the operational readiness and efficiency of surface combatants. Digital twins provide advanced capabilities for proactive identification by continuously monitoring onboard systems, analyzing performance data, and predicting potential failures before they occur.

- **Continuous Monitoring**: Digital twins integrate data from onboard sensors, monitoring systems, and historical performance data to provide real-time insights into the health and condition of critical components and systems. By continuously monitoring parameters such as temperature, pressure, vibration, fluid levels, and operational status, digital twins detect deviations from normal operating conditions and identify potential issues as they arise.
- **Anomaly Detection**: Using advanced analytics algorithms, digital twins analyze sensor data to detect anomalies and deviations from expected behavior. By comparing current data with historical patterns and performance benchmarks, digital twins identify abnormal trends, patterns, or behaviors indicative of potential issues or impending failures, enabling proactive intervention before problems escalate.
- **Predictive Analytics**: Digital twins leverage predictive analytics techniques to forecast potential failures and anticipate maintenance requirements based on historical performance data, degradation patterns, and operational conditions. By analyzing trends, correlations, and patterns in sensor data, digital twins predict the remaining useful life of components and systems, enabling operators to plan maintenance activities, allocate resources, and schedule downtime proactively to prevent costly failures and minimize disruption to operations.
- **Early Warning Alerts**: When anomalies or potential issues are detected, digital twins generate early warning alerts, notifications, or alarms to alert operators to the presence of a problem. These alerts provide timely information and actionable insights to enable operators to investigate, diagnose, and address issues before they escalate, minimizing the risk of unplanned downtime and mitigating the impact on mission performance.
- **Root Cause Analysis**: Digital twins facilitate root cause analysis by providing detailed insights into the underlying factors contributing to potential issues or failures. By analyzing sensor data, system behavior, and operational parameters, digital twins enable operators to identify the root causes of problems, assess their impact on overall system performance, and implement corrective actions to address underlying issues and prevent recurrence in the future.
- **Optimized Maintenance Planning**: Based on predictive analytics and early warning alerts, digital twins enable operators to optimize maintenance planning and scheduling. By prioritizing maintenance activities according to the severity and urgency of potential issues, digital twins help operators allocate resources effectively, minimize downtime, and maximize system availability while ensuring mission readiness and operational efficiency.
- **Continuous Improvement**: Digital twins support a culture of continuous improvement by providing feedback mechanisms and performance metrics for evaluating proactive identification strategies. By analyzing the effectiveness of early warning alerts, predictive analytics, and maintenance interventions,

digital twins enable operators to refine their proactive identification approaches, optimize response strategies, and enhance operational resilience over time.

Proactive identification of potential issues to prevent costly downtime is essential for maintaining the operational readiness and efficiency of surface combatants. Digital twins enable operators to continuously monitor onboard systems, detect anomalies, predict failures, and implement proactive interventions to address issues before they escalate. By leveraging real-time monitoring, predictive analytics, early warning alerts, and optimized maintenance planning capabilities, digital twins empower operators to minimize downtime, reduce maintenance costs, and ensure mission readiness and operational effectiveness in dynamic and challenging maritime environments.

5. EMPOWERING DECISION-MAKING PROCESSES

5.1 UTILIZING DATA-DRIVEN INSIGHTS FROM DIGITAL TWINS FOR INFORMED DECISION-MAKING

Utilizing data-driven insights from digital twins for informed decision-making is a powerful capability that enhances operational efficiency, effectiveness, and resilience in surface combatants. Digital twins provide real-time monitoring, predictive analytics, and simulation capabilities that enable naval operators to make informed decisions based on comprehensive data analysis.

- **Real-time Monitoring and Situational Awareness**: Digital twins integrate data from onboard sensors, monitoring systems, and external sources to provide real-time insights into the operational status, performance, and environment of surface combatants. By continuously monitoring parameters such as position, speed, heading, temperature, pressure, and sensor readings, digital twins enhance situational awareness and enable operators to make informed decisions based on current conditions and evolving threats.
- **Predictive Analytics and Forecasting**: Digital twins leverage predictive analytics techniques to forecast potential outcomes, assess risks, and anticipate future events based on historical data, trends, and simulation models. By analyzing sensor data, performance metrics, and environmental conditions, digital twins predict system behavior, operational performance, and mission outcomes, enabling operators to proactively plan and optimize strategies for mission success and operational effectiveness.
- **Scenario Analysis and Simulation**: Digital twins facilitate scenario analysis and simulation of different operational scenarios, including routine operations, emergencies, and combat situations. By simulating various scenarios and assessing their potential impact on mission objectives, digital twins enable operators to evaluate different response strategies, contingency plans, and courses of action, ensuring readiness and resilience in dynamic and uncertain maritime environments.
- **Optimized Resource Allocation**: Based on data-driven insights from digital twins, operators can optimize resource allocation and decision-making processes to maximize efficiency and effectiveness. By analyzing performance data, operational constraints, and mission priorities, digital twins help operators allocate resources, personnel, and assets effectively, ensuring the optimal use of available resources while minimizing risks and maximizing mission success.
- **Risk Assessment and Mitigation**: Digital twins support risk assessment and mitigation by identifying potential risks, assessing their likelihood and impact, and recommending risk mitigation measures based on data-driven insights. By analyzing sensor data, historical performance data, and simulation results, digital twins enable operators to prioritize risks, implement preventive measures, and develop contingency plans to minimize the impact of adverse events on mission objectives and operational effectiveness.
- **Continuous Improvement and Learning**: Digital twins facilitate continuous improvement and learning by providing feedback mechanisms and performance metrics for evaluating decision-making processes and outcomes. By analyzing the effectiveness of decisions, identifying areas for improvement, and incorporating lessons learned into future operations, digital twins enable operators to enhance decision-making capabilities, adapt to changing conditions, and improve operational performance over time.

Utilizing data-driven insights from digital twins for informed decision-making enhances the operational effectiveness, efficiency, and resilience of surface combatants. By leveraging real-time monitoring, predictive analytics, scenario analysis, and optimized resource allocation capabilities, digital twins enable operators to make informed decisions, optimize strategies, and ensure mission success in dynamic and challenging maritime environments.

5.2 ENHANCING SITUATIONAL AWARENESS AND MISSION PLANNING CAPABILITIES

Enhancing situational awareness and mission planning capabilities is crucial for ensuring the success of surface combatant operations. Digital twins offer advanced capabilities to improve situational awareness and streamline mission planning processes through real-time monitoring, predictive analytics, and simulation.

- **Real-time Monitoring**: Digital twins integrate data from onboard sensors, surveillance systems, and external sources to provide real-time insights into the operational environment. By continuously monitoring parameters such as position, speed, heading, weather conditions, and sensor readings, digital twins enhance situational awareness by providing operators with up-to-date information on the surrounding maritime domain.
- **Geospatial Visualization**: Digital twins enable geospatial visualization of the operational environment, allowing operators to visualize the location, movement, and status of friendly and enemy forces, as well as other relevant objects such as navigational hazards, maritime traffic, and environmental features. By overlaying real-time sensor data on digital maps and charts, digital twins provide operators with a comprehensive understanding of the operational landscape, enhancing situational awareness and decision-making.
- **Threat Assessment and Analysis**: Digital twins leverage predictive analytics algorithms to assess threats, identify vulnerabilities, and analyze potential risks based on real-time data and historical trends. By analyzing sensor data, intelligence reports, and simulation results, digital twins help operators identify potential threats, assess their likelihood and impact, and develop strategies to mitigate risks and ensure mission success.
- **Scenario Simulation and Analysis**: Digital twins facilitate scenario simulation and analysis of different mission scenarios, enabling operators to evaluate the potential impact of various factors on mission outcomes. By simulating different courses of action, environmental conditions, and adversary behaviors, digital twins help operators assess the feasibility, effectiveness, and risks associated with different mission plans, enabling informed decision-making and optimized mission execution.
- **Optimized Route Planning**: Digital twins support optimized route planning by analyzing environmental conditions, operational constraints, and mission objectives to identify the most efficient and effective routes for surface combatants. By considering factors such as sea state, weather conditions, traffic patterns, and threat levels, digital twins help operators plan safe and efficient routes that minimize risks, conserve resources, and maximize mission effectiveness.
- **Mission Rehearsal and Training**: Digital twins serve as valuable tools for mission rehearsal and training, allowing operators to simulate mission scenarios, practice response strategies, and train personnel in a realistic virtual environment. By replicating the operational environment and simulating various mission scenarios, digital twins enable operators to enhance readiness, decision-making skills, and coordination abilities, ensuring mission success in actual operational situations.

Enhancing situational awareness and mission planning capabilities through digital twins enables surface combatants to operate more effectively, safely, and efficiently in dynamic and challenging maritime environments. By providing real-time monitoring, predictive analytics, scenario simulation, and optimized route planning capabilities, digital twins empower operators to make informed decisions, optimize strategies, and ensure mission success in a wide range of operational scenarios.

5.3 OPTIMIZING RESOURCE ALLOCATION AND OPERATIONAL EFFICIENCY

Optimizing resource allocation and operational efficiency is critical for maximizing the effectiveness and success of surface combatant operations. Digital twins offer advanced capabilities to streamline resource allocation processes and improve operational efficiency through real-time monitoring, predictive analytics, and simulation.

- **Real-time Monitoring and Data Integration**: Digital twins integrate data from onboard sensors, systems, and external sources to provide real-time insights into the operational environment and the status of onboard resources. By continuously monitoring parameters such as fuel consumption, power usage, equipment performance, and mission objectives, digital twins enable operators to make data-driven decisions about resource allocation and utilization in real-time.
- **Predictive Analytics and Forecasting**: Digital twins leverage predictive analytics techniques to forecast resource requirements, anticipate operational demands, and optimize resource allocation strategies. By analyzing historical data, trends, and simulation models, digital twins predict future resource needs based on current operational conditions, mission objectives, and environmental factors, enabling operators to proactively allocate resources and optimize efficiency.

- **Dynamic Resource Optimization**: Digital twins enable dynamic optimization of resource allocation based on changing operational requirements and mission priorities. By continuously analyzing real-time data and predictive analytics outputs, digital twins adjust resource allocation in response to evolving threats, mission objectives, and environmental conditions, ensuring that resources are allocated efficiently and effectively to support mission success.
- **Scenario Simulation and Analysis**: Digital twins facilitate scenario simulation and analysis of different resource allocation scenarios, allowing operators to evaluate the impact of different allocation strategies on operational outcomes. By simulating various scenarios, including different mission profiles, resource constraints, and threat levels, digital twins help operators identify the most effective resource allocation strategies to optimize operational efficiency and mission effectiveness.
- **Optimized Maintenance Planning**: Digital twins support optimized maintenance planning by predicting equipment failures, forecasting maintenance requirements, and scheduling maintenance activities proactively. By analyzing sensor data, historical performance data, and predictive analytics outputs, digital twins enable operators to prioritize maintenance tasks, allocate resources, and schedule downtime in a way that minimizes disruption to operations and maximizes system availability.
- **Efficient Mission Planning and Execution**: Digital twins facilitate efficient mission planning and execution by providing operators with real-time insights into resource availability, mission requirements, and operational constraints. By integrating data from multiple sources, including sensor data, intelligence reports, and mission objectives, digital twins enable operators to plan missions more effectively, allocate resources efficiently, and optimize mission execution for maximum efficiency and effectiveness.

Optimizing resource allocation and operational efficiency through digital twins enables surface combatants to operate more effectively, safely, and efficiently in dynamic and challenging maritime environments. By providing real-time monitoring, predictive analytics, scenario simulation, and dynamic optimization capabilities, digital twins empower operators to make informed decisions, allocate resources efficiently, and ensure mission success in a wide range of operational scenarios.

6. BOLSTERING MISSION EFFECTIVENESS

6.1 LEVERAGING DIGITAL TWINS TO ENHANCE MISSION READINESS AND EFFECTIVENESS

Leveraging digital twins to enhance mission readiness and effectiveness is crucial for ensuring the success of surface combatant operations in dynamic and challenging maritime environments. Digital twins offer advanced capabilities to optimize mission planning, resource allocation, operational efficiency, and situational awareness, enabling operators to make informed decisions and execute missions with maximum effectiveness.

- **Real-time Monitoring and Situational Awareness**: Digital twins integrate data from onboard sensors, surveillance systems, and external sources to provide real-time insights into the operational environment. By continuously monitoring parameters such as position, speed, weather conditions, and sensor readings, digital twins enhance situational awareness, enabling operators to respond quickly and effectively to changing conditions and emerging threats.
- **Predictive Analytics and Risk Assessment**: Digital twins leverage predictive analytics techniques to forecast potential risks, assess threats, and anticipate operational challenges based on historical data, trends, and simulation models. By analyzing sensor data, intelligence reports, and environmental conditions, digital twins help operators identify potential risks, assess their likelihood and impact, and develop strategies to mitigate risks and ensure mission success.
- **Optimized Resource Allocation**: Digital twins support optimized resource allocation by analyzing operational requirements, mission objectives, and resource availability to identify the most efficient and effective allocation strategies. By integrating data from multiple sources and leveraging predictive analytics, digital twins help operators allocate resources such as personnel, equipment, and supplies in a way that maximizes mission effectiveness while minimizing risks and resource constraints.
- **Dynamic Mission Planning and Execution**: Digital twins facilitate dynamic mission planning and execution by providing operators with real-time insights into mission requirements, resource availability, and operational constraints. By integrating data from multiple sources, including sensor data, intelligence reports, and mission objectives, digital twins enable operators to adapt mission plans and allocate resources dynamically in response to changing conditions, ensuring mission success and operational effectiveness.
- **Scenario Simulation and Analysis**: Digital twins enable scenario simulation and analysis of different mission scenarios, allowing operators to evaluate the potential impact of various factors on mission outcomes. By simulating different courses of action, environmental conditions, and adversary behaviors,

digital twins help operators assess the feasibility, effectiveness, and risks associated with different mission plans, enabling informed decision-making and optimized mission execution.

- **Continuous Improvement and Learning**: Digital twins support continuous improvement and learning by providing feedback mechanisms and performance metrics for evaluating mission outcomes and identifying areas for improvement. By analyzing the effectiveness of mission plans, resource allocation strategies, and operational decisions, digital twins enable operators to refine their approaches, optimize performance, and enhance mission readiness and effectiveness over time.

Leveraging digital twins to enhance mission readiness and effectiveness empowers surface combatants to operate more effectively, safely, and efficiently in dynamic and challenging maritime environments. By providing real-time monitoring, predictive analytics, scenario simulation, and dynamic optimization capabilities, digital twins enable operators to make informed decisions, allocate resources efficiently, and ensure mission success in a wide range of operational scenarios.

6.2 ADAPTIVE STRATEGIES ENABLED BY REAL-TIME SIMULATION AND ANALYSIS

Adaptive strategies enabled by real-time simulation and analysis are pivotal for surface combatants to respond effectively to evolving threats, changing environments, and dynamic mission requirements. Digital twins provide advanced capabilities to simulate different scenarios, analyze real-time data, and dynamically adjust strategies to optimize mission outcomes.

- **Dynamic Threat Response**: Digital twins continuously monitor sensor data, intelligence reports, and environmental conditions to assess threats and anticipate adversarial actions in real-time. By simulating different threat scenarios and analyzing potential threat vectors, digital twins enable operators to develop adaptive response strategies that dynamically adjust defensive measures, countermeasures, and tactics to mitigate threats and ensure mission success.
- **Flexibility in Mission Planning**: Digital twins facilitate flexibility in mission planning by simulating different mission scenarios, evaluating operational constraints, and adjusting mission plans in real-time. By analyzing sensor data, resource availability, and mission objectives, digital twins enable operators to adapt mission plans dynamically, reallocating resources, adjusting routes, and modifying objectives to respond to changing conditions and emerging requirements while optimizing mission effectiveness.
- **Optimized Resource Allocation**: Digital twins support optimized resource allocation by analyzing real-time data, assessing operational requirements, and dynamically adjusting resource allocation strategies. By continuously monitoring parameters such as fuel consumption, power usage, and equipment performance, digital twins enable operators to allocate resources efficiently, reallocating assets, personnel, and supplies as needed to maximize mission effectiveness while minimizing risks and resource constraints.
- **Adaptive Command and Control**: Digital twins facilitate adaptive command and control by providing real-time insights into mission status, operational conditions, and strategic objectives. By integrating data from multiple sources and simulating different command scenarios, digital twins enable commanders to make informed decisions, adjust tactics, and allocate resources dynamically to respond to changing threats, evolving missions, and unforeseen contingencies while maintaining operational effectiveness and mission success.
- **Continuous Risk Assessment**: Digital twins support continuous risk assessment by analyzing real-time data, assessing threats, and dynamically adjusting risk mitigation strategies. By simulating different risk scenarios and analyzing potential consequences, digital twins enable operators to identify emerging risks, evaluate their likelihood and impact, and implement adaptive risk mitigation measures to minimize risks and ensure mission success.
- **Operational Resilience**: Digital twins enhance operational resilience by enabling adaptive strategies that anticipate and respond to disruptions, failures, and contingencies in real-time. By simulating different operational scenarios and analyzing system behavior, digital twins enable operators to develop adaptive response strategies that optimize system performance, mitigate risks, and ensure mission continuity in the face of unforeseen events or adversarial actions.

Adaptive strategies enabled by real-time simulation and analysis through digital twins empower surface combatants to respond effectively to changing conditions, emerging threats, and dynamic mission requirements. By providing real-time insights, dynamic scenario simulation, and adaptive decision support capabilities, digital twins enable operators to develop flexible, responsive strategies that optimize mission outcomes, enhance operational effectiveness, and ensure mission success in diverse and challenging maritime environments.

7. FOSTERING INNOVATION AND ADAPTABILITY

7.1 ADAPTABILITY OF DIGITAL TWINS TO EVOLVING OPERATIONAL REQUIREMENTS

Fostering innovation and adaptability is essential for the continued success and relevance of surface combatants in an ever-changing maritime landscape. Digital twins play a crucial role in this endeavor by providing a flexible and adaptable platform that can evolve to meet evolving operational requirements.

- **Agile Development and Iterative Improvement**: Digital twins enable agile development processes, allowing naval organizations to rapidly prototype, test, and iterate new technologies, systems, and operational concepts. By providing a virtual environment for experimentation and simulation, digital twins facilitate rapid feedback loops, enabling continuous improvement and innovation in response to evolving operational needs and emerging threats.
- **Integration of Emerging Technologies**: Digital twins serve as a foundation for integrating emerging technologies such as artificial intelligence, machine learning, autonomy, and advanced sensor systems into surface combatants. By providing a common framework for data integration, analysis, and visualization, digital twins enable seamless integration and interoperability of diverse technologies, enhancing operational capabilities and fostering innovation in naval engineering and operations.
- **Modularity and Scalability**: Digital twins are inherently modular and scalable, allowing naval organizations to customize and expand their capabilities to meet evolving operational requirements. By integrating data from different systems and sensors, digital twins support a holistic view of the surface combatant's operational environment, enabling operators to adapt and scale their digital twin solutions to address new missions, operational challenges, and technological advancements.
- **Adaptive Decision Support**: Digital twins provide adaptive decision support capabilities, enabling operators to make informed decisions in real-time based on dynamic operational conditions and evolving mission requirements. By analyzing real-time data, predicting potential outcomes, and simulating different scenarios, digital twins empower operators to adapt strategies, allocate resources, and optimize mission plans to achieve mission success in changing and uncertain environments.
- **Continuous Learning and Optimization**: Digital twins support continuous learning and optimization by providing feedback mechanisms and performance metrics for evaluating operational effectiveness and identifying areas for improvement. By analyzing simulation results, mission outcomes, and operational feedback, digital twins enable naval organizations to refine their strategies, optimize resource allocation, and enhance operational performance over time, ensuring continuous adaptation and innovation in surface combatant operations.
- **Collaboration and Knowledge Sharing**: Digital twins facilitate collaboration and knowledge sharing among naval organizations, industry partners, and research institutions, enabling the exchange of ideas, best practices, and lessons learned. By providing a common platform for data sharing, simulation, and analysis, digital twins foster collaboration across different domains and disciplines, driving innovation and fostering a culture of adaptability and continuous improvement within the naval community.

Digital twins foster innovation and adaptability in surface combatants by providing a flexible and adaptable platform for agile development, integration of emerging technologies, modularity and scalability, adaptive decision support, continuous learning and optimization, and collaboration and knowledge sharing. By enabling naval organizations to evolve and adapt to changing operational requirements, digital twins ensure that surface combatants remain effective, resilient, and capable of meeting the challenges of the future maritime environment.

7.2 COLLABORATION AND KNOWLEDGE SHARING FACILITATED BY DIGITAL TWIN ECOSYSTEMS

Collaboration and knowledge sharing facilitated by digital twin ecosystems play a vital role in enhancing innovation, efficiency, and effectiveness across naval organizations, industry partners, and research institutions. Digital twin ecosystems provide a collaborative platform for sharing data, expertise, and insights, enabling stakeholders to work together seamlessly to address complex challenges and drive continuous improvement in surface combatant operations.

- **Data Sharing and Integration**: Digital twin ecosystems enable stakeholders to share data from various sources, including onboard sensors, external systems, and historical databases, to create comprehensive digital representations of surface combatants and their operational environments. By integrating data from multiple sources into a unified platform, digital twin ecosystems provide a holistic view of the maritime domain, enabling stakeholders to collaborate on data-driven analyses, simulations, and decision-making processes.

- **Interoperability and Standardization**: Digital twin ecosystems promote interoperability and standardization by adopting common data formats, protocols, and interfaces that enable seamless integration and exchange of information among different stakeholders and systems. By adhering to industry standards and best practices, digital twin ecosystems facilitate collaboration across organizational boundaries, ensuring compatibility and interoperability among diverse technologies and platforms.
- **Collaborative Modeling and Simulation**: Digital twin ecosystems enable stakeholders to collaborate on modeling and simulation activities, allowing them to develop and refine digital representations of surface combatants, operational scenarios, and mission objectives collaboratively. By providing shared access to simulation tools, data repositories, and analytical capabilities, digital twin ecosystems facilitate collaborative analysis, experimentation, and optimization of operational strategies and tactics.
- **Knowledge Sharing and Best Practices**: Digital twin ecosystems serve as repositories for knowledge sharing and best practices, enabling stakeholders to document and disseminate lessons learned, best practices, and innovative solutions across the naval community. By providing collaborative platforms for sharing insights, experiences, and expertise, digital twin ecosystems foster a culture of continuous learning and improvement, driving innovation and efficiency in surface combatant operations.
- **Cross-Disciplinary Collaboration**: Digital twin ecosystems facilitate cross-disciplinary collaboration among naval engineers, operators, analysts, and researchers, enabling stakeholders with diverse backgrounds and expertise to collaborate on complex challenges and solutions. By providing collaborative environments for interdisciplinary teams to work together, digital twin ecosystems foster creativity, innovation, and synergy, leading to more effective and holistic approaches to addressing operational requirements and challenges.
- **Industry-Academia Collaboration**: Digital twin ecosystems promote collaboration between industry partners and academia, enabling stakeholders to leverage the latest research, technologies, and methodologies to enhance surface combatant operations. By providing collaborative platforms for industry-academia partnerships, digital twin ecosystems facilitate the transfer of knowledge, expertise, and technology innovations from research institutions to practical applications in naval engineering and operations.

Collaboration and knowledge sharing facilitated by digital twin ecosystems are essential for driving innovation, efficiency, and effectiveness in surface combatant operations. By providing collaborative platforms for data sharing, interoperability, modeling and simulation, knowledge sharing, cross-disciplinary collaboration, and industry-academia partnership, digital twin ecosystems enable stakeholders to work together seamlessly to address complex challenges, drive continuous improvement, and ensure the readiness and effectiveness of surface combatants in dynamic and challenging maritime environments.

8. CONCLUSION

The transformative potential of ship design digital twins lies in their ability to revolutionize naval engineering and maritime operations by integrating advanced technologies such as AI, data analytics, and sensor networks. These technologies enhance the capabilities of digital twins, enabling them to provide unprecedented insights, predictive capabilities, and operational efficiencies across various aspects of surface combatant optimization. AI algorithms enhance the intelligence and decision-making capabilities of ship design digital twins. AI techniques such as machine learning enable digital twins to analyze vast amounts of data, identify patterns, and learn from past experiences to improve simulation accuracy, predict vessel behavior, and optimize operational strategies. AI-powered digital twins can autonomously adapt to changing conditions, optimize performance, and support real-time decision-making, enhancing operational efficiency and mission effectiveness.

Data analytics techniques play a crucial role in extracting actionable insights from the vast amount of data generated by ship design digital twins. Advanced analytics algorithms analyze sensor data, simulation outputs, historical performance data, and external environmental data to identify trends, anomalies, and optimization opportunities. Data analytics enable naval engineers and operators to assess performance, predict failures, optimize designs, and optimize operational strategies, ultimately improving vessel efficiency, safety, and reliability.

Sensor networks form the backbone of ship design digital twins, providing real-time data on various aspects of vessel behavior, performance, and environment. Sensor data from onboard systems, external sources, and environmental sensors are integrated into the digital twin to enhance simulation accuracy and fidelity. Sensor networks enable continuous monitoring of critical parameters, early detection of anomalies, and proactive maintenance strategies, ensuring the reliability, safety, and readiness of surface combatants in diverse operational conditions.

In summary, the integration of advanced technologies like AI, data analytics, and sensor networks enhances the transformative potential of ship design digital twins by augmenting their capabilities in simulation, analysis, prediction, and decision support. By leveraging these technologies, digital twins empower naval engineers and operators to optimize vessel design, performance, and operation throughout the ship's lifecycle, ultimately enhancing mission effectiveness, operational efficiency, and maritime security. As the naval industry embraces digital transformation, ship design digital twins are poised to play a central role in shaping the future of naval engineering and maritime operations.

As digital twin technology evolves, it will enable more sophisticated simulations and predictive analytics, leading to enhanced mission effectiveness. Digital twins will provide naval operators with real-time insights into vessel performance, enabling them to optimize tactics, strategies, and resource allocation for mission success.

Future advancements in digital twin technology will lead to more accurate predictive maintenance capabilities. By leveraging AI and machine learning algorithms, digital twins will anticipate equipment failures, optimize maintenance schedules, and improve overall vessel reliability, reducing downtime and operational costs.

Digital twins will play a crucial role in the development and deployment of autonomous and *Unmanned Surface Vessels* (**USVs**). By simulating autonomous behaviors and interactions with manned vessels, digital twins will facilitate the integration and coordination of mixed manned and unmanned fleets, enhancing operational flexibility and capability.

Future digital twins will offer comprehensive lifecycle management capabilities, integrating design, construction, operation, maintenance, and decommissioning phases. By providing a holistic view of the vessel's lifecycle, digital twins will enable more efficient resource allocation, decision-making, and optimization strategies, maximizing the value and longevity of naval assets.

One of the primary challenges in implementing digital twin technology in naval operations is the integration and standardization of data from diverse sources. Naval vessels are equipped with a wide range of sensors, systems, and platforms, each generating vast amounts of data. Ensuring interoperability, compatibility, and consistency of data across different systems and platforms is essential for the effective implementation of digital twins.

Digital twins rely on interconnected networks and data exchange, making them vulnerable to cybersecurity threats and privacy breaches. Safeguarding sensitive information, protecting against cyber-attacks, and ensuring data integrity and confidentiality are critical challenges in deploying digital twins in naval operations, requiring robust cybersecurity measures and protocols.

Naval vessels are complex systems with numerous interconnected components and subsystems, making the development and implementation of digital twins challenging. Ensuring the accuracy, reliability, and scalability of digital twin models to accurately replicate the behavior and performance of physical vessels across diverse operational scenarios is a significant challenge that requires advanced modeling, simulation, and validation techniques.

Adopting digital twin technology entails cultural and organizational changes within naval organizations. Embracing a data-driven mindset, fostering collaboration across different departments, and promoting innovation and adaptability are essential for successful implementation. Overcoming resistance to change, building trust in digital twin technology, and ensuring user acceptance and proficiency are critical challenges that require effective leadership and change management strategies.

Overall, while digital twin technology holds immense promise for enhancing naval operations, addressing challenges related to data integration, cybersecurity, complexity, and organizational change is essential for successful implementation. By overcoming these challenges and leveraging the transformative potential of digital twins, naval organizations can unlock new levels of efficiency, effectiveness, and resilience in maritime operations.

In conclusion, digital twins represent a pivotal advancement in naval engineering and maritime operations, poised to shape the future of surface combatants in profound ways. By providing dynamic, real-time simulations of physical vessels and integrating advanced technologies such as artificial intelligence, data analytics, and sensor networks, digital twins offer unparalleled capabilities for optimizing design, enhancing performance, and maximizing mission effectiveness.

Digital twins enable naval engineers and operators to gain deeper insights into vessel behavior, predict performance, anticipate issues, and optimize operational strategies throughout the ship's lifecycle. From design optimization and predictive maintenance to mission planning and training simulations, digital twins offer a comprehensive framework for improving efficiency, reliability, and resilience in maritime operations.

As naval organizations embrace digital transformation, digital twins will play a central role in driving innovation, fostering collaboration, and enhancing adaptability within the naval domain. By leveraging the transformative potential of digital twins, naval organizations can achieve greater operational efficiency, effectiveness, and mission success, ultimately ensuring maritime security and superiority in an evolving and dynamic maritime environment.

In essence, digital twins represent not only a technological advancement but also a paradigm shift in how surface combatants are designed, operated, and maintained. By harnessing the power of digital twins, naval organizations can unlock new levels of performance, agility, and capability, shaping the future of surface combatants and advancing the goals of naval engineering and maritime operations in the 21st century and beyond.

9. REFERENCES

1. AGGARWAL, A., *"The Birth of AI and the first AI Hype Cycle"*. 2018

2. BENAYAS-AYUSO, A., and PEREZ FERNANDEZ, R., (2019). "What should shipbuilding expect from the CAD/CAM systems of the future?". *The Naval Architect magazine*. April. pp. 28-31.

3. CAPEK, K., "R.U.R. (Robots Universales Rossum)". *Books Mablaz,* 2017

4. MUÑOZ, J A., and PEREZ FERNANDEZ, R., "CAD tools for designing smart ships in the world of the Internet of Things*", SMART SHIP 2017*, 24 and 25 January. 2017

5. PEREZ FERNANDEZ, R., BENAYAS-AYUSO, A., and PEREZ-ARRIBAS, F., "Data management for smart ship or how to reduce machine learning cost in IoS applications", *SMART SHIP*, 23 and 24 January. 2018

6. PEREZ FERNANDEZ, R., and PÉTER COSMA, E., "The use of CAD systems to manage modularity in multi-role warships", *Conference Warship*, 2019.

7. TURING, A M., "Computing Machinery and Intelligence", *Mind 49*, pp.433-460. 1950

10. AUTHORS BIOGRAPHY

Rodrigo Perez Fernandez FRINA, PhD, holds the current position of Software Engineering Senior Director at SIEMENS Digital Industry Software.

Leonie Wong holds a Bachelor of Engineering (Infomechatronics) from Queensland University of Technology. She currently holds the position of Senior Account Director at SIEMENS Digital Industries Software based in Melbourne, Australia.

Marc Shipley, BA (Hons) (Northumbria University), holds the position of Strategic Lead for key clients in Maritime for Siemens Digital Industries Software based in England, United Kingdom

THE IMPORTANCE OF DOCKING FACILITY QUALIFICATION IN THE POST DEFENCE STRATEGIC REVIEW ERA

M de Cure and **M Blamires**, Department of Defence, Australia

SUMMARY

In the modern era, naval platforms are continually becoming larger, more complex, and significantly more expensive. The resource investment required to design, build, and maintain naval platforms has made it such that warships and other craft can take years or even decades to go from blank paper to deployable capability. Due to this resource requirement, it is important to ensure that the platforms are protected, as best as possible, from inadvertent damage.

The Defence Strategic Review and National Defence Strategy have emphasised the importance of every vessel in the fleet, with Australia's national strategy moving to that of deterrence as the primary objective. Use of the Royal Australian Navy's docking facility qualification program will minimise the risk that sea days, or whole platforms, are lost to avoidable accidents involving docking infrastructure. This will be especially vital as Australia moves into performing maintenance on Australian nuclear powered submarines.

1. INTRODUCTION

The dry-docking of a ship, boat, or submarine is necessary to carry out periods of maintenance. It is a high-risk activity that requires unique considerations, including the stability of the vessel, the cradling/land support arrangement, the transfer mechanism if the vessel is to be removed from the dock or slip, and extreme environmental loadings, such as high winds and earthquakes.

The docking facilities used by the Royal Australian Navy (RAN) are strategic assets that enable sovereign control and conduct of warship maintenance, and because of this should be viewed as a capability that needs to be correctly resourced and maintained. As such, they are to meet as a minimum, an Australian Defence Force (ADF) Maritime Materiel Requirement Set (MMRS) for facility qualification. There is also a requirement on the System Program Office (SPO) to conduct due diligence on the docking facilities being used by the class or classes of vessel they manage, to ensure the risks of the docking evolution have been mitigated and minimised.

A SPO is an organisation within the Australian Department of Defence, charged with managing through-life aspects of ADF vessels, usually on an individual vessel or class of vessels basis. This allows for specialist knowledge and a narrow focus to be maintained for each of the classes of vessels that are managed.

To provide assurance that the facilities used for the docking of RAN vessels have appropriate equipment and systems in place, each facility is assessed by the Directorate of Navy Engineering (DNE) against the MMRS as part of the facility qualification program. The program assesses the equipment, systems, processes, and personnel of a facility; to provide confidence that they meet RAN requirements and therefore, that the facility is acceptable for use by ADF vessels.

Facility qualification is not certification, which is the independent verification by an appropriate third-party, that the equipment and infrastructure of a facility is capable of functioning to its design capabilities. Certification however, contributes valuable objective evidence against the requirements for facility qualification, that the facility is adequate for its intended use.

By working together, DNE and the SPO can ensure that facility qualification is maintained, and that due diligence for the conduct of docking evolutions is completed. This enables the RAN to be assured that their vessels and people are at the least possible risk of damage or injury, and also that the risk of an RAN vessel damaging a docking facility is minimised.

2. HISTORY

History has unfortunately shown that accidents do occur during the docking of vessels, with two notable incidents having occurred at Australian facilities during the docking of RAN vessels. In 1997, HMAS *Gawler* was damaged during a docking using a shiplift; following the failure of a support cable, causing the platform to become unstable and the vessel to tip and eventually slide into the water. Partially because of this incident, the requirements for docking and slipping were written in 2002, as part of the wider ADF MMRS.

A second incident in 2008, took place as HMAS *Diamantina* was being docked on a floating dock, which began to take on water, resulting in a list of approximately 6 degrees.

Figure 1: 1997 - HMAS *Gawler*

Figure 2: 2008 - HMAS *Diamantina*

Fortunately, there were no injuries as a result of either incident and there was no damage to *Diamantina*; however *Gawler* sustained extensive damage, resulting in a lengthy and expensive repair period.

Following the 2008 incident, a facility qualification program was set up as part of the Navy's Centre for Maritime Engineering. However, it was not until 2010 that dedicated personnel were assigned to the facility qualification role fulltime.

Facility qualification is a Supplier Quality Safety Assurance program which provides the RAN with a common approach of assessing that docking facilities operated by Defence or other commercial operators are suitable and safe.

3. STRATEGIC VALUE OF THE FACILITY QUALIFICATION PROGRAM

Physical failure during a docking evolution can result in damage to, or the loss of, a vessel, or in human terms can cause injury or death; with prevention of such incidents being the focus of the facility qualification program. However, failure should not only be defined as physical failure, but also as the failure of a facility to meet the requirements to achieve and maintain facility qualification; a failure of this sort requires significant additional work to conduct risk assessments for each individual evolution.

Failure of the Commonwealth and the dockyard to successfully implement and manage the facility qualification program could lead to major issues relating to the equipment, systems, processes, and personnel of a facility being overlooked and leading to a situation that could negatively impact a successful and safe evolution. The up-front assessment of a facility lowers the risk of an incident occurring to the asset and/or associated personnel, and ensures resourcing is properly planned and implemented.

In addition to the cost of damage to an ADF vessel, an incident during a docking/undocking evolution can also negatively impact the RAN's strategic capability, scheduling, and reputation. There will also be impacts to the docking facility itself, which can have flow-on effects to RAN capability through maintenance delays for other vessels, but also to the commercial viability of the facility if commercial work is lost.

Facility qualification provides the RAN with an assessment independent to that of a SPO or facility, and when combined with the due diligence that should be conducted by each SPO, makes decision making on the use of a facility for the docking of ADF vessels more robust. Without this common approach, all SPOs would be required to carry out their own risk assessments for each docking and undocking. In addition, there would be instances where multiple SPOs, when using the same facility, would be required to carry out risk assessments concurrently, but separate to each other. An overarching facility qualification program reduces this potentially resource intensive burden on individual SPOs, as well as providing a uniform approach to facility assessments.

However, the success of the facility qualification program is only as good as the time, effort and resources put into it, from both the Commonwealth and facilities. Backing by SPOs is also critical in the success of the program, as they have responsibility for selecting the facilities used for the docking of ADF vessels. If the SPOs do not adhere to the requirement for the use of a qualified facility, and make use of a facility that has not been awarded facility qualification, then there is little incentive for facilities to engage in the process, and increases the risk of an incident occurring during an evolution, and represents a failure of process.

In addition, the need for facilities to fully engage in the program is required to ensure that the most efficient use of resources is achieved for both parties. Facilities are required to provide a comprehensive application for assessment, which includes a thorough description of their facility as well as appropriate objective evidence from third-party engineering authorities to support any claims made about the facility and its equipment, systems, processes and personnel. This will stop the facility needing to provide similar information to each SPO when a risk assessment is being conducted.

The Commonwealth is pleased to note that there been no physical failures during docking/undocking evolutions in a qualified facility since 2009. However, it is important to remain vigilant to ensure that complacency does not lead to a reduction in resources being applied to these strategic assets, which could lead to failures in the future.

4. TYPES OF DOCKING FACILITIES

There are several types of docks in Australia used by the RAN. The most common types are:

4.1 GRAVING DOCK

This is an excavated basin enclosed at one end and accessed by a vessel at the other. Once a vessel has entered the flooded dock and has been positioned, a caisson or gate is put in place to seal the area and the water is removed. As the water level lowers, the vessel will come to rest on pre-arranged blocks on the dock floor. At the completion of works, the vessel is re-floated and the caisson removed, allowing the vessel to depart.
The Captain Cook Dock at Garden Island, Sydney is an example of a graving dock.

Figure 3: STS *Young Endeavour* and HMAS *Arunta* in CCGD - LSIS David Cox – RAN

4.2 SLIPWAY

Also known as a Marine Railway, this type of facility involves a bogie arrangement that is run down rails which have been laid on an incline, to below the low-tide mark. The vessel is positioned over the cradle arrangement mounted on the bogies and winched up the incline until it is clear of the water, where it can be worked on or transported to a working bay. During an unslipping, the bogies and vessel are lowered until the vessel reaches the water, where it can be re-floated.

Tropical Reef Shipyard in Cairns is an example of a slipway.

Figure 4: The launch of the first Landing Helicopter Dock at the Navantia Slipway in Spain – RAN Media Library

4.3 FLOATING DRY DOCK

Is a water-tight, floating structure that is partially submerged by flooding water into ballast tanks, whereby a vessel can enter. The vessel is positioned within the dock and the water pumped from the tanks, raising the dock and setting the vessel onto pre-arranged blocks, providing a dry working area. Alternatively, once raised, a vessel can be transferred from the floating dock onto a hardstanding for work to be conducted.
The Australian Marine Complex Common User Facility, in Henderson, Western Australia, contains a floating dock and self-propelled modular transporters (SPMTs) to move vessels to hardstandings ashore.

Figure 5: HMAS *Perth* docked on Floating Dry Dock *Yargan* - CPO David Connolly - RAN

4.4 SHIPLIFT

Is a platform that is lowered into the water by a set of synchronous winches. The vessel is positioned over the platform, which is raised to allow the vessel to come to rest on pre-arranged blocks clear of the water, providing a dry work area. Again, if required, the vessel can be transported ashore to a working bay.
The Australian Naval Infrastructure Common User Facility in Osborne, South Australia operates a shiplift.

Figure 6: Nuship *Hobart* docked on Australian Naval Infrastructure – Air Warfare Destroyer Alliance

4.5 TRAVEL LIFT

Is a horseshoe-shaped, wheeled vehicle that is driven out along a pair of finger wharves and a number of lifting straps lowered below the water. The vessel then enters, the straps are raised and the boat is carried to the working bay at a hardstanding ashore.
Norship Marine and Austal Cairns in Queensland both operate travel lifts.

Figure 7: Norship Marine Travel Lift, Cairns Queensland – Norship

5. POLICY

The subsequent investigation into the *Gawler* and *Diamantina* incidents identified that a deficiency existed in the application of the docking requirements during the maintenance periods, and that contractual documents did not reference the expected RAN standard.

An outcome of the investigations recommended that all contracts for the docking of RAN vessels be reviewed immediately, to ensure that the requirements of the MMRS were invoked and followed. This recommendation was implemented via a directive from Head Maritime Systems in February 2008 and written into Defence policy in 2015.

The policy states that Defence personnel must use and adhere to the MMRS for the qualification of docks for use by ADF vessels. This policy applies to all Defence personnel charged with managing the through-life aspects of ADF vessels, with a number of stakeholders having responsibilities within the program:

- Commander Joint Logistics – the policy owner;
- Facility Management – submission of an application for facility qualification and the management of the qualification throughout the validity period; including notifying the Commonwealth of any changes that have occurred at the facility during that time;
- SPOs responsible for ADF vessels – ensure that prior to using an Australian dock, the facility has been assessed and awarded facility qualification;
- Directorate of Navy Engineering – assessment of applications against the MMRS for facility qualification and re-qualification; as well as assessing any changes that may have occurred during the facility qualification period; and
- Director General Engineering - Navy – the overall delivery of the facility qualification process, including the formal authorisation of the use of Australian docks with ADF vessels.

6. PROCESS

6.1 ADF MARITIME MATERIEL REQUIREMENT SET

The RAN Maritime Materiel Requirement Set is based on the US MIL-STD-1625, *Department of Defense Standard Practice: Safety Certification Program (SCP) for Drydocking Facilities and Shipbuilding Ways for US Navy Ships*, which is used by the United States Navy to assess docking facilities used by their vessels.

The MIL-STD contains mandatory inspection checklists for all structures and systems applicable to the facility type. Due to the smaller throughput of vessels in Australian facilities when compared to those in the US, the RAN MMRS is less prescriptive and does not require mandatory checklists to show compliance. The MMRS involves the overall assessment of the equipment, systems, processes and personnel of a facility and allows the facility management to determine the appropriate objective evidence required to demonstrate compliance to each requirement.

6.2 APPLICATION FOR FACILITY QUALIFICATION

The RAN MMRS specifies the minimum requirements for the safe docking of ADF ships, boats, and submarines in facilities operating within Australia and its territories and is therefore to be applied for all Department of Defence or contractor facilities.

The Directorate of Navy Engineering (DNE) is the designated Engineering Authority for the qualification assessment of facilities, with the Director General of Engineering - Navy (DGENG-N) the Executive Authority responsible for awarding facility qualification.

- A request for facility qualification is to be made through facility management, via an application to DNE. The application is to detail how the facility meets each of the RAN requirements and to contain appropriate objective evidence in support. A document, generated by DNE, and available to all facilities, provides guidance on how to compile and submit a comprehensive qualification application;

- The application will be assessed by DNE and be followed up by a site visit, to view the facility as well as carry out a review of the maintenance management system;

- If the facility is found to meet the RAN requirements, a recommendation on the suitability of the facility will be made to DGENG-N; and

- The validity period of facility qualification is on a case-by-case basis, but is generally five-years. A mid-qualification review is carried out at the half-way point of the qualification period, to assess what, if any changes, have taken place since facility qualification was awarded, and if these changes have had an impact on the facility meeting the RAN MMRS.

6.3 RE-QUALIFICATION

If a facility requires qualification beyond the stated expiry date, facility management is to make a submission for re-qualification. The submission is to confirm that the information provided during the previous qualification submission is still current or identify any changes that have occurred. The re-qualification submission is assessed by DNE, to ensure that the facility continues to meet the RAN MMRS and is acceptable for continued use by ADF vessels.

6.4 QUALIFICATION VS CERTIFICATION

It is often asked why certification of a facility via a class society, for example, is not adequate for RAN assurance. Classification Societies can provide validation that the design, construction and maintenance of docking facilities are compliant to a set of rules they have established and maintained.

However, class certification involves only the structural aspects of a facility and does not take into consideration the other key areas of the RAN MMRS, when certifying a facility.

The facility qualification assessment carried out by DNE extends beyond the structural scope covered by Classification Societies to include:

- adequacy of the facility for its given purpose;
- maintenance of the facility;
- operational, emergency and security procedures for the docking/undocking of RAN vessels;
- qualifications and experience of key personnel undertaking docking activities; and
- quality procedures that meet the requirements of ISO 9001.

While class certification is not sufficient in isolation for facility qualification, it is considered one form of objective evidence used as part of the facility qualification assessment.

6.5 VALUE OF THE FACILITY QUALIFICATION PROGRAM

Investigations into the two RAN incidents found that the contractual documents between the RAN and the docking facilities did not call up the expected standard for docking requirements. In addition, issues leading to these incidents were found to be in-line with the requirements of the MMRS, and that to prevent future incidents a facility qualification assessment for all docking facilities used with Australian Defence Force vessels, is necessary.

A robust facility qualification process increases the likelihood that RAN capability is maintained, repaired and deployed on time, every time; with a qualified facility reducing the risk to Navy to so far as is reasonably practicable. A qualified facility addresses the areas most likely to result in an incident during docking. Therefore, to prevent future incidents the areas most likely to contribute are assessed, to ensure they are being managed to a low risk.

The areas assessed as part of the facility qualification program view the facility within a holistic framework, much broader than individual class certifications, they include:

6.5 (a) Adequacy of the facility, an assessment:

- of the material condition of the equipment and infrastructure to be used with ADF vessels;
- that regular inspections and certification of the equipment and infrastructure are being carried out by appropriate, third-party engineering authorities;
- that appropriate objective evidence is available for the material condition of the equipment and infrastructure, as well as suitability for on-going use to design capabilities; and
- that there is timely attention to inspection findings/recommendations/requirements made via third-party engineering authority, including damaged equipment and negative results following testing.

6.5 (b) Maintenance, an assessment that:

- a facility has a preventative maintenance management system and regime in place;
- the maintenance regime has a basis in OEM recommendations, Australian or other appropriate standards;

- maintenance tasks are being carried out as required and as scheduled;
- maintenance is being carried out by a suitably qualified workforce;
- deferred maintenance, for any reason, is tracked and any negative outcomes that may occur are understood;
- an established mechanism is in place for ensuring emerging work and feedback on maintenance issues, particularly results of testing, are captured and addressed in a timely manner;
- an established mechanism is in place to disseminate maintenance feedback between the operators and maintenance management; and
- the use of appropriate materials is being monitored and applied.

6.5 (c) Operational procedures and check sheets, an assessment that a facility:

- has in place procedures and check sheets for the operational processes during docking/undocking; and
- has in place procedures and check sheets for critical items prior to docking/undocking, e.g. watertight integrity.

6.5 (d) Emergency and Security procedures and check sheets, an assessment that a facility:

- has in place emergency and security procedures or check sheets for any issues that may occur during docking and undocking, including contact with emergency services;
- ensures that critical docking equipment is isolated and cannot be operated by unauthorised personnel during the lay period; and
- carries out regular emergency drills.

6.5 (e) Key docking personnel, an assessment that a facility has:

- personnel qualified and experienced in the type of docking equipment being used, examples being the Dockmaster, or mobile transporter operators;
- in place back-up personnel for each of the key positions; and
- a training regime for additional personnel under the supervision of qualified and experienced staff, to ensure continuity of operation.

7. OUTCOMES

The primary desired outcome from facility qualification is the avoidance of further mishaps with Defence naval assets. In the current shipbuilding and sustainment reality, avoidable damage to existing assets has the potential to blow out finely tuned maintenance schedules. This would result in an ongoing capability loss as the industrial base fight to catch up with the schedule, and platforms are delayed in their return to service.

The RAN's primary requirement from the enabling functions of the enterprise is materially capable warships and submarines being available when required to conduct tasks as directed by the government of the day. This requirement is underpinned by Defence Seaworthiness, which sets out the framework for safe and capable naval platforms. It is through this lens that the support organisation approaches its work in providing naval platforms for service.

Safe and efficient docking and undocking of naval assets enables the ongoing drumbeat of maintenance, and subsequently, the RAN's operational capability to be maintained. The qualification process is conducted to minimise the risk of damage caused during the docking or undocking evolution, through lack of training or currency of personnel, or lack of maintenance leading to partial or total failure of the docking equipment.

The first and most fundamental obligation of Defence Seaworthiness is the structural integrity of the naval platform. Protecting the structural integrity from degradation from minor or major damage during the docking or undocking evolution is vital given the cost and time associated with repairing an existing asset or building a new asset to replace it.

Facility qualification provides the first and most fundamental element in the risk minimisation strategy. By only docking at facilities that are properly operated and maintained, and that use suitably trained and experienced personnel, Defence is taking a large step towards ensuring the events of *Gawler* and *Diamantina* are not repeated.

8. FACILITY QUALIFICATION AND THE COLLINS SYSTEM PROGRAM OFFICE – A CASE STUDY

The Collins SPO implementation of the facility qualification program has been shown to be best practice in a number of areas.

8.1 AVAILABILITY OF MULTIPLE FACILITIES

The qualification of twelve facilities around the country have allowed the Collins SPO to have the flexibility required to dock submarines at a facility to best support the strategic requirements of the fleet. The Collins SPO is able to dock in South Australia, Western Australia and New South Wales as part of the facility qualification program, knowing that a level of due diligence has already been undertaken. This assurance can then be used as a baseline for the SPO to make a decision on a docking.

8.2 COMMON APPROACH TO FACILITY QUALIFICATION

The use of the facility qualification program allows the Collins SPO to make a decision on the use of a facility knowing that all facilities have been assessed using a common process and a common set of requirements. Again providing a baseline that the SPO can build on for their assurance, without the need for resource intensive risk assessments for each evolution.

8.3 RAPID SOLUTIONS TO EMERGING NEEDS

A qualified facility has a set of parameters for which it has been assessed and agreed to, including equipment, infrastructure, procedures, and personnel, a SPO is to use a facility only up to the maximum limit of these parameters. If a facility is required to be used in excess of the agreed parameters, collaboration between all stakeholders is required to collate the additional objective evidence required for assessment against the MMRS prior to use.

A situation arose in December 2021 where a Collins-class submarine was required, at short-notice, to be set-down on a hardstanding area that had been assessed and awarded qualification for use with Frigate-class vessels. The maximum load acknowledged for the hardstanding was less than that required for the submarine and therefore its use was not permitted under the facility qualification.

Knowledge of the facility qualification program within the Collins SPO, as well as the relationships that had previously been established between the key stakeholders, allowed a rapid solution to this situation. The Collins SPO and Facility Qualification Cell of DNE, worked with the Facility Management and Maintenance Contractor to identify the information and objective evidence required to provide assurance that the setting-down of a Collins-class submarine on the hardstanding would not damage the vessel or cause injury to the personnel involved in the movement. In addition, assurance was also required that the work carried out would not damage the hardstanding.

The involvement of the stakeholders and their understanding and commitment to the facility qualification program allowed the approval required for the setting-down of the submarine to be achieved in less than three weeks and provided the Collins SPO with assurance that the hardstanding could be used successfully; with a low risk of damaging the submarine, causing injury or damaging the hardstanding and therefore negatively impacting the on-going maintenance of the Frigate fleet.

Following the successful setting-down, the scope of that facility's qualification was updated to include the revised maximum loads to accommodate a Collins-class submarine, allowing the hardstanding to be used for future submarine movements.

9. BEST PRACTICE

It is the aim of the facility qualification program to have the level of engagement and collaboration it has with the Collins SPO, with all Systems Program Offices. The relationship between the facility qualification program and the Collins SPO has developed due to the involvement the SPO has with day-to-day issues that impact the qualification of a facility, as well as the SPOs understanding of the facility qualification process; what it strives to achieve, and how it works to provide a program that supports through-life activities.

The experience with the Collins SPO in general, and during the use of the Frigate hardstanding in particular, highlighted the need for an adequate level of resourcing within all stakeholder organisations and the importance of the correct implementation and management of the facility qualification program. This resourcing is required for the submission and assessment of a facility for initial qualification, as well as subsequent re-qualifications and any short-notice changes required during the qualification period.

Once this relationship has been established, it requires work to ensure that it is not lost; especially with the movement of personnel through SPOs and facilities in particular. Adequate resourcing and importance placed on the facility qualification program within SPOs, facilities, and contractors is required to maintain the relationships and collaboration crucial to the success of facility qualification.

These relationships and collaborations also provide DNE with local support and eyes on-the-ground, which allows for a better knowledge of what is taking place in facilities around the country, in-between DNE visits.

10. FUTURE

The Surface Fleet Review has recommended a significant increase in the number of front-line combatants for the RAN, changing the future combatant fleet from twelve vessels to twenty-six. As a result, the requirement for docked maintenance will likely double, pending the maintenance cycles of the new vessels.

This combined with the increase in size of the submarine fleet, from six Collins class vessels to at least eleven nuclear powered vessels (three Virginia, and eight SSN-AUKUS), represents a significant increase in the required qualified docking capacity to ensure sufficient availability to conduct scheduled and unscheduled maintenance on the expanded RAN fleet. The increase in fleet size will require an increase in the capacity of maintenance facilities, and likely the construction of new docking facilities to handle the throughput required to keep the fleet well maintained and at sea.

This situation is further exacerbated by the fact that nuclear powered submarine maintenance is currently conducted in graving docks by both the United States and the United Kingdom. There are advantages and disadvantages to the conduct of maintenance in a graving dock; with the submarine not needing to be lifted out of the water, the risk of it being dropped is almost eliminated. However, conducting work in a graving dock is likely to be less efficient than conducting it in a maintenance hall, as it limits the number of submarines that can be docked at one time, and purpose built maintenance infrastructure cannot be built within a dock.

Given there is only one graving dock currently in commission in Australia that is large enough to accommodate an SSN, being at Garden Island in NSW, this would suggest that new graving docks will need to be commissioned and qualified to meet the RAN's ongoing maintenance requirements, if the use of graving docks is the preferred solution.

If the decision is made to go with an alternative to a graving dock for nuclear powered submarine maintenance, these facilities will need to be properly maintained and licensed. This nuclear licensing will be conducted by an independent regulator, whose sole purpose will be to ensure nuclear safety. This will provide a large part of the risk reduction puzzle to ensure events similar to *Gawler* and *Diamantina* do not occur to a nuclear powered submarine.

When considering the future docking requirements for the nuclear powered submarines, these docking facilities will need to be nuclear rated and licensed. This additional design rigour and scrutiny is required for seismic events to ensure nuclear safety for the rare occurrence where there is an unforeseen significant environmental event, such as a large earthquake. This will add layers of complexity in construction and during qualification to ensure that the nuclear requirements are met, and add additional requirements on the maintenance that is conducted on the facility itself.

11. CONCLUSIONS

Ensuring that incidents similar to *Gawler* and *Diamantina* do not occur for Navy's future surface fleet and nuclear powered submarines is paramount. The ongoing capability and environmental impacts of not correctly qualifying facilities, and an incident occurring, are far too great to be allowed to happen.

The robust qualification process developed by Navy Engineering Branch, which has been implemented with the help of the Collins Submarine Program in the submarine domain, ensures that materiel standards of docking facilities is maintained. The continued use of and adherence to, the qualification process will ensure the risk to vessel seaworthiness, and subsequently Navy capability, from docking evolutions is reduced so far as is reasonably practicable.

As the continuous naval shipbuilding program ramps up and new docking facilities are brought online to support nuclear powered submarines, the qualification process will support the safe docking and undocking of RAN vessels. In order to maintain the strategic sovereign capability that is provided by Australia's docking facilities, all involved parties need to approach qualification with the correct mind-set and sufficient resources, to minimise the risk such that incidents of the past are not repeated.

12. ACKNOWLEDGEMENTS

The authors would like to thank Danielle Hodge for providing the impetus for this paper. We would also like to thank our respective Directors General and management lines for their support during the writing process.

13. AUTHORS' BIOGRAPHY

Matt de Cure holds the current position of Senior Naval Architect at the National Shipbuilding and Sustainment Group within the Department of Defence. The Author is responsible for design acceptance review and material assurance engineering for the Collins Class submarines as part of a larger team. The Author's previous experience includes facility qualification inspections and survey work on submarines and support assets.

Megan Blamires holds the current position of Qualification Manager - Docking and Slipway Facilities at the Directorate of Navy Engineering within the Department of Defence. The Author is responsible for the management of the facility qualification program, including the assessment of docking facilities against the ADF Maritime Materials Requirement Set for facility qualification. The Author's previous experience includes certification of materials and processes within the commercial aerospace industry.

RADICAL WARSHIP PLATFORM CONCEPTS – TESTING AND ACCEPTANCE

M W Hood BEng MSc CEng FRINA MAPM MIET CDipAF RCNC - Nova Systems, UK

SUMMARY

This paper examines the shape of future naval fleets and how they could contain radical new surface warship platform concepts. It addresses the drivers for moving from evolution to revolution in warship design and how, when faced with radical new concepts, we can test, and accept them using modern technologies.

1. INTRODUCTION

At Warship 2015 the author presented a paper called "The Shape of Fleets to Come", reference 1. That was a historical review of naval fleets and the step changes and evolutions through history. It ended with a view of the future, looking to 2040 and concluded with a diagram of a future amphibious multi-hull, Figure 1.

The concept in this paper is based on lessons learned from the amphibious landing ships and the challenges of fleet management inside Navy Command. The hypothesis is that it is no longer sufficient to just design a warship, the whole eco-system must be considered across all Defence Lines of Development (TEPIDOIL), or Fundamental Inputs to Capability (FICs).

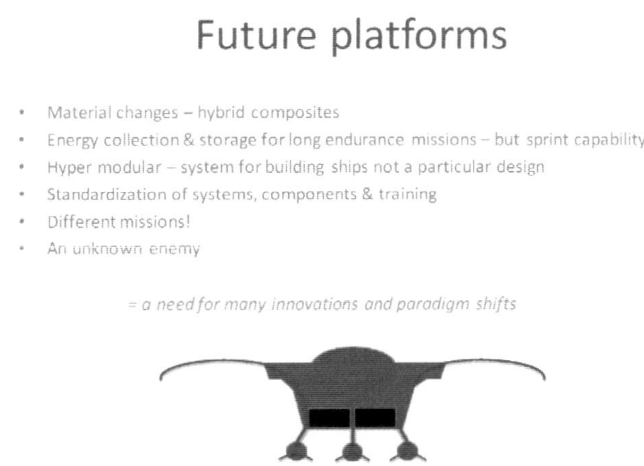

Figure 1: The view of future platforms presented in 2015

In 2016 the author was developing ideas for ships to deliver humanitarian aid with a concept of "Peaceships". The problem was that the UK Government, even the Overseas Development Agency did not see it as their role. The Navy saw humanitarian aid delivery as a secondary role even though the Atlantic Patrol Tasking (North) had become a Bay Class Royal Fleet Auxiliary (RFA) (like HMAS CHOULES) or River Class Offshore Patrol Vessel on standby in the Caribbean for hurricane season. We see RFA CARDIGAN BAY acting as a sea base at the Eastern End of the Mediterranean ready to deliver aid into Gaza at the time of writing.

2. THE CHALLENGE

Warships have not generally been required to carry heavy loads, but with the advent of asymmetric threats and the need for multi-role platforms to counter the rapidly changing nature of threats a different approach is needed. The theme of the Warship 2019 conference was multi-role vessels and several papers, including that of David Andrews, reference 2, further inspired this concept. Since then, there has been much development of containerised solutions and we are now in an era of offboard systems, in particular drones. Previous research by the author reviewed the success of modularity, looking back to the Danish STANFLEX concept and others. Modularity is very much with us today, and the common module of choice is an ISO shipping container, normally in the 20-foot robust variant. Current warfare in the Black Sea, Red Sea and Eastern Mediterranean requires much larger capacity for missiles, munitions, and the delivery of logistics. There is also a pressing need to respond urgently to rapidly developing crises. This drives a need for new types of platforms optimised for offboard systems, modular solutions, adaptability for multiple missions and the ability to sprint into action whilst also being able to loiter for long periods in areas of strategic interest as a deterrent. Today there is a need to deliver all the following:

- Large numbers of defensive missiles
- Vast quantities of ammunition for artillery and deployed land forces
- Clean water and food for displaced civilians
- Medical and humanitarian aid
- Intelligence, surveillance, reconnaissance, and information operations
- Armour, vehicles and heavy military equipment and troops to an unprepared shore
- Offboard autonomous systems in the air, surface, and underwater domains

There is also a pressing need to deal with non-combatant evacuations and recovery of troops and to render assistance to climate emergencies that are occurring more frequently around the globe. A lean, lithe frigate or destroyer just can't carry the payload required to really help.

Current warship procurement processes take too long and surface combatants are very expensive. A £1.3Bn ship to carry 35 missiles that can be exhausted in a few minutes with cheap drones does not sound sustainable. Can it be done differently? This paper explores a radical solution for a family of extremely modular ships and looks at what we would need to do to test and accept such radical new platforms into service with global navies.

3. MODERN CONFLICT

What innovative technologies do we need to include to address the challenges of modern conflict? How will we overcome swarms of small, cheap drones in the air, on the surface and under the surface? There has been a step change in the threat environment in recent years, particularly in the volume of attacks against surface units. We have not been fighting a warship-on-warship war since arguably 1941. The need to strike from the sea, see reference 3, has evolved from aircraft carriers through land attack missiles such as Tomahawk (TLAM) to the deployment of Royal Marines with the increasing need to engage prior to a conflict beginning and by having a visible presence to deter. TLAM and similar strike weapons were intended as a weapon of coercion but no longer wield the power of intent because of their frequent use in modern conflict. Attack systems today and tomorrow will be mostly cheap drones and uncrewed vehicles.

Nick Childs writing on the Britain's Future Navy, reference 4, addresses each type of ship required and looks at the balance of the fleet (Ch 10). Discussing the Royal Fleet Auxiliary (RFA) Bay Class Landing Ships Dock Auxiliary he identifies "*They have huge amounts of space to carry equipment and stores. They lack a helicopter hangar, but that could be added. Some critics say they are critically short on power. They are not pretty, but they are certainly imposing, due largely to an astonishing tower block of superstructure*". Swan Hunter's won the competition with a variant of the Rotterdam class, developed into the Enforcer class. The procurement team pushed to build a class of 4 relatively cheap ships that would be extremely useful and more available than the two Landing Platform Docks HM Ships ALBION and BULWARK. The dock ship concept remains extremely versatile and suits modern warfare with offboard systems extremely well. The Maritime Afloat Replenishment and Support (MARS) programme was to have a variant for multi-role, that was delayed and will have to await the Multi-Role Support Ship competition.

A surface combatant now needs to take the fight to the enemy, wherever they may be in the world, the combat radius sought is from 50nm out to 100nm from the ship with a desirable range from base of 750nm to 1000nm to deliver effect. That 'effect' used to be cruise missiles but might now be offboard drones that can loiter closer to the objective and react when instructed. A single warship is no longer the fighting unit, it is an operating base for offboard systems, most of which will be uncrewed (see the UK Defence Drone Strategy, reference 5). A more unified approach across services is demanded, as is a more unified approach across allied forces. The concept of interchangeability is being discussed more in NATO conferences than interoperability. Future warships need to carry and deploy sensors, weapons and troops from other nations' navies and wider armed forces. War now has many more shades of sub-threshold conflict and deliberately incited civil wars make use of militias to de-stabilise an area for broader and less obvious political gain. Lawrence Freedman in his book 'The Future of War – A History', reference 6, concludes with his view that "*War therefore has a future. It can make an appearance wherever there is a combination of an intensive dispute and available forms of violence*".

And to fail to plan for an unknown war against a little understood or unrecognised foe is to plan to fail. However, only five nations can afford an aircraft carrier navy and not many more can afford amphibious ships. The challenge for smaller nations is to operate useful ships, without presenting a threatening posture to their neighbours. This allows them to contribute in times of crisis, Bangladesh have contributed to every UN Peacekeeping force for the last three decades and are developing a very useful navy that they describe as three dimensional, i.e. with helicopters.

The concept presented here is deliberately aimed at providing platforms of substantially lower cost than first-rate warships. These can be operating bases or mother ships for offboard systems and drones of various sizes. Existing and operating safely within the ocean environment will remain the primary challenge for any ship, see reference 7.

In the RUSI Journal of 2022 the issue of water shortages as a cause of conflict is cited, particularly with reference to Iraq, reference 8. Intervening with fresh water making equipment is likely to become a need for warships in the future, just delivering water is insufficient. The LSDAs were deliberately fitted with huge additional water making capacity for this reason, but in future it may be more appropriate to offload containers of water purification or desalination equipment to be deployed ashore together with electrical generators – all in ISO containers. So, what are the requirements for a versatile warship?

4. KEY REQUIREMENTS

There is a pattern of requirements for a warship. CHAT GPT was tested in November 2023 to find out how good Artificial Intelligence (AI) was at this task. It was asked: *What are the ten key user requirements for a future ship designed for security and humanitarian relief in the event of war?* Avoiding asking about a warship directly as this would be rejected. In a fraction of a second the internet returned the requirements in Figure 2.

Security Measures: Robust defense systems to ensure the safety of the ship and its occupants during wartime.

Humanitarian Facilities: Dedicated spaces for medical facilities, storage of relief supplies, and areas for treating and accommodating displaced persons.

Versatility: A design that allows for quick adaptation between security and humanitarian roles, ensuring the ship can respond effectively to changing circumstances.

Advanced Communication: State-of-the-art communication systems to coordinate both security and relief efforts efficiently.

Sustainability: Integration of eco-friendly technologies, such as renewable energy sources and waste management systems, to minimize the environmental impact during extended missions.

Modular Design: A modular structure that allows for easy upgrades, maintenance, and adjustments to meet evolving security and humanitarian requirements.

Rapid Deployment: Quick deployment capabilities for timely response to crises, with features like high-speed propulsion systems and efficient loading/unloading systems.

Adaptive Technology: Incorporation of cutting-edge technologies like artificial intelligence, drones, and autonomous systems for enhanced security and operational efficiency.

International Collaboration: Interoperability with vessels from different nations and compatibility with international relief organizations to facilitate collaborative efforts.

Crew Well-being: Consideration of living conditions for the crew, including comfortable accommodations, recreation areas, and mental health support services for extended missions.

Figure 2: CHAT GPT version of key requirements for a future ship

Research was focused on these things and how original, radical, or innovative solutions might be presented. A key theme was how solutions could be tested and accepted if you move too far from an evolutionary approach.

5. SWAN CONCEPT

In 2015, inspired by work in Navy Command Headquarters over 2½ years, the author developed a concept referred to obliquely in reference 1. So, what is a SWAN? A Small Waterplane Area Nonohull. That is nine hulls. Each fulfils a function and is a separate unit that can be designed, built, maintained, and potentially operated separately. They are normally connected together. This is a combination of a Small Waterplane Area Twin Hull (SWATH), a trimaran, a hydrofoil, and a Surface Effect Ship (SES).

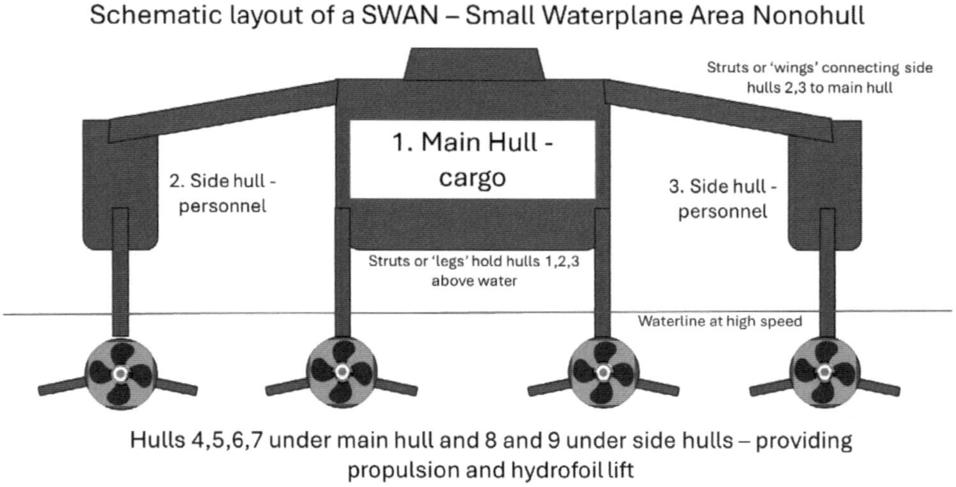

Figure 3: Schematic layout (section view) of the nine hulls of a Small Waterplane Area Nonohull

The function of each hull can vary, and different configurations are likely to be adopted. The key concept is multiple separate hulls interfaced to each other, rather than a monohull that is heavily integrated.

Table 1. Hull purpose and modes of operation

SWAN hulls

Hull No	Purpose	Characteristics	Notes
1	**Main hull** for carrying cargo and connecting together the parts of the ship	Large hull providing strength to connect the other parts together and with large open spaces for storage and flow of cargo including a through dock	Through dock is based on a canal, rather than a pumped ballast system like a Landing Platform Dock (LPD)
2,3	**Side hulls** provide spaces for personnel and sustainment of people	The side hulls of a trimaran that distance personnel from hazardous cargo in the main hull. Separated by 'wings' that can also provide lift at high speed.	Optimised for flow of personnel
4,5,6,7	**Sub hulls** provide propulsion and lift through buoyancy and hydrofoil lift. May be azimuthing or have rudders	Relative size and buoyancy depends on mission speed profiles. May be able to retract for restricted water operations.	Can be changed during the life of the ship. May only have two under main hull.
8,9	**Side sub hulls** provide propulsion and lift through buoyancy and hydrofoil lift	Alternative layout options available. Could be designed as independent craft	May separate and operate independently as surface or underwater vehicles

At low or zero speed hulls 1,2 and 3 are floating. These are partially supported by buoyancy from the sub surface hulls 4-9. Once forward speed commences with propulsion from the sub surface hulls the hydrofoils provide lift and raise the floating hulls above the water. Table 3 describes the modes of operation. Weight reduction is a key driver for this concept and the main hull also provides aerodynamic lift.

Table 2. Modes of operation

Modes of operation

Mode	Configuration	Purpose
Stationary	Floating in displacement mode with hulls 1,2,3 in the water. Sub hulls 4-9 provide dynamic positioning	For offload and onload at sea or loitering
Low forward speed	Some lift provided by hydrofoils reduces draft of above water hulls. Propulsion provided by sub hulls 4-9	For manoeuvring or surveillance operations
High speed	Lift provided by hydrofoils carries hulls 1,2 and 3 clear of the water. Sub hulls 4-9 foiling.	For transit and sprint

Sprint capability is the key driver for this design concept. The need to respond to an incident quickly over long distances, potentially trans-oceanic, is the primary driver for this concept. This drives the need for foiling clear of the top of the waves. The need for speed in a warship is an ancient characteristic that has been reduced as a requirement today due to cost and the introduction of high-speed missiles and aircraft. However, high speed sprint allows the ship to respond to incidents to gain military advantage or to save lives, speed can also help avoid trouble and evade threats. Responding within 12 or 24 hours saves lives of those injured, if the ship doesn't arrive for three to five days there will be many more casualties and disease will have set in. To pursue this concept, you must want to respond quickly to emerging incidents. Whether for your own benefit or others.

This concept requires high power density, which we now have available in gas turbines. Hybrid configurations of propulsion and storage enable new concepts that include electric and diesel propulsion and other fuels. In hydrofoil mode these craft should be able to move at 40 – 60 knots extending their radius of response to 960-1440nm in a day. Further work on the propulsion concept is being addressed for another paper to be presented at INEC along with marine engineering colleagues.

Loiter and offload are other key features of this concept. The ability to load and offload at sea and operate as an amphibious ship enables a huge range of mission types. With the advent of autonomous offboard systems this has become much more important. These ships are designed to load and offload surface and underwater vehicles including large and heavy ones. The key to this operation is a through dock, or canal in the centre of the main hull. The multi-hull configuration enables the buoyancy to be moved around to provide stability that can overcome free surface effects. Lock gates provide the means of loading and offloading craft. The general principle is to onload at the bow and offload at the stern. This allows craft to be gathered with a low forward speed of the ship, this may extend to small boats, or uncrewed craft not under power. Offload at the stern provides protection from waves and shields the daughter craft from hostile enemy action.

This canal approach will be faster than ballasting using pumps. Much of the volume of water in the canal is intended to be 'scooped' up by the ship using forward speed to fill the canal system. The canal will vary in size across the family of SWAN ships. Around 3m wide in the smallest increasing to around 8m wide to take a Landing Craft Utility sized craft.

Segregation for survivability is a key feature of this SWAN concept. Moving the propulsion motors into separate pods and having many of them enables survivability through duplication of propulsive power. Each hull will have internal watertight and fire or blast bulkheads and may have foam filled sections to prevent flooding. The primary benefit is to build separate hulls for people and cargo. Each hull should be able to survive independently and operate to some extent if disconnected. Modular design will also enable swapping or cannibalisation of modules between hulls and between different craft in the SWAN family.

A family of SWANs is covered in this concept with different sizes, see figure 4. Different configurations can be achieved with the same construction modules. The smallest craft, dubbed the Cygnet, should fit into the larger craft with 8m wide docks. Lengths of craft are selectable and the sub hulls providing buoyancy and power might be changed through life.

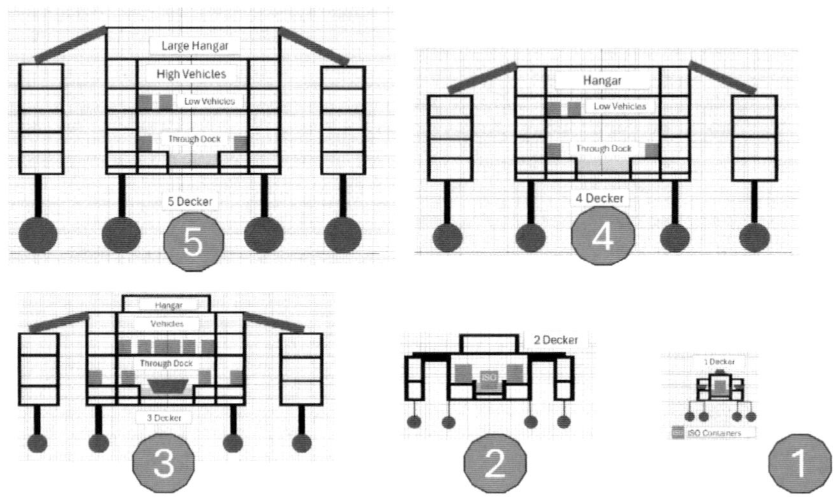

Figure 4: Family of SWAN concept craft, the small 1 Decker is the cygnet in the family

Cygnet – the one deck craft. This small craft is designed to carry up to three 20-foot ISO containers in barges (possibly inflatable), vehicles, or uncrewed craft, that fit within the through dock. It also carries 2 boats and is intended for a mission crew of 8 persons with two operators. This is conceived as a 24m long by 8m wide craft that can fit inside the dock of the larger craft in the family. Some renders of the early concept are shown in figures 5 and 6. The main hull acts as an aerofoil to provide additional aerodynamic lift during high-speed transit.

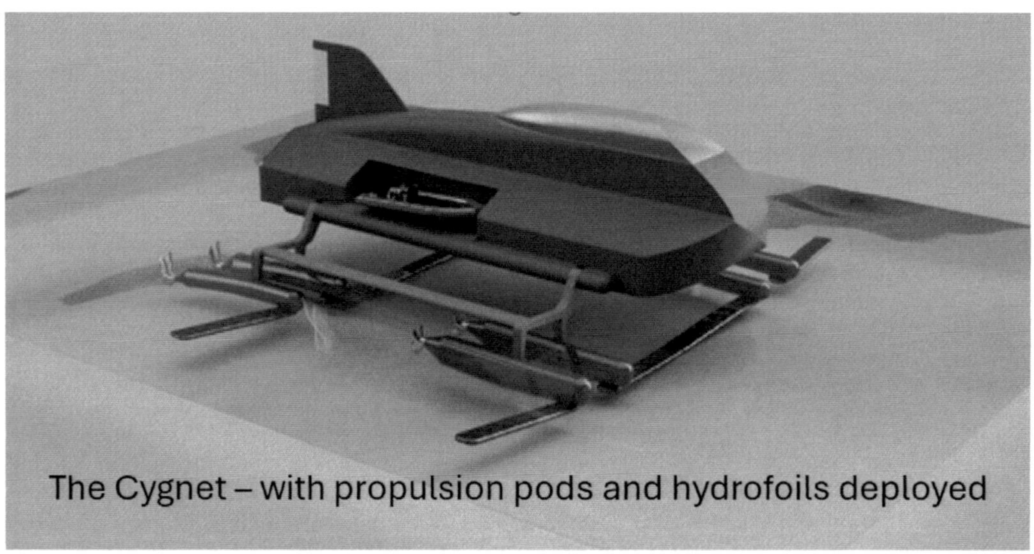

Figure 5: 1-Decker Cygnet render showing pods and foils deployed and sea boats. This craft is c.24m long

Thanks to Mark Storer of MES Consulting for assistance with CAD design and rendering of this concept.

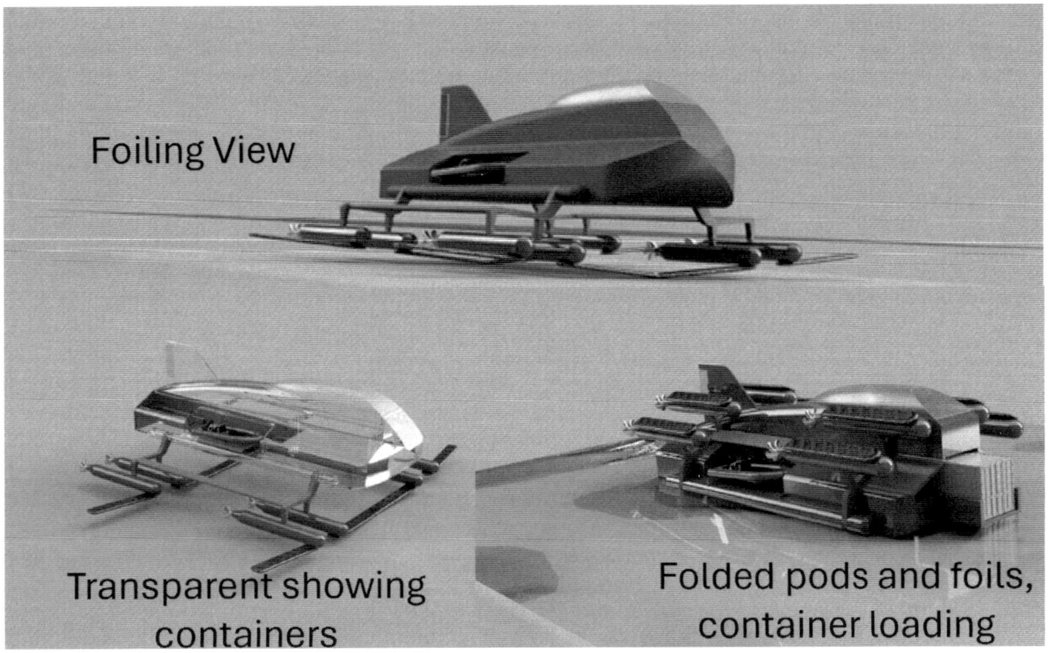

Figure 6: Renders of Cygnet craft showing different aspects including stowage of three 20' ISO containers

Hyper-modular construction is intended across this family of craft. Wherever possible the systems and equipment are to be included in ISO containers and 'plugged in' to the craft. This includes the following:

- Mission modules
- Some electrical generation and possibly storage
- Water making and wastewater treatment
- Refrigeration and stores
- HVAC equipment and controls
- Firefighting systems including mist or fog systems
- Computing and communications
- Secondary accommodation for passengers
- Medical facilities so that they can be donated ashore

Modules will be able to be swapped during life and even during operations. Consequently, there is a need to move loaded ISO containers at sea. This will of course be restricted by ship motions, but it is intended that this type of craft has inbuilt overhead gantries for controlled movement of containers. Containers will be able to be moved around the ship, not just as cargo but as operational systems. Interfaces for containers placed around the ship will allow exchange.

Hybrid composite construction is intended to achieve the light weight. This is sandwich panel construction developed from work the author carried out in 1989 at the Ecole Nationale des Sciences et Techniques Avancées (ENSTA) with the French and Belgian navies. Recently developed materials including fire-resistant structural foams and carbon fibre with fire retardant resins enable a whole new approach. This does require the design to be started from the ground up with composite construction in mind. Standard external module sizes for the sandwich panels are intended with selectable sandwich cores for different parts of the ship, for example acoustic insulation, thermal insulation, or ceramic and Kevlar armour.

Included within the thickness of the hybrid composite panels are ducts for electrical cabling and pipework. Deck panels will incorporate scuppers and drains and deckhead panels will have air extraction for HVAC, lighting, smoke and heat detection and firefighting spray systems built into the structure. Smart panels are intended with sensors and local controllers for the systems included within the composite panels. Computer controllers will be built into the structural panels to record the lifecycle and events to which each panel is subjected. This data can be used for maintenance and decisions about life expiry of panels based on their actual behaviours, such as vibration or thermal transfer.

Evolution through life is a key objective for this concept. By separating the hulls, designing for modular construction, and incorporating many systems and equipment into containers these craft will be able to adapt. All mission systems will be modular and carried rather than integrated. By including the ability to move ISO containers from watercraft up through the vessels to the flight deck or hangar these ships will also be able to carry and operate UAVs and weapon systems. It is intended that some mission systems such as defensive missiles or directed energy weapons will operate from the upper deck of the larger ships.

Electric ship principles will be applied. This is currently developing but is intended to be a Direct Current (DC) architecture for distribution with generation from a range of prime movers combined and possibly using rotating machines that double as gyroscopes to provide motion smoothing when moving fast in foiling model. The emergence of DC digital motors with high power enables this approach with storage in batteries, capacitors and possibly flywheels to enable peak-lopping. The expectation is that these ships will spend a large portion of their lives loitering with occasional sprints to respond to a crisis.

It is understood that this concept is radical – that is deliberate. When we as a community are presented with radical new solutions what do we do? Let us consider the emergence of turbines with Mr Parsons' Turbinia, or hovercraft, hydrofoils and surface effect ships that were prototyped in the 1950s and 1960s. They have come of age and a few brave nations have exploited these fast lightweight approaches. But who is brave enough to commission them, accept them into naval service or sign off a safety certificate?

6. MODELLING AND SIMULATION

This concept will go nowhere unless we can digitally model and simulate the solutions. The mathematics of the design must balance across:

- Mass and Lift
- Thrust and Drag
- Energy for endurance and speed
- Strength and deflections

At Nova Systems we carry out Digital Engineering and Model Based Systems Engineering, almost always targeted at a customer's system or concept. We model systems at different stages of the lifecycle, often working on developing requirements, conducting verification and validation (V&V) and operational test and evaluation (OT&E). We conduct detailed studies of motions in six degrees of freedom and computational fluid dynamics, often between two moving bodies. However, in the main we do not design ships. So, this concept will only 'take off' if you and others like the ideas presented here enough to consider collaborating on the development of new ship types.

A key test to build a business case for a new capability is to identify and evaluate the operational and financial benefits. There may also be time benefits and political benefits that come from this approach. Nova Systems evaluates military capabilities across all Defence Lines of Development (DLODs) or Fundamental Inputs to Capability (FICs). The architecture of this new capability needs to be developed in the Unified Architecture Framework (UAF), taking over shortly from the NATO Architecture Framework. This enables all aspects of the capability to be modelled rigorously in advance of build.

We produce digital twins of equipment, in particular the interactions between aircraft or drones and ship flight decks. The Digital Twin consortium, part of the Object Management Group has published their Periodic Table of currently 61 types of digital models. This is expanding with each version and now covers Data Services, Integration, Intelligence, User Experience, Management and Trustworthiness. See https://www.digitaltwinconsortium.org. Our community are all going to need to adapt and interface our computer design systems and supporting data to achieve a much greater level of interoperability and through life support. Many organisations have mature Product Lifecycle Management (PLM) systems in place, but we are still in an era of a multitude of independent digital models that are often recreated, resulting in duplication of effort, even within companies. We need data interoperability and interchangeability that can be sustained through life. Amazing new technologies are available for this.

Nova Systems Model Based Systems Engineering (MBSE) team are addressing this challenge and smoothing the way to multiple organisations collaborating on complex systems at different parts of the lifecycle.

7. BENEFITS

Applying a digital engineering approach allows one to investigate and review concepts more thoroughly before committing to building them. In this case of the SWAN what are the key capability benefits? Measures of Effectiveness (MOE), and Measures of Performance might include those shown in Table 4.

Table 3. Capability Measures of Effectiveness and Performance

Requirement	Measure of Effectiveness	Measure of Performance
Security	Ability to defend the ship from different threat types at various ranges	Analysis of threat arcs and effective radii of offensive and defensive missiles or guns
Humanitarian Facilities	Capacity for evacuation of casualties and delivery of aid	Number of personnel recovered by category of casualties. Tonnes of aid delivered.
Versatility	Range of mission types and score for effectiveness of each type	No. of different mission equipments that can be carried. Simplify to no. of ISO containers.
Sustainability	Endurance of craft including crew food and fuel consumption	Duration of operations per tonne of food and fuel. Emissions per mile and operating day
Modular Design	Proportion of reusable modules across the SWAN family or fleet	No of mission modules that can be carried and distributed. No of system modules
Rapid Deployment	Time to respond to a military or humanitarian crisis	Time to configure for each mission. Operating radii in 12hrs, 24hrs, 48hrs. Effective ranges.
Adaptive Technology	No. of equipment or systems that can evolve through life v fixed for ship life	Time required to change each mission and system type to respond to new technologies
International Collaboration	No. of nations participating in interoperable SWAN programme	No. of nations participating. No. of ships. No. of interchangeable modules. Interop score.
Crew Well Being	Reductions in motion sickness and injury, illness and fatigue	No. of crew unable to work in heavy weather. No. of crew injured per year. Retention.

The benefits of this radical approach are that the lifecycle of the systems, missions, weapons, and many other aspects can be broken away from the lifecycle of the platform. Taking a collaborative approach to design and construction of modules in many locations can dramatically change the speed of acquisition. A digital approach to design, acceptance and testing supported by digital modelling and simulation should also help to reduce acceptance time for a first of class.

8. TESTING AND ACCEPTANCE

Testing these craft will primarily be concerned with the more radical aspects. The structures and modules can all be tested separately, and systems integration facilities constructed ashore and even distributed at different locations. By taking an approach of digital integration via exchange of recognised data messages the ship systems can be tested before they are brought together within the ship. The physical and communications interface for many being the ISO container interface.

An integrated test evaluation and acceptance plan (ITEAP) will still be required. However, with the accuracy of digital design and Computer Aided Manufacturing, including 3-D printing of some components there will be much less variability in the construction than with traditional steel warship construction. This enables a different approach.

Cloud Towing Tanks can now be used to model hydrodynamics. This will be crucial, but the essential element of avoiding the waves proposed in this concept dramatically reduces complexity. Foiling at high speed does not scale well in a towing tank. There will be challenges to confirm the control parameters required, but it is evident right up front that real time computer controlled foilborne ride control systems will be required. This is normal for all modern vehicles.

Electrical systems and digital controllers will also need thorough testing and acceptance in offsite facilities. With a data interfacing approach, rather than tight integration between systems, it will be possible to test systems together across wide area networks. A progressive approach to acceptance has been proven in most of our projects.

Our experience with aircraft and air systems addresses an extremely thorough but more risk-based approach to certification and acceptance. This radical platform type will certainly require a prototype of at least the smallest craft. That in turn can provide acceptance evidence to support the larger members of the family. It is normal for air platforms to enter service with limited operating envelopes that expand through life. The same approach is proposed here with constrained speed, power, payload, and weather limitations applied, particularly to launch, recovery, onload and offload of offboard systems early in the lifecycle. The Bow Tie approach to safety is recommended, as used by the Military Airworthiness Authority, we have achieved good successes in pushing boundaries and assisting senior responsible owners and duty holders to understand the residual safety risks that they are signing for.

Sensors and weapons testing can largely be divorced from the life of the ship. Indeed, cooperative use of sensors and weapons across allied nations must be a key goal. These ships could carry novel weapons. The ability to separate crew from hazardous cargo or weapons creates a wonderful opportunity for furnishing these craft with extremely powerful weapons systems, whether those are directed energy or high-powered missiles (or possibly satellite launchers).

9. SAFETY AND REGULATION

High-speed foiling is beyond the capability of human controllers. These platforms require multiple automated systems for stability, 'flight control', power distribution and many other aspects. Operations will be automated wherever practicable, and oversight, diagnosis and analytics of operational data will be carried out ashore. First line maintenance aboard will be significantly reduced by a deliberate approach to 'repair by exchange' of modules.

Each module will require its own lifecycle, type approval, maintenance checks and inspections for this concept to work. A data-driven approach to product lifecycle management is essential. Different parties are likely to maintain responsibility for different modules, some may be operated on a service basis with ownership remaining with an OEM or leasing agent, much like 'power by the hour' is now common in commercial aviation.

The ship safety will be based around the High-Speed Craft Code, reference 9. But with a fundamental need to address people living on board. Materials selection within the hybrid composites and fire protection will be critical. Blast radii from hazards, in particular munitions, is mitigated by the fundamental arrangement of the multiple hulls. High speed reduces vulnerability to attack by some threats but creates other hazards. The ISO containers for any weapons systems will need be constructed as magazines.

Escape arrangements must be considered carefully. It may be acceptable to evacuate to one of the side hulls and for this to be detached from the main hull to provide a safe haven, rather than moving into life rafts or lifeboats. Redundancy of power systems and propulsion is inherent within this design concept. This should also enable much greater recovery of wasted energy in the forms of heat and motions. The use of a primarily DC storage and distribution approach to electricity also mitigates some hazards but creates others.

The challenge with this concept is that it moves a long way from a traditional monohull. Basing the approach on commercial Class will require a different approach to risk compared with commercial shipping. These craft are not well suited to commercial operation. Indeed, the primary requirement is to respond rapidly to threats and incidents. Therefore, the scope of consideration of the safety case may need to be wider than just the ship.

The safety case for operation of armed and unarmed offboard systems in all environments will require particular care. The use of Ship Vehicle Operating Limits based on Ship Helicopter Operating Limits (SHOL) is something that Nova Systems is developing for drones and exploring with very advanced modelling and simulation.

To take on this radical approach requires a cautious approach to regulators. A pressing need for new warship types exists and an approach that avoids acquisitions of twenty years and a billion pounds is required. By designing for a multi-national, multi-role, multi-type, multi-supplier ecosystem this concept addresses those challenges.

10. CONCLUSIONS

This paper provides a deliberate challenge to current warship naval architecture. Forcing your body through the waves is something that nature evolved away from many millions of years ago. Humans, in particular naval architects, may have got it wrong by persisting with surface transit. However, we are clearly wedded to the concept of buoyancy as a cheap way of transporting goods, so this paper is unlikely to impact commercial shipping.

The thesis that a radical design concept requires a radical approach to test, evaluation, and acceptance may be flawed. That is up to you as colleagues and professionals. This institution exists to promote advances in science, technology, and shipbuilding. This concept offers an approach that pushes those boundaries.

Since this concept is the author's individual research rather than a company sponsored design, please recognise that the opinions expressed should not reflect on Nova Systems. It remains the author's goal to see ships of this nature applied to disaster relief missions where the response time to an event occurring in an unknown place at an unknown time provides the key driver for speed. Over 10,000 people a year are killed by natural disasters and whilst these ships can't save those who are killed by the initial event, navies can and do help to save the lives of those who are injured, trapped, or left without clean water, shelter, or functioning infrastructure. There is a pressing need for internationally managed 'peaceships' as well as next generation warships.

These ships speak to the deterrent effect of being able to rapidly respond and to surprise a hostile entity. The most significant warlike feature of these ships is "*surprise*" (to quote Monty Python). Others will never know what these ships are capable of because they can be filled with non-descript boxes that can be deployed to other units, to locations ashore and could contain a myriad of different capabilities. As these ships evolve through their lives the modularity will enable them to keep up with the pace of technology and respond to different issues that threaten our sea lines of communication and our interests around the globe. They could also adapt to oceanographic changes forecast in reference 7 that will driver wider climate change and political tension, resource competition, displacement of people, and war.

11. REFERENCES

1. HOOD, M W., 'The Shape of Fleets to Come', *Warship 2015: Future Surface Vessels, RINA*, 2015.

2. ANDREWS, D., 'The Multi Role Combatant – Jack of All trades, Master of None', *Warship 2019: Multi-Role Vessels, RINA*, 2019.

3. BALLANTYNE, I., 'Strike from the Sea', *Pen and Sword Books*, ISBN 1-84415-059-3, 2004.

4. CHILDS, N., 'Britain's Future Navy', *Pen and Sword Books*, ISBN 978-1-84884-291-5, 2012.

5. UK Ministry of Defence, 'Defence Drone Strategy – the UK's Approach to Defence Uncrewed Systems', UK Ministry of Defence, 22 February 2024 Defence Drone Strategy - the UK's approach to Defence Uncrewed Systems - GOV.UK (www.gov.uk).

6. FREEDMAN, L., 'The Future of War – A History', *Penguin Random House*, ISBN 978-1-846-14749-4, 2017.

7. CZERSKI, H., 'Blue Machine – How the Ocean Shapes Our World', *Penguin Random House*, ISBN 978-1-9117-0910-7, 2023.

8. SCHULMAN, S., "Iraq's Water Plight – The Drought Between Two Rivers", *Royal United Services Institute Journal*, Issue 2/2022.

9. IMO, '2000 HSC Code – International Code of Safety for High-Speed Craft, International Maritime Organization, ISBN 987-92-801-4240-2, 2008 edition.

12. AUTHORS BIOGRAPHY

Matthew Hood holds the current position of Head of Practice at Nova Systems in the UK. The Author is responsible for expanding capability in supporting the management of defence systems. The Author's previous experience includes training through the Royal Corps of Naval Constructors and design, acceptance and support of warships and submarines. He has also spent many years supporting aircraft, rail systems and C4ISTAR systems.

NAVAL CREW SIZE AND HABITABILITY – WHERE IS THE FUTURE?

M O'Connor and **D Bong**, Taylor Bros Marine Pty Ltd, Australia

SUMMARY

Habitability standards on naval surface ships have changed over the years with an increased focus on crew welfare, space and well-being. These standards drive a reduction in crew size. Naval crewing policy however has not changed in the same manner as the commercial shipping industry with the introduction of more automated equipment. Crew sizes have been maintained and on certain vessels increased. Taylor Bros has been involved in numerous installations and studies where modern crew habitability standards have been applied while the increased vessel capability has driven a crew increase rendering crew habitability standards impossible to meet. Any crew increase has a flow on effect to other ships auxiliary systems and design aspects, while any crew decrease will have the same flow on effect but in the reverse direction. Such an effect can have some major impacts on ship capability such as ship displacement, range and particularly endurance.

1. INTRODUCTION

Naval Habitability standards have changed substantially over the last 20 to 30 years with an increase focus on crew welfare. These standards quantitively provide increased space for crew members in berthing, recreational and working areas. However, such space is not often available as the naval surface combatant design footprint has not changed. Thus, to meet habitability standards either the crew berthing spaces needs to increase which will provide the adverse effect of reducing vessel capability in other areas or crew numbers need to decrease.

Based on real experience, however, neither vessel capability nor crew numbers are reduced. Instead, crew habitability standards are compromised. It is worth viewing the evolution in the commercial shipping world, as due to vessel equipment automation, neither crew habitability standards or vessel capability are compromised. Crew numbers are reduced.

2. HABITABILITY AND ACCOMMODATION RULES

The Royal Australian Navy (RAN) has 2 major habitability standards relevant to the fleet. These are:

- **Def (Aust)5000 Vol. 08 Part 01**: Accommodation Requirements
- **DST-Group-TR-3550**: A Revised Maritime Physical Accommodation Guidance for the Royal Australian Navy

As the majority of Australia's fleet is designed outside of Australia, these standards are rarely adhered to or in compliance for newly designed vessels. However, any modifications or work done during operation of the vessels these standards are referenced.

Taylor Bros has a lot of experience in (and ongoing attempts in) retrofitting existing designs to meet these standards, and understands the implications these standards have on vessel design and capability.

Commencing with the Def (Aust)5000 Vol. 08 Part 01 as being the specified standard for habitability across the fleet, some important context and citations of the standard are included below:

General: 'The RAN's primary mission is to be prepared to fight and win in the maritime environment either alone or as an element of a joint or combined force. The Navy is dependent upon its shipboard personnel to accomplish this mission. Navy shipboard life contains a number of elements that may have a considerable bearing on the physical and psychological wellbeing of the ship's crew. These include separation from family and friends, intermittent danger, boredom, forced contact with others, and the unpredictable nature of the sea. The RAN recognizes the positive impact that appropriate habitability criteria and design practices can have on mitigating these factors.'

Capability Aspects: 'Good habitability standards are essential in ensuring that a ship's crew performs at its maximum potential. It can be shown from research and anecdotal evidence that a ship's crew that is happy with their living conditions will be more motivated, productive, proactive and perform their duties to a higher standard for a longer period.'

Safety: 'Safety is of paramount importance and should be considered at all times in the design and implementation of surface ship habitability. Habitability directly and indirectly influences the safety of the crew and the platform. A ship that is well designed from a habitability view point will minimize the crew's exposure to hazards such as excessive noise and vibration, falls, motion induced interruptions, exposure to large variations in climatic conditions. Good habitability standards will directly lead to increased motivation and awareness which will in turn have a positive effect on safety.'

Consequences of Poor Performance or Hazard: 'Poorly designed and outfitted Habitability spaces will lead to a crew that is unhappy with their quality of life. This will have several short and long term impacts. This will directly impact on crew safety, quality of life, recruitment and retention, crew performance and morale.'

From the above overview direct from the navy's primary standard on ship habitability, habitability is not only important from a crew morale and quality of life perspective, but it is intertwined deeply in ship capability and crew safety.

The overarching rules in the Def (Aust) 5000 standard which dictates the number of crew is the berthing requirements per cabin. These are summarised below for each rank of crew member.

Captain/Commanding Officer suites shall comprise of:

- A combined day/dining cabin of 19 square metres (approx.),
- A sleeping cabin of 7 to 9 square metres (approx.), and
- An ensuite compartment.
- The combined day/office, dining cabin, sleeping cabin, and ensuite ablution facilities shall be fitted out to the same requirements as the corresponding Flag Officers apartment.

Cabin - Head of Department (HOD):

- HOD cabins shall have 8 to 9 square metres deck area. These cabins are for use by Heads of Departments and should be suitable for interviews, meetings, and office work.

Officer Cabins - Single Berth:

- Officer cabins shall have 6 to 7 square metres deck area depending on the users rank and responsibility.

Officers Cabin – Double Berth:

- Officer cabins of double berth cabins shall have 8 to 10 square metres deck area. The allowances for furniture shall be one outfit for each occupant with the exception of washbasins, which are provided on the basis of one per two occupants.

Senior Sailors Accommodation.

Warrant Officers:

- Warrant Officers are entitled to a single berth cabin of between 5 to 6 square metres deck area. Single cabins should be the designers aim.

Chief Petty Officers:

- Chief Petty Officers shall be provided with multi-berth sleeping cabins. Two berth cabins should be the designers aim.
- Two tier bed berths shall be provided, aligned in the fore and aft direction.

Petty Officers:

- Petty Officers shall be provided with multi-berth sleeping cabins. Four berth cabins should be the designers aim.
- Two tier bed berths shall be provided, aligned in the fore and aft direction.

Junior Sailors:

- Deck area for Junior Sailors sleeping spaces shall be based on 1.6 square metres per person messed, not including bunk/locker area. The maximum number of Junior Sailors located in each sleeping cabin shall be twelve. The minimum number shall be six
- Two tier bunks shall be provided, aligned in the fore and aft direction. The sleeping cabin shall be fitted with ensuite compartments.

Defence Science and Technology (DST) Group provided guidance in 2015 regarding navy habitability standards in the stipulated documents below:

- **DST-Group-TR-3550:** A Revised Maritime Physical Accommodation Guidance for the Royal Australian Navy
- **DST-Group-TR-3564:** Revised Anthropometry Guidance for the Royal Australian Navy

In 2015 an Anthropometric Survey of the Royal Australian Navy (ASRAN) was conducted providing comprehensive female and male body size and shape data of the Royal Australian Navy operational workforce. Details of this study can be found in document DST-Group-TR-3564.

The RAN's anthropometric data is essential for accurately aligning and designing the physical dimensions of compartments, workspaces, systems, or equipment with those of the users. This ensures precise fitting to the diverse range of individuals within the RAN population. The RAN anthropometric data in DST-Group-TR-3564 serves as the basis for evidence based human factors engineering design guidance, as outlined in DST-Group-TR-3550. This document was developed to tailor design solutions to optimise the fit between the RAN population and HMA Surface Ships and Submarines, and their systems, subsystems, and facilities. The main purpose of this guidance is to provide future and modified naval vessels with physical habitability requirements that reflect the physical and social needs of the personnel that use it. The end goal is to optimise performance, health and safety, quality of life, and satisfaction.

Key areas addressed in the document in regards to naval habitability design include:

- General body space requirements
- Workstation requirements
- Bunk requirements
- Passageway requirements
- Mess seating requirements
- Water closets and showers

All of the above areas of study provide design guidance for acceptable size and space requirements for ships' berthing areas and bunks, ships passageways, ships offices, workstations and consoles, mess seating and water closet and shower requirements.

Further details of the methodology and application examples of these DST standards can be found in Annex 1.

When considering the broader context of ship design, particularly for RAN's fleet, it becomes evident that many design elements are brought or influenced by international standards and practices. This is as seen in the case of the Hobart Class Destroyers being based on the Spanish F100 design and the SEA5000 Hunter Class Frigates, which are based on designs from T26 Frigate and the Royal Navy's anthropometric data. While drawing from these external sources can offer valuable insights and best practices, there is a clear imperative to prioritize the collection and utilization of anthropometric data specific to the RAN's operational workforce. This data provides invaluable insights into the physical dimensions and requirements of RAN personnel, ensuring that ship designs are optimized for the individuals who will be using them. By incorporating this data into the design process, naval engineers and architects can better tailor accommodations, workspaces, and equipment to suit the needs of RAN personnel, ultimately enhancing safety, performance, and overall operational efficiency.

3. CREW SIZES AND EXAMPLES OF INCREASED CREW SIZE

Table 1 provides an overview of the crew size for a range of RAN current and future surface combatant fleets. Specifically, Taylor Bros experience has been with the Hobart class Destroyer vessel where a crew size of 30 additional personnel to the baseline design was specified.

To note also but not specifically related to Surface Combatants, Taylor Bros has been recently involved in an increase in crew size on the AOR vessels of an additional 10 personnel and the Armidale Class patrol vessels of an additional 8 personnel.

The rationale behind the increased crew size of the destroyer vessels was an increase of training capacity for personnel, particularly during helicopter operations as this already provided a maximum capacity of the ship's crew.

Table 1 below indicates the crew size for a range of RAN Surface combatants. The overall number of crew has not drastically changed over the last 20 to 30 years. The table also shows the proposed increase in bunk size for the Hobart Class Destroyers for an additional 30 personnel.

Table 1. Crew size overview per RAN Surface Combatant Ship Class.

	ANZAC Class Frigate	Hobart Class Destroyer	Proposed Increase Hobart Class Destroyer	Hunter Class Frigate
CO	1	1	1	1
Officers	21	22	22	30
CPOs	25	29	29	30
POs	50	62	62	30
Junior Sailors	65	72	72	117
Total	163	186	186	180
		Spare Staff		
CO		1	1	
Officers	5	11	11	5
CPOs	5	17	32	5
POs	5	10	25	8
Junior Sailors	1	9	9	10
Total	16	48	78	28
Total Berths	**178**	**234**	**264**	**208**

4. IMPACT OF CREW SIZE ON SHIP DESIGN AND AUXILIARY SYSTEMS

It is certainly not negligible the impact crew size has on other ship systems and the total ship design and vessel capability.

By adding 30 extra personal to the Hobart Class Destroyer the following direct influences on the ship capability and ship auxiliary systems were affected:

- The endurance of the vessel reduced from 34 to 30 days.
- The time taken for the slowest group to embark the vessel from working stations to embarkation stations increased by 31 seconds.
- An additional 2 life rafts were needed.
- Additional washer and dryer capacity is needed in both the communal laundry and self-service laundries. Self-service washer and dryer demand increased by approx. 1.5 kg per berthing zone

(changing the size of each stacked washer and dryer from 6kg capacity to 8kg capacity) and communal dryer capacity increased to 13kg.

- Potable water storage capacity was needed to increase by approx. 1400l to achieve a required storage capacity of 1.5 days.
- The ships HVAC system does not meet ISO standards for the stipulated amount of fresh air entering certain cabins due to the increase in the number of berths.
- Recreational space (Mess area and facilities) does not meet the minimum requirements for seating needed for the number of crew.
- The number of toilets, showers and wash basins in junior and senior sailors berthing does not meet the minimum requirement for numbers per berth.
- Stowage capacity for senior sailor's cabins does not meet the minimum amount of $0.75m^3$ per person.
- Galley size and capacity does not meet the minimum requirements.
- Food stowage capacity does not meet the requirements for the stipulated crew size and vessel endurance. Thus, the vessel endurance is compromised.

Berthing arrangements was one major area where the current baseline design does not meet the Def (Aust) 5000 habitability requirements. However, by increasing the number of berths in certain compartments the habitability of the crew has moved further away from acceptable standards. This is shown in table 2 below with the berthing area analysis of the baseline design and the proposed upgraded design for the Hobart class destroyers.

The deck area requirement is a minimum of $1.6m^2$ per person messed, not including bunk and locker area.

It should also be noted to comply with the need to increase crew capacity 3-tier bunks have been specified over 2-tier bunks. This is in direct contradiction to the Def (Aust)5000 standard and the DST-Group-TR-3550 standard. These documents both recommend 2-tier bunks to be used rather than 3-tier bunks.

Table 2. Berthing area analysis Hobart Class Destroyer.

Cabin	Cabin Name	Initial Arrangements				Design Arrangements				No. of Bunks Increase
		Deck Area, Excluding Furniture (m^2)	No. of Bunks	Deck Area per person (m^2)	Compliant to DEF (AUST)	Deck Area, Excluding Furniture (m^2)	No. of Bunks	Deck Area per person (m^2)	Compliant to DEF (AUST)	
3-172-0-L	Junior Sailor's Berthing N-2	7.44	9	0.83	NO	-	9		-	0
3-173-2-L	Junior Sailor's Berthing N-1	7.62	9	0.85	NO	-	9		-	0
3-164-3-L	PO's Berthing N-1 (F)	8.83	8	1.10	NO	9.16	10	0.92	NO	2
3-156-1-L	PO's Berthing N-4 (M/F)	7.61	8	0.95	NO	7.10	10	0.71	NO	2
3-158-2-L	PO's Berthing N-3	7.98	8	1	NO	7.72	9	0.86	NO	1
3-164-2-L	PO's Berthing N-2	5.78	4	1.45	NO	5.52	6	0.92	NO	2
3-50-1-L	Ratings Berthing N-9 (M/F)	6.71	9	0.75	NO	6.74	9	0.75	NO	0
3-57-1-L	Ratings Berthing N-6 (F)	6.87	9	0.76	NO	6.91	9	0.77	NO	0
3-51-2-L	Ratings Berthing N-8	6.67	9	0.74	NO	6.96	9	0.77	NO	0
3-58-2-L	Ratings Berthing N-7	6.14	9	0.68	NO	6.41	9	0.71	NO	0
2-50-3-L	CPO's Cabin N-5	5.75	4	1.44	NO	5.57	6	0.93	NO	2
2-50-4-L	CPO's Cabin N-6	5.79	4	1.45	NO	5.68	6	0.95	NO	2
2-57-2-L	CPO's Cabin N-4 (F)	5.72	4	1.43	NO	5.56	6	0.93	NO	2
3-36-2-L	PO's Berthing N-11	7.90	6	1.32	NO	6.94	9	0.77	NO	3
3-44-4-L	PO's Berthing N-9	8.05	6	1.34	NO	7.30	7	1.04	NO	1
3-44-1-L	PO's Berthing N-7	6.19	4	1.55	NO	5.70	6	0.95	NO	2
3-37-1-L	PO's Berthing N-10	8.90	8	1.11	NO	8.64	9	0.96	NO	1
2-44-0-L	CPO's Cabin N-9	5.30	4	1.33	NO	5.82	6	0.97	NO	2
2-44-1-L	CPO's Cabin N-8	5.30	4	1.33	NO	5.82	6	0.97	NO	2
2-172-6-L	CPO's Cabin N-3	5.50	4	1.38	NO	5.82	6	0.97	NO	2
2-172-0-L	CPO's Cabin N-2	5.50	4	1.38	NO	5.87	5	1.17	NO	1
2-172-1-L	CPO's Cabin N-1	5.50	4	1.38	NO	5.87	6	0.98	NO	2
3-44-2-L	PO's Berthing N-8	5.10	6	0.85	NO	5.09	7	0.73	NO	1
		TOTAL	**144**			**TOTAL**	**174**		**TOTAL**	**30**

5. CREW SIZES IN THE COMMERCIAL SECTOR

An analysis and overview of current crew sizes in the commercial shipping sector is worth noting. It is acknowledged that commercial ships have a much different mission than navy vessels, but the influence of equipment automation and crew size over the years should not be discounted.

It is fair to say in the commercial sector crew size has halved over the last 30 years. with the increased use of automation and the common use of unmanned machinery spaces. Studies are being conducted to further reduce crew size but as of today the commercial sector has minimum manning requirements dictated by both the IMO and flag state authorities.

Many commercial ships are single mission non-complex ships with little comparison with a navy vessels multi-role capability and complex ship systems. Some vessels however provide a closer comparison where commercial vessels are multi-role with various complex ship systems are employed that require some specialised crew competence to operate.

One vessel is the RSV Nuyina, a new icebreaking research and supply vessel owned by the Australian Antarctic Division. This vessel has various specialised ship missions from carrying specialised equipment and scientists to Antarctica as well as conducting specialised scientific research.

The vessel's principal particulars are below in Table 3.

Table 3. RSV Nuyina principal particulars.

Length OA:	160.3m
Beam Moulded:	25.6m
Depth to Main Deck:	19.2m
Draught:	9.2m
Displacement (Full load):	25 500 tonnes
Speed Max:	16+ knots
Range @ 12 knots:	16000+ nm
Icebreaking @ 3knots:	1.65+ m
Silent Operation:	8 knots
Endurance:	90+ days
Crew:	32
Passengers:	117

The RSV Nuyina has over 20 different modes of operation on its complex hybrid twin screw propulsion system. The propulsion system can be used in electric mode, mechanical mode, or a combination of both in either booster mode for increased speed or power for icebreaking or increased power generation where the PTO electric motors can be used as generators. The vessel also has 3 stern thrusters and 3 bow thrusters with a DP3 rated dynamic positioning system providing excellent manoeuvrability.

Equipment is designed to function in the extreme weather of Antarctica and also tropical summer conditions of 45 deg air temperature. This necessitates a complex HVAC system on the vessel. For instance the vessel has 2 chilled water systems supplying 11 air handling units and 47 fan coil units on board to control the air temperature of various specialised compartments throughout the vessel.

The vessel has a doctor and medical facility, fitness centre, theatre and various specialised scientific equipment that can be deployed to conduct specialised scientific research. These can be deployed via the use of an aft A-frame, moon pool handling system, Fwd. deployment boom and drop keels.

The crew size for the RSV Nuyina is 32 which is broken down in Table 4 below:

Table 4. RSV Nuyina crew size.

RSV Nuyina Crew Size	
Master	1
Deck officers	6
Engineering officers	7
Integrated Ratings	12
Stewards/cooks	6
Total	**32**

The RSV Nuyina does have berthing capacity for 116 additional personnel and 1 doctor but these are not considered part of the vessels crew to operate the vessel.

In determining the appropriate crew size for a commercial vessel generally the ship owner undergoes a crewing evaluation. For international vessels this is outlined in IMO Resolution A.1047(27) Principles of Minimum Manning. For Australian flagged vessels Marine orders 504, Schedule 1 clause 6(4) lists the factors that must be considered when undergoing a crewing evaluation.

It is important to state that based on the well-developed IMO convention for the Standards for Training Certification and Watchkeeping (STCW) crew competence is standardised and the level of competence needed by the crew for shipborne tasks and levels of responsibilities is stipulated. This makes the important task of calculating crew numbers in the vessel crewing evaluation much easier.

Australian Maritime Safety Authority (AMSA) has developed some guidance in performing a crewing evaluation with an overview of the factors that effect crewing numbers below.

AMSA Overview of crewing evaluation: The owner of the vessel must determine the level of appropriate crewing through a process referred to as 'crewing evaluation'. Appropriate crewing may change depending on the tasks and the kind of operation being carried out.

Marine order 504, Schedule 1 clause 6 (4) lists the factors that must be considered when carrying out a crewing evaluation:

- Tasks and activities performed in addition to the safe navigation of the vessel, and the demands they impose on the master and crew
- Risk of fatigue of the master and crew
- Number of people to be carried on the vessel and the effectiveness and timeliness of arrangements for passenger monitoring, taking into account that the master of the vessel must be able to find out the number of passengers on board the vessel at any time
- Design characteristics of the vessel, including its general arrangements, machinery and equipment
- Qualifications and competencies of the master and crew, including circumstances where only the master holds mandated engineering qualifications
- Competency needed to use technological aids to safety and navigation fitted, in addition to the mandatory requirements (e.g. fire safety systems, remote engineering monitoring and diagnostics, electronic communication and navigation equipment, closed circuit TV)
- Area of operation of the vessel and expected conditions, for example weather, climate, and water temperatures
- Duration of the voyage
- Requirements for the vessel's emergency preparedness, including the vessel's emergency plan and evacuation arrangements
- Maintenance requirements of the vessel, and its machinery and equipment
- External support available to the vessel

6. SHIP OPERATION IN THE COMMERCIAL SECTOR AND CULTURAL DIFFERENCES

It should be noted and explained the large cultural difference between the navy and commercial ship operators. Commercial ship operators have managed to successfully streamline technical and commercial roles and responsibilities of ship management as well as ship side and shoreside management into one organisation.

The naval structure is bureaucratic in nature where there are clear lines of management and organisational roles and responsibilities. Multiple organisations would manage a naval vessel from ship's crew to shoreside staff to repair and maintenance personnel. These roles and responsibilities can be more centralised in the commercial world where one organisation is responsible for multiple functions, allowing a far greater level of efficiency in operations. Only certain specialised tasks would be sub-contracted out to a 3rd party organisation giving the ship manager far greater autonomy of the ship's operation. This also allows the crew to perform more maintenance tasks of vessel equipment with only major equipment breakdowns or maintenance tasks being sub-contracted out. Navy crew are performing far less maintenance tasks than commercial ships with most of the equipment maintenance being sub-contracted out to equipment manufacturers or representatives.

Bureaucracy is evident in the breakdown of a navy vessel crew compared with a commercial vessel. A navy vessel has clear lines of management and roles and responsibilities which has not greatly changed over the years with the change in equipment technology. A commercial ships lines of management have become much flatter with greater roles and responsibilities placed on fewer personnel.

In summary a commercial ships crew have far greater levels of responsibility and autonomy than in the naval sector. Commercial ships crew have become more multi-functional and ship's crew perform a greater level of maintenance tasks.

A breakdown of both a typical naval surface combatant crew organisational chart is contained below and compared with a typical commercial ship breakdown. What is not shown on the naval chart is the further breakdown of ship's crew into Chief Petty Officers, Petty Officers and Able Seaman. What is strikingly different is the levels of management are much less on a commercial vessel. Also, the number of personal for each level is much less on a commercial ship.

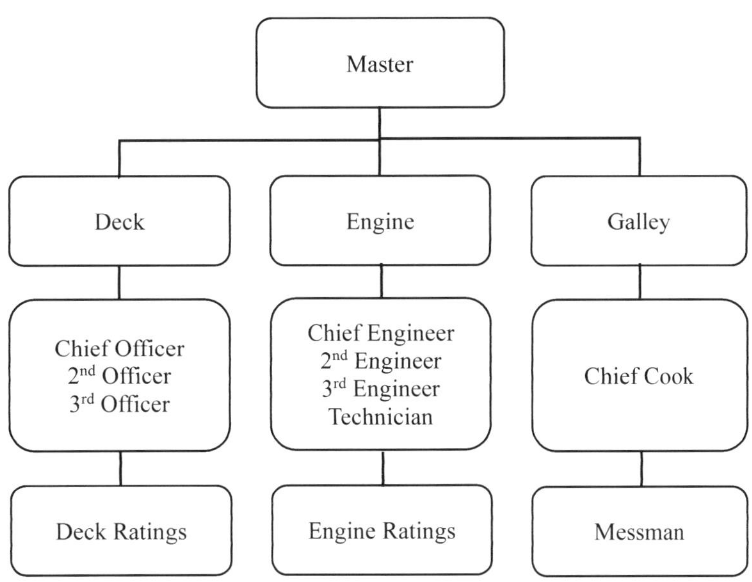

Figure 1: Typical merchant ship crew breakdown.

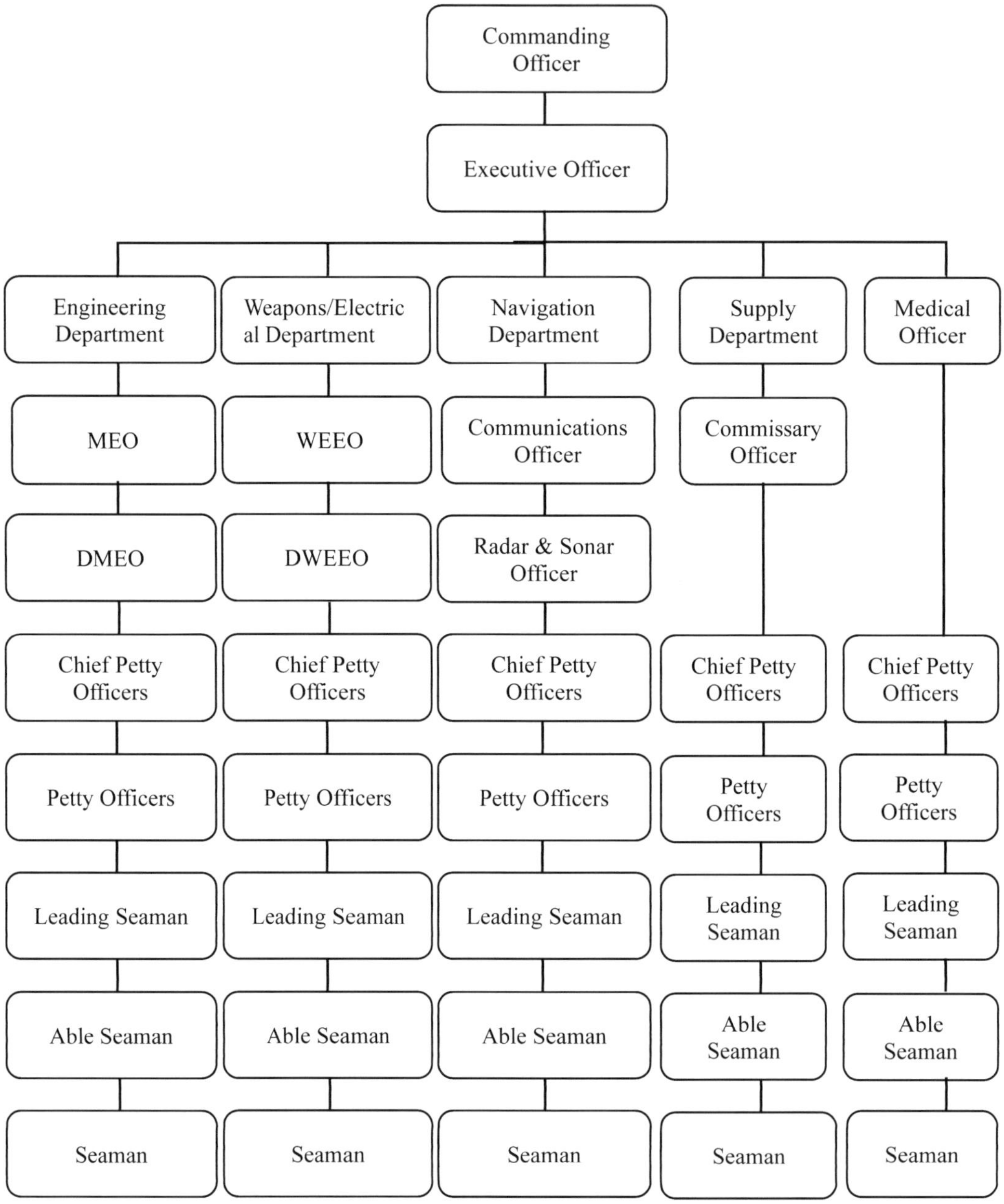

Figure 2: Typical RAN surface combatant crew breakdown. (Indicative)

7. DESIGN DIFFERENCES IN THE COMMERCIAL SECTOR COMPARED WITH WARSHIPS

There are some major ship mission philosophical differences when designing commercial ships compared with Navy ships. A naval vessel and in particular a surface combatant is a nations means of defence and crew onboard, damage control systems and procedures onboard are oriented towards saving the ship, rather than in the commercial world where the loss of life is the priority rather than the loss of the ship.

Thus, crew on a naval vessel take greater responsibility in damage control activities. This includes firefighting, damage control from flooding and Chemical, Biological, Radiological and Nuclear Defence (CBRND) and general preparation for warfare.

A naval vessel will have a greater level of communication equipment onboard as a surface combatant has a command centre where it can be part of a fleet of ships and exercise some form of command and control. This function requires a high level of specialised communication equipment to communicate within a battlefleet and home to the flag state nation. Top side design is also very important for a surface combatant as various sensors and weapons systems for both defence and attack as well as specialised radars and communications equipment need to be carefully placed on the top deck to ensure there is no interference.

Specialised communications and warfare officers and crew is thus needed on naval vessels and this aspect of the vessel design requires a higher manpower than a commercial crew.

Due to this nature of a warship crew size is often seen as an enhancement of vessel capability as the crew have not been replaced in their ability to save the ship in a wartime or damage control scenario and fight a battle.

No doubt damage control activities have become more automated over the years, such as the use of automated watertight doors, centralised and fixed firefighting systems, IPMS Systems where ship control as well as damage control activities can be controlled from multiple work stations and compartments throughout the ship.

If naval warfare and damage control activities can be performed as equally effective with less personnel than currently used is an aspect of Naval ship design that needs further consideration. It is unclear if this is an area that has been greatly studied but it certainly needs further examination of the risk-benefit of moving naval habitability in line with modern standards and its effect on the ability of a warship to better defend itself in a number of damage control circumstances.

In summary ships functions such as vessel navigation and operation are unchanged between naval and commercial vessels. However, command and control, damage control and warfare specifics of a naval vessel require a larger crew. It is not clear however weather this need for a larger crew has been quantified in the context of today's modern and automated onboard equipment.

Also, when designing a new surface combatant and integrating habitability standards into the vessel requirements from the outset some interesting characteristics are observed that are explained in the next chapter.

8. REALISTIC CHANGE PROPOSALS AND THE IMPACT IT WOULD HAVE ON SHIP CAPABILITY

This study does not propose a reduction in crew size based on any crewing evaluation. It does recommend a crewing evaluation be conducted that is better reflected with the needs and benefits of a modern naval surface combatant.

This study does however propose a surface combatant configuration that meets current naval habitability standards and highlights the influence this has on the vessel capability and the design aspects of the vessel and its auxiliary systems.

Firstly, returning to the Hobart Class Destroyer study conducted. Working within the confines of the as-built cabins the table in Annex 2 proposes the number of berths per cabin that will meet the navy standards for habitability.

This is an over 50% reduction in crew size for the senior sailors and below. As all officer's accommodation meets the naval standards this was not scrutinised and kept as per the baseline design.

Total crew numbers have reduced from 234 to 125 personnel in order to meet the navy habitability standards.

What does this crew reduction do for other ship systems and the general design and capability of the vessel.

A summary is contained below:

- The Endurance of the vessels increases from 33 to 62 days. Or the food storage capacity needed to meet the current endurance reduces from 117m^3 to 65m^3.
- The Potable water storage capacity needed and water making capacity reduces by a factor of 2.

- Domestic hot water consumption reduces by a factor of 2, reducing hotel load by approx. 180kW.
- The grey water capacity and storage tanks reduces by a factor of 2.
- The ships sewage treatment plant can be reduced by a factor of 2.
- The HVAC system can be simplified with the capacity of Air Handling Units and Fan Coil Units reduced and the capacity of supply fans reduced.
- The Laundry capacity can be reduced by a factor of 2 for both communal laundries and self-service laundries.
- Galley size and capacity is reduced by a factor of 2.
- Messed area and dining area space requirements are reduced.

It has not been quantified but the vessel light ship weight will certainly also be reduced. This can lead to reduced running costs with less fuel & provisions needed to meet the vessel mission, or vessel range and capability can be increased with capacity for increased fuel and payload (in the form of combat systems) due to the reduced light ship weight.

Crew numbers certainly have such a vast impact on the surface combatant design that such a design may lead to a rethink of how navies project such a capability.

With less crew vessel capability can be greatly increased on the same platform or for the same capability the vessel size being both displacement and/or dimensions can be greatly reduced. This can allow navies to better optimise their fleets also by providing the same capability with either more ships or less ships than currently utilised.

Also, it's worth noting the operating costs of a vessel are greatly influenced by crew size and any major change will have a substantial effect on vessel running costs.

9. CONCLUSION

The RAN habitability standards for surface ships are modern, well detailed and specific to RAN personnel. These standards being the Def(Aust)5000 Vol.8 Part 01: Accommodation Requirements and DST-Group-TR-3550: A Revised Maritime Physical Accommodation Guidance for the Royal Australian Navy.

Any attempt to meet these standards involves a compromise.

If these standards are intended to be met then they need to be incorporated into the design from the outset and any attempt to retrofit such standards into existing designs based on experience is not possible without major ship wide design implications.

Naval crewing policy is currently not aligned with such standards and they seem to have diametrically opposed objectives that need major design compromises for successful integration.

A comparison with commercial ships and crewing policy provides valuable insight. Although it is noted the major design differences and objectives of naval and commercial vessels it is unclear this is truly reflective of the vast difference in crew numbers between the two. Due mainly to vessel automation and an increase in crew autonomy with fewer layers of management the commercial shipping world has managed to reduce crew size without compromising vessel capability and maintaining high crew habitability standards.

This is an area the RAN has not achieved and it is unclear if any major studies into crewing policy have envisaged to reduce the crew size of its surface combatants in an attempt to provide compliance with modern habitability standards.

A reduction in crew size can certainly provide both ship wide and fleetwide capability increases if incorporated in the design from the outset. Ship auxiliary services are greatly reduced and vessel range, endurance & combat system capability can be increased on the same platform, while providing crew a safer environment with a higher standard of living giving them a better opportunity to perform at their maximum potential.

10. REFERENCES

1. Australian Government Department of Defence., 'Def (Aust)5000 Vol. 08 Part 01: Accommodation Requirements. *Defence DRN Maritime Materiel Requirements Set*, 4TH March 2008

2. PONTON, K., "A Revised Maritime Physical Accommodation Guidance for the Royal Australian Navy", *Maritime Division Defence Science and Technology Group*, October 2018.

3. PONTON, K., HUF, S., GAMBLE, D., COLEMAN, J., FURNELL, A., COCKSHELL, S., FLETCHER, K., 'Revised Anthropometry Guidance for the Royal Australian Navy'. *Maritime Division, Land Division and Weapons and Combat Systems Division Defence Science and Technology Group*, 2019.

11. AUTHORS BIOGRAPHY

Michael O'Connor currently is the Design Manager at Taylor Bros Marine Pty Ltd. The author is responsible for oversight in all Marine Engineering and Naval Architecture related tasks within Taylor Bros. Taylor Bros have designed, manufactured and installed the entire accommodation facilities for the Hobart Class Destroyers and Sea 1180 Offshore Patrol Vessels. The author has over 20 years of Marine based design experience with experience in the design and upgrade of naval amphibious vessels, patrol vessels and surface combatants.

Dean Bong currently serves as a Project Engineer at Taylor Bros Marine Pty Ltd. The author is tasked with overseeing the planning, execution, and management of engineering projects pertaining to naval vessels and maritime systems. The author's previous experience spans various areas relevant to this paper, including conducting ergonomic studies on the SEA1180 Offshore Patrol Vessel, contributing to accommodation outfitting engineering projects within the RAN fleet, and managing aspects such as project, configuration, and quality management.

ANNEX 1: ERGONOMIC STUDY IN ACCORDANCE WITH DST GROUP STANDARDS

As outlined in DST-Group-TR-3550, the default accommodation target range is 5th percentile female to 95th percentile male (or central 90%) population. This target range also aligns with international standards such as MIL-STD-1472G (US standard) and Def Stand 00-251 (UK standard) and serves as the benchmark for SEA1180 Offshore Patrol Vessels and SEA5000 Hunter Class Frigates. While in most cases, the 5th percentile female will capture the 5th percentile male, and the 95th percentile male will capture the 95th percentile female. However, there are a few anthropometric dimensions for design where females are larger and males are smaller. In such instances, female and male data shall be examined and used separately for precision and accuracy.

Percentile data is most effectively used when one dimension needs to be considered for a single design point. For example, when determining the height of a structure that has no other constraints, stature is the dimension to consider, and reference to the percentile data can be made. When multiple dimensions are relevant to a design point, the percentile data should not be used or added together, instead a multivariate approach to design is required. For example, if designing a doorway, both stature and width (bideltoid breadth or forearm-forearm breadth) are necessary to consider simultaneously. Using 95th percentile male data for both stature and bideltoid breadth would only accommodate 90% of that male population. As more dimensions that are included, the level of accommodation decreases when using 95th percentile data. Therefore, a multivariate data is to be used where multiple anthropometric body dimensions are considered key and are interrelated in the design point. An overview of the assessment process of using univariate or multivariate data is illustrated in Figure 3.

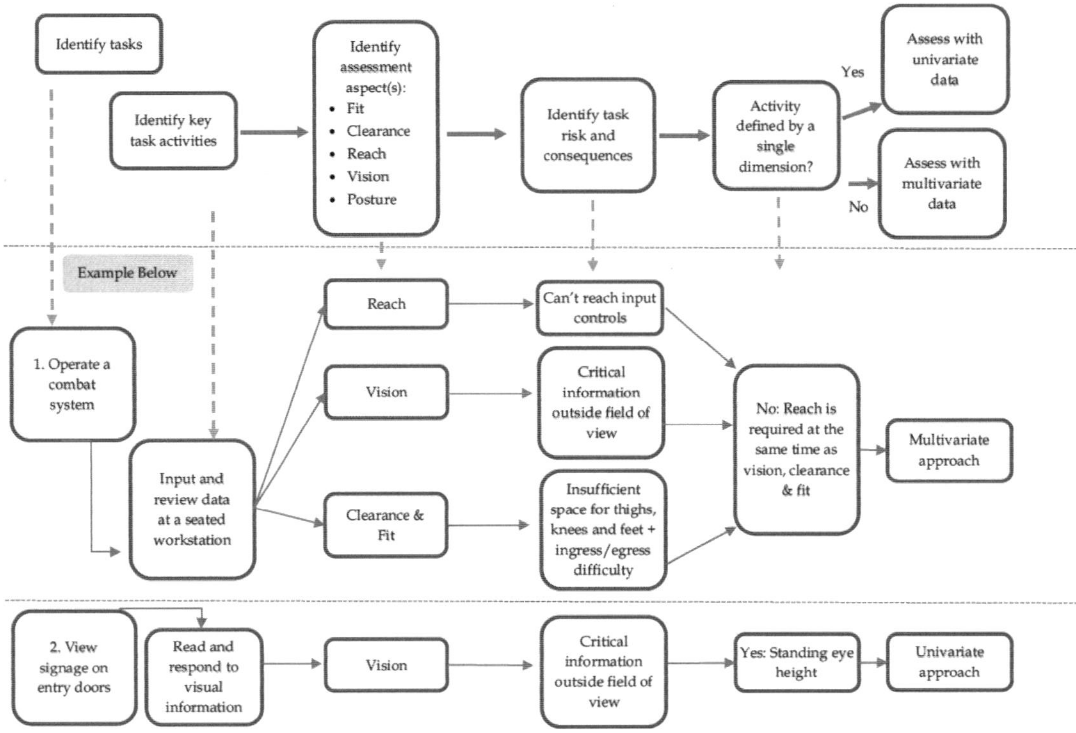

Figure 3: Assessment process of using univariate or multivariate data as per DST-Group-3564.

Anthropometric data is typically collected on semi-nude participants, as is the case with the 2015 ASRAN data. Consequently, it is essential to consider additional allowances including:

- Secular trend – refers to the generational changes in dimensions that can occur over time.
- Personal equipment and clothing correction factors – refers to the additional volume that clothing and equipment or other encumbrance normally worn adds to a dimension.
- Dynamic movement – refers to the additional space required for normal posture and movement when performing a task.
- General allowances and clearances –includes physical and psychological comfort measures, as well as facilitating efficient ingress and egress.
- Subtractions – to account for cases such as the compression of the seat cushion and postural slump in individuals.

There are four key principles of applied anthropometric design requirements:

- Design for the smallest – applies primarily to the application of physical force and reach distances.
- Design for the largest – applies primarily to clearances.
- Design for the average – applies when adjustability is not feasible in the design.
- Design for the range – applies when determining the amount of adjustability required, ideal for accommodating a range of users.

Taylor Bros was engaged by Luerssen to conduct an ergonomic study on the workstations on the SEA1180 Offshore Patrol Vessel (OPV). Workstations include those on the bridge, operations room, and machinery control room. The study focuses on verifying that specified workstations complies with the design guidance and criteria outlined in DST-Group-3550, identifying any system design issues, and recommend any design alternatives. The assessment of each workstation's physical habitability considered five physical aspects; fit, reach, clearance, vision and posture. This assessment was guided by the four principles mentioned above, utilising the anthropometric (univariate and multivariate) data alongside the allowances.

The findings revealed significant non-compliance with DST-Group-3550, particularly concerning the fit within the target accommodation range. Issues included chairs not meeting height and width requirements, inadequate forward and backward pan, improper positioning of chairs in relation to the desks or consoles, and inability to reach the far end of the console while seated. In several workstations, insufficient clearance hindered the rotation of chairs, impeding user ingress and egress. Of the most notable and crucial concern is the Helmsman's console on the bridge, failing to meet SOLAS Regulation 22 regarding the visibility of the bow and horizon while seated.

The study made recommendations to optimise workstation ergonomics in line with DST-Group-3550 and to accommodate the target population as far as reasonably practical, with allowances for size increase over time for future crew (secular trend), clothing, cushion compression and dynamic movement. The recommendations involved implementing an alternative chair model that is better suited to the target accommodation range, modifying chair height adjustability, adjusting console height, installing deck rails for chair clearance, and optimising desk dimensions for improved reach and ergonomics. While these recommendations were implemented, the goal of accommodating the target range of the population was hindered by the limited availability of commercial off-the-shelf chairs. Furthermore, constraints related to economic feasibility restricted the feasibility of custom-made chairs and significant alterations to workstation design, such as console height adjustability.

It is apparent that the crew sizes is evolving and this has a number of implications on the compliance with navy habitability standards. Upon examination (in accordance with DST-Group-3550) of the proposed addition of crew in the Hobart Class Destroyer, it was also found that 3-berth bunks do not meet the vertical separation requirement, leading to a recommendation for 2-berth bunks instead. This recommendation contrasts with the standard practice across all in-fleet surface ships, where 3-berth bunks are commonly used in junior sailor cabins and even in some senior sailor cabins.

ANNEX 2: HOBART CLASS DETROYER BERTHING COMPLIANCE STUDY

Table 5. Hobart Class Destroyer berthing in compliance with Def (Aust) 5000.

Compartment	Compartment Name	Initial Arrangements				Design Arrangement in compliance with Def (Aust) 5000			
		Deck Area, Excluding Furniture's (m^2)	No. of Bunks	Deck Area per person (m^2)	Compliant to DEF (AUST)	Deck Area, Excluding Furniture's (m^2)	No. of Bunks	Deck Area per person (m^2)	Compliant to DEF (AUST)
3-172-0-L	Junior Sailor's Berthing N-2	7.44	9	0.83	NO	7.44	4	1.86	YES
3-173-2-L	Junior Sailor's Berthing N-1	7.62	9	0.85	NO	7.62	4	1.90	YES
3-164-3-L	PO's Berthing N-1 (F)	8.83	8	1.10	NO	8.83	4	2.21	YES
3-156-1-L	PO's Berthing N-4 (M/F)	7.61	8	0.95	NO	7.61	4	1.90	YES
3-158-2-L	PO's Berthing N-3	7.98	8	1	NO	7.98	4	1.99	YES
3-164-2-L	PO's Berthing N-2	5.78	4	1.45	NO	5.78	4	1.45	YES
3-155-2-L	PO's Berthing N-5	4.18	6	0.70	NO	4.18	2	2.09	YES
3-50-1-L	Ratings Berthing N-9 (M/F)	6.71	9	0.75	NO	6.71	4	1.68	YES
3-57-1-L	Ratings Berthing N-6 (F)	6.87	9	0.76	NO	6.87	4	1.72	YES
3-51-2-L	Ratings Berthing N-8	6.67	9	0.74	NO	6.67	4	1.67	YES
3-58-2-L	Ratings Berthing N-7	6.14	9	0.68	NO	6.14	2	3.07	YES
3-63-0-L	Junior Sailor's Berthing N-3	4.53	9	0.50	NO	4.53	2	2.27	YES
3-63-2-L	Junior Sailor's Berthing N-4	4.56	9	0.51	NO	4.56	2	2.28	YES
3-63-4-L	Junior Sailor's Berthing N-5	7.23	9	0.80	NO	7.23	2	3.61	YES
2-50-3-L	CPO's Cabin N-5	5.75	4	1.44	NO	5.75	2	2.87	YES
2-50-4-L	CPO's Cabin N-6	5.79	4	1.45	NO	5.79	2	2.90	YES
2-57-2-L	CPO's Cabin N-4 (F)	5.72	4	1.43	NO	5.72	2	2.86	YES
3-36-2-L	PO's Berthing N-11	7.90	6	1.32	NO	7.90	4	1.98	YES
3-44-4-L	PO's Berthing N-9	8.05	6	1.34	NO	8.05	4	2.01	YES
3-44-3-L	PO's Berthing N-6	7.41	8	0.93	NO	7.41	4	1.85	YES
3-44-1-L	PO's Berthing N-7	6.19	4	1.55	NO	6.19	4	1.55	YES
3-37-1-L	PO's Berthing N-10	8.90	8	1.11	NO	8.90	4	2.23	YES
2-44-0-L	CPO's Cabin N-9	5.30	4	1.33	NO	5.30	2	2.65	YES
2-44-1-L	CPO's Cabin N-8	5.30	4	1.33	NO	5.30	2	2.65	YES
2-44-3-L	CPO's Cabin N-7	5.30	4	1.33	NO	5.30	2	2.65	YES
2-44-2-L	CPO's Cabin N-10	5.49	4	1.37	NO	5.49	2	2.74	YES
2-35-0-L	CPO's Cabin N-11	5.30	4	1.33	NO	5.30	2	2.65	YES
2-35-6-L	CPO's Cabin N-12	5.40	4	1.35	NO	5.40	2	2.70	YES
2-172-6-L	CPO's Cabin N-3	5.50	4	1.38	NO	5.50	2	2.75	YES
2-172-0-L	CPO's Cabin N-2	5.50	4	1.38	NO	5.50	2	2.75	YES
2-172-1-L	CPO's Cabin N-1	5.50	4	1.38	NO	5.50	2	2.75	YES
3-44-2-L	PO's Berthing N-8	5.10	6	0.85	NO	5.10	2	2.55	YES
		TOTAL	**201**			**TOTAL**	**92**		

DESIGNING THE MULTI ROLE COMBAT VESSEL INTEGRATED FULL ELECTRIC PROPULSION SYSTEM: FOSTERING COLLABORATIONS FOR SUCCESSFUL DESIGN

M S Wibawa, **T W Teo**, and **J S Pua**, Defence Science and Technology Agency (DSTA), Singapore, **K P Auyong**, Republic of Singapore Navy, Singapore, **R Edlund** and **K Jansson**, Saab Kockums, Sweden, **C D Schmidt** and **C Müller**, Odense Maritime Technology (OMT), Denmark

SUMMARY

Over 18 months, an on-site, fully integrated project management team consisting of engineers and managers from Singapore, Sweden, and Denmark, as well as naval officers from the Republic of Singapore Navy, completed a Class-certified basic design of Singapore's new Multi Role Combat Vessel (MRCV). As a new ship class for the Republic of Singapore Navy, the MRCV is designed to be a mothership for unmanned systems and comes with an extensive ability to incorporate future upgrades. Traditionally, complex new build naval design programmes are associated with long lead times and schedule delays. However, with a focus on several key management elements such as transparency and collaboration, the team navigated various challenges successfully to deliver the basic design within a challenging timeline. By exemplifying the design development of the MRCV Integrated Full Electric Propulsion system, the paper explains the required management and engineering applied by the project team to fulfil the vision of the Singapore MRCV.

NOMENCLATURE

DSTA	Defence Science and Technology Agency
HSDG	High Speed Diesel Generator
IFEP	Integrated Full Electric Propulsion
IPMT	Integrated Project Management Team
MRCV	Multi Role Combat Vessel
MSDG	Medium Speed Diesel Generator
OMT	Odense Maritime Technology
RSN	Republic of Singapore Navy
SK	Saab Kockums

1. INTRODUCTION

Large Naval programmes are by definition complex and they are more than often suffering from both schedule and cost overruns [1] [2]. This is often attributed to the uncertainty in the requirements due to the ever-changing security landscape over the course of the programme, the unknown missions that the vessel might need to undertake over the lifetime of the vessel, 20 to 30 years in the future, and the rate of technology change. We have seen in recent times how the COVID-19 pandemic and the Russian invasion of Ukraine changed the world as we know, disrupting supply chains and putting inflationary pressure on the economy. We have also seen how the warfare tactics and strategies are changing due to the proliferation of drone technology in warfare. Hence, it is imperative that future surface combatants are designed with the flexibility to be able to adapt to the changing threat and security landscape, while still remaining affordable within the limited defence budget.

The Singapore MRCV is designed with such flexibility and adaptability in mind, acknowledging that change is expected. As a mothership for unmanned systems, it is well placed to capitalise on the fast-developing drone technologies. In order to be flexible and adaptable, margins have been built into the design [3]. One of them is provisioning sufficient power for future systems via the selection of IFEP system, which is discussed in this paper.

Besides the engineering development of the vessel itself, in such a complex programme it is important to pay attention to the management of the programme as well. The MRCV basic design had many challenges due to the high number of stakeholders and the nature of international cooperation. In addition, there were time and budgetary pressure due to slow down during COVID-19 and the ensuing inflation thereafter, which happened right at the inception of the programme. In an effort to improve the work efficiency and collaboration to meet the ambitious schedule, the IPMT agreed from the onset to implement some fundamental changes to the traditional systems engineering approach: agile way of working [4], blurring of organisation boundaries to cultivate trust and adopt transparency [5]. This affected how the technical specifications were used, the way 'reviews' were held, and the effort put into truly integrating all stakeholders into the IPMT. These aspects are discussed in the later part of this paper.

2. GENESIS OF THE MRCV DESIGN

The MRCV is a new ship class in the RSN to replace the aging Victory-class Missile Corvettes, being designed as a mothership for unmanned systems with an extensive ability to support unknown future upgrades. The MRCV would be the first ship in the RSN to be equipped with IFEP system, and one of the first combatants in the world that is equipped with full diesel engine driven IFEP system.

In the beginning, there were two concepts. The OMT concept was based on the scaled up Danish Iver Huitfeldt class frigate in order to house the specified payload and to ensure sufficient future growth capabilities. The Iver Huitfeldt class is recognised for its robust and cost-efficient design, a legacy from OMT's experience in commercial shipbuilding which was successfully transferred to naval design. It was also used as a basis for the Arrowhead 140 under construction to Royal Navy and others [6]. The concept from SK stemmed from a different type of naval tradition, being primarily involved in stealthy, compact and lightweight designs such as HMS Visby class.

While traditionally Sweden and Denmark are rivals, the two companies realised the potential to complement each other, both in terms of conceptual design solutions and inherent company skills. OMT on one hand, had successfully promoted a cost efficient and proven concept that was based on the Iver Huitfeldt design. SK on the other hand had experience of cooperation with DSTA for more than 30 years including the basic design for the Littoral Mission Vessel project in 2013 – 2014. As a part of the Swedish defence company SAAB AB, SK would also benefit from an extensive knowledge base including weapons and sensors integration as well as signature management. Eventually, OMT and SK initiated communication and gradually established a cooperation that finally grew into a partnership with focus on a common design for the MRCV programme. See Figure 1.

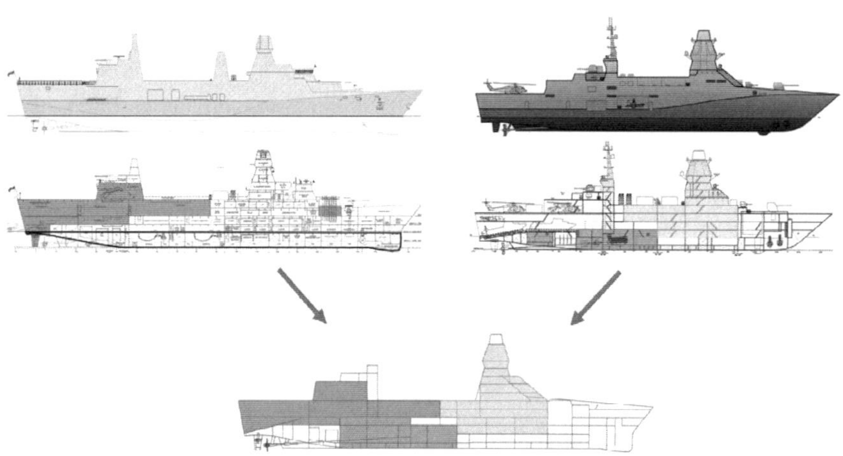

Figure 1: The merging of the concepts

The merged MRCV design inherited features from both concepts. From the OMT side it inherited the hull lines, structural steel design and payload handling systems. From the SK side, the initial contributions were the propulsion arrangement, weapons/sensors/signature integration, and carbon fibre reinforced plastic design of mast and superstructure.

Despite their differences, the two companies already found a common platform for sharing design ideas that complemented each other in the initial commercial phase. By focusing on the different strengths of the two design houses, responsibility for the design could be appropriately distributed. Later, some parts of the design were exclusively developed by one or the other party, while other parts were developed jointly. The propulsion system in particular required very close cooperation as external suppliers were involved. The common understanding of the technical challenges combined with the extremely close collaboration method in IPMT made it possible to develop the design in a very short time.

Unknown to the two companies, while they were busy merging their concepts, the DSTA project team was deliberating a change in the propulsion system from the initial CODLAD system to the IFEP system. Soon thereafter, a change in requirement for MRCV with IFEP system was issued.

3. SELECTION OF MRCV PROPULSION SYSTEM

3.1 KEY CONSIDERATIONS

Operational flexibility to prepare for future uncertainties was one of the key considerations. Increasingly, advanced combat systems like Multi-Function Radars (MFR) need more electrical power for increased performance, and concomitantly more cooling capacity than before. The Royal Navy, as an early and heavy adopter of electrification – started with Type 23 frigates, and more recently with the Type 45 and QEC – has led the way in showing the flexibility that this entails for warships, by allowing easy upgrade and reroute of electrical power (e.g. reduction of speed to shift power from propulsion to combat loads) when required, to suit the different payloads that might come aboard [7].

An additional operational benefit is to reduce the underwater radiated noise. The IFEP design allows decoupling of main diesel engines from the shaft line, eliminating the need for a complex and noisy gearbox to manage the multiple inputs into the shaft. It also allows the main engines' vibrations to be isolated from the shaft, leading to a quieter ship. IFEP also means that, as battery technologies mature, it may be possible to use large battery banks for hotel and propulsion loads. These can be designed to be incorporated into the system to allow the vessel to loiter without running any engines, thus greatly reducing the noise.

Defence budget for a small country is always limited, every dollar must be spent prudently. Being one of the cost drivers onboard the ship, the decision on the propulsion system for MRCV was deliberated for a long time. The initial requirement given was for a CODLAD system to reap some benefit of electrification, while keeping within the resource constraints as the conventional wisdom was an IFEP system would cost more in terms of capital cost. As more information was obtained from the industry, however, it became clear that IFEP system for MRCV did not necessarily cost more than a CODLAD system as seen in Table 1.

Besides the operational benefits and budgetary constraints, supportability was also a key consideration. Most navies with electrified warships use gas turbines to generate peak power. However, in the RSN's context, this would mean setting up an entirely new logistics and support chain, which is not supportable. RSN has found it imperative to leverage on established logistics chains, whether commercial or military, and thus relies on the continuing use of diesel engines.

Manpower is also a real constraint for small country like Singapore, so the RSN has always had to find innovative solutions through technology to overcome these constraints while constantly punching well above its weight. Having only limited number of crew and with projected decline in the population, maintenance workload could quickly become a burden. In an effort to keep maintenance man-hours to a minimum, MSDG could have lower maintenance due to reduced number of cylinders and allows for better optimisation of the running hours, when compared to HSDG.

Given the above considerations and the operating profile, medium speed diesel engine driven IFEP solution without gearbox emerged as the most attractive propulsion system for MRCV. However, there were still risks on the technical feasibility of such IFEP solution within the specified boundary, even though shipborne electrification technologies are mature, driven by the Cruise and Offshore industries. See Table 1 below for the comparison.

Table 1. Assessment of propulsion design

Criteria		**CODAD (Mechanical)**	**CODLAD (Hybrid)**	**IFEP (Diesel Electric)**
Performance	Relative fuel Consumption	Baseline	–	+
	Underwater radiated noise	Baseline	+ +	+ + +
Cost	Capital cost	Baseline	– – –	– –

Criteria		CODAD (Mechanical)	CODLAD (Hybrid)	IFEP (Diesel Electric)
	Lifecycle cost incl. maintenance and fuel	Baseline	–	Baseline
Future Growth	Power available at cruise	Baseline	+	+ + +
	Space margin	Baseline	– –	– – –
	Weight margin	Baseline	–	– – –
Workload for operator	Annual Maintenance Workload	Baseline	+	+ + +

3.2 FEASIBILITY OF IFEP ON MRCV

The DSTA project team had to work closely with SK and OMT teams to establish the technical feasibility as the owner of the hullform. A few factors which contributed to the feasibility of the IFEP on MRCV are summarised as follows:

- A favourable operating profile which limits the required power generation
  As the MRCV is designed to be a mothership for unmanned assets, there is no need for the mothership itself to perform high-speed intercepts when UAVS and USVs can perform these tasks. This reduces the need for the vessel to be light with high power to weight ratio, thus obviating the need for gas turbines in the design. This allowed the exploration on the use of MSDGs and appropriately sized electrical motors.
  .
- A favourable ship size to fit the motors and switchgear
  Due to the need for the vessel to embark all the daughter craft and mission modules, the vessel has to be sufficiently large. However, these crafts are typically stored above the waterline - leaving the underwater volume of the ship available for the electrical switchgear and power generation equipment. As for the 400 tons increase in displacement compared to the CODLAD, it would result in an increased draught of about 0.2 meters which was concluded to be feasible with regards to speed performance, stability and future growth weight margins.

- Availability of local support
  Singapore being a maritime hub was also beneficial for the RSN, as most Original Equipment Manufacturers (OEMs) have established offices and maintenance networks based out of Singapore. This ensured a level of supply and logistic resilience – a lesson reinforced by the COVID-19 pandemic situation where global trade nearly ground to a halt and support from overseas was not possible. Hence, with an established network of OEMs with local support, and maturing electrification technologies offered by multiple OEMs, RSN found it suitable to foray into electrification.

- Cost Benefits
  By going from CODLAD to IFEP cost savings was gleaned from the elimination of a complex and expensive gearbox. Although there is cost increase in the electrical motor and power distribution systems, overall there is a slight reduction in capital cost over the CODLAD solution.

- Maintenance Reduction
  The major benefit was significant reduction in the crew maintenance load of more than 15%. Besides the reduction from the elimination of the gearbox, there was also significant reduction from running hour optimisation between the MSDG and HSDG.

4. DIESEL GENSET AND ELECTRICAL CONFIGURATION

Initially SK wanted to take the approach to optimize for low weight given their heritage in designing and producing smaller, lightweight vessels such as HMS Visby class. As gas turbine was not an option, an option to use all HSDG was proposed. When considering what was available on the market in terms of HSDG, the maximum available power was about 3 MW_e per genset, which resulted in a total of ten gensets. This was not acceptable to the DSTA project team as the maintenance profile would exceed what could be supported.

On the other hand, the DSTA project team proposed to use all MSDG option to minimise the maintenance load, but this was not feasible from the weight and space aspect. There was also concern on the damaged stability due to the length of the MSDG, but the DSTA project team proposed to split the diesel engine and the alternator in two adjacent compartments, connected via shaft through a bulkhead seal adapted from the cruise ships design to overcome the damaged stability. SK and OMT agreed to proceed based on this split genset configuration.

The proposed way forward was therefore a combination of HSDG and MSDG.

At this stage the MRCV had two engine rooms that were located forward and aft in two separate fire zones. As different combinations of HSDG and MSDG to achieve total power of about 30 MW_e were evaluated it was finally narrowed to two options: combination of three MSDG and two HSDG (3M + 2H) option which was the baseline proposal from SK and combination of two MSDG and four HSDG (2M + 4H) option. The 3M + 2H option would result in asymmetric arrangement with one MSDG + one HSDG in the aft engine-room and two MSDG + one HSDG in the forward engine-room, whereas the 2M + 4H option will result in symmetrical arrangement of one MSDG + two HSDG in each engine room.

Another consequence of changing to IFEP was that a low-voltage AC or DC system would not be feasible due to limitations in the short circuit resilience. The low voltage system was therefore replaced by a 6.6 kV medium-voltage AC system. With the alternators requiring less head room than the diesel engines, it was thereby possible to add a split deck level on top of the alternator rooms where the main switchboards could be located, in favourable proximity to the alternators. OMT had to make structural changes to increase the deck height though, as the 6.6 kV switchboards were taller.

4.1 3M + 2H CONFIGURATION

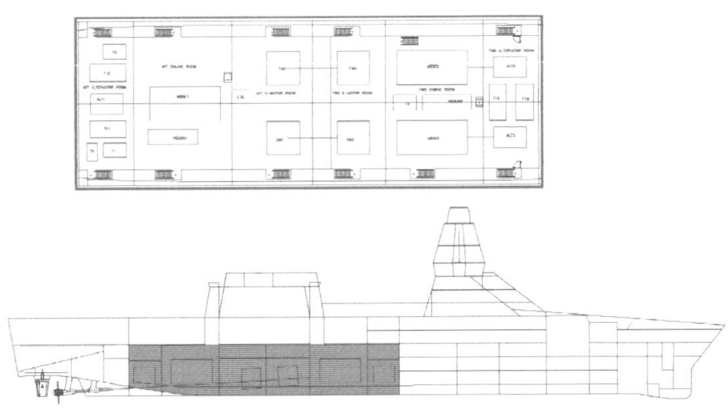

Figure 2: 3M + 2H arrangement

The arrangement of the forward engine room turned out to be challenging due to the size of the MSDG. There was considerable risk to fit 2 MSDG and 1 HSDG in the forward engine room as shown in Figure 2. The total power was just below 30 MW_e with power density of 75 kW/ton.

The asymmetry was not limited only to the physical arrangement, but also affected the electrical configuration. Due to the asymmetry, the forward 6.6 kV main switchboard would support 63% of the total load, as opposed to only 37% in the aft 6.6 kV main switchboard. This presented an issue for scenario where the two switchboards are required to run in open bus tie configuration in action station, to fulfil the requirement to minimise impact to operation in case of damage to one main switchboard. See figure 3 below.

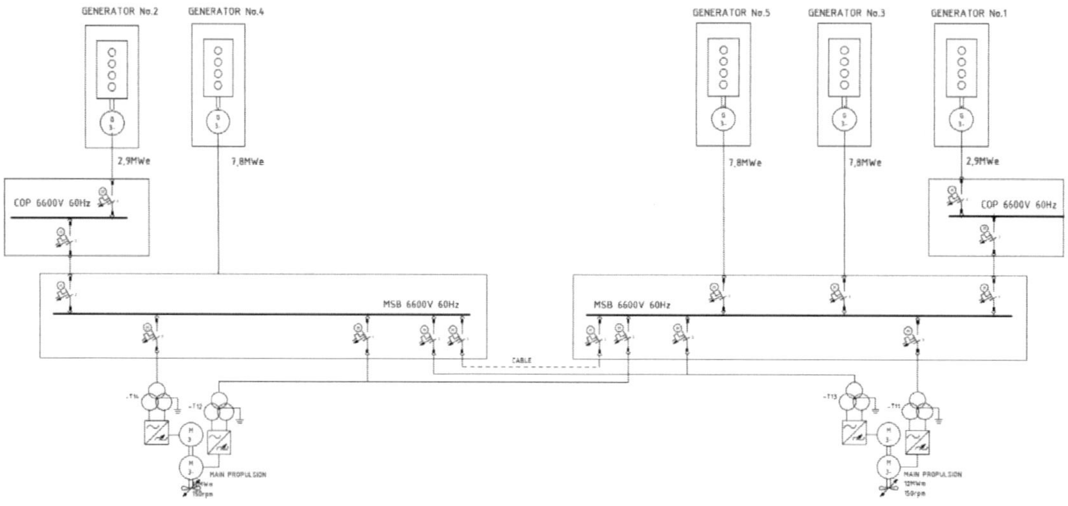

Figure 3: 3M + 2H Single Line Diagram

Besides the arrangement and the power distribution, the maintenance consideration was the key criteria. Hence, the IPMT worked together with the potential engine OEM to identify the maintenance tasks required to be done by the crew and computed the maintenance cost for the life cycle of the vessel. The computation was based on the operation profile of the vessel and the prevailing quotation cost of OEM's manpower and spares in 2022. Based on the computation for the 3M+2H configuration, it was estimated that 1617 man-days of RSN manpower effort would be incurred over the vessel lifecycle of 30 years.

4.2 2M + 4H CONFIGURATION

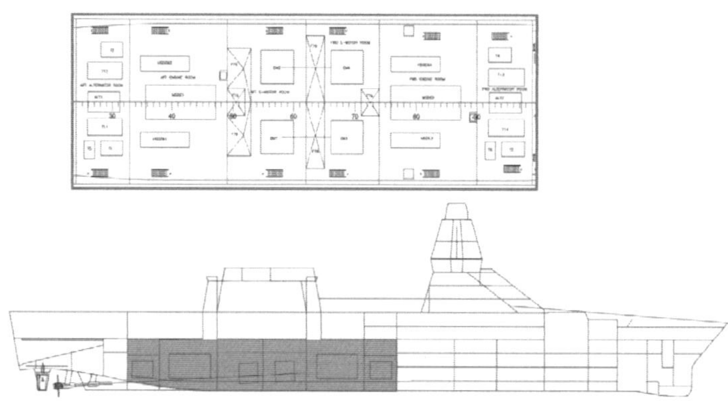

Figure 4: 2M + 4H arrangement

The 2M + 4H configuration was lighter as the two HSDG are much lighter than one MSDG. Total available power is slightly higher too, hence the much higher power density of 90 kW/ton. The arrangement in the forward engine room is less crowded and more manageable, although the aft engine room was slightly more crowded due to the additional one HSDG, as shown in Figure 4. Being symmetrical, it was also easier to manage for the weight and stability.

Electrically the configuration was also symmetrical with each 6.6 kV main switchboard supporting 50% of the total load, hence there was no issue to run the open bus tie configuration that was not possible in the 3M + 2H configuration. See Figure 5 below.

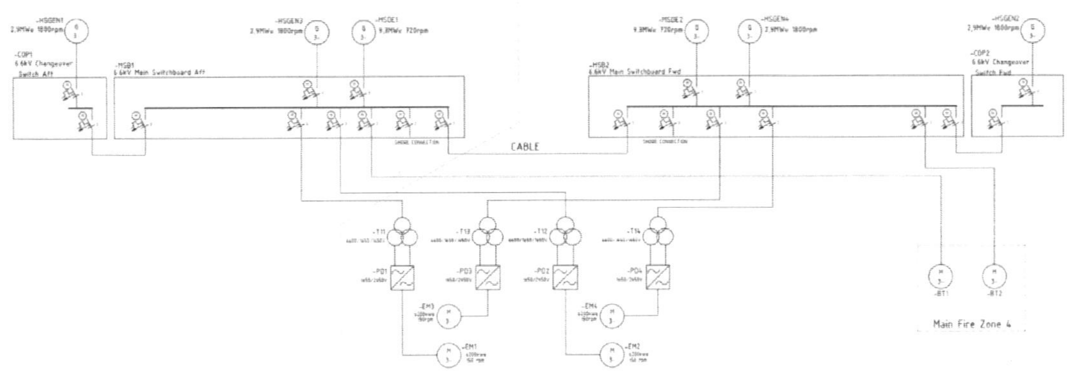

Figure 5: 2M + 4H Singe Line Diagram

The maintenance computation for the 2M + 4H configuration estimated that 1384 man-days of RSN manpower effort would be incurred over the vessel lifecycle of 30 years. The significant saving (15 %) on the RSN maintenance manpower achieved in the 2M + 4H configuration was mainly due to the lower running hours and hence the maintenance of the four HSDG. Although the running hour of the MSDG was higher compared to the 3M + 2H configuration, the increase in maintenance hour was negligible as it still did not reach the engine overhaul limit, due to the much higher Mean Time Between Overhaul of the MSDG.

4.3 CONCLUSION TO THE GENSET AND ELECTRICAL CONFIGURATION ASSESSMENT

The outcome of the assessment is summarised in Table 2.

Table 2. Assessment of Gensets Configuration

Criteria		**3 MSDG + 2 HSDG**	**2 MSDG + 4 HSDG**
System Performance	Power capacity	~29MW	~30MW
	Power to weight ratio	75 kW/ton	90 kW/ton
	Engine Room Arrangement	Substantial risk of arrangement in the forward engine room	Sufficient space in both engine rooms
	Electrical Power distribution flexibility	Asymmetric loading of aft (37%) and forward (63%) MSB resulted in limited Vessel operation in action station (i.e. open bus tie).	Symmetric loading of aft (50%) and forward (50%) MSB.
Maintenance (for life-cycle of 30 years)	RSN Manpower for maintenance	1617 Man-days	1384 Man-days
	Contractor Service and Spare Cost	100%	103%
Capital Cost		100%	109%

The 2M + 4H Config was selected for the IFEP configuration due to the significant (~15%) saving of RSN maintenance manpower, despite slightly higher overall maintenance and capital cost. Furthermore, the 2M + 4H configuration has better system performance, which the 3M + 2H configuration cannot match and there is no work around to overcome the discrepancies (i.e. the limitation in action station, space constraint to accommodate two MSDG in the forward engine room and additional weight of 60 tonnes). The reduction in the weight of the propulsion system also enables the IPMT to explore weight increase in other areas to offset the cost.

5. KEY SUCCESS FACTORS

As can be seen from the above sections, just highlighting the propulsion system design iteration, many stakeholders were involved in the design of the MRCV who contributed various requirements and solutions. For such large and complex programme, as can be expected for any future surface combatants, an open communication and transparency among all the stakeholders were crucial in ensuring the success of the programme. From the start of the basic design, the IPMT was implemented, and an emphasis was put on initiatives that could create an agile and collaborative environment between the key stakeholders. The main elements of the approach are described in the following sub-sections.

5.1 THE LETTER KILLS, BUT THE SPIRIT GIVES LIFE

The most common approach to technical specification is to treat it as the bible to the project, where each engineer will read the technical specification, interpret it in his/her own way, and work out the solution, sometimes blindly following the text without understanding the intent. This was not the approach taken by the leadership team from the IPMT, as they recognised that the text could easily be misinterpreted. Instead, a requirements review session was conducted at the start of the design with all the relevant engineers to clarify the spirit of the text, the interpretation, and the approach to the solution. In this way, the leadership team was able to set clear direction for each engineering discipline and ensure that everyone had clarity of the end goal of what the project was trying to achieve.

Emphasis was also placed more on the desired end outcome rather than the written text itself. In essence, the technical specification formed the foundation on which the design was built upon, rather than a design that was built according to. This principle is well embodied by the house shown in Figure 6. Utilising the technical specifications in this way, there was more flexibility in the design, and the designers were free to propose the best solution that were able to meet the end goal even if it deviated from the text of the specifications. In fact, based on the understanding that the fundamental requirements was one of the parameters to tune in complex projects [8], the DSTA project team would constantly encourage the designers to propose alternative solution, if that could better meet the MRCV vision. This might come across as counter intuitive, but worked very well for the basic design.

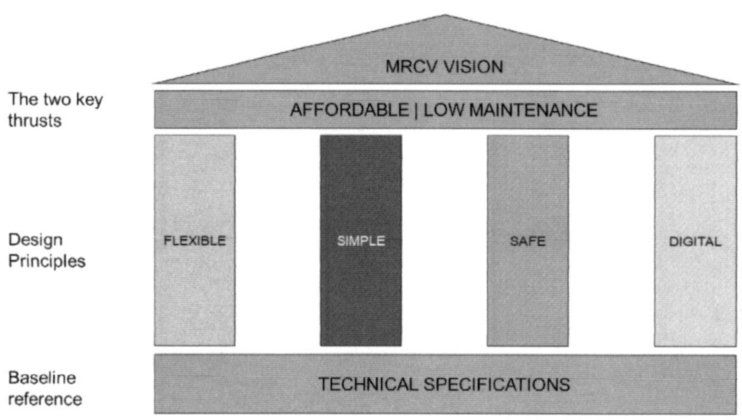

Figure 6: The Building Blocks of MRCV Vision

5.2 DISCIPLINE OF PEOPLE INTEGRATION

The term systems integration is well known in engineering circles, especially when one is managing a large, complex programme similar to the design of the MRCV. The discipline of people integration, however, is seldom heard of if at all. In the MRCV basic design, however, the people integration was given the utmost priority, on equal importance with systems integration.

Heading up a program that was intrinsically technically complex by itself, could be further complicated if we added several different nation specific cultures, various business cultures and finally various historical experiences.

SK and OMT brought different pieces of the puzzle to the table. Formally OMT was a sub-contractor to SK. During the start-up of the program, however, it was decided that the two companies would aim to treat each other as colleagues rather than customer/supplier. As a consequence, the two company formed an integrated team were engineers as well as project managers within the programme could come from either organisation. This was a leap of faith as it involved the parties maintaining the balance between monitoring the contract obligations versus working fully integrated. During the lessons learned it was found that the mix between Swedish and Danish culture, properly managed, was very successful.

During the whole period of the basic design, SK made available dedicated project office facility in Karlskrona, Sweden where the various stakeholders were co-located to ease communication [9]. The common project office enabled an efficient working environment for collaboration between the customer, end-users and basic designers and was characterised by an open-door culture. It enabled an agile interaction between the stakeholders, where design input and feedback could be quickly exchanged. The fact that design decisions in most circumstances, could be made on-site, did drive a fast decision making in general, resulting in an accelerated design process. When needed, the on-site team would facilitate fast interaction with their home offices, which also saved valuable time in the decision making and approval process. All in all, the common project office served as a catalyst for an efficient and quality-minded working environment.

Besides common physical office, SK also set up shared digital workspace with project specific servers, shared network for 3D CAD application, and secure file transfer mechanism among the IPMT members, designed and configured to meet the high standards of cyber security required for a naval project. Project specific computers and SAAB e-mail addresses were issued to the members of the IPMT and the lead engineers. This facilitated file and information sharing between customers and designers, hence faster design turnaround time, as most of the work was done digitally.

After all the above efforts, one could envision that the communication would naturally occur. Surprisingly, the engineers displayed some hesitation in interacting with the customers initially. Review of the matter revealed that a large portion of the reluctance was based on mental projections of decades old previous experiences. This was brought to the attention of the leadership team of the IPMT. An agreement was made to foster a "safe" environment without hostility, enabling trust and transparency. Initially, a simple measure was to rename "reviews" to "workshops", where the focus is to collectively solve engineering problems, rather than assigning blame. After approximately 4 months into the project, the engineers started to initiate dialogue directly with whom they best needed advice from rather than looking to which organisation they belonged to.

5.3 SMALL, QUICK ITERATIONS VS. BIG BANG REVIEW

The IPMT adopted agile methodology with short sprints and fast design iterations in between. The basic design phase was roughly divided into three phases of six months each, divided naturally by seasonal breaks in summer and winter, which set the project tempo. There were also clear goals and set of deliverables at the end of each sprint, which helped to align expectations between the stakeholders. Having the customers on-site with frequent workshops had the added advantage that the customers were well aware of all the design statuses, issues, as well as good alignment of the design within the IPMT. As a result, the traditionally lengthy and arduous milestone design reviews transformed into open and constructive discussions with some focus on the next upcoming phase. Formally, the milestones still served the purpose of presenting the design in its current state and to further document comments and actions if any. Nevertheless, it was the small, quick design iterations of the IPMT that enabled very efficient formal design review milestones.

5.4 CREATING CONSENSUS IN DECISION MAKING

Key to the good progress of every design project is availability of information, as well as timely decision making. Fortunately, the top management of the participating organisations empowered the project team to take most of the design decisions the to ensure good progress of the project. This mandate considerably reduced the design timeline and gave a sense of empowerment to the project team. Mirroring the overarching vision of the technical specification, the project management adopted a "Business-Savvy" approach [10], never losing sight of the customers vision of the project thereby enabling well balanced priorities during execution.

A central tool was the Main Decision and Assumptions List (MDAL) that was initiated after the main ship requirements and design characteristics were settled. Although the tool was new to the DSTA and SK project teams, the MDAL is generally accepted within the engineering industry and has been used by OMT with success on previous projects. The list served two overall purposes: firstly, to capture the assumptions made by the basic designers when design-critical information was unavailable, and secondly, to document design decisions made jointly by the IPMT. The list was reviewed weekly during progress meetings for awareness of all stakeholders and decisions could be taken at project level when necessary. The MDAL allowed the IPMT to progress the design on a known and agreed basis, thus enabling a transparent design process with acceptable risk levels.

Another important tool was the list of Vendor and Authority Furnished Information to track the status of all equipment- and system-information received on the project. During the design process, the list was actively baselined to monitor the status of received and missing information that was critical to progress the design. Where critical information was missing, the way ahead would be discussed and agreed, and often times this results in entry into the MDAL.

Besides specific tools mentioned above, a central part of creating consensus in decision making was the philosophy of transparency within the IPMT. An example of transparency was the multi-disciplinary task forces that were created to handle design issues with critical impact on product quality and project timeline. The setup of the task forces enabled open, fast paced communications between the members of the IPMT, which sometimes also draw on experts from other partners to approach the issues. Customers were involved in the work of the task force and informed of the progress on a daily basis. As a result of the fast-acting nature of these task forces, the IPMT was provided with input for decision-making in a shorter time frame than usual.

6. CONCLUSIONS

The basic design of the MRCV has been successfully completed, as shown in one aspect via the development of the IFEP system in this paper, through the highly successful collaboration between Swedish, Danish, and Singaporean engineers. While it started from an unlikely alliance, fuelled by historical baggage and cultural differences, the leadership of the IPMT managed to overcome the initial suspicion and mistrust to create an environment of trust and unity to deliver a successful programme, as demonstrated by the following record of events:

The very first call for an engineering meeting was sent out two weeks after the arrival of the customer on-site team in Karlskrona. The engineers from SK were reluctant to meet the customer and refused to show up. Offered rationale was that "We have nothing to show yet", and "The customer may be disappointed". That was the situation when the project started. Fast forward 18 months. When the on-site team announced their plan to return to Singapore, the reaction from the same initially "hesitant" engineers, was now of frustration as "the work cannot be completed without the customers". The view on the cooperation with the customer had undergone a complete 180 degrees turn!

In the same way that the will and the ability to collaborate in a transparent environment was matured and strengthened through the project, so was the will and the ability to jointly find optimal solutions for the complex ship that the MRCV was. All parties have proposed solutions that have been reversed, accepted or rejected, ultimately resulting in a highly refined design enriched by the knowledge and experience of efficient ship design, cost awareness and optimised maintenance regime contributed by the overall IPMT.

7. ACKNOWLEDGEMENTS

The views expressed in this paper are solely those of the authors and not necessarily reflect those of the organisations. The authors would like to acknowledge the contributions given by our colleagues and industry partners for the successful basic design, and the Republic of Singapore Navy for funding this programme.

8. REFERENCES

1. POWELL, M., "Big Defence Projects are Usually Late and Over Budget – Here's What We Can Learn From the Build-up to WW2", https://www.port.ac.uk/news-events-and-blogs/blogs/security-and-risk/big-defence-projects-are-usually-late-and-over-budget-heres-what-we-can-learn-from-the-build-up-to-ww2, 2023.

2. ZINK, T., "Strategies for Planning Complex Systems Development", *PMI Global Congress*, 2015.

3. ANDREWS, D., "The Multi Role Combatant – Jack of All Trades, Master of None?", *RINA WARSHIP 2019*, 2019.

4. PMCollege, "PMOs Move Toward Agility", PMCollege white paper series, 2017.

5. AHIAGA-DAGBUI, D.D., TOKEDE, O., MORRISON, J., and CHRINSIDE, A., "Building High-performing and Integrated Project Teams", *Engineering, Construction and Architectural Management,* 2020.

6. JOHNSON, J., and HOWARD, M., "Type 31 Frigate: Complex Warship Design for a Dynamic Operational Environment", *International Naval Engineering Conference*, 2022.

7. COREY, C., CASTLES, G., WILGRESS-PIPE, C., and MOON, H., "Electric Propulsion for Modern Naval Vessels", *MADEX 2021*, 2021.

8. COVEY, G. and FABER, G., "Writing Better Technical Specifications", *Appita Journal*, 2003.

9. ZENUN, M M N., LOUREIRO, G., and ARAUJO, C S., "The Effects of Teams", *Co-location on Project Performance', Complex Systems Concurrent Engineering*, 2007

10. PMCollege, "The Business-Savvy Project Manager", *PMCollege white paper series*, 2012.

9. AUTHORS BIOGRAPHY

Martin Sulaiman Wibawa holds the current position of Senior Programme Manager at DSTA. He is currently responsible to deliver a new training vessel for the Singapore Navy. He was previously the Senior Programme Manager in charge of the MRCV design and was posted to Sweden to lead the basic design. A recipient of the DSTA Scholarship, he graduated with a Bachelor of Science in Engineering (Naval Architecture and Marine Engineering) degree from the University of Michigan, Ann Arbor, USA, in 2008. He further obtained a Master of Science (Naval Architecture and Marine Engineering) degree from the Massachusetts Institute of Technology, USA, in 2010.

Teo Teck Wee holds the current position of Principal Engineer at DSTA. He is responsible for the naval architecture and structural aspects of the MRCV and was posted to Sweden to assist in the development of the basic design. His previous experience includes the design of the green harbour patrol craft for the Singapore Maritime Port Authority, and more than 10 years of experience within the maritime industry prior to joining DSTA. He holds a BEng (Hons) in Naval Architecture from University of Newcastle Upon Tyne, UK.

Pua Jiunn Siong holds the current position of Principal Engineer at DSTA. He is responsible for the design of the propulsion systems and auxiliary systems for the MRCV. His previous experience includes the propulsion systems and automation systems design onboard the Formidable class Frigate. He holds a BEng (Hons) Mechanical Engineering (Design) from Nanyang Technological University.

Auyong Kok Phai is an officer in the Republic of Singapore Navy and holds the current position of Deputy Director, MRCV Project Office. He is responsible for ensuring that the ship design is coherent with the operational intent. His previous experience includes serving as Commanding Officer of RSS Tenacious, as well as staff officer in various operational and planning capacities.

Robert Edlund holds the current position of Senior Project Manager at Saab Kockums. He is responsible for being the Prime Project Manager for the MRCV Basic Design. He has a long history of project management roles for Singapore programmes stemming back to the Submarine programme in 1997. He was also the prime Project Manager for the Basic Design of the Littoral Mission Vessel for Singapore.

Kristian Jansson holds the current position of Chief Engineer for surface ships at Saab Kockums. He is responsible for development and coordination of ship design and has been active in the MRCV programme since early on in the project. He holds a MSc in Naval Architecture from Royal Institute of Technology in Stockholm.

Christian Dyreborg Schmidt holds the current position as Chief Project Officer at Odense Maritime Technology. During the MRCV project, he led the OMT design team and supported the design development as an integral member of the IPMT. His previous experiences include, in addition to a naval architectural background, 35 years of shipbuilding experience and was initially at Odense Steel Shipyard technically responsible for the design of the Iver Huitfeldt class.

Christoffer Müller holds the current position as Project Manager at Odense Maritime Technology. During the MRCV project, he functioned as the Design Manager of the OMT organisation and facilitated the design development as an integral member of the IPMT. He has a naval architectural background and 7 years of experience within the maritime industry.

ENHANCING MARITIME ASSET STRUCTURAL MANAGEMENT: WHAT CAN NAVAL AND CIVILIAN FLEETS DO?

M T Williamson and **M N F Zailani,** Floating Solutions Consulting, Australia

SUMMARY

The use of digital visualisation and modelling has made significant progress in the last few years; there is now a vast amount of data that can be generated and interrogated using a wide range of digital methods. Effective display, use and manipulation of this data can be costly and may not provide a return on investment. However, harnessing this data in a manner that enables structural optimisation to be simpler and more accessible can be a powerful enabler for through life support.

This paper reviews practical methods of managing hull structures through life and how these can be optimised. The similarities between naval and certain civilian assets are highlighted. The volume and type of data that needs to be generated to effectively optimise the management of structures through life is discussed. Finally techniques to visualise the generated information in a meaningful way, for all stakeholders, using existing generic software are presented.

1. INTRODUCTION

Floating structures such as Floating Production, Storage and Offloading (FPSO) units and naval ships operate in some of the most challenging marine environments on earth. These assets face relentless marine corrosion, dynamic sea stresses, and complex regulatory requirements, necessitating a robust structural integrity management strategy. Effective management of these complex structures not only ensures safety and regulatory compliance, but can have significant cost and risk reduction benefits, as well as increasing platform availability and prolonging operational life.

To date, structural integrity management techniques have been codified into rule-based methodologies based on years of accumulated experience and knowledge. These rules and standards serve the maritime industry well however for the most part are fairly prescriptive and remain generic to a vessel class across which there can be significant differences in design detail and operational environment. Whilst offering robust assurance for hull integrity, these historical approaches will rarely be the most efficient for a specific asset and operating conditions. For bespoke naval and high value civilian assets, tailored and more efficient in-service structural maintenance can lead to significant improvements in operability and availability whilst reducing costly and intrusive maintenance.

Recent advances in analysis, inspection and digital recording techniques mean that there can now be a vast amount of information to process. Although this is invaluable to optimise structural maintenance, the limitation is how effective human decision makers are in digesting and making the correct decisions based on the information. This paper seeks not only to highlight what information needs to be generated to optimise through life structural maintenance but also how to use existing generic commercial software to effectively manage and present the data to facilitate the decision-making process.

2. MANAGEMENT OF STRUCTURES IN SERVICE

2.1 OIL AND GAS AND NAVAL ASSETS COMPARED

Engineering support for structures during the in-service phase can be tailored depending on the type of asset [1]. When considering Figure 1, it can be seen that naval assets have many attributes similar to oil and gas assets. Given the significant costs and implications of platform down time as well as the difficulties and high costs associated with finding suitable facilities and skilled workforce for inspection and repairs, an increased level of engineering optimisation can provide significant value.

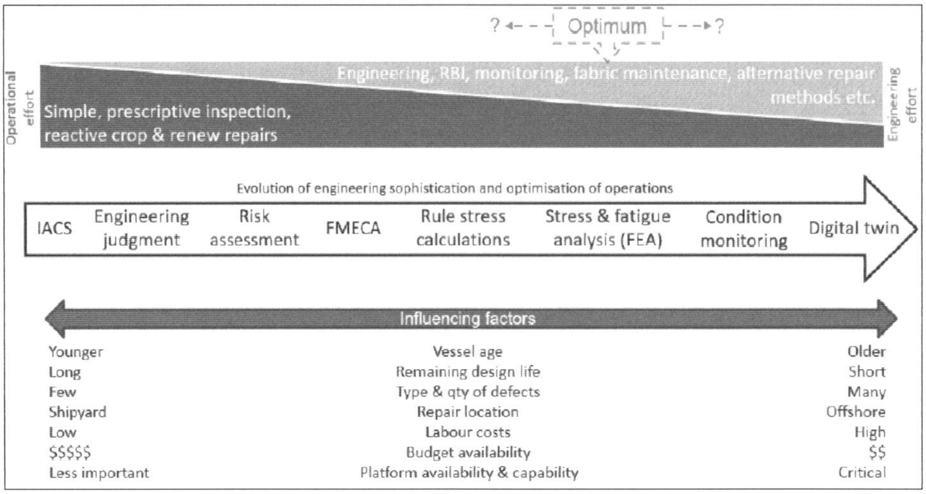

Figure 1: Tailoring of the level of engineering depending on influencing factors

2.2 REPAIR AND OTHER RISK MITIGATION TECHNIQUES FOR STRUCTURAL DEFECTS

Once defective structure has been identified (this could be corrosion, pitting, cracks, fractures, buckling or other issues), there are a number of actions that can be taken to mitigate and manage risk. Table 1. summarises some techniques currently available to address defects.

Table 1. Structural defect risk mitigation techniques - advantages and disadvantages

Technique	Description	Advantage	Disadvantage
Crop and renew	Involves cutting out the damaged or defective area and replacing it "like for like" with new material that is as per the original scantlings.	Standard 'Rule' based approach. Well understood and easy to plan. Minimal engineering input or understanding of the structure in question required. Low risk option.	Can be expensive and intrusive. Can have significant impacts on operational capability and availability during repair. Absence of good inspection data can result in an excessively large repair. May not fix the issue if it was a design deficiency.
Reinforce	Will vary depending on the specific requirements of the vessel and the areas being reinforced. Common approaches include adding additional stiffeners or brackets. In other cases, composite reinforcement can be considered.	Can facilitate increase in capability. Can address deficiencies in original design. Can enable introduction of advanced materials (e.g. composites).	Same as "crop and renew". May need significant calculations and longer approval time to justify. Can impact weight and stability. May introduce new points of failure.
Liquid loading restrictions	Often used for stability reasons, from a structural point of view, mandating maximum or minimum fill levels of tanks can reduce the structural loads on tank boundaries or on the entire vessel hull girder.	No physical impact on vessel. Can be used to manage both stability and structures.	Can be operationally quite restrictive. A flexible but effective restriction can require large numbers of permutations and combinations to be calculated. Can add to decision making burden of the crew.
Change of service / operational restrictions	Limiting allowable sea states / motions / accelerations by reducing speed and / or operation in areas with excessive conditions can	No changes to the structure are required. Weather routing and similar techniques are well established.	Can be logistically challenging if a vessel needs to remain close to a port of refuge. Analysis to define limitation can be resource intensive.

Technique	Description	Advantage	Disadvantage
	relieve loads on structures (e.g. restricted to North Arabian Gulf or the Northwest Shelf of Australia).	Existing areas of restrictions are well defined (e.g. Great Lakes or similar).	There may be a disconnect between the limitation (distance from port) and the driving physical parameters (e.g. wave height)
Monitor	Using various techniques from strain gauges to physical and remote inspections, highlighted areas of interest are regularly checked for changes (e.g. monitoring the growth of cracks in non-critical structure).	Regular real time information on a defect can be invaluable for effective decision making. Large numbers of mature techniques exist (strain gauges, digital photos, NDT etc.)	Monitoring does not contribute to the structural strength of the part merely reports on its status. Large amounts of data produced can be difficult to process. Multiple inputs may be required to be effective (e.g. strain gauge results, loading condition and sea state simultaneously).
Structural reassessment	Design calculations and or analyses are revised. This may be due to a change in service, updates to design assumptions (e.g. metocean data), or for older assets to the advent of more sophisticated and accurate assessment methods.	Previously built in redundancy can be retired (e.g. for commercial reasons limited number of plate thickness used at build). Advances in engineering theory, analysis methods and computing power can mean that conservatism can be reduced.	Engineering resource required can be intensive and time consuming. Depending on the extent of analysis, an update in the applicable rules may occur (this may affect other areas such as lifesaving appliances, electrical design or more).

3. METHODS TO OPTIMISE STRUCTURAL DEFECT MANAGEMENT

When considering how best to address structural defects, multiple steps are typically required. There will be some type of inspection information, a level of engineering assessment and then some concrete action taken such as monitoring or repair. All the data produced through these processes needs to be managed and accessible in a meaningful and easy to understand manner. The sections below discuss some of the methods available for each of these steps.

3.1 APPROPRIATE INSPECTION DATA

Common practice when conducting inspections is to limit the scope and details recorded to those mandated by Class [2]. These requirements will often only be sufficient to establish the location and worst-case severity of a defect but not necessarily its extent to the degree required to carry out engineering assessment or design an effective repair.

The consequence of limited information when performing calculations can mean for example that a "worst case' or conservative thickness needs to be assumed. This can mean that failure may be suggested when better representative thickness data could in fact demonstrate adequate strength.

When a defect needs repairing, accurate details need to be known regarding its extent. In this way the right size repair can be specified as opposed to one that is too large or too small. If insufficient inspection data is available, conservative assumptions must be made and a larger more intrusive repair than is necessary may end up being specified. Lack of sufficient inspection data can also lead to surprises and late changes to repair details which can be very disruptive to a repair campaign cost and schedule.

It is therefore very important to specify inspection requirements for defects to ensure not only adequate information for Class, but also for the engineers who will assess the defect and the team who will carry out the repair. Typical information required using corrosion as an example includes: sufficient thickness measurements to determine average representative thickness, clearly defined boundaries of the defect as well as distance to the nearest identifiable bulkhead / frame or stiffeners. Figure 2 provides and example of the level of detail that can aid structural optimisation.

Figure 2: Example of detailed annotation of anomalous plate

3.2 RULE ASSESSMENT OF DEFECTIVE AREAS

At build, numerous considerations go into the selection of structural sections and plate thicknesses. Final scantlings often exceed those required by design rules and calculations. For example, plate thicknesses are often rationalised to reduce inventory and simplify the build process. Material availability, excessive rolling tolerances or corrosion margins may also result in as built scantlings being greater than the minimum required. In these cases, it may not be necessary to replace a defective structure with "like for like" scantlings.

In some cases, especially for example if the defect is away from the peak shear or bending moment locations of a span, the reduced structural capacity brought about by the defect may be acceptable, with no repair required.

3.3 FINITE ELEMENT ANALYSIS (FEA)

Finite Element Analysis (FEA) is a mature analysis tool that is widely used. However, its greatest use is normally at the design stage Referring to Figure 1, it is suggested that FEA can be used as part of a more optimised strategy for managing through life structural integrity for assets like naval vessels.

Optimal use of FEA can be through:

- Identification of structural weaknesses: This can enable more asset specific targeted and optimised inspections.
- Design optimisation: Reviewing the design through life, can enable the design to be optimised for in service conditions as opposed to the design drivers at build (which may have been most efficient production, ease of procurement pull through from a previous design or similar).
- Better assessment of defect risk: High quality FEA based on latest material state and up to date assumptions on loading can provide good insight into the level of risk posed by a defect, group of defects or pattern of degradation. This enables better decision making when deciding the most appropriate risk mitigation actions to carry out.
- Repair optimisation: Where there are limitations and constraints on the extent of a potential repair, a number of repair scenarios can be run and checked structurally, particularly where structural compliance appears close to the acceptable limit.

- Providing evidence to demonstrate safety and compliance (Class and regulator). Especially if any proposed repairs indicate a departure from the as built scantlings, a refined and optimised structure may only be justifiable through the use of FEA.

3.4 FAILURE MODES AND EFFECTS CRITICALITY ANALYSIS (FMECA)

FMECA as codified in military and civilian standards [3], [4] is a thorough, step-by-step approach designed to identify potential failure modes within a system, assess the effects of those failures on the system's operation, and prioritise each loss based on its criticality.

The following are considered: failure mode severity, frequency of the failure occurring, detectability of the failure (both before and after it has occurred) and actual impact on safe operations, the environment or the ability of a naval asset to prosecute the Command Aim. In this way FMECA can play a crucial role in identifying and prioritising failure modes that could impact mission-critical systems. In a naval context, this could enable the Command to focus repair and maintenance efforts on systems that are most vital to operational readiness and combat effectiveness. The use of FMECA therefore lends itself very effectively to the optimisation of structural maintenance or quantifying to senior management and the Command the consequences of a defect and the options to address it.

An additional benefit is that FMECA is not a one-time analysis but a cyclical process that evolves with the asset. As new information becomes available, the FMECA analyses can be updated, further optimising any decisions on repairs or asset deployment.

3.5 TYPES OF DATA GENERATED

When using these inspection and engineering assessment methods to optimise defect risk mitigation, large amounts of data of various sorts are generated. Table 2 summarises some of the types of data that can be generated.

Table 2. Types of data generated in support of structural optimisation

Inspection reports	Inspection reports offer a firsthand account of the current state of structural elements. These reports document visual inspections, measurements, and notes on any observed defects or abnormalities. The data collected in inspection reports serve as a foundation for subsequent analyses and are pivotal in prioritising repair efforts.
Survey reports	Survey reports extend beyond routine inspections, providing a holistic overview of the asset's condition. They consolidate findings from various inspections, non-destructive testing, and evaluations. Survey reports are instrumental in identifying trends, patterns, and potential areas for optimisation in structural management.
As-built drawings	As-built drawings are necessary to understand the original design intent. Comparing these drawings with the current state of the structure helps identify modifications, deviations, or areas where wear and tear are more pronounced. This understanding is crucial for optimising repairs in alignment with the asset's original specifications.
UTM data	Ultrasonic Thickness Measurement (UTM) data offers insights into the thickness of structural components. Changes in thickness over time can indicate corrosion or material degradation. Integrating UTM data into structural analyses facilitates predictions regarding the remaining useful life of components, aiding in the optimisation of repair schedules.
Emailed reports	Communication is vital in structural management. Emailed reports containing textual descriptions, observations, and recommendations serve as a means of conveying critical information between stakeholders. Integrating this form of qualitative data into the decision-making process ensures that insights from on-site personnel are considered in the overall structural optimisation strategy.
Change drawings / shipyard workpacks	Over the lifespan of maritime assets, changes are inevitable. Change drawings (or shipyard workpacks) document modifications, renovations, or upgrades made to the original structure. Understanding these changes is essential for accurate structural assessments and optimising repairs that align with the current state of the asset.
Calculation sheets	Calculation sheets provide the quantitative foundation for structural analyses. They contain mathematical computations related to stress, strain, load-bearing capacities, and other critical factors. Incorporating these calculation sheets into the optimisation process ensures that repairs are visually guided and supported by rigorous engineering analyses.

FEA reports	FEA reports can be long and detailed to satisfy Class or other requirements. Such information as boundary conditions, load conditions, meshing, validation, convergence coupled with the provision of sub models are essential to provide the evidence behind any optimisation.
FMECA / risk assessments	FMECA and the associated risk assessments necessitate the generation of large databases detailing the various considerations for each anomaly. These give the reasoning behind the chosen risk mitigation for any defects. As such are normally fundamental to the justification to regulators or other interested parties.

3.6 DATA MANAGEMENT

Optimisation of defect risk mitigation requires more information than simple traditional crop and renew repairs. This is both in terms of the technical input and the documentation required to justify the more sophisticated approaches. The opportunity comes from being able to effectively take advantage of the vast amounts of information and present it in a meaningful and easy to understand manner to enable quick and confident decisions to be made. The challenge is managing all the data generated.

Currently there are a number of tools available in order to manage the information associated with an asset. Normally any organisation will have a range of information management tools, however what tends to happen is that a mix of tools and data repositories exist or end up being created. In general, current tools can be split into two main groups, flexible and inflexible tools.

3.6.1 Flexible tools

These are normally generated from generic industry standard software that lend themselves well to customisation. For example, many organisations will have either Excel or Access databases to track certification or drawing documentation.

Table 3. Advantages and disadvantages of flexible data management tools.

Advantages	**Disadvantages**
Readily available Easy to customise Little or no training required Cheap Simple to use	Easy to lose version control of the data and tools Can struggle when vast amounts of data added Can handle single data sets (e.g. list of drawings or analysis), but are limited for multiple related data sets (e.g. relating analysis to locations or photographs may be challenging) 3D visualisation and interaction difficult to programme Multiple 'sources of truth'

3.6.2 Inflexible tools

Depending on the size of an organisation, large and bespoke sets of software may be used. These might be originally conceived for accounting or programme management reasons. In other cases, these may have been developed by Classification Societies to manage the survey cycle. Unfortunately, due to their rigid nature, users often end up creating separate spreadsheets or similar to present, manage or assess the data contained within.

Table 4. Advantages and disadvantages of inflexible data management tools.

Advantages	**Disadvantages**
Very effective at managing their target data set Difficult to create uncontrolled versions Are able to handle large data sets	Sometimes difficult to reconfigure for different assets (e.g. design changes have to be implemented by another office in a different time zone) Can be expensive with access limited by licensing issues Are vulnerable to limited or ending support Can require significant IT support to setup and maintain with new versions Do not lend themselves to integration across different sets of software

4. REQUIREMENTS FOR A TOOL TO MANAGE STRUCTURAL OPTIMISATION

To successfully develop or select a tool to support the structural optimisation process, the following is a suggested list of requirements that can be tailored to an individual organisation's needs.

4.1 IT CAPABILITIES

- Remotely accessible through the internet
- No complex licensing or burdensome IT setup requirements
- Licensing limited to standard and readily available software (e.g. Microsoft packaged software)
- Simple but secure access protocols to ensure no barriers to key stakeholders
- Low bandwidth requirements so as to be easily accessible remotely through low bandwidth internet connections
- Multiple simultaneous user access and data input possible

4.2 STRUCTURAL / ENGINEERING INFORMATION (CENTRALISED REPOSITORY FOR ALL STRUCTURAL DATA)

- Vessel build drawings
- Through life update / conversion drawings
- Repair drawings
- Repair workpacks
- Photographs
- Inspection and survey reports
- Condition monitoring results
- NDT reports
- Approval letters (Class and Regulator)

4.3 DATA INPUT

- Multiple data input methods available
- Single input and updates of individual anomalies
- Mass input of data through transfers of batches
- Flexible input fields to be able to receive multiple sources of information
- Tabular or graphical input formats
- Ability to attach / link drawings /images to individual anomalies

4.4 GRAPHICAL USER INTERFACE

- 3D visualisation of vessel
- Zoomable to local areas
- Defect symbols in geometrically correct locations
- Defect symbols to indicate types of defect (corrosion, deformation, pitting etc.)
- Defect symbols to indicated status of repair (repaired and date, pending repair with accepted drawings etc.)
- All data related to a defect to be linked to the defect (e.g. structural drawings, inspection reports etc.)
- Dashboards available to show graphs charts or other items required to manage the anomalies
- 3D visual interface for data input or tabular database lists

5. CASE STUDY – FPSO-A: PANDEMIC SHIPYARD STRUCTURAL PROJECT MANAGEMENT

5.1 BACKGROUND AND DEVELOPMENT OF TRUHULL

During the COVID-19 pandemic, an FPSO normally operating in Australian waters was in Singapore for her scheduled five yearly dry docking. The pandemic's immediate impact was a significant reduction in on-site resources. Travel restrictions effectively barred the deployment of engineering teams and consultants to the shipyard, disrupting the traditional workflow. Two critical issues compounded this scenario:

- Scattered information sources: The absence of a unified data repository led to reliance on disparate, often inconsistent human-centric sources of information.
- Visual inspection challenges: The inability to conduct physical inspections raised concerns about accurately identifying and assessing defects.

Amid these challenges, using recently developed commercially available software (e.g. PowerBi) some naval architects with a suitable disposition to handling and manipulating code were able to develop a fit for purpose asset specific structural integrity software package (TruHULL). This was designed to be simple and flexible, to enable seamless project management, effective defect detection and classification and repair decision making without the need for a physical presence.

5.2 BENEFITS AND IMPACTS OF TRUHULL

Key features and benefits of TruHULL:

- Centralized Information Hub: By acting as a single source of truth, TruHULL eliminated the inefficiencies of scattered data sources, ensuring all stakeholders had access to consistent and accurate information.
- Remote Project Management Capability: The software facilitated effective project management, enabling stakeholders to monitor progress, make informed decisions, and communicate effectively, all from remote locations. Importantly the metrics were not hours or spend but rather number of anomalies and whether they had been addressed.
- Advanced 3D Visualization: TruHULL's 3D visualisation tools allowed for detailed remote inspection of ship defects, providing a level of detail previously achievable only through physical examinations. This helped in particular with understanding the criticality of recorded defects.

Impact of Implementing TruHULL:

- Continuity of Operations: Despite the severe restrictions imposed by the pandemic, TruHULL facilitated the continuation of the shipyard campaign, ensuring that essential repair and maintenance activities could proceed without significant disruption.
- Enhanced Project Management Efficiency: The software's capabilities led to improved efficiency in project execution, with stakeholders able to access real-time updates, make timely decisions, and maintain a clear overview of the project status.
- Increased Stakeholder Confidence: The transparency and accuracy provided by TruHULL improved confidence among all parties involved in the shipyard campaigns, from the ship owners to the engineering teams and consultants.
- Data Management: The 3D graphical interface and links to relevant information improved the management of the large amounts of data required to be processed.

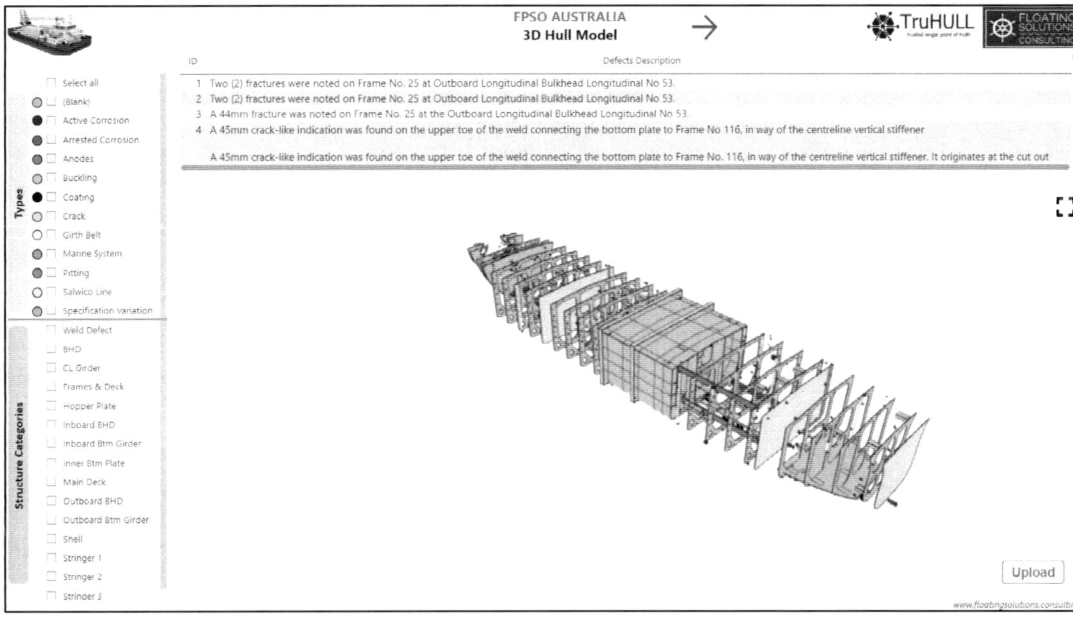

Figure 3: 3D Visualisation with filtering functionality

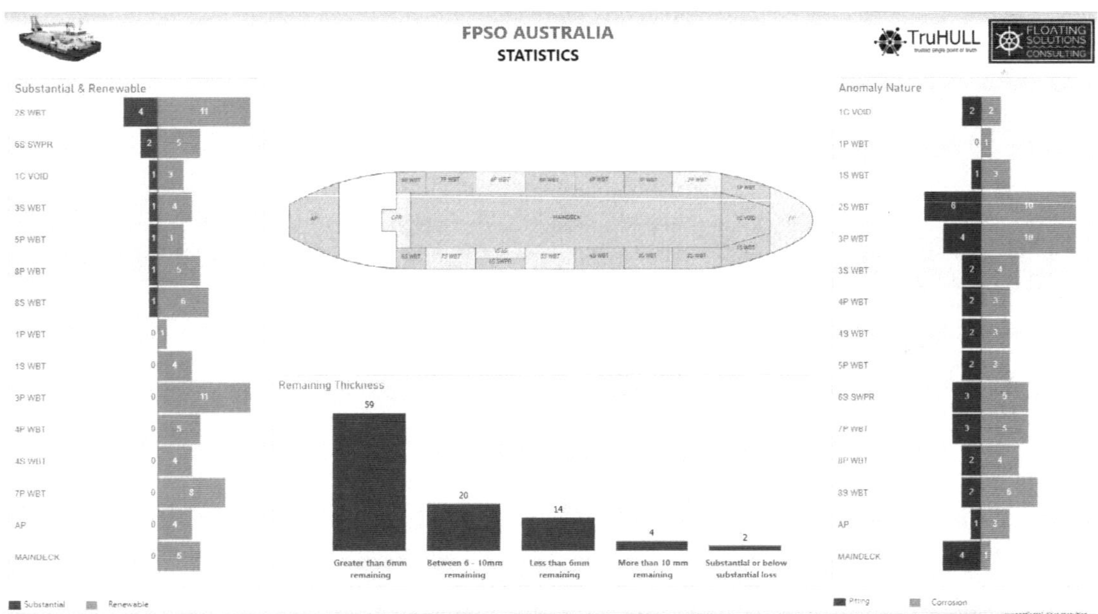

Figure 4: 2D Interactive Dashboard

6. CONCLUSION

High value oil and gas assets share many characteristics with naval vessels particularly in terms of their complexity and the significant impact of taking them out of service to effect repairs. Structural integrity optimisation through careful scrutiny of defects, detailed consideration of optimum risk mitigation methods and any proposed repairs can increase operability, availability and capability.

This structural optimisation can be completed using tried and tested techniques more normally deployed at the concept and design stage. However, to effectively deliver structural optimisation requires the gathering of significant amounts of data and the generation of large amounts of information. This is both to quantify the structural performance but also to document the calculations that reflect the material state and operational status of the asset in question. Although daunting in its range and volume, this information can be effectively managed using modern generic software to create tools to not only marshal and collate the data, but also act as effective visual aids to the engineers and presentation mediums to decision makers and senior management.

7. REFERENCES

1. WILLIAMSON., et al, "Lessons learnt from Oil and Gas: Practical opportunities to increase operability and capability through optimising inspection, engineering and repairs", *IMC 2022*, Sydney

2. LR "Shipright Design and Construction Procedure for Ship Units", 2014

3. MIL-STD-1629A, Procedures for performing a failure mode, effects and criticality analysis

4. BS EN 60812:2006 Analysis techniques for system reliability - Procedure for failure mode and effects analysis (FMEA)

8. AUTHORS' BIOGRAPHIES

Matthew Williamson is a Principal Naval Architect with Floating Solutions Consulting in Perth Australia where his role is as Lead – Client Delivery. Matthew is a Chartered Naval Architect with a career that has spanned Oil and Gas, Defence, and Commercial Shipbuilding.

As a Professional Engineer with a proven record in technical delivery and project management in the Marine Field. Experience includes the full range of projects from conception and bid through contract negotiation, detail design, build and in service. This has been whilst working for suppliers, prime contractors, regulators and clients internationally and within Australia. This understanding of the full lifecycle of marine projects is now being leveraged with Floating Solutions Consulting to deliver optimised inspection and repair scopes for marine assets.

Nabhan Zailani is a Naval Architect with Floating Solutions Consulting based in Malaysia and is also responsible for leading FSC's Digital System and R&D. Nabhan is a Chartered naval Architect from the Engineering Council and a Professional Engineer from the Board of Engineers Malaysia with broad experience in ship design, classification, consulting, construction, and inspection.

Nabhan has expertise in risk and engineering-based assessment of Floating Offshore naval architecture facilities, having been involved in ship repair and shipbuilding for commercial, Floating Production Storage and Offloading (FPSO), military and para-military ships of various types, sizes, and conditions.

Warship 2024: Future Surface Combatants
18th – 19th June 2024
Adelaide, Australia

BUILDING AUSTRALIA'S SOVEREIGN SHIP DESIGN CAPABILITY: LESSONS LEARNT FROM NAVANTIA AUSTRALIA'S LMV-M DESIGN

N Jayarathne and **J Coleiro**, Navantia Australia, Australia

SUMMARY

This paper examines the development of Australia's sovereign ship design capability using Navantia Australia's Littoral Manoeuvre Vessel – Medium design as a case study. First, it discusses how to develop a well-defined roadmap for the development of the capability that aligns with the requirements. Second, when, and how to involve the key stakeholders in the development process. Third, how, and where to invest in research and development to ensure that the design contains the cutting-edge technology. It then discusses the key stakeholders involved, the key milestones achieved, the key elements of the capability, and the lessons learnt that can be applied to the future. The paper concludes by discussing the future of Australia's sovereign naval ship design capability and the opportunities and challenges that lie ahead.

1. INTRODUCTION

Australia's latest strategic and geopolitical circumstances have made clear the importance and urgency for the nation to further develop sovereign ship building, design, and sustainment capability. There is recent evidence that the importance of this capability has been realised due to the release of an Independent Analysis of Navy's Surface Combatant Fleet [1]. This review detailed the Australian Governments path forward for a larger and more lethal surface combatant fleet and commitment to a plan for sovereign activities.

With a forward-looking approach, Navantia Australia (NAUS) has always understood the importance for Australia to establish this sovereign design capability. Since establishing an office within Australia in 2006, NAUS has developed a successful working partnership with the Commonwealth of Australia (CoA) through both design and sustainment activities on multiple Royal Australian Navy (RAN) platforms. This paper will focus on the specific design activities and processes which are recommended to be followed by Australian companies when designing vessels for Australia. NAUS' Kodal Littoral Manoeuvre Vessel – Medium (LMV-M) design will be used as a case study to detail this.

The LMV-M vessel is a bespoke design that was purposefully engineered following best practices at NAUS's Melbourne Naval Design and Engineering centre (DEC). It was designed against the Australian Army's LAND 8710 operational and technical requirements. The vessel was optimised to both littoral and riverine operational environments while meeting open ocean voyage requirements for extended distances. The design work concluded with lessons learnt, which are detailed within this paper. It is hoped these will be utilised by future Australian naval ship designers for the benefit of the nation's sovereign capability.

2. CURRENT SOVERIGN CAPABILITY

Australia's current naval sovereign design capability is limited and has been recognised as such for some time. There are several factors which have caused this, however, the largest is likely the common approach by the nation to procure foreign Off the Shelf (OTS) designs and have vessels built overseas. This approach has been deemed more risk adverse, less complex, and hence more cost effective though is rarely of benefit to the local industry. OTS designs only get limited input from local companies and yards for modification work to suit Australian conditions and requirements. This approach has significantly hindered capability as it has not provided indigenous companies with the required funding or time to develop the skills, resources and supply chains needed for sovereign capability.

The lack of capability and/or utilisation of the capability in existence is particularly evident when considering the number of ships in service with the RAN that are fully locally designed. To bolster national security and economic prosperity, we as a nation should be aiming to have RAN and army ship fleets which have been completely designed, constructed, and sustained locally.

3. DESIGN DISCUSSION

The following subsections detail critical activities which enable sovereign ship design and lessons which have been learnt in achieving a sovereign design.

3.1 DESIGN ROADMAP

Naval ship design activity is a multi-disciplinary effort, thus a systems engineering approach is vital for a successful design. A well-defined design roadmap with input from all disciplines involved is critical to the success of a design project. A design roadmap is considered a visual representation of the project path that will generally include a high-level overview of tasks involved whilst providing information on a timeline, dependencies, milestones, and resourcing. It is important to note that the roadmap does not have to be low-level to be considered well defined. Its purpose is to provide clarity to a team, as well as promote communication and cohesion. Information management including what information is required, how to find the information, how to use the information must also be considered in the roadmap development. Due to the magnitude of the overall design project, it must be divided into manageable parts within the roadmap, each of which has a disciplinary focus [2].

For Australian ship designs, all aspects of a Request for Tender (RFT) first must be well understood before a roadmap is constructed. The RFT is a document release by the CoA and typically provides specific details about the design which is requested along with (but not limited to) a range of ship characteristics, relevant requirements, operational concepts, contractual conditions, known hazards, and checklists. When these are well understood, design milestones, deliverables, risks, assumptions, and schedule can be appropriately defined to aid in creating a roadmap for the design team to follow. It should be noted that once these are defined, they should be reviewed regularly to track progress and update items as required as unforeseen changes can occur.

For the LMV-M design roadmap development, senior and executive figures within NAUS were utilised and drew upon their previous experience and knowledge. They understood the resources available and their skillsets across the varying disciplines of engineering. This included in house engineers with electrical, mechanical, structural, and naval architecture backgrounds. There was also an understanding on the milestone dates, critical work required, relevant points of contact and what the overall hierarchy in the team was. The roadmap generated was visual and high level with major headings used to describe the type of work which was to be completed over the course of the project. Headings covered items such as 'development of requirements and concepts of operation', 'design work' by discipline of engineering, 'industrial partner and supplier responses' and through to the 'RFT delivery'. This roadmap was communicated to the team and made accessible to all in a local drive which assisted in the team's overall clarity on the project.

3.2 KEY STAKEHOLDER INVOLVEMENT

Key stakeholders have a vested interest and/or influence in the outcome of a project. In the context of Australian ship design, key stakeholders can vary but typically include CoA representatives, Australian Defence Force (ADF) personnel and design partners/suppliers. Having knowledge of who the key stakeholders are and what they require is important to ensure the project and/or design is given every opportunity to succeed. Key stakeholders must be engaged from the start of a design and communicated with effectively throughout the duration of a project to ensure they have complete clarity. This communication can be through progress meetings or general progress emails sent out at a given time interval.

For the LMV-M design, NAUS identified the key stakeholders for the project within an internal meeting. The team identified the CoA, design partner UGL and the Australian Army as the key stakeholders for this design. However, due to it being a tender process, contact with the CoA and Australian Army was limited to clarification meetings only. Design partner UGL, who were selected as the design's sovereign ship builder, were involved in major aspects of production planning and design, formal design, and approval reviews. Contact with UGL occurred regularly and via multiple channels. These being email, phone and in person meetings.

3.3 RESEARCH AND DEVELOPMENT INVESTMENT

Research and Development (R&D) investment provides companies the chance to be innovative and forward-thinking. Currently, Australia only spends approximately 1.68% of their GDP on innovation. This is well behind other major R&D nations such as the United States, Japan and Germany who all spend around 3% of their GDP on such endeavours [3].

In regard to ship design, innovative approaches must be taken by Australia to ensure a ship will have the latest features and cutting-edge technology at commissioning and for the years it will be in service. This is particularly challenging as upon completion of a design, there will typically be a significant lag period before a ship is commissioned due to the building and testing activities required. For this reason, it has always been NAUS' approach to design for the future. In the case of the LMV-M design, areas best suited for innovation were first identified and the design team was given time to investigate cutting edge technologies and ways to incorporate these. The areas deemed most suitable were around vessel autonomy, modular integration for items on the operations deck and the Unmanned Aerial Vehicle (UAV) integration.

In relation to vessel autonomy, the design team investigated technology which could enable autonomous capability in a navigation capacity. Ultimately it was found there are options for technology to assist with controlling (course and velocity), guidance (path planning and following), estimation (state of vessel) and obstacle detection. The appropriate architecture was able to be incorporated into the system design and gives the option of fitted for but not with capability of remote control, semi-autonomous navigation, or complete autonomous navigation. In remote mode, navigation control could be carried out by an operator on land or in a mothership. Semi-autonomous capability results in the system assisting the operator when making decisions by providing speed and course propositions. A full autonomous mode could be achieved and would result in the operator not needing to make any decisions regarding navigation. This feature has the capability to be deactivated at any time and was not intended to completely replace crew.

The operations deck needed to be capable to serve multiple purposes during various operations and the team identified modular systems are a particularly good way to achieve this. This is as modular systems have the ability to be rearranged, removed, or replaced easily depending on the operation. The team therefore gave the vessel the ability for twenty-foot equivalent unit containers to be installed on the operations deck. Containers could house additional weapons, medical facilities and/or sensors depending on the operation. Additionally, the operations deck featured a landing spot for an UAV. The integration of the UAV was in line with the future of operations where they can provide an economical and safe way for surveillances and attack capabilities.

3.4 KEY DESIGN MILESTONES

Design milestones can be tailored and usually differ between individual projects, teams, and companies. Typically, the key milestones are identified when planning the project along with their associated estimated dates of completion. In ship design, milestones can feature anything from the completion of a concept hull, predicting power requirements and all the way to the completion of the design. The LMV-M design team identified several key milestones throughout the design process, three of which are presented within this paper.

Firstly, achieving an optimised hull form for competing operational requirements. These being beach landing, riverine manoeuvre, cargo carrying and open ocean voyages. For riverine manoeuvres, the overall size of the vessel had to be relatively small as there is typically little space in riverine systems. The vessel then had to be large enough to meet cargo carrying requirements and have good seakeeping characteristics in open ocean conditions. For beaching operations, the hull benefits from having a flat bottom and shallow draft which directly competed with the cargo carrying operation where the draft would be deeper and the ocean-going operation where the hull should cut through the waves to limit slamming (not flat). To overcome these competing operations, the design team decided to put a weighting of importance against each hull form parameter and determine its effects on meeting all criteria and operations. Through doing this the team managed to achieve a design which was able to operate in all competing operational profiles in a safe and effective manner.

Secondly, the completion of model experiments to validate Computational Fluid Dynamics (CFD) calculations for performance (power and seakeeping) was a major milestone in the design process. To complete this, a scaled model of the hull and appendages (1:10) had to be produced and a towing tank program undertaken with various environmental conditions simulated. The testing was able to show the desktop exercises agreed with the experimental results for resistance, power, and seakeeping motions. This gave NAUS additional assurance in their calculation methods, which will continue to be used for future designs, as well as assurance regarding compliance for the vessel against the relevant RFT and classification society requirements.

Finally, the submission of this design to the CoA was a key milestone as it was NAUS' first fully Australian designed warship. Months of works with all branches of the company involved culminated in a design which could have underpinned the Army's future littoral manoeuvre capability. The design was also particularly important in a broader sense as one released to the public, it showed the nation that NAUS has and will continue to have the capability to be a sovereign designer for the CoA.

3.5 LESSONS LEARNT

Reflections on lessons learnt on the completion of designs can provide opportunities for growth and development on personal, company and government levels. However, the real value on this reflection comes when it is time to implement changes based on the lessons in future projects. In building a sovereign naval capability, every effort should be made to reflect on lessons learnt by companies after design programs as it will enable a more efficient process for future programs. It is particularly important to evaluate lessons learnt as when building new capability, teams are often crossing unchartered territory and do not have experience to draw upon.

Drawing upon lessons learnt from our allies across the globe with great naval design and building experience such as Spain will be beneficial for our nation's capability, however, the most effective way to learn will be through completing designs locally. Proof of this is the lessons which NAUS LMV-M design team have listed at the conclusion of the design process. These were documented for future learnings and some major points are presented below.

- A concept arrangement should be developed as early as possible which encapsulates as many of the design considerations as possible before refining it in later design sprints.
- Developing a concept before without understanding all requirements presents risks and inefficiencies as there is hesitance to move away from the initial design.
- The importance of spending time to identify the key design drivers early in the design phase should not be overlooked. For the LMV-M design, these included (but weren't limited to) beach water depth, air draft, speed, range, sea state, cargo capacity, crew/endurance.
- Don't artificially constrain the vessel design based on perceived limits such as rule set 'trigger points' – the implications are significant.
- A weight and moment control system should be set up against the hull hydrostatic parameters to ensure concept designs do not end up being significantly overweight or out of trim. This is important as it would identify issues before significant analysis of such things gets completed.
- Identification of a line between what is being procured and what is being designed should be done as early as possible.

The lessons listed were identified to have mainly been centred around the initial stages of the vessel design process. In particular the importance of understanding the key design drivers, developing a concept design early and limiting artificial constraints. This highlights that more time could be spent in planning and understanding requirements before commencing work. This is challenging as there can be time constraints and an eagerness to begin, but ultimately the finished design will benefit from more time in this phase.

4. FUTURE OUTLOOK

With proper investment and favourable decision making, Australia's sovereign design capability has the potential to expand and develop rapidly. Multiple companies within Australia, including NAUS, are already taking steps to increase their design capability through additional resources, partnerships and training though need to be engaged and considered more frequently for 'clean sheet' designs. This will be critical for our sovereign development as it will allow the industry to gain valuable experience and learn the lessons that can only come from completing tasks. The engagement of local companies will likely come at a higher financial cost and risk profile then alternative foreign OTS solutions; however, it would be in the best interest of the nation and its security to take this step now due to our latest strategic direction.

Australia's surface fleet is in an expansion phase. The CoA's latest commitment to expand the RAN's surface combatant fleet to numbers that have not been seen since the end of World War II, as well as growing now the local shipbuilding industry makes the perfect time for the CoA to prioritise future design sovereignty.

5. CONCLUSION

The development of sovereign ship design capability is critical for national security and economic prosperity. Australia's current capability is limited however there are opportunities for expansion and growth with favourable decision making from the Australian Government. Indigenous companies such as NAUS are taking steps to increase their design capability but do need to be engaged for 'clean sheet' designs for the nation to begin on the road to complete design sovereignty.

NAUS's LMV-M vessel design was a significant milestone in the development of Australia's sovereign ship design capability. It is a bespoke design that was purposefully engineered following best practices at NAUS's Melbourne Naval Design and Engineering centre. It was designed against the Australian Army's LAND 8710 operational and technical requirements. The completion of the design allowed NAUS to reflect on the critical activities required to build a sovereign capability and provided many key lessons.

For sovereign designs, the importance of establishing a well-defined roadmap should not be overlooked. A design roadmap should include a high-level overview of tasks involved whilst providing information on a timeline, dependencies, milestones, and resourcing. For NAUS' LMV-M design, senior and executive figures within the company were utilised and drew upon previous experience and knowledge to define a roadmap. The roadmap listed items such as development of requirements and concepts of operation, design work by discipline of engineering, industrial partner, and supplier responses and through to the RFT delivery.

Key milestones for sovereign designs can vary by company and the baseline from which they are beginning but are important to identify for a team to work towards. For NAUS' LMV-M design, some key milestones which were identified were the optimisation of the hull form against competing requirements, validation of CFD results against experimental data and the completion of design.

Reflecting on lessons learnt from ship design activities are vital for the nations naval industry to grow and be efficient. Upon completion of NAUS' LMV-M design, lessons were documented and reflected upon. The lessons listed highlighted the importance of planning and understanding what is required. NAUS's biggest learnings from the design were to ensure there is an understanding of the key design drivers, developing a concept design early and limiting artificial constraints.

6. ACKNOWLEDGEMENTS

The authors of this paper would like to acknowledge the NAUS Kodal LMV-M design team. In particular, Peter Goodin, Dylan Barnett and Dylan Dwyer who provided valuable support and insightful information through the development of this paper.

7. REFERENCES

1. Commonwealth of Australia, 'Enhanced Lethality Surface Combatant Fleet – Independent Analysis of Navy's Surface Combatant Fleet', www.defence.gov.au/about/reviews-inquiries/independent-analysis-navy-surface-combatant-fleet , 2024.

2. LAMB, T., 'Ship Design and Construction', *SNAME*, 2003.

3. Australian Academy of Technological Sciences and Engineering, 'A future made in Australia needs to be built on Australian research and development', www.atse.org.au/news-and-events/article/a-future-made-in-australia-needs-to-be-built-on-australian-research-and-development, 2024.

8. AUTHORS BIOGRAPHY

James Coleiro is a naval architect with six (6) years of experience in the Australian maritime industry. He graduated from the Australian Maritime College at the end of 2017 and has since gained exposure to small recreational craft design as well as naval sustainment and innovation design activities through his various roles across the country. Currently, James is working as a naval architect within the Innovation team at Navantia Australia and is assisting in the development of cutting-edge designs and technology for the naval defence sector.

Nirman Jayarathne is a naval architect with 18 years of experience in various roles in the maritime industry, including ship management, commercial and naval ship design, marine surveying, lecturing, commercial and defence research, and leading innovation projects. He completed his PhD in Ship-to-Ship Interaction in 2018 at the Australian Maritime College. Currently as the Innovation Lead at Navantia Australia, he is responsible for driving the development of new and innovative technologies and solutions for the naval defence sector. His research interest focuses on application of AI for naval ship and system designs and ship hydrodynamics.

DISRUPTIVE INNOVATIONS FOR NAVAL VESSELS: A CASE STUDY OF THE NAVANTIA AUSTRALIA KODAL UNCREWED LANDING CRAFT

N Jayarathne and **J Coleiro**, Navantia Australia, Australia

SUMMARY

Navantia Australia (NAUS) has developed a Kodal Uncrewed Landing Craft (ULC), a high-speed, high-payload concept vessel which can operate in a wide range of maritime environments and perform various missions. ULCs can offer a number of advantages over traditional landing crafts, including increased survivability, reduced cost, and enhanced capabilities. This paper discusses NAUS's experience in developing the Kodal ULC concept and provides valuable insights into the challenges and opportunities of disruptive innovation in naval ship design. Additionally, the development journey of the ULC by NAUS is examined, focusing on the key drivers and enablers, the key technological challenges that NAUS has faced, and the potential operational impact of Kodal ULCs on naval warfare. The paper concludes by discussing the future of disruptive innovations for naval vessels and the opportunities and challenges that lie ahead.

1. INTRODUCTION

Disruptive innovations have played a significant role throughout history in the design of naval vessels. They have provided navies with new and once unimaginable capabilities and ultimately shaped the way naval warfare has been conducted. However, in more modern times, the promotion and uptake of such innovations in a naval context has been slow or even non-existent due to risk adverse and legacy approaches by government and defence force agencies. This has in turn reduced the willingness of companies to develop or present new and disruptive designs. Recent technological advancements and geopolitical circumstances are shifting the sentiment however and there is now an increased appetite from agencies for disruptive innovations and technology.

Uncrewed Surface Vessels (USV) are a product of disruptive innovation and have the potential to revolutionise the future of naval warfare. They can provide navies with many distinct advantages including the ability to operate in hazardous environments, conduct missions that would be too difficult, dangerous, or expensive for manned vessels, and provide enhanced lethal capability. Around the globe, navies have been taking an increased interest in USVs as well as the disruptive new technologies which enable their operation. Recent evidence of interest and investment in uncrewed capability is seen through the latest commitment by the Commonwealth of Australia's (CoA) to acquire six (6) Large Optionally Crewed Surface Vessels (LOSVs) [1], which as the name suggests, will have the option to operate autonomously without crew.

Conventional crewed landing crafts are versatile and have typically provided navies with the capability to deliver personnel, equipment, vehicles, and supplies to shore based locations with no port facilities. Most significant navy fleets across the globe, including the United States, United Kingdom, and Australia currently feature several crewed landing craft of small to medium size. Their strategic importance is well recognised and has been for some time, particularly after their positive influence in historical global conflicts such as World War I and II. Based on such history and knowledge of geography, it can be reasonably assumed that landing craft will play a pivotal role in future warfare operations and therefore new and innovative designs and concepts, like Navantia Australia's (NAUS) Kodal Uncrewed Landing Craft (ULC), should be developed to match the needs and technology of the future. The ULC concept was developed by NAUS at their Melbourne Naval Design and Engineering Centre (DEC) with the intention for it to act as an amphibious warfare capability extender. It has been developed against a set of expected future Australian Defence Force (ADF) requirements and operational profiles. Innovative approaches were taken by the design team to develop the concept, and this yielded a fully autonomous craft which can reach high speeds and has a high payload capacity. Furthermore, the Kodal ULC has cutting edge technology integrated which can allow interoperability with existing Royal Australian Navy (RAN) platforms and also teaming capability to allow the formation of sea trains, bridges or jetties making it extremely versatile. This paper explores the concept further with emphasis on disruptive innovations.

2. DISRUPTIVE INNOVATION AND IDENTIFICATION

Disruptive innovation is formally defined as the process where a company is able to successfully challenge established incumbent businesses with their new products and services [2]. Disruption is said to have successfully occurred when the product or service becomes mainstream and is therefore widely used. In a naval ship context, there have been many examples of this occurring. Some of the more notable disruptions which have occurred over history include the design and integration of artillery weapons on ships, communication system equipment and the design of U-Boats and aircraft carriers.

Noting the strategic importance of landing craft and uncrewed surface vessels moving forward, NAUS identified an opportunity to combine the two and develop a new disruptive variant which provides capabilities that align with expected future missions. To fully understand what would be considered disruptive however, NAUS first had to conduct a study as to what designs and technology currently exists, what other businesses are considered well established with quality landing craft designs, and if the market would require and/or embrace such a design. Additionally, to assist in determining areas in which a naval landing craft could benefit from disruptive innovation, NAUS chose to utilise a thought-provoking technique labelled SCAMPER. The SCAMPER technique was developed by Bob Eberle in 1971 and uses a set of directed questions from its acronym to resolve a problem or meet/identify opportunities [3]. The points below highlight what each letter in the SCAMPER acronym stand for as well as examples of what the NAUS Kodal ULC team identified when considering each against a traditional landing craft.

- S – Substitute (e.g., substitute crew and human decision-making on-board through autonomous operation),
- C – Combine (e.g., combine underwater cameras with sensors and berthing areas to suit modular connectivity),
- A – Adapt (e.g., new design to have adaptable features such as underwater recording and cargo recovery),
- M – Modify (e.g., modify existing designs to increase payload capacity, decrease drag and add modular ability),
- P – Put to other use (e.g., utilise new vessel to form bridges and jetties, act as a power barge and/or helideck),
- E – Eliminate (e.g., eliminate the need for different platforms in a fleet due to extreme versatility of new design),
- R – Reverse/Rearrange (e.g., rearrange decks to allow for an enclosed cargo hold as opposed to exposed cargo deck).

3. KODAL ULC DEVELOPMENT

3.1 KEY DESIGN DRIVERS

Key design drivers are usually formed from an understanding of customer requirements, design/system requirements and application drivers [4]. In a naval vessel design context, customer requirements essentially guide the designing team to what type of vessel is needed and what they want for it to have. Design and system requirements are specifications for the vessel and its subsystems onboard such as principal particulars or its payload capacity. Application drivers are lesser known but focus on the 'how' and are generally describing what the vessel needs to have onboard for it to do what the customer would like. The Kodal ULC team understood the design drivers from the applicable aforementioned areas and managed to put together an extensive list of what the key drivers were. The following list highlights some, but not all, of the key design drivers identified.

- The Kodal ULC shall fit within the docking area of the RAN Landing Helicopter Dock (LHD). Additionally, it should have the capability to act as a replacement vessel for the RAN LHD Landing Craft (LLC) which currently docks inside the LHD,
- The vessel should have fully autonomous operation and self-learning capability through technology integration,
- The vessel should have limited moving parts and its features should be of simplistic designs to limit potential failures and therefore the need to return to port (i.e., increased endurance),
- The vessel should have enough versatility to meet the ADF needs in the year 2040,
- The vessel shall have adequate capacity and stability to transport specific cargo and army vehicles (cargo and vehicle types not disclosed). Additionally, it should be able to autonomously load and unload the cargo,
- Emissions during operation are to be limited, but ideally zero,
- The vessel is to have classification society approval for Uncrewed Surface Vessels (USV).

Ultimately, the key design drivers listed created a framework which could be followed and referred to during the development process. To promote innovative designs however, the team remained opened and investigated various technologies and features which were not specifically driving the design. This is considered essential in innovation as constraints are a well-known killer of creativity.

3.2 TECHNOLOGICAL CHALLENGES

Incorporating cutting edge technology or technology which is not typically used provides many challenges, particularly as the best way to utilise or integrate it has not necessarily been discovered. In the development of the Kodal ULC, identifying technology which could enable autonomous operation for all missions and systems presented the greatest challenge. Autonomous technologies are quite new and for successful integration onto naval vessels, many different factors needed to be considered. Three of the challenges faced relating to autonomous technology are presented below.

Firstly, selecting the appropriate technology and developing the required algorithms for the vessel to be capable of self-learning presented significant challenges. The self-learning model needed to ensure the vessel can use measured data from previous missions to improve performance without any intervention from humans. The design team therefore had to determine what data is relevant to enable learnings in specific areas and what conclusions the system could draw. It was critical to overcome this challenge as without a self-learning model, there would still be a heavy reliance on human involvement.

Secondly, developing algorithms and selecting sensors which would allow vessels to be able to team to create a platform had a unique set of challenges as external factors such as environmental conditions had to be considered. Environmental factors such as wind speed, wave height, and tidal surges all had to be accounted for along with how they are impacting the movement of other vessels which they are teaming with. The algorithm therefore had to be capable of understanding likely impacts of specific conditions and then be able to successfully locate and lock onto sensors on the nearby vessel to which they should mate with.

Finally, the Kodal ULC had to have the ability to communicate with a control station as well as other surrounding ULC vessels when working as a team. Therefore, the team was faced with the challenge of creating a vessel and system which is smart enough to process data and relay it to either the control station or other vessels. If working as a team, the vessels required the ability to use the information of the surrounding vessels to adjust their behaviours. Thus, the team had to prioritise the Robotic Teaming application development in the ULC development roadmap.

3.3 WARFARE OPERATIONAL IMPACT

Naval warfare describes the tactics of military operations conducted on, under or over the sea and it has a major influence on the outcomes of historical conflicts [5]. There has been significant evolution in such warfare and how it is conducted over history. From using rams as the primary weaponry in ancient times, to the introduction of guns and submarines in more recent times. Now, in the 21st century, technological advancements and innovative designs are shaping how naval warfare will be conducted in the future.

NAUS's Kodal ULC concept incorporates cutting edge technology and innovative features which have the potential to revolutionise naval warfare operations. It is expected the Kodal ULC will have its greatest impact on warfare through its autonomous operation which eliminates the need for humans onboard. This is a significant change to current warfare tactics as it ensures navies can avoid endangering human life but still achieve missions. If operators prefer to have human input for the operation of the vessel, it can be done safely from a command centre on shore or on a mother ship.

The versatility of the Kodal ULC also has potential to significantly impact naval warfare operations. The design enables navies to achieve missions which previously could not have been possible entirely or not possible with just the single platform. Some key missions which could be achieved by the vessel (apart from traditional landing craft capability) and impact warfare tactics are as per the points below. Note not all features and mission profiles could be disclosed due to the commercial sensitivity of the design.

- Act as a fuel replenisher for nearby vessels,
- Carry ammunition to dangerous locations without risking human life,
- Connect to form a bridge for vehicles to cross or form a jetty for vessels to dock,

- Carry drones to locations where surveillance is required,
- Transfer vital supplies such as food and portable water.

Finally, with its minimal maintenance and zero crew requirements, the endurance of the Kodal ULC will force opposing navies to rethink their strategy and resourcing. Currently naval ships need to return to port to gather provisions for crew, refuel and/or get maintenance at certain time intervals which can leave a nation, or waters vulnerable. The Kodal ULC has addressed such issues by ensuring its onboard systems requires minimal maintenance through simplistic designs as well as a reduced number of moving parts. Additionally, with no crew onboard, the vessel does not need to return to port to gather provisions which will significantly increase endurance during operations.

4. FUTURE OUTLOOK

The future is never certain; however, it is clear that governments and navies will need to continue to consider and embrace concepts of disruptive innovation and technology if they wish to achieve future naval superiority. Disruptive innovation has already had a significant impact on the way naval warfare is conducted and being planned for in the 21st century and should technological advancements continue at a rapid pace, disruptive innovation for naval vessels will likely be provoked and enabled. This is essential as naval vessels are going to require new and innovative features to address threats which will be encountered in the future. There is no guarantee on what such threats will be, only, that they will not all be the same as today.

To further enable future disruptive innovation for naval vessels, the maritime industry must now seize the opportunity at hand and invest in their people and give them the time and resources to produce designs and products which develop capability and disrupt the industry. If future widespread investment occurred, it would promote competition which would in turn drive innovative designs. It is important to note that disruptive innovation will only be worthwhile if it can provide navies a new or enhanced capability. Not all disruptive innovations are worthwhile in the naval context.

A major challenge moving forward will likely be around ethics in disruptive innovations as they have increasingly been based upon the use of uncrewed vessels and autonomous operations in naval warfare. Questions exist over if it would be fair to operate such systems against those who do not have access to it, as well as if it would be suitable to use them to directly harm humans. Currently there is no right answer, however, there will need to be more discussion on this topic moving forward by the relevant law makers.

5. CONCLUSIONS

The development of the Kodal ULC is a significant step in the case for disruptive innovations in naval ship design. The design features cutting edge technology to enable complete autonomous operation while it also has an operational versatility that is unmatched from current naval vessels. The autonomous operations will provide navies with many distinct advantages over traditional landing craft designs. Namely, the ability to operate in hazardous environments, conduct missions that would be too difficult, dangerous, or expensive for manned vessels, and provide enhanced lethal capability.

To achieve advantage in naval designs, innovative and disruptive approaches have become a necessity. There are many methods to identify areas for disruptive innovation in existing products or designs, however NAUS utilised the SCAMPER method in the design of the Kodal ULC. The SCAMPER method provides a thought-provoking series of questions and enabled the design team to move forward with several ideas for disruptive innovation.

Key design drivers create a framework for a design team to follow in the design of any product. They enable an understanding of what the customer requirements, design requirements and application drivers are which in turn provides a clear picture of what the end product will look like. The Kodal ULC team understood the design drivers from the aforementioned areas and managed to put together an extensive list of what the key drivers. To promote innovative designs however, the team remained opened and investigated various technologies and features which were not specifically driving the design. This is considered essential in disruptive innovation as constraints are a well-known killer of creativity.

There were several technological challenges faced through the design process for the Kodal ULC. Most were identified to be around the integration of autonomous technologies. First, the team was challenged in developing the required algorithms for the vessel to be capable of self-learning. Second, developing algorithms and selecting sensors which would allow vessels to be able to team to create a platform was challenging as it had to consider external environmental conditions. Finally, it was a challenge to provide the vessel with the ability to communicate with a control station as well as other surrounding ULC vessels when working as a team.

The Kodal ULC has the potential to revolutionise naval warfare operations. The vessel's ability to operate autonomously, achieve a variety of different missions and also have a high endurance would force opposing navies to rethink their strategy and resourcing moving forward. Whilst the future is not guaranteed, with warfare operations changing through designs such as the Kodal ULC, it is clear that governments and navies will need to continue to consider and embrace concepts of disruptive innovation and technology if they wish to achieve future naval superiority.

6. ACKNOWLEDGEMENTS

The authors of this paper would like to acknowledge and thank the ULC design team including Luke Baxter, Sam De Vincentis, Lachlan Stammers, Charlie Williams, Wilson Njoroge, Saskia Milne, and Peter Goodin for their continual support provided with the Kodal ULC design.

7. REFERENCES

1. Commonwealth of Australia, 'Enhanced Lethality Surface Combatant Fleet – Independent Analysis of Navy's Surface Combatant Fleet', www.defence.gov.au/about/reviews-inquiries/independent-analysis-navy-surface-combatant-fleet , 2024.

2. CHRISTENSEN, C., RAYNOR, M., and McDONALD, R., 'What Is Disruptive Innovation?', https://hbr.org/2015/12/what-is-disruptive-innovation , 2015.

3. SERRAT, O., 'The Scamper Technique', https://www.researchgate.net/publication/318018918_The_SCAMPER_Technique , 2017

4. HEEMELS, WPMH., 'The Use of the Key Driver Technique in the Design of Copiers', https://www.gaudisite.nl/ICSSEA2006_HeemelsEtAl_KeyDrivers.pdf , 2006

5. GILANI, A., 'Naval Warfare: A Journey from Ancient Galleys to Autonomous Weapons and the Role of International Law', https://www.dlpforum.org/2023/10/11/naval-warfare-history-international-law/ , 2023

8. AUTHORS BIOGRAPHY

James Coleiro is a naval architect with six (6) years of experience in the Australian maritime industry. He graduated from the Australian Maritime College at the end of 2017 and has since gained exposure to small recreational craft design as well as naval sustainment and innovation design activities through his various roles across the country. Currently, James is working as a naval architect within the Innovation team at Navantia Australia and is assisting in the development of cutting-edge designs and technology for the naval defence sector.

Nirman Jayarathne is a naval architect with 18 years of experience in various roles in the maritime industry, including ship management, commercial and naval ship design, marine surveying, lecturing, commercial and defence research, and leading innovation projects. He completed his PhD in Ship-to-Ship Interaction in 2018 at the Australian Maritime College. Currently as the Innovation Lead at Navantia Australia, he is responsible for driving the development of new and innovative technologies and solutions for the naval defence sector. His research interest focuses on application of AI for naval ship and system designs and ship hydrodynamics.

CHALLENGES IN FUTURE WARSHIP DESIGN: THE MULTI-MISSION DRONE CARRIER

R Pinto da Costa, Damen Naval, The Netherlands

SUMMARY

In this paper, the conceptual design process of a multi-mission drone carrier warship will be reviewed. This paper follows as a revision and update concerning a conceptual design project undertaken within Portugal's Instituto Superior Técnico aimed at developing innovative warship concepts as well as supporting design tools for new-generation surface combatant design. Within this project various challenges were faced in process and software development which will be further elaborated within the contents of this paper. Additionally, various learnt lessons have been incorporated into the developed solutions in order to improve upon the existing toolset. This toolset was tailor built for the aforementioned project and has since undergone a scaling-up process in order to increase the range of its applicability and potential to obtain further innovative concept designs.

1. INTRODUCTION

In modern conflicts, a surge in the use of naval unmanned assets has been observed. These systems and products have shown unprecedented operational success in both strike and reconnaissance operations coupled with a low buy-in cost for acquisition and deployment when compared to other market alternatives [1]. The state of the art of warships designed to deploy these systems has also consequently seen an increase in development [2]. As the modern naval battlefield evolves, so do the products that end-users make use of for their operational objectives – within which unmanned systems have recently taken a leading role in.

One of the most striking recent examples of this fact has been the widespread usage of surface loitering munitions in the Ukraine War. In 2024 alone, the Magura V5 USV's developed within Ukraine have been responsible for the sinking of 3 different warships – each of which has a highly reduced acquisition cost estimated to be approximately 273,000 US$ [3]. The use of this unmanned surface vehicle, along with a myriad of other unmanned assets, have been a hallmark of the Ukraine war and consequently prompted many navies to reflect upon their approach to modern naval warfare. Additionally, the state of the art of warships designed to deploy these systems have seen a surge in development in recent times [2], with many high profile projects focusing on the design of drone carrier ships as well as the incorporation of these assets into conventional warship designs. Despite this, the state of the art of drone carriers is still limited to a reduced range of built ships – a lot of which are either reconverted pre-existing vessels or ships solely dedicated to scientific research purposes.

With the above considerations in mind, in the academic year of 2022-2023, a project was developed within the Centre for Marine Technology and Ocean Engineering (CENTEC) aimed at developing methodologies, processes, best-practices and a toolset for the concept design of multi-mission drone carriers – carried through the lens of the concept design of two warships catered towards two distinct end users [4]. The findings from this work were published and discussed in various forums from which new conclusions, improvements and modifications to the existing work were later implemented and realized. This paper provides a summary of the previous work and discusses the various challenges faced throughout the project, at each individual stage, and how these were addressed by the author – along with still ongoing efforts to develop upon the already established work. The paper is thus divided into the following sections:

Chapter 2 and 3 – Literature and state of the art review: a brief overview of ship design procedures was developed, the differences between commercial and military ship design were elaborated, upon and a study of the the selected process methodologies and toolsets used by institutions for warship concept design was conducted.

Chapter 4 – Operational Requirement Definition: The ship's operational concept was defined based on its intended use, perceive likely threats, environmental conditions, benchmarked systems and other factors that will lead to a final set of operational requirements that will guide the concept development process throughout its entirety.

Chapter 5 – System Definition Process: The ship's systems will be selected in order to ensure that the vessel complies with its initially defined operational requirements. These will include all of the unmanned vehicles, platform systems, naval payload and other relevant systems for each configuration of the multi-mission drone carrier.

Chapter 6 – Ship Synthesis Model: A tailor-made concept design toolset was developed in the form of a ship synthesis model. This model makes use of non-linear optimization algorithms in order to find optimal concept solutions based on a set of design objectives. This synthesis model underwent various revisions throughout the course of the project and further underwent a validation process in order to ensure the validity of the obtained results.

Chapter 7 – Concept Results and Discussion: The resulting concept designs obtained through the aforementioned process will be discussed, critically reviewed and further refined.

Chapter 8 – Conclusions and Future Challenges: The project results and process will be further elaborated upon and discussed. Future improvements to the chosen process and synthesis model will be proposed as well as possible future applications of this process and toolset.

2. LITERATURE REVIEW

A ship's design is an inherently complex process. It requires a broad range of expertise across diverse scientific domains together with an iterative process of ideation, analysis, and refinement until an adequate solution is achieved. Typically, analytical-empirical methods – often derived from past designs – are employed [5]. Originally introduced in [7], the Ship Design Spiral [6] is among the earliest depictions of the contemporary ship-design process. The many iterative phases that go into designing a ship are displayed in this design spiral. However, several researchers have addressed this method's limitations and proposed other solutions that make up the state of the art of modern ship design [8]. Currently, the three most influential branches of ship design methodologies include: System-based design, Requirements Elucidation and System Engineering focused processes.

The most recent of the previous methodologies, are the System Engineering focused processes. This approach orients itself on the realization of successful systems by focusing on the definition of customer needs and required functionality in the early design stages, documenting requirements and then proceeding with design synthesis, verification and validation – oriented throughout with the initially documented requirements in mind [9][8]. This entails the decomposition of larger complex systems into their sub-subsystems which will then be further decomposed into components. At the component level, design processes are independent and based on the initial project requirements. A discussion of maritime applications of this methodology can be seen in [10] where it's recognized that system engineering places engineering at the core of management processes but that the methodologies should be tailored to specific projects.

Naval ship design comes with its own unique set of challenges when compared to commercial ship design – particularly due to a larger focus on design for operational capability. To address this, modern navies have traditionally made use of a point-based serial engineering process [11]. This method, although very commonly adopted throughout the industry, has shown to be inefficient in increasingly more complex naval design challenges. This methodology often leads to frequent redesigns due to the lack of expertise and information transfer between stages. This process has evolved in consideration that naval ship design is actually a "system of systems" type design challenge, in which the interaction and integration of various complex systems gives rise to new challenges in concurrent design methodologies [12].

A key part of this process that also deserves a detailed analysis is the design toolsets used throughout the concept design development of naval vessels. Although the initial stages of operational requirements definition and system selection all proceed with relatively small variations between institutions, the same is not true for the later stages of initial sizing and preliminary design (with a particularly emphasis on the toolsets used for this purpose). One of these toolsets is ASSET, a ship synthesis model developed by NAVSEA. This synthesis model is tailor-made to be used on concept design stages of naval vessels [13]. ASSET has been used widely in US Navy projects, however, history with the software has shown that it has various problems related to its robustness, crashes and computational time. ASSET data model and input configurations have also shown to be limited in applications, particularly in weight space and electric power estimations as well as in design flexibility for non-conventional ship designs [13].

Another challenge associated with this project is that the current literature regarding the early-stage design of drone carriers is still highly limited – principally due to the still emerging nature of these naval ships [13][15]. One of the few available works for drone carrier concept design can be seen in [15] in which a vessel designed to carry a fleet of unmanned vehicles was conceptualized through the use of architecturally oriented early-stage ship design processes. Within this thesis, decision-making and ship concept design tools were developed – intending on providing a holistic means of integrating aspects of a fleet of unmanned vehicles into early stage mothership design. At the point of time of this research, the concept of a "drone carrier" or "mothership" type surface combatant was yet to be solidified, but the reference project still provides valuable insights into the development process of similar ships.

Final considerations should also be made regarding how to implement and manage innovation within the context of concept design. To do so, it's possible to resort to C-K (Concept-Knowledge) Theory [16]. Within the context of this theory, designed systems are determined by a form (design synthesis), function (functional analysis and allocation) and performance (requirements and analysis). Each of these parameters provide insight into which elements are conceptual and which are based on established knowledge. This leads to a distinction between two "spaces": the concept space (C-Space) and the knowledge space (K-Space). The C-Space includes all concepts that do not yet have knowledge established within the K-Space. Concepts are generally regarded as necessary departure points within the conceptual design process, since without these the design is reduced to an optimization or small incremental improvement problem.

With the above literature in mind, the process for this project was thus defined. The project will start with an operational concept and requirements development process as per [12]. This stage is followed by a system definition stage, in which the various systems for the vessel will be selected – this process was later integrated into the ship synthesis model itself. A ship synthesis model, similar to the US Navy's ASSET software, will be then used to further elucidate the concept design and perform relevant calculations for the ship's initial sizing, crewing, resistance estimation, operational capability among other data, using UCL [17] guidelines and making use of non-linear optimization algorithms to find optimal designs for selected end-users. These design proposals would then be further elucidated and modelled using CAD software, from which the initial synthesis model would be readjusted.

3. STATE OF THE ART

Within the state of the art of this type of surface combatant warship, various cases were considered and compiled within a common database. These included examples from NATO member states as well as other modern navies such as the Chinese PLA Navy. These examples included: Zhu Hai Yun (PLA Navy). TCG Anadolu (Turkish Navy), NRP D. João II (Portuguese Navy), UXV Combatant (BAE Systems) and Octopoda 500 (Mauric).

Throughout this research, it was observed that the current state of the art of drone carriers is exceedingly limited when compared to other common warship typologies. Despite this, some design trends could be clearly observed and noted regarding these vessel's configurations, typical naval payloads and dimensions. The main distinction lies in the differences between drone carriers conceptualized to accommodate a large amount of unmanned aerial vehicles and ones designed to mainly operate surface and underwater vehicles. The first of these are generally characterized by large hangar spaces and flight decks, while being generally larger than their counterparts. The second however, are generally smaller in size with hangar spaces and flight decks designed to mostly accommodate VTOL UAVs.

Since the inception of this project, the state of the art of this type of ships has shown little advances. The only notable exception has been the reconversion of a containership into a drone carrier-like surface combatant by Iran – the Shahdid Bagheri [18]. On the other hand, unmanned assets have been increasingly implemented in naval warfare. Just in March of 2024, the US 3rd fleet concluded Integrated Battle Problem (IBP) 24.1 off the coast of Southern California. This exercise, through analysis, simulation, prototyping and demonstration, focused on employing unmanned systems in naval warfare [19]. In addition to this, in January of the same year, the Defense Innovation Unit (DIU) of the US Military put out a solicitation for companies to submit pitches for small unmanned surface vehicles that may tie into the Pentagon's Replicator initiative – an effot to acquire new kinetic, lethal USVs that can be employed, preferably, in a western Pacific context [20]. This telegraphs a clear interest in the further development, acquisition and integration of these assets into modern naval forces and thus signalling the importance of also developing naval ships that leverage these technologies.

4. OPERATIONAL REQUIREMENTS

The first part of the concept development process for this vessel was defining the requirement set which would later guide and baseline the entirety of the concept development. This is a multi-stage process and ones of the most important ones within the context of a concept design development. Generally, this process began with the definition of a core operational concept in which mission requirements were defined for a set of operational roles which the ship should be able to undertake. These were later compiled into a User Requirements Document (URD) and requirements database, in which only the most stringent requirements within the entire mission range for the vessel were considered. A summary of this process can be seen below in Figure 1:

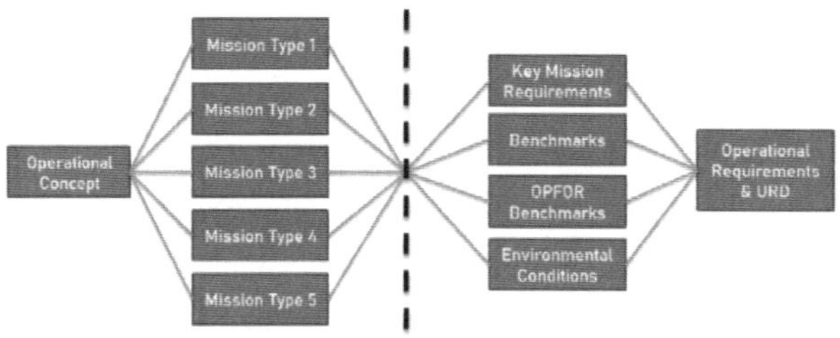

Figure 1: Operational requirements definition process [4]

4.1 INITIAL CONSIDERATIONS

Some initial considerations were taken into account, taking into account the general purpose of the developed concept as well as its intended end user. These will include:

End-user(s): the concept is aimed at NATO member states as end-users. These are states with a need to rapidly integrate unmanned capabilities into their fleets at scale. End-users will make use of the vessel in both wartime and peacetime scenarios, and it should be designed as such.

Area of operations: during peacetime, the vessel will operate principally in the end-user's EEZ – which is assumed to be located in the North Atlantic for project development purposes. In conflict and wartime, it'll be assumed that the ship will most likely need to be able to sail in the Baltic, Norwegian and Barents Sea. Occasionally, in extreme conditions, the ship should be made to make use of the navigational passageways (such as the Northwest Passage) in the Arctic Ocean in close-to-shore operations.

Deployment: the ship shall be able to deploy year-round in patrolling missions within the end user's EEZ as well as in NATO peacetime operations – deployed to missions such as disaster relief or search and rescue. Combat deployments will occur when in crisis or transition to war. These will include littoral strike and raiding operations targeting hostile country's infrastructure, naval assets or military personnel or task force support operations such as ASW, ASuW, MCM or AAW support to NATO fleets.

Concept success conditions: In peacetime, the ship should be able to endure long mission times characterized by high system utilization. The vessel should be able to provide reconnaissance in as large of an area as possible with the use of its unmanned systems while guaranteeing high versatility and low operational costs. Crewing demands should be kept to a minimum through the choice of modern systems and widespread usage of automation. As a combat system, the warship should be able to traverse high-risk areas while remaining mostly undetected/inconspicuous, quickly deploy a large amount of unmanned systems and then successfully extract while guaranteeing high mission success rates combined with low system loss rates for retrievable unmanned systems. Acquisitions costs should be minimized considering the innovative nature of the concept, in order to reduce the financial risk associated with such a venture, as well as reducing the financial consequences of losing the ship itself in combat as it will likely operate outside of conventional task forces in high-risk deployments during wartime. In task force support operations, the ship should be able to be fitted with various configurations allowing it to provide differentiated support to naval task forces in the submarine, air, mine-countermeasures and surface domains. Additional capabilities regarding disaster response operations should also be considered.

4.2 MISSION RANGE

Further elucidating the ship's operational concept, a mission range was defined for the vessel, from which mission specific requirements were derived. This mission range includes: littoral strike and raiding operations, task force support (ASW, ASuW, AAW, MCM), search and rescue and disaster relief operations. Each of these were detailed separately and mission specific operational requirements were thus defined, these would consider, among other design drivers, the following considerations: mission specific requirements, likely OPFOR and threats, current market solutions used as benchmarks, end-user specific naval ship requirements, environmental aspects among other miscellaneous considerations that may be translated into ship requirements.

An example of one of these missions will be littoral raiding and strike operations. In these missions, the vessel will sail into hostile zones while remaining undetected throughout. Thus, the ship's own hull and superstructure should be designed in order to reduce its radar signature and systems should be selected taking into account the objective to decrease the system's detectability. When nearing designated targets, the ship will sail along in close to shore operations while deploying unmanned strike assets at scale – including USVs, UAVs and UUVs – before returning to friendly controlled waters. This should be done underway, in the shortest amount of time possible and retrieving any designated systems that the ship has deployed that are non-expendable. If detected, the ship should be able to elude opposing forces through the use of its own self-defence weapon systems, combat UAVs and USVs as well as its own maximum speed. Initial operational requirements were thus derived considering the above considerations.

With this summary mission description in mind, and in order to define additional relevant operational requirements, both NATO and non-NATO ships designed for this exact purpose as well as the likely threats the ship will face while deployed were used as initial benchmarks. Operational requirements that will ensure the ship is able to successfully complete these types of missions were thus defined, including: a maximum speed of 35 knots, the capability to deter small naval patrol ships through the use of unmanned systems and defensive weaponry (e.g.: combat UAVs or a VLS system with a range of over 5,500 meters), the ship's unmanned systems and manned craft should be able to be deployed outside of coastal defence systems' range (22 km, considering likely threats), anti-aircraft measures should be equipped on the ship along with a radar system able to detecting threats at a range of 300 km, the hull and superstructure shall be constructed with non-magnetic materials or a MCM payload should be equipped on the vessel.

Finally, the operational area in which the drone carrier will operate in will also results in its own set of requirements. For this project, the operational areas considered for the drone carrier were: The North Atlantic Ocean, during peacetime operations, and the Barents, Norwegian, Baltic, Bering and North seas when in transition to crisis or wartime (with rare but possible crossings into the Laptev and East Siberian Seas). A set of environmental restrictions pertaining to these operational areas were also thus listed and included in the final requirement set. A particular consideration should be made to class guidelines [21] associated with requirements derived from the presence of floating ice in some of the above operational areas.

The process was repeated for all operational roles, and the resulting requirements were then compiled into a database. From within this database, the most stringent requirements were compiled and assigned in order to define the final concept requirements. Initially, this was developed manually. As the project developed however, this process has been automated by defining a range of different mission sets, operational areas and likely threats and allowing the user to simply select the relevant operational roles for the vessel, end-user and threats from which the developed toolset will automatically derive operational requirements.

An important consideration within this process is that only publicly available data was used. It's important to acknowledge both the possibility for inaccuracies in publicly reported data as well the fact that opposing forces and threats will evolve with time. With this in mind, since the first publication pertaining this project [4], this process has been readjusted by adding margins to derived operational requirements. This is done to both account for inaccuracies in publicly available data as well as to attenuate incremental innovation by opposing forces. This margin is based on recent historical evolution of naval assets on the side of these perceived threats. These were added principally in operational performance relevant requirements such as weapons and detection range, maximum speed and payload capacity. Additionally, systems currently in development (including concepts) were thus also included in this database. This process is inexact, but serves as an attenuating factor in order to future-proof the conceptualized vessel.

5. SYSTEM DEFINITION PROCESS

In this stage, the ship systems will be further elucidated and selected through an analysis of already existing market alternatives, benchmarks from modern naval vessels, considering NATO member state suppliers and each equipment's functionality. This decision process was initially undertaken manually, in which the following systems were selected: hull, propulsion type, sensors and effectors, energy production, unmanned vehicles and other auxiliary platform systems.

Taking into account the multi-mission concept for the ship, as well as the requirement for the ship to be reconfigured at port in order to be deployed for different roles, the selected systems were divided into two groups: fixed and variable systems. The first of which include systems which will stay consistent in type and quantity throughout all the generated concept iterations and include systems such as: propulsion type, sensors, hull type, sonar, platform auxiliary equipment, among others. Meanwhile, the variable payload will include all systems that will vary both within the same concept iteration (through reconfiguration of the vessel itself) as well as between concept iterations (generally through varying payload size). These will include systems such as: manned and unmanned vehicles, containerized mission modules and naval weapons/payloads.

The drone carrier's mission configurations were also developed with the several roles it should be able to change to perform in mind. Ten mission configurations were conceived in all, including search and rescue, disaster response, task force support (anti-submarine, anti-aircraft, anti-surface, and mine countermeasures), and littoral strike (low to high-risk, amphibious and non-amphibious). Two of the three forms of modularity specified in the Royal Navy's maritime modularity concept are taken into account in these configurations [22]: integral and installed modularity. An example of one of these variable payload configurations can be seen below in table 1:

Table 1. Task force support configurations' payloads [4]

	Task Force Support			
Unmanned System	**MCM**	**AAW**	**ASW**	**ASuW**
HERO-400EC	25+	50+	25+	50+
HERO-900	20+	40+	20+	40+
HERO-1250	15+	30+	15+	30+
Kratos XQ-58	4+	4+	4+	4+
Schiebel Camcopter	4+	0	0	4+
Tekever AR3	10+	0	0	10+
Loitering USV	0	0	0	4+
Protector USV	4+	4+	2+	3+
Zodiac MACH II MM	2+	2+	2+	2+
Swarmdiver USV	10+	0	5+	0
Echo Voyager	0	0	1+	0
Loitering UUV	0	0	5+	5+
SDV Mk9	0	0	0	2+
SDV Mk8	0	0	0	2+
Modular TEU	4+	0	4+	0

One of the main challenges faced throughout this process was the impact that the system selection process would later have on the resulting concept design outputs. Performing this process manually would later restrict the range of innovative solutions that would come from the model itself. As an example, one of the initially selected radar system for the vessel was the Thales APAR Block2. This selection served as a fixed input into the synthesis model and all subsequent iterations of the concept would make use of this radar model. However, there are other market alternatives that also fulfil the ship's operational requirements – despite the initial selection of the previous radar system as the ideal one for the ship. Proceeding in this way, with fixed system decisions, would disregard all possible ship iterations that would make use of different radar systems – there being a possibility that the choice of a different, initially seemingly less adequate radar system, would lead to a better solution when taking into account information obtained at later stages. To overcome this, this manual process was later replaced by a toolset connected to the operational requirements database. This toolset, through a set of decision gates, selects a range of adequate systems for the vessel and forms configurations used as input in the ship synthesis model. This results in a wider range of concept solutions based on these different configurations. The updated process can be seen summarized below in Figure 2:

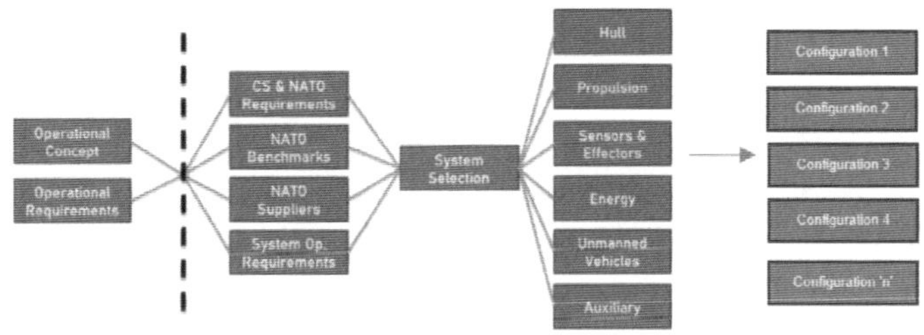

Figure 2: System selection process diagram

6. SHIP SYNTHESIS MODEL

Within this stage, a ship synthesis model was developed that would allow the transition from a concept of operations and selected systems into a concept design proposal. This model was based on adapted methods and calculations from [23][24][25] as well as Royal Navy and UCL methodologies standardized within [17]. The model is made up of a series of concurrently running modules which, through the input of restrictions based on the operational requirements and system selection process, performing various relevant calculations, and by iteratively varying the ship's main particulars, will output a final concept solution which is optimized for a certain end-user. These modules include:

Database: Compiled database of warships from the past 25 years, with similar mission profiles to the designed drone carrier concept. This module is used as a reference point for a later comparison process of the obtained concept solutions;

Ship Details & User Interface: Within this module, the project variables for the ship synthesis model are inputted. The module also serves as the main display for the final concept solution information which is extracted from other modules.

Hull Form Coefficients: Performs calculations and estimates regarding the ship's hull form coefficients, including: block coefficient (Cb), waterplane coefficient (Cwl), midship coefficient (Cm), and prismatic coefficient (Cp);

Volume Calculations: Performs volume estimates for the vessel, which will include both the hull and superstructure volumes based on [17];

Resistance Estimation: Estimates ship resistance, based on Holtrop & Mennen's method extension of 1982 [26], for both the ship's service and maximum speeds;

Propulsion, Fuel & Range: Dimensions the main propulsion installation and performs calculations to obtain how much fuel the ship will need to comply with its operational requirements;

Initial Stability Analysis: Initial intact stability analysis is performed as per [17] and compares results with modern NATO warship requirements;

Mission Payload Profiles: This module will serve as a database for the vessel's naval payload and configurations. It includes information regarding all of the ship systems including: weight, volume, deck area, overall dimensions, cost and other relevant technical details which are input from the system selection database;

Crew: Performs an estimate of the ship's required crew. Crew is divided into functional departments; whose personnel requirements vary with each ship iteration. Calculations are obtained regressively from Royal Navy (UK) data as well as specifications from installed systems;

System Groups: All of the ship's systems are divided into functional groups that make up the full ship. For each system group a set of individual systems are detailed and calculations are performed to obtain the weight, volume and acquisition cost of each individual equipment. Estimates are also obtained for operational capability details – e.g. the amount of unmanned vehicles the ship is able to carry for each iteration. The formulas used to estimate the above details are extracted from [17] or from publicly available data. The following system group modules were included: hull, personnel, ship systems, propulsion, electrical power, naval payload, variables;

Costing: The ship's acquisition cost is estimated in US$ based on data from other modules;

The synthesis model, making use of each separate constituent module, creates various iterations of ships through the use of non-linear optimization algorithms (in this case, a GRG non-linear algorithm [27]) until a final iteration, which complies with all of the project restrictions, operational requirements and optimization goals, is obtained. This iteration will serve as the output for each concept solution. A summarized diagram of this process can be seen below in Figure 3:

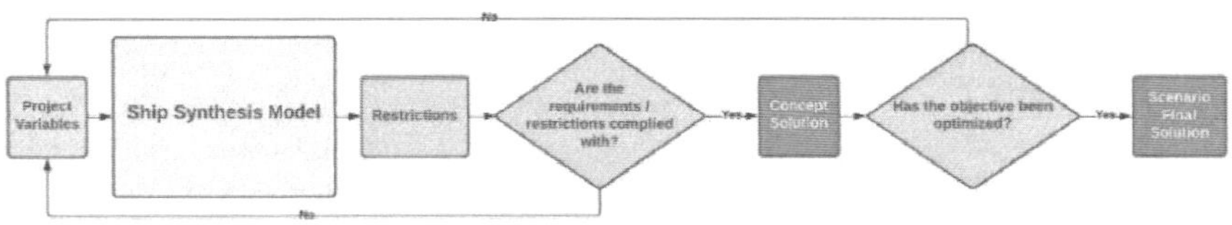

Figure 3: Ship synthesis model simplified process diagram [4]

This model was later validated, making use of existing NATO warship examples. The main objective of this validation process was to ensure that all of the individual modules were working correctly and that the model itself was providing realistic concept design results. Various ships' data was input into the model which then performed calculations for this warship. These model results were then compared with the real case data. This process was repeated for various existing examples of modern warships that fit parts of the drone carrier's operational concept, were it was observed that all variations fell within a 1% to 6% range – adequate margins considering the context of a conceptual design toolset [4].

Following the validation process of the ship synthesis model, final concept solutions were developed. To do so, project variables were selected, which include: length overall, beam, depth, draught and Cb. These were chosen considering that they will be the ones that have the largest impact on the final concept design and will be used by each module to determine the final ship's characteristics. The project restrictions will be related to the operational concept and requirements defined in the operational requirements and system definition chapter – which will be set for all mission configurations and operational scenarios.

Following the selection of project variables and restrictions, the optimization goals will need to be defined within the synthesis model. Two scenarios were developed, leading to two different concept solutions. The first scenario would be an optimization of the ship's acquisition cost – leading to a ship that could comply with the operational requirements and system payloads while retaining the lowest possible acquisition cost. The second scenario would be to create a product that would, within a pre-established acquisition cost, maximize a set of capability indicators (defined by the size of the ship's naval payload for each mission configuration).

This synthesis model would evolve substantially since the first publication pertaining the described work. This would include further refining of the derived formulas used in each module as well as the optimization of the model as a whole to allow for faster computing times. At the time of publication, the OpenSolver software is currently being used in order to run different non-linear optimization algorithms within the model. Despite this, it was observed that the model would converge to the same concept solutions even when using different optimization algorithms – leading to the conclusion that the relatively simpler GRG non-linear algorithm is adequate. An important consideration in the development of this program however is that a large percentage of the formulas used, particularly ones regarding the system groups and crewing, were derived from Royal Navy data. Thusly, it's important to understand that concept results will vary between end-users and that this data should also be updated to match the current state of the art in order to derive adequate results.

7. CONCEPT RESULTS AND DISCUSSION

Two distinct concept solutions were thus obtained. When comparing these to case studies that made use of UCL formulas, it's possible to observe that the drone carrier's payload volume fraction resides on the lower end range of aircraft carriers while the enclosed volume fraction is closer to a small to medium-sized frigate with payload densities lower than most warship examples but being closest to aircraft carriers. These findings were shared by both concept iterations output from the ship synthesis model. The main results for both of these iterations can be seen below, in Table 2 and Table 3:

Table 2. Scenario 1 main results

SSM Module Results: Scenario 2		
Details	**Results**	**Units**
LOA	110.9	m
Beam	18.5	m
Depth	18.5	m
Draft	4.6	m
Freeboard	6.1	m
Cb	0.579	
Max. Speed	35	knots
Service Speed	12	knots
Propulsive Power	70	MW
LBP/B	6.01	
LOA/T	24.13	
B/T	4.02	
Displacement	5 241	t
Cost	0.331	[B] US$

Table 3. Scenario 2 main results

SSM Module Results: Scenario 1		
Details	**Results**	**Units**
LOA	182.4	m
Beam	18.6	m
Depth	18.6	m
Draft	4.7	m
Freeboard	5	m
Cb	0.406	
Max. Speed	35	knots
Service Speed	12	knots
Propulsive Power	76	MW
LBP/B	9.83	
LOA/T	38.63	
B/T	3.93	
Displacement	6 245	t
Cost	0.374	[B] US$

Both concept solutions, considering the output of the ship synthesis model, comply with the initially established operational requirements as well as restrictions associated with the selected systems and end-user goals. It was thus possible to consider that the concepts were adequate proposals – within the context of a concept design. It's important to consider however, that these results provide only a reference for possible successful solutions. In practice, the output information will vary slightly as the design is further elucidated and refined. In order to gauge how these results would vary, both ship concept iterations were further elucidated and subsequently modelled in 3D in order to both further develop the concept designs as well as ensure the validity of the obtained results. The first of these developed models can be seen below in Figure 4:

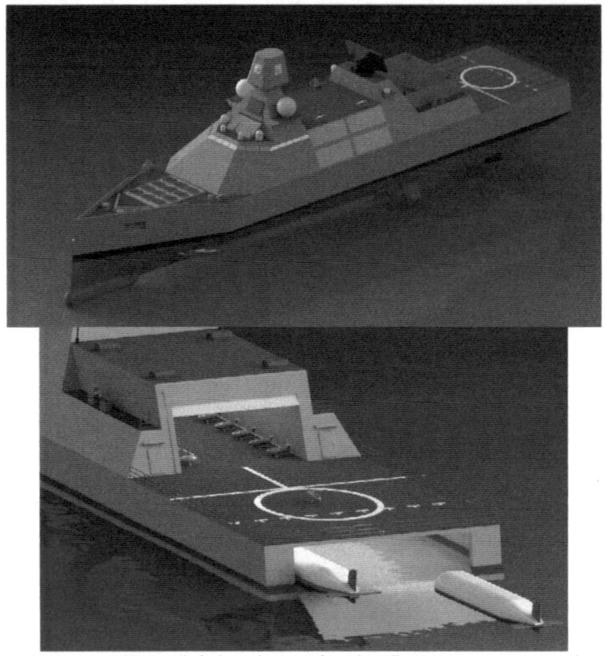

Figure 4: 3D model developed for the first concept scenario

While further refining the concept designs, it was observed that small adjustments would need to be made to the formulas that were developed within the ship synthesis model. While most formulas used, that could be further validated through the use of CAD software, fit into acceptable ranges of deviation when transitioning into basic design stages, some required small adjustments. One of the principal adjustments made concerned the payload distribution formulas – including ones used for the containerized modules as well as the UAV payload prediction. These formulas were thus further refined considering the effect of the superstructure shape in reducing the available payload space as well as ergonomic factors in both the hangar and containerized module mission bays. These adjustments would have a relatively larger impact on the available payload space present in the smallest range of vessel iterations.

8. CONCLUSIONS AND FUTURE CHALLENGES

Within this project, the initial design stages of a drone carrier ship were performed along with the development of a tailor-built ship synthesis model. This was done through an initial definition of a process methodology, followed by a review of the current state of the art of drone carrier warships. Concluding these initial studies, an operational concept and subsequent requirements were defined. This process was later improved upon by the use of a custom built database and by the addition of margins to account for future threat developments. From here, a system definition process was undertaken where key warship systems were selected. This process was later automated through the use of a market available systems database and the programming of a set of decision gates which would select valid system configurations considering the operational concept of the vessel. With this process concluded, a ship synthesis model was developed. This toolset would be used to obtain the first concept solutions, and would use key insights, data and calculations obtained from UCL warship design guidelines [17]. This model additionally underwent a validation process. The final results were then analysed (as seen in the previous chapter) where the concept solutions were both determined to be compliant with the design requirements and could thus be considered successful concept solutions and ready for refinement. These solutions, were later developed and 3D models of the concept ships were developed from which the design formulas use within the synthesis model were further refined.

Further challenges and improvement upon this work are still underway. These currently include the further refinement of the ship synthesis model – as well as decreasing the large computational power requirements currently serving as bottlenecks to the model. Additionally, this model is currently being used to make single optimized iterations of concept warships when considering single ship operational requirements. This framework can further be expanded upon by taking into account fleet operational requirements instead of single ship requirements. The model is thus evolving in order to be used to optimize a small fleet of ships (with variable numbers and ship configurations) or to optimize a single ship concept while taking into considering pre-existing ships within an end-user's fleet.

10. AKNOWLEDGEMENTS

This work contributes to the Strategic Research Plan of the Centre for Marine Technology and Ocean Engineering (CENTEC), which is financed by the Portuguese Foundation for Science and Technology under contract UIDB/UIDP/00134/2020.

This work is entirely carried out in the author's personal title and does not reflect the views and opinions of Damen Naval.

9. REFERENCES

1. KAHN, L., 'All-out drone war in Ukraine points to future', https://www.channelnewsasia.com/world/ukraine-invasion-all-out-drone-war-points-future-3203156, Retrieved 2 February 2023.

2. RAND Corporation, 'A New Direction for China's Defence Industry', 2022.

3. MILITARNYI, 'Україна розкрила деталі про ударні морські БПА', https://mil.in.ua/uk/news/ukrayina-rozkryla-detali-pro-udarni-morski-bpa/, Retrieved 20 March 2024.

4. COSTA, R., VENTURA, M., 'Concept Design of a Drone Carrier', *Transactions of MARTECH*, 2024.

5. COSTA, R., 'Challenges in the Design of a Multi-Mission Surface MUS (Project VENT-Sup EN)', *SNAME Maritime Convention*, 2019.

6. ANDREWS, D J., PAWLING, R., 'A Case Study in Preliminary Ship Design', *IJME RINA Trans. Vol. 150, Part A1, 2008. Discussion and Authors'*, 2008.

7. EVANS, J H., 'Basic Design Concepts', *Journal of the American Society for Naval Engineers*, 71: 671-678, 1959.

8. BRUINESSEN, T V., SMULDERS, F., HOPMAN, H., 'Towards a Different View on Ship Design: The development of ships observed through a social-technological perspective', *TU Delft*, 2013.

9. INCOSE, 'Systems Engineering Handbook', International Council on Systems Engineering, 2006.

10. GRIETHUYSEN, W., 'Marine Design, can systems engineering cope?', *Naval Engineers Journal*, 39(2): 233-245, 2000.

11. SOBEK, D., WARD, A., LIKED, J., 'Toyota's Principles of Set-Based Concurrent Engineering', *MIT Sloan Management Review* 40 (2): 67–83, 1999.

12. KWAK, H M., 'Early-Phase Design of Naval Ship Using Systems Engineering and Interface Management', *Seoul National University*, 2006.

13. KERNS, C., BROWN, A., NEU, W., XIROS, N., 'Naval Ship Design and Synthesis Model Architecture Using a Model-Based Systems Engineering Approach', *Virginia Polytechnic Institute and State University*, 2011.

14. BAUER, J., BOUDREAU, B., BRYANT, C., CASIAS, J., DEMEYER, A., DUBOIS, H., GASPERETTI, F., MACK, T., NOWLIN, B., ZANE, T., 'TSUNAMI Maritime Threat Response Ship', *Naval Postgraduate School*, 2006.

15. KOURIAMPALIS, N., 'Applying Queueing Theory and Architecturally-Oriented Early Stage Ship Design to the Concept of a Vessel Deploying a Fleet of Uninhabited Vehicles', *University College London*, 2019.

16. HATCHUEL, A., WEIL, B., 'C-K design theory: An advanced formulation', *Research in Engineering Design, Volume 19*: *181-192*, 2009.

17. UCL, 'Ship Design Procedure & Data Book', *UCL Department of Mechanical Engineering, Naval Architecture and Marine Engineering. University College London*, 2016.

18. USNI, 'Iran Building Drone Aircraft Carrier from Converted Merchant Ship, Photos Show', https://news.usni.org/2023/01/03/iran-building-drone-aircraft-carrier-from-converted-merchant-ship-photos-show, Retrieved 20 March 2024.

19. DVIBS, 'Integrated Battle Problem 24.1 Concludes', https://www.dvidshub.net/news/466700/integrated-battle-problem-241-concludes, Retrieved 20 March 2024.

20. USNI, 'Pentagon Puts Out Call for Swarming Attack Drones That Could Blunt a Taiwan Invasion', https://news.usni.org/2024/01/30/pentagon-puts-out-call-for-swarming-attack-drones-that-could-blunt-a-taiwan-invasion, Retrieved 20 March 2024.

21. DNV, 'DNV-RU-SHIP Pt.6 Ch.6: Cold Climate', 2016

22. Royal Navy, 'Maritime Operating Concept (MarOpC): Maritime Modularity Concept', 2023.

23. VENTURA, M., 'Ship dimensioning in the initial design', *Developments in Maritime Transportation and Exploitation of Sea Resources*, 2014.

24. D'ALMEIDA, J., 'Arquitectura Naval – O Dimensionamento do Navio', *Prime books*, 2009.

25. NATO, 'ANEP-41 Ship Costing', 2006.

26. HOLTROP, J., MENNEN, G., 'An approximate power prediction method', *International Shipbuilding Progress: 166–170*, 1982.

27. ARORA, S., 'Introduction to Optimum Design (Third Edition)', *Chapter 13 – More on Numerical Methods for Constrained optimum design, Pages 533-573*, 2012.

10. AUTHORS BIOGRAPHY

Rui Nuno Pereira Pinto da Costa holds the current position of Verification & Validation Lead at Damen Naval. The Author is responsible for leading a team of systems engineers tasked with all verification and validation of end-user operational and technical requirements for a wide portfolio of naval products, mostly including surface combatant warships. The Author's previous experience includes a 10-year career in the Portuguese Navy, in which the author has undertaken various operational deployments as well as projects with a focus on surface combatant ship design, mid-life upgrades, through life support and technical management. In addition to his professional history, the author has also had a wide range of published works within the fields of naval architecture, naval autonomy and warship concept design.

OPTIMISING THE FUTURE NAVY SURFACE FLEET: A SYSTEMS-BASED APPROACH

T M Lee and **P J Marshall**, AMOG, Australia

SUMMARY

After a dip over the coming decade, the Future Surface Fleet of the Royal Australian Navy (RAN) will be substantially enlarged, with a focus on distributed and enhanced lethality in order to meet the strategic goal of providing security for sea lines of communication and maritime trade in Australia's region of strategic military interest. This increase in the fleet size creates complexities around producibility, maintainability and manning that must be addressed to support Navy in meeting its fundamental operational objective: to fight and win at sea.

Meeting the challenge of increasing the surface fleet size while maintaining operational effectiveness requires consideration of constraints around the industrial shipbuilding workforce and shipyard capacity; the training pipeline and lag to grow crew numbers; impacts of increased automation, increasing systems complexity, and reduced crewing on ship operations and maintenance; the demand for interoperability between Australian and allied assets across all four operational domains (space, air, sea and land) and continued operational effectiveness of other assets such as replenishment, patrol, amphibious and auxiliary forces.

This paper will examine how these factors may be balanced, and how a systems-based approach may be used to evaluate the influence of these constraints to optimise strategies for ensuring operational effectiveness during surface fleet enlargement.

1. INTRODUCTION

Australia and the world have entered a period of increasing geopolitical uncertainty. As the post-cold war era of global hegemony of the United States of America has receded, and along with it, the established liberal democratic order [1]. The world has entered a new period of what has been termed "global disorder" [2]. The resulting deterioration in Australia's strategic circumstances was recognised in the recent *National Defence: Defence Strategic Review* (DSR), and the requirement for the Australian Defence Force (ADF) to contribute to collective security in the Indo-Pacific Region was reaffirmed [1]. The DSR considered all operational domains of the ADF and the strategic context within which it operates and examined the force design and structure. A key change articulated in the DSR was a change from a balanced force structure to a 'focused, integrated force' with a demand for 'the minimum viable capability'. Within the maritime domain, a priority element in achieving this is

> "*Enhancing Navy's capability in long-range strike (maritime and land), air defence and anti-submarine warfare requires the acquisition of a contemporary optimal mix of Tier 1 and Tier 2 surface combatants, consistent with a strategy of a larger number of smaller surface vessels.*" [1].

The method of achieving this enhanced surface force capability was further refined via an independent analysis of the surface force of the RAN in the *Enhanced Lethality Surface Combatant Fleet* review [3]. Key recommendations of this analysis, as accepted by the Australian Government, were:

- Continue with the procurement of 6 Hunter Class frigates under SEA 5000 Phase 1;
- Upgrade the 3 Hobart Class destroyers Aegis combat system to Baseline 9 configuration;
- Acquire 6 Large Optionally Crewed Surface Vessels (LOSVs) to provide distributed lethality and increased long-range strike capability;
- Accelerated acquisition of 7 to 11 general purpose frigates, to address a capability gap that may emerge as Anzac Class frigates reach end of life; and
- Develop a force of 25 minor war vessels, principally of Evolved Cape Class Patrol Boats supplemented by 6 Arafura class Offshore Patrol Vessels.

The enhanced lethality surface combatant fleet structure that would emerge from these changes in the medium term would comprise 9 Tier 1 combatants (3 Hobart Class destroyers and 6 Hunter Class frigates) and 7 to 11

Tier 2 combatants (the general purpose frigate of a design to be selected). The 6 LOSV would further bolster the Tier 1 force in a longer timeframe. This Future Surface Fleet has the potential to be a potent force.

The RAN Future Surface Fleet would further include 2 Canberra Class LHD, and would be supported by a number of replenishment and logistics vessels, and operate closely with the amphibious craft force being procured under project LAND 8710, as well as the previously mentioned minor war vessels. And of course this is supplemented by aviation assets, the commercial ships of the National Support Squadron purchased over the last ten years such as ADV Reliant, and the evolving foray into autonomous vessels.

The result is a potentially substantial increase in the size of the RAN combatant fleet (currently 3 Hobart Class destroyers and 7 Anzac Class frigates). This substantially enlarged fleet introduces a number of challenges around building, maintaining and operating the additional ships within the ongoing constraints of build and sustainment naval and industrial capacity.

2. A SYSTEMS MODEL OF THE NAVAL ENTERPRISE

The Naval Enterprise required to build, sustain and crew the Naval Fleet can be conceptualised as a system of systems. A representation of this system of systems is shown in Figure 1. Whilst the Naval Feet is rightly at the centre of the Naval Enterprise, it relies upon and interacts with the enterprises for shipbuilding, sustainment, and crewing, and is otherwise supported and enabled by Australia's industrial base and the other Fundamental Inputs to Capability (FICs). In the context discussed here, the sensor, combat and weapons systems of a warship exist within the "Shipbuilding Enterprise" at the integrated level.

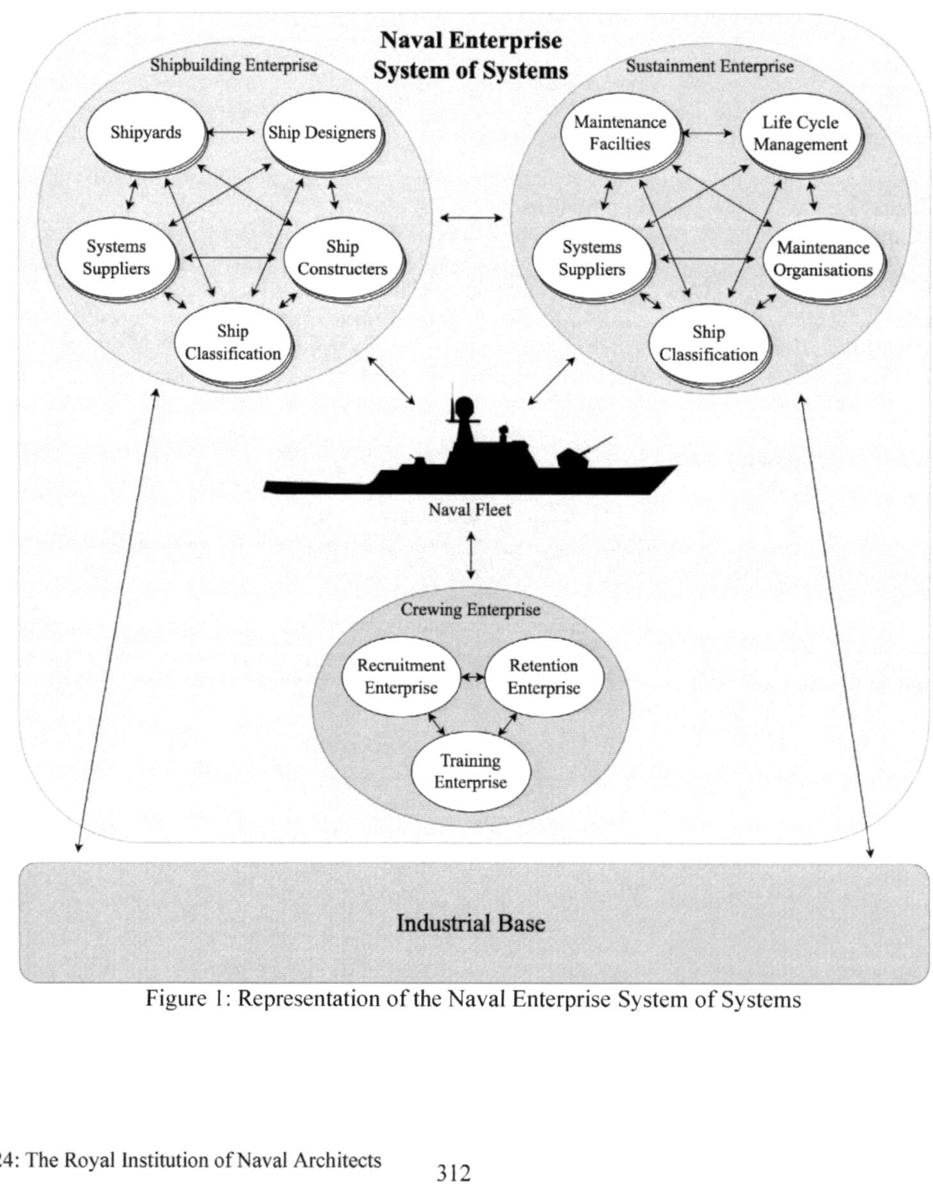

Figure 1: Representation of the Naval Enterprise System of Systems

Each of the three principal enabling enterprises (Shipbuilding Enterprise, Sustainment Enterprise and Crewing Enterprise), are themselves a "system-of-systems", and all three are critical to delivering operational naval capability. Each of these supporting enterprises has its own requirements, constraints, interactions with systems within their boundaries, and interactions with other systems (including many not represented here). There is significant co-dependence between each and all systems within the enterprise, and as a result, shortcomings in the performance of any of the enabling enterprises will have a deleterious effect on the delivery of operational capability. Overall performance of the fleet cannot exceed the worst performing of the three enabling enterprises. Accordingly, optimisation of the Naval Fleet requires achieving a balanced performance of all three enabling enterprises, such that it is not throttled by a poorly performing enabling enterprise, nor that resources are wasted on improving performance of an enabling enterprise over and above the level of Naval Fleet performance the other enabling enterprises can sustain.

3. THREE OPTIMISATION CHALLENGES

3.1 BALANCING THE WORKFORCE

Shipbuilding and ship sustainment activities draw upon similar resources to achieve their outputs. Shipbuilding and ship maintenance draw upon similar technical workforces (such as welding, electrical, mechanical, painting trades) and professional workforces (i.e. engineers, managers, schedulers, purchasing). Similar facilities are utilised (dry docks, wharf sides, ship lifts, transfer systems, slipways etc.) and similar supply chains (equipment, subsystems, major systems suppliers) engaged. As such, for a given industrial base within Australia, the capacity of each of these resources must be shared between the Shipbuilding Enterprise and Sustainment Enterprise. The resource base can be grown, but such growth will often be limited by developing a suitable base of experience for supervisory roles and is best achieved as gradual rather than rapid growth. Further, whilst the two enterprises use similar resources, the specific skills vary between shipbuilding and sustainment, and as such there is a temporary frictional dip in overall capacity when changing resources (in either direction) between the Shipbuilding Enterprise and Sustainment Enterprise.

The first optimisation challenge faced is developing a balance between the tempos of shipbuilding and sustainment, such that workforces may be maintained for each enterprise.

Avoiding the "valley of death" between periods of ship construction has been an acknowledged critical driver for the success of Australian shipbuilding policy, commencing with the 2016 Defence White Paper and 2017 Naval Shipbuilding Plan [4], [5].

Whilst the intent here is to focus on the technical drivers for fleet optimisation, the third principal enabling enterprise, the Crewing Enterprise, must be mentioned. Whilst the Crewing Enterprise is less dependent on the same resources as the Shipbuilding Enterprise and Sustainment Enterprise (albeit with a good deal of cross fertilisation, particularly as both shipbuilding and sustainment provide employment opportunities for former Navy personnel), the balance between tempo of new ship construction, operational availability of existing ships between maintenance periods, and the supply of trained crews from the Crewing Enterprise also needs to be balanced. Providing financial resources to the Shipbuilding and Sustainment Enterprises to increase capacity to a tempo that outstrips the Crewing Enterprise will result in a mismatch between available ships and available crew, and therefore waste resources of the overall Naval Enterprise.

Naturally this balance becomes even more important as Australia commences construction of the SSN under AUKUS, and again must be balanced against the demands of the Life of Type Extension (LOTE) of the existing Collins Class, accepting that many (but not all) elements of the submarine present their own unique ecosystem.

3.2 DESIGN EVOLUTION AND DESIGN OBSOLESCENCE

Acquisition of new surface combatants for the RAN has, over much of the last 40 years, commenced with the selection of a new ship design to meet a particular operational requirement, followed by the establishment of a shipyard, reconstitution of a workforce, and then construction of the ships, largely in isolation from consideration of the selection of the ship design and the construction location for preceding or subsequent ship classes. Typically, by political mandate and arguably without comprehensive understanding of the implications, the design has come from overseas, and therefore brings its own unique ecosystem. As a result, for much of this period, the RAN fleet has consisted of classes of ship with relatively low levels of commonality, in aspects ranging from national origin (and with it, underlying) design philosophy, major subsystems, and even through to construction locations. As a result, each new class of ships has presented something of a revolution in design compared to the preceding class, rather than evolution.

The benefit of this approach is that each ship class can, at least notionally, be selected to optimally meet the requirements and reap the benefits of the best of available technology. The drawback is that each new class of ship presents a learning curve in terms of design philosophy, construction methods and shipyard culture, supply chain and equipment, and in many cases, also working up a new construction workforce following the "valley of death" from the preceding class. Substantial changes in equipment and systems to meet Australian WHS legislation, our operational requirements such as interoperability, and at times organisational preference, also introduces additional complexity in shipbuilding and sustainment, and increases the complexity for crewing.

The second optimisation challenge is therefore balancing evolution vs revolution in design.

Fortunately, this challenge is not binary, and there are opportunities to evolve a design whilst introducing new systems and technology. Strategies that can be employed include:

- Pull through of evolved system designs into new ship designs to mitigate the magnitude of change.
- Ordering of ships classes in batches, with each batch built to a common configuration but with targeted design modifications and system upgrades only implemented between batches.
- Ordering larger numbers of the same platform to increase the level of commonality across the fleet.

A further example of how this challenge could be addressed is where a new ship design is needed to fulfil a requirement, but there is no suitable domestic in-production design from which to evolve a solution. An approach could be to:

- Adopt a foreign in-production design and put a batch into production with minimal (i.e. essential only) modifications. This reduces design risk associated with an un-built design and allows a degree of familiarity to be established with the design and construction principals of the base ship.
- Simultaneously, evolve the ship design for a second batch, with a greater degree of "pull through" of systems that have been identified as critical to maintain fleet-wide commonality. In the context of the RAN, this may be modifications or replacement of the combat system to utilise an indigenously developed common tactical interface or radar system as has occurred with the mandate of the SAAB 9LV system, Lockheed Martin's Aegis system and the CEA CEAFAR radar across multiple current platforms.
- Once sufficient of the second batch ships are in service, conduct a retrofit program on the first batch to bring them up to second batch standard, and thereby reduce the sustainment and training burden of within class configuration differences.

Opposing such initiatives is the Government directed mandate for 'No Change' in the SEA3000 General Purpose Frigate program. This is not to criticise that mandate – it has been applied to ensure expeditious delivery of warfighting capability and mitigate the capability gap, but when applied in isolation and not in balance with other elements of the system, it is likely to introduce considerable other adverse outcomes.

A guiding principle for this optimisation challenge will be the requirement to have a warfighting capability in operation. Compromise on the level of capability will be required to ensure that in future the fleet-in-being gains the increased lethality that is sought, within the constraints of forecast reduced availability of the existing fleet (notably progressive retirement of the Anzac class frigates, and extended maintenance of the Hobart Class destroyers for the Aegis upgrade program), concurrently with a deteriorating international situation. A fleet-in-being exerts its influence by fact of its existence, and so increasing the warfighting capability of the fleet sooner changes the strategic effect sooner compared to a fleet still in the planning or design stages. The counterpoint is that too great a subjugation of warfighting capability to schedule will risk investing in insufficient warfighting capability, and the risk of bringing a proverbial "knife to a gunfight".

3.3 AUTOMATION VS COMPLEXITY

The Future Navy Surface Fleet plans include the acquisition of the 6 LOSV. These vessels are described as:

> "*six Large Optionally Crewed Surface Vessels (LOSVs) with 32 Vertical Launching System cells, providing enhanced lethality through additional multi-domain strike capacity and directly increasing survivability, lethality and endurance. This investment will increase distributed fleet lethality with a lower cost and crewing impact.*" [3]

It is clear that the LOSV will need to integrate a high degree of automation in order to operate in the uncrewed mode. This will extend to operation of the essential ship platform systems as well as the combat and weapons systems. This will be a step forward in terms of automation compared to conventional ships, however it can also be expected that the Future Surface Fleet will also incorporate increasing levels of automation over existing ships. This may be expected to have positive consequences for the Crewing Enterprise, as crew requirements should be reduced on an aggregate per-vessel basis.

However, increased automation will also result in increased system complexity especially in relation to control systems. The increased system complexity will in turn increase the requirements for maintenance placed on the Sustainment Enterprise acknowledging that this increased control systems functionality should bring with it an increased level of self-diagnosis perhaps akin to that currently offered in most cars available in Australia today. This will be further exacerbated as reduced crewing will reduce the capacity for organic maintenance whilst at sea.

This leads to our third optimisation challenge, balancing automation against complexity.

4. CONSIDERATIONS FOR RESOLVING THE OPTIMISATION CHALLENGES

Continuing to think of the Naval Enterprise as a system-of-systems is an important precursor to considering methods for addressing the optimisation challenges posed above. This helps keep in mind that changes in any one of the individual systems to improve output may be ineffective at best if other systems are not similarly adjusted, or may even counteract the improvement by reducing output elsewhere.

However the types of optimisation challenges posed are not easily resolved solely with mathematical or theoretical models. A large part of the system response is not deterministic and will depend on the range of cognitive and emotional responses to change. However considering the general techniques that are applied in the engineering field for optimisation may be fruitful when addressing this situation.

In truth however, there will regularly be call for decisions based upon good judgement rather than on pure engineering merit. It is therefore important that the engineering analysis considers all factors and the implications on other systems within the enterprise.

Of the three optimisation challenges posed, the third, balancing automation against complexity, is the most tractable to directly solve by engineering analysis. Approaches to assessing the whole of life-cycle balance between managing the operational effectiveness of a system (including system availabilities and the risk of failing to achieving an operational effect from a system) and the costs of operation (including maintenance overhead, and limitations on system availability) using reliability analysis techniques have been described, and indeed employed using quantitative models for the Hobart Class destroyers. [6], [7] These techniques, utilising Reliability Block Diagram (RBD) models of the end-to-end contributors to enabling a function (i.e. extending beyond single system models) can consider labour resource availability, system utilisation by mission type, lifecycle usage phasing (e.g. upkeep and deployment), maintenance activity types, failure modes, and limitations on resources for equipment repair. By considering all these aspects, it is possible to understand the drivers of change to mission lengths, increased utilisation or maintenance regimes, in the form of labour hours, component downtime and system availability. In the context utilised for the Hobart Class destroyers, where this modelling was performed during the sustainment phase, it has been principally used to guide decision making regarding resourcing of maintenance activities and upkeep cycle planning. If brought forward into the design stage however, these techniques can be used to guide design decisions for automation of systems.

Developing high-level system models of the enabling enterprises is proposed as a method of examining the response of each enterprise to changes in constraints and resourcing, in turn the impact of these changes on the Naval Enterprise. These high-level models will themselves be a synthesis of multiple lower level models; for example and simplistically, it may be a model of the potential increase in shipyard capacity over time (in terms of length of weld that may be completed per year) based on models of rates of growth of welder workforce, taking into account the ability to organically train, qualify and supervise additional welders, and the level of construction activity required to sustain this. The models may, by necessity, be only a broad indication of the general behaviour of a system, as the response of the actual system can be expected to be stochastic, as each system involves people, who do not respond deterministically. Other models (such as the reliability analysis-based models described above) may be more robust and less subject to variance from the prediction. None the less, it is suggested that such models, used with multivariate analysis techniques to examine overall system behaviour, would support the optimal development of the Future Navy Surface Fleet, by allowing testing of different scenarios as a guide to how the outcome in the real world may be influenced by changes in the constraints and resourcing on different elements of the Naval Enterprise, or indeed how changing the approach used to conduct differing elements of the system (i.e. implementing a new business practice) may impact the Future Navy Surface Fleet.

The systems model of the Naval Enterprise will be an imperfect representation of reality, this is inherent in any model of a system. Nor will the impact of policy changes on the actual Naval Enterprise be able to be rapidly evaluated, and the outcomes changed as readily as can be done with a model. This is however also the justification for use of a systems-based modelling approach, as the impact of changes cannot be rapidly evaluated in the real world, and indeed the perturbation inherent in change takes time to settle such that the effect can be measured. Consideration of second and sometimes third order impacts will also be of importance in assessing overall impact. Accordingly, overly rapid change in an enterprise can easily result in "chasing one's tail" and systematic instability, making changes in direction before the prior course change has taken effect will be counterproductive. Testing assumptions in the model can be instructive to develop the collection of policy settings that are applied in the real world.

5. CONCLUSIONS

The creation of the Future Surface Fleet of the RAN, as outlined by the DSR and the *Enhanced Lethality Surface Combatant Fleet* review, will be a challenge to the Naval Enterprise of Australia. Achieving the outcome will require balancing competing priorities, and just a few of these challenges are described in this paper. However the process for making decisions on these challenges may be guided by conceiving of the naval enterprise, and the principal enabling enterprises, the Shipbuilding Enterprise, Sustainment Enterprise and Crewing Enterprise, as a system of systems. These can be decomposed into constituent systems, for which representative models may be created. Using these system models, the broad response of the enterprise to changes in inputs, constraints and resourcing can be evaluated, as a support to the decision-making process for creation of the Future Surface Fleet. By this method, the strategies for achieving the enlargement of the Future Surface Fleet may be optimised, so that the operational effectiveness of the Future Surface Fleet is maximised within the constraints that will be faced by the Naval Enterprise.

8. ACKNOWLEDGEMENTS

We would like to acknowledge the support of our AMOG colleagues Dan Washington, Eric Thompson, and Ben Clark for the inspiration they have provided for preparation of this paper, principally through many years of ad-hoc discussions of the abstract challenges involved in shipbuilding, ship sustainment, ship crewing, and ultimately fighting the ship at sea – often taking place after working hours over many shared bottles of wine.

9. REFERENCES

1. Australian Government, "National defence: defence strategic review," Canberra: Commonwealth of Australia, 2023.

2. STIGLITZ, J., "We are in process of creating 'global disorder,' says Nobel Laureate Stiglitz," Sep. 05, 2022. Accessed: Jun. 12, 2024. [Video]. Available: https://www.youtube.com/watch?v=uVR6oHbY4Z4

3. Australian Government, "Enhanced lethality surface combatant fleet: independent analysis of Navy's surface combatant fleet," Canberra: Commonwealth of Australia, 2024.

4. Australian Government, "2016 Defence White Paper," Canberra, ACT: Commonwealth of Australia, 2016, p. 191. [Online]. Available: https://www.defence.gov.au/about/strategic-planning/defence-white-paper

5. Australian Government, "Naval Shipbuilding Plan," Canberra, ACT: Commonwealth of Australia, 2017.

6. LEMERANDE, T., "Optimising Availability, Capability, and Affordability across the Fleet: A Total Life Cycle Management Approach for Improving Seaworthiness," presented at the *World Conference on Engineering Asset Management*, Brisbane, Australia, Jan. 2018.

7. BOUSTEAD, N., et al., "Achieving Lifecycle Seaworthiness on The Hobart-Class Guided Missile Destroyer (DDG) Through Reliability Centred Maintenance Analysis And Probabilistic Risk Assessment," presented at the *International Maritime Conference*, Sydney, Australia: Royal Institution of Naval Architects, 2022.

9. AUTHORS BIOGRAPHY

Tim Lee is a Fellow of the Royal Institution of Naval Architects and is the Chief Operations Officer at AMOG. The Author oversees the engineering delivery of work by the AMOG group, an Australian specialist engineering consultancy. Tim has managed the system safety programs of several defence procurement and upgrade programs across the maritime, land and air domains.

Rear Admiral Peter Marshall AM, RAN (Retd.) is the Non-Executive Chair of the AMOG group of companies, as well as a principal of Pacific Management Partners. Peter consults widely on strategic matters to Defence and related industries. Prior to this, Peter served for 33 years in the Royal Australian Navy, including roles as Chief Naval Engineer and Head of Maritime Systems, overseeing the sustainment of the Navy fleet.

MACHINE LEARNING APPROACH TO WAVE ENVIRONMENT AND HULL STRESS PATTERN IDENTIFICATION FOR A PATROL VESSEL

T Magoga, **M Mogeke**, and **P Dennis**, Defence Science and Technology Group, Australia, **L McLeod** and **M Herandi**, Monash University, Australia, **I Drummen** and **R Hageman**, Maritime Research Institute Netherlands, The Netherlands

SUMMARY

Hull monitoring systems have been deployed to support the structural management of defence vessels, by providing a high-volume of ship response data. The data can be exploited in data-driven approaches such as machine learning, to learn how to classify data as belonging to a pattern within the dataset. This paper presents a study on the identification of patterns in the relationship between hindcasted significant wave height and wave period and hull stresses in a coast guard cutter, through the development and application of a machine learning (ML) framework. Of three candidate supervised ML models explored, the CatBoost model was found to be the most effective. The SHapley Additive exPlanations (SHAP) library was used to determine the relative importance of the interactions between input variables on the output. Patterns in the wave parameters that correspond to increased stresses were successfully identified, indicating that further development of the framework is warranted. The paper concludes with recommendations to aid in structural assessment of defence vessels.

NOMENCLATURE

β	Ship heading relative to principal wave direction (°)
λ	Wavelength (m)
H_s	Significant wave height (m)
L	Ship length (m)
T_e	Wave encounter period (s)
T_p	Wave period at spectral peak / peak period (s)
V_s	Ship's speed over ground (kn)
V_w	Wave velocity (m/s)

1. INTRODUCTION

The hull of a ship experiences millions of stress cycles during its lifetime. These stress cycles can cause fatigue cracking, primarily at welded joints [1]. Fatigue cracking can drive increases in ship repair costs that potentially exceed ship valuation, and contribute to structural failure [2]. The ability to assess the fatigue life of a ship is crucial in establishing design criteria and predicting maintenance and repair costs over a ship's lifetime [3].

Defence and patrol vessels are purpose-built, may operate in harsh ocean conditions, and must demonstrate operational reliability [1, 2]. These factors have driven the development of hull monitoring systems to record stresses experienced in naval hulls. Such systems can provide a large volume of operational and performance data, presenting an opportunity to apply 'data-driven' approaches to automate the discovery of specific wave environments that result in high stresses, and potential fatigue-critical conditions, in a ship's hull.

Machine learning (ML) is a subset of artificial intelligence algorithms that aims to identify and understand patterns within a given dataset. The classification algorithms learn how to form decision boundaries to classify or predict data as belonging to a sub-group within the dataset [4]. Some example applications of ML algorithms are data mining, predicting or classifying elements of a population, or determining the sub-structure of a dataset [4].

ML is innately data-driven; thus, compared to physics-based approaches, there is a smaller onus on the user to have expert knowledge of the physics that govern a specific ship's interactions in varying wave environments [5]. Although the use of ML to predict the structural performance of a ship in varying ocean conditions has been sparse [6], there is precedent for the use of ML techniques to identify or improve ship performance. Kim and Jeong [7] created a supervised learning model to predict optimal routes for naval vessels operating concurrently in an area. The model was trained on a fictional fleet of vessels with optimal routes in a pre-defined ocean area. Gao and Shi [8] demonstrated a clustering method applied to vessel positional data to identify crew decision-making in relation to collision avoidance and route planning. The authors were able to identify similar segments between vessel routes that reflected decision-making from crews, to maintain course and avoid other vessels.

In the context of ML models, explainability refers to the ability for humans to understand why a model produced a given set of predictions. Interpretability refers to the deeper meaning behind the predictions [12]. The explainability of ML models is crucial for understanding the possible trends contained in the dataset, and to validate the predictions against understood physical principles. This can also help to define the range of conditions in which the ML model can reliably be used.

This paper presents an exploratory study on the development and application of a ML framework, based on a full-scale hull monitoring dataset. The primary aim of the study is to identify ML models within the supervised learning approach that can relate patterns in wave environments to hull stresses in a naval ship. A secondary aim is then to determine which of the developed models is most effective, and to explain the model's identification of significant wave patterns. Although interpretability is an important aspect of ML applications, this study is focused on explainable models – to understand why the patterns are unique and important in the context of specific wave environments resulting in higher structural stress.

Section 2 of the paper presents the ML and model explanation methods, and the dataset used in the study. The dataset comprises full-scale hull stresses, ship speed and ship position coupled to wave hindcast data. Section 3 describes the training and comparison of different ML models to evaluate the best performing model, use of this model to demonstrate the importance of the different input parameters, and identification of patterns in wave environment parameters associated with relatively large structural stresses. Section 4 discusses the main findings and recommendations from the results, and outlook for continued development of the ML framework. Conclusions are presented in Section 5.

2. MATERIALS AND METHOD

2.1 SUPERVISED MACHINE LEARNING

Supervised learning is a subset of ML that uses datasets with explanatory labels. For classification algorithms, these labels identify which class or group each data point in the training set belongs to [10]. Supervised learning, specifically neural networks and tree-based algorithms with labelled datasets, have been widely used as a prediction tool in applications such as computer vision, voice recognition and natural language processing [11]. However, supervised learning algorithms are inherently labour-intensive because of the requirement that the dataset must be satisfactorily labelled. To circumvent this issue, neural networks that have already been trained on a large general dataset may be used [11]. However, for atypical datasets, use of pre-trained networks is unlikely to be feasible.

2.1 (a) Regression Decision Trees and Random Forests

Decision trees are non-linear ML models that map an independent variable to a dependent variable [12]. Regression decision trees aim to determine the most appropriate variables and values at which to separate data points. The class to which each data point belongs is known in advance. Splitting criteria are determined by their relative impact on reducing prediction error. They achieve this by first choosing a variable and continually splitting data points into two smaller groups, based on their value of the chosen variable. This splitting is repeated until termination, which occurs when stopping criteria are met. The criteria can include reaching a maximum tree layer depth, or minimum samples per leaf. This process is then repeated for all variables in the dataset, unless the tree prematurely stops [13].

Random forests are combinations of decision trees, each tree using a bootstrapped sub-sample of the dataset. Bootstrapping is a method that generates small samples (subsets) by randomly sampling a larger dataset. As random forests are combinations of decision trees, each using a bootstrap sub-sample, they are less prone to overfitting [14].

2.1 (b) Gradient Boosted Models

Gradient boosted models aim to improve performance beyond that of a regular random forest. Whereas regular random forests take the average of the decision trees, gradient boosted models build fixed-size trees to reduce the prediction errors of the previous decision tree [15]. These decision trees are referred to as weak learners because they are constrained to be shallower, and have poor individual predictive ability. However, by combining these weak learners, gradient boosted random forests are able to reduce bias and encapsulate complex trends that exist within a dataset [16].

2.1 (c) Shapley Values

Lloyd Shapley developed a framework for calculating the contributions of individual players in a group game scenario in which players cooperate to generate a surplus [17]. The Shapley value for each player represents the deserved share of the surplus, based on their relative contribution. This theory underpins the ML explainability library developed by Lundberg and Lee [18], called SHapley Additive exPlanations (SHAP). The SHAP library calculates each feature (or independent) variable's quantitative impact on a model's prediction (or dependent variable). SHAP treats individual features of a dataset as 'players', and the difference between a model's prediction for a given variable/data point, and the average value of that specific variable as the 'surplus'. A high SHAP value corresponds to a high model output (prediction) for the given data point (relative to the model's baseline prediction), and vice versa.

The SHAP library is used to address this study's secondary aim – to explain model predictions of structural stresses in various wave environments.

2.2 SUMMARY STATISTICS

Summary statistics are used to ascertain the ML model's ability to predict the measurements.

The bias is defined in Equation 1 (where y is the variable, and the subscripts m and o refer to model and observations respectively):

$$\text{bias} = \overline{y_m - y_o} \qquad \text{Equation 1.}$$

The Root Mean Square Error (RMSE) of predicted values is calculated using Equation 2:

$$\text{RMSE} = \sqrt{\overline{(y_m - y_o)^2}} \qquad \text{Equation 2.}$$

The scatter index (SI) is the normalised RMSE by the average observed value (dimensionless), as per Equation 3:

$$\text{SI} = 100\% \times \frac{\sqrt{\overline{\left((y_m - \overline{y_m}) - (y_o - \overline{y_o})\right)^2}}}{\overline{y_o}} \qquad \text{Equation 3.}$$

2.3 FULL-SCALE SHIP DATA COUPLED TO WAVE HINDCAST DATA

The United States Coast Guard (USCG), in collaboration with Maritime Research Institute Netherlands (MARIN) and other industry partners, are conducting a rigorous monitoring campaign on the *USCGC Stratton* (Figure 1a). The motivations for this project were the USCG cutter's unique superstructure design, and the operation of these types of ships in harsh ocean conditions [19].

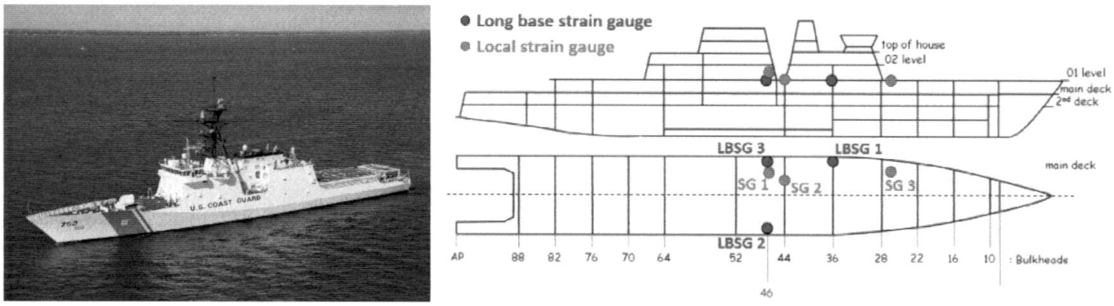

a. b.

Figure 1: a. *USCGC Stratton* [1], and b. schematic of strain gauge locations [20].

The *USCGC Stratton* dataset consists of statistics (for 30-minute windows) of speed over ground (V_s) and stress measurements from strain gauges installed at the locations shown in Figure 1b. In the present paper the analysed parameters (the prediction variables) are the maximum stresses at strain gauges SG01 and SG03, collected during a deployment of approximately 2000 hours. The stresses are axial in the longitudinal direction. The stress recordings have been filtered with a low-pass filter applied to obtain the wave-only component of the signal.

The monitoring system dataset is augmented with wave hindcast data from the Copernicus Marine Environment Monitoring Service (CMEMS) [21] for the ship's geolocations during the deployment. This hindcast data includes significant wave height (H_s), wave period at spectral peak or peak period (T_p) and principal wave direction. Thus, the ship encountered wave angle (β), or heading, can be calculated. The angle β equal to 180° is head seas.

The feature variables, or features, used to train the models are H_S, β and the ratio between the length of the ship to the wavelength, L/λ. The ratio L/λ is a constructed feature, calculated using Equation 4 where T_e is the wave encounter period, V_w is wave velocity equal to the acceleration due to gravity divided by the wave period (T_p) in radians per second:

$$\frac{L}{\lambda} = \frac{L}{T_e(V_w - V_s cos\beta)} \qquad \text{Equation 4.}$$

The ratio L/λ is used as a feature to improve the performance of the trained models, and to validate the supervised learning approach against understood physical phenomena. L/λ equal to 1 is a typical design condition that corresponds to the largest wave-induced loading [22].

Figure 2a and 2b present scatterplots of H_s vs maximum measured stress at SG01 and SG03 respectively.

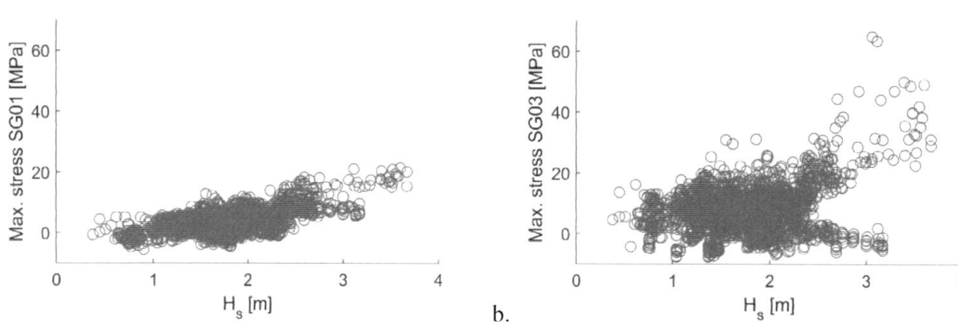

a. b.

Figure 2: a. Scatterplot of H_s vs maximum measured stress at SG01, and b. at SG03.

Figure 3a and 3b show the distribution of the maximum measured stress with heading β at SG01 and SG03 respectively. The asymmetry depicted is largely due to the strain gauges being located off the vessel centreline.

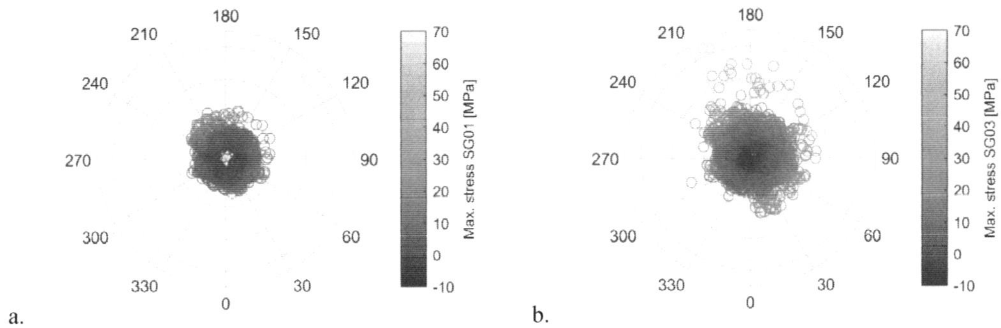

Figure 3: a. Polar scatterplot of β vs maximum measured stress (radius) at SG01, and b. and at SG03. β = 180° represents head seas.

Figure 4a and 4b present scatterplots of maximum measured stress vs L/λ for SG01 and SG03 respectively.

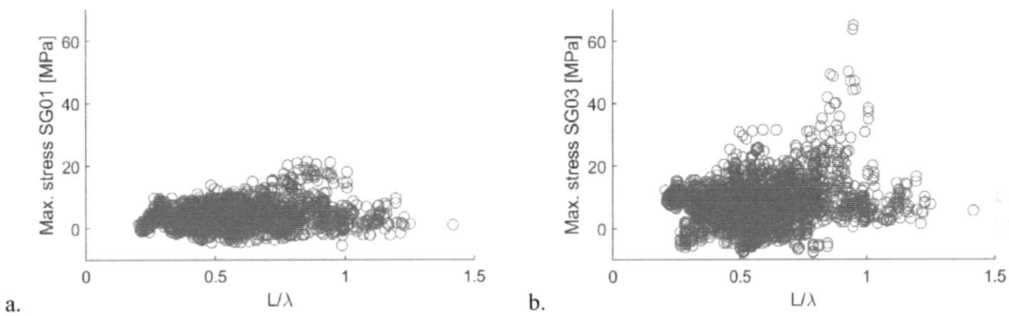

Figure 3: a. Scatterplot of L/λ vs maximum stress at SG01, and b. maximum stress at SG03.

3. ANALYSIS AND RESULTS

The analysis steps are given in Figure 5. The SHAP library is used to generate summary, dependence and stacked force plots for the predictions (steps 3 and 4).

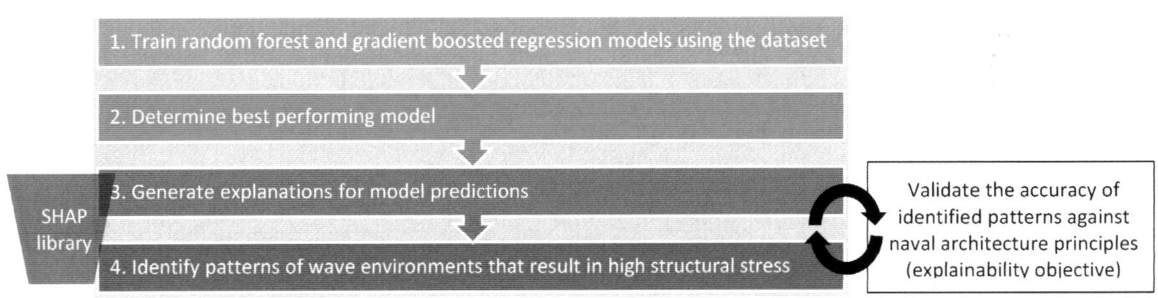

Figure 4: Analysis steps

3.1 TRAINING OF SUPERVISED LEARNING MODELS

Regression random forest models implemented in the ML library scikit-learn RandomForestRegressor and gradient boosted models XGBoost [23] and CatBoost [24] were trained to predict the maximum SG01 and SG03 stresses. 75% of the dataset was randomly assigned as the training set and 25% as the testing set.

Hyperparameter tuning was carried out for each model type and for each predicted variable. Hyperparameter tuning is an optimisation technique to improve the accuracy of the models. Hyperparameters are 'external' to the model; they express higher-level properties of the ML algorithm itself, as opposed to characteristics relating to the dataset. One technique to find optimal hyperparameter values is grid search, whereby the parameter values are listed in a grid, all parameter value combinations are assessed, and the final optimal parameters are found [25]. Using this method, hyperparameters minimising training error (RMSE) for a model were used.

3.2 EVALUATION OF BEST PERFORMING SUPERVISED LEARNING MODEL

The training and test RMSE values of the maximum stress at SG01 and SG03 for the tree-based model types are compared in Figure 6a and 6b, respectively. The XGBoost and Random Forest models produced large differences between the training errors (low) and test errors (high). This is indicative of model overfitting, in which a model learns how to predict data in the training set but is then unable to generalise its predictions on unseen data. The CatBoost models yielded similar training and test losses, suggesting that overfitting is not present, and provided predictions with the lowest test error on unseen data. Therefore, the CatBoost model was selected for further use.

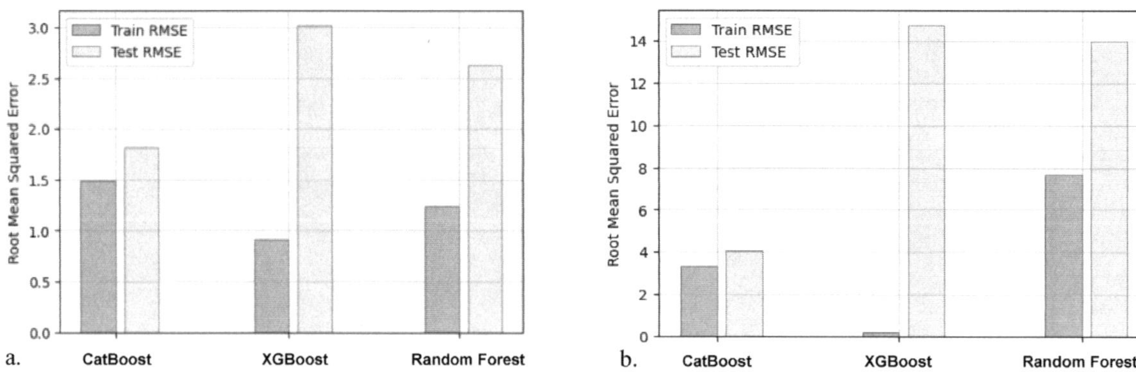

Figure 5: RMSE of maximum stress (in MPa) for training and testing of each model type for a. SG01 and b. SG03.

Based on the whole dataset, Figure 7a and 7b present scatterplots of the measured versus the CatBoost models' predicted maximum stresses at SG01 and SG03, respectively. Also shown in these figures are the bias, RMSE and scatter index of the predictions. Although the bias is negligible and the RMSE results are small relative to the range of measured stresses, the scatter indices are 43% and 46%, respectively. This indicates that while the model generally performs well, there is substantial scatter in some of the predictions of the measured maximum stress; for example, at SG03, the model underpredicts some of the ~30 MPa maximum stress values by approximately a third.

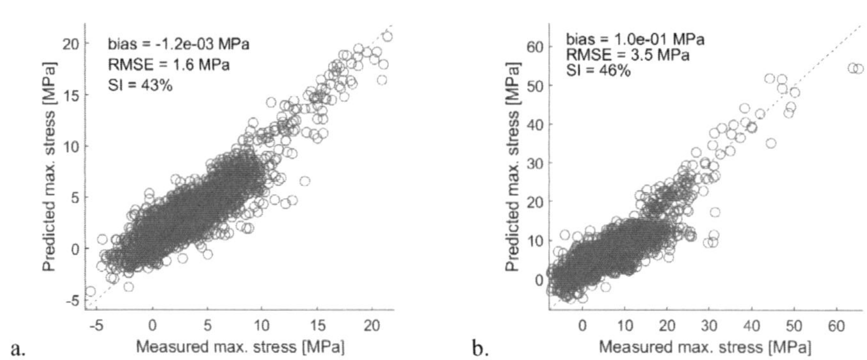

Figure 6: Measured vs CatBoost predicted maximum stress at a. SG01 and b. SG03.

3.3 EXPLAINABILITY

SHAP summary plots are used to illustrate the importance of each feature on the model predictions. A positive SHAP value and a positive feature value indicates a direct positive correlation. If the low and high feature values are interspersed at a similar SHAP value, it is indicating that the feature value is not well correlated (positive or negative) with the model prediction. Figure 8a and 8b show the SHAP summary plots of the effect of each feature variable on model predictions for maximum SG01 and SG03 stress, respectively.

a.

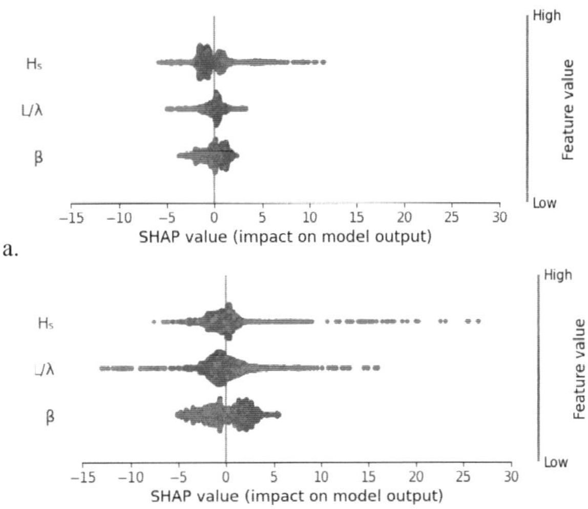

b.

Figure 7: SHAP summary plot of wave environment effects on CatBoost model predicted maximum stress for a. SG01 and b. SG03.

At both strain gauge locations, the range of SHAP values is largest for H_s indicating that this feature has the largest influence on the predictions. Further, the larger the wave height the larger the predicted stress. L/λ has a relatively greater impact on predictions for SG03 (SHAP values between -15 and 15) when compared to SG01 (SHAP values between approximately -5 and 3). The effect of β on predictions at both structural locations is relatively minor.

SHAP dependence plots are used to illustrate the relationship between each wave environment variable and its highest correlated feature on the predicted variable. Figure 9a and 9b show the predictions of SG01 and SG03 maximum stresses, respectively, based on the combination of H_S and L/λ. Below H_S of 2.4 m, each model behaves similarly with L/λ. However, above H_S of 2.4 m the predictions separate by L/λ. This observation is reflected in the scatterplots of H_S versus measured maximum stress in Figure 2; at H_S above 2.4 m there are also separated trends in the data.

Figure 10a and Figure 10b present the SHAP dependence plots of β effects on the predicted maximum stresses for SG01 and SG03, respectively. The SHAP values tend to be slightly positive in bow quartering seas and slightly negative in stern quartering seas; this is expected as, in general, longitudinal stresses are likely to be higher in bow quartering seas. These observations agree with the scatterplots of β vs maximum measured stresses (Figure 3).

Figure 11a and 11b show the SHAP dependence plot of L/λ on the predicted maximum stress for SG01 and SG03, respectively. Similar to that observed from Figure 9, there is a peak in the predictions for L/λ between 0.8 and 1 and H_S greater than 2.4 m. However, for H_S less than 1.2 m, the SHAP values are negative despite L/λ being in the range of 0.8 to 1.

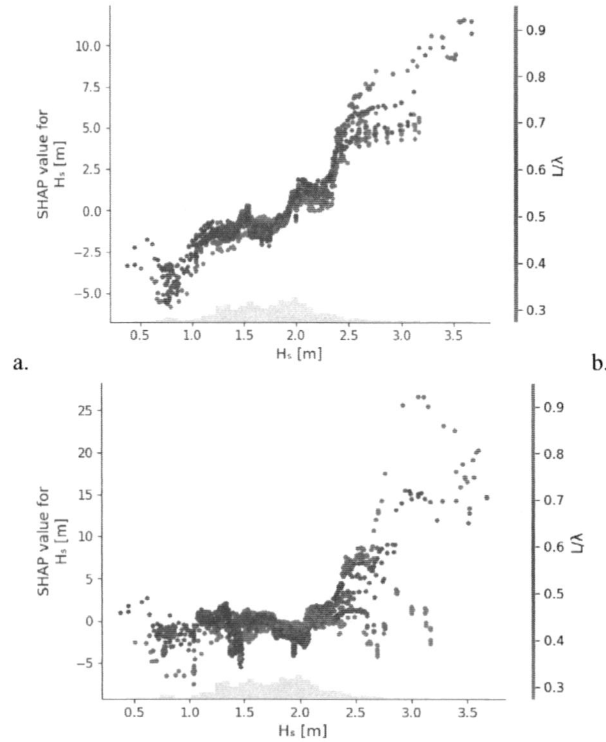

Figure 8: SHAP dependence plot of H_S effects on predicted maximum stress for a. SG01 and b. SG03.

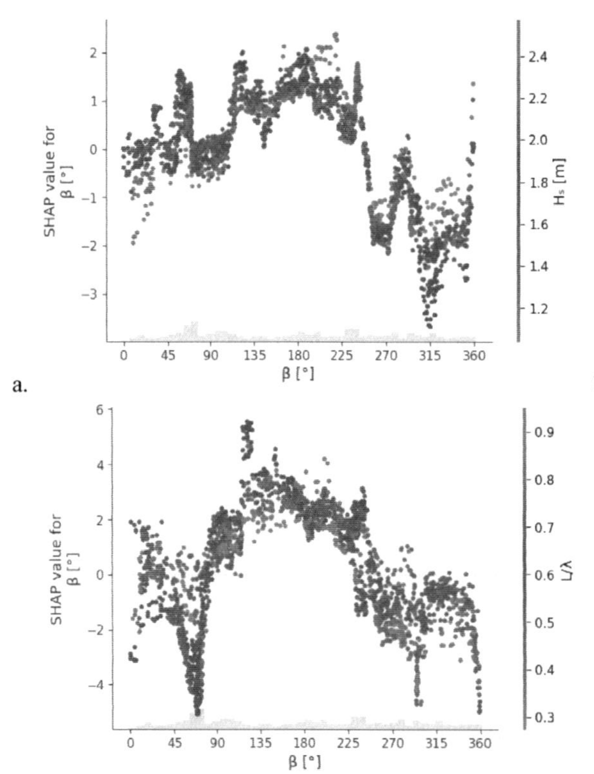

Figure 9: SHAP dependence plot of β effects on model's predicted maximum stress for a. SG01 and b. SG03

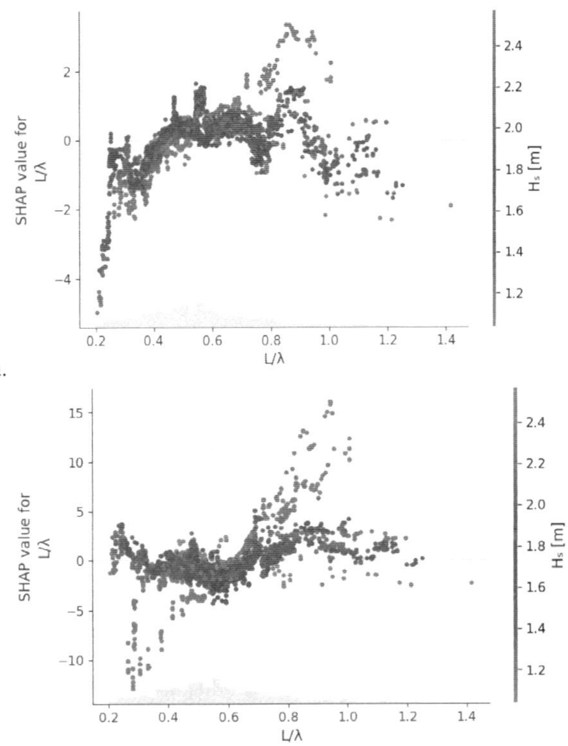

a. b.

Figure 10: SHAP dependence plot of L/λ effects on model's predicted maximum stress for a. SG01 and b. SG03.

3.4 PATTERNS OF WAVE PARAMETERS THAT RESULT IN HIGH STRUCTURAL STRESS

A stacked force plot enables a form of supervised hierarchical clustering, by sorting data points horizontally based on the leaf order obtained from the clustering of their SHAP values [26]. Data points with features that share similar individual SHAP values (similar effects on model predictions) are grouped together. This tool is used to identify combinations of wave environment variables that result in relatively high maximum stresses.

Figure 12 shows a SHAP stacked force plot for the predictions of maximum SG01 stress. Data points are grouped by wave environment. For example, the cluster on the far left (H_s > 2.5 m, L/λ > 0.71) elicits the largest maximum SG01 stress grouping. In comparison, the cluster on the far right (H_s < 1.8 m) has the smallest magnitude stress predictions.

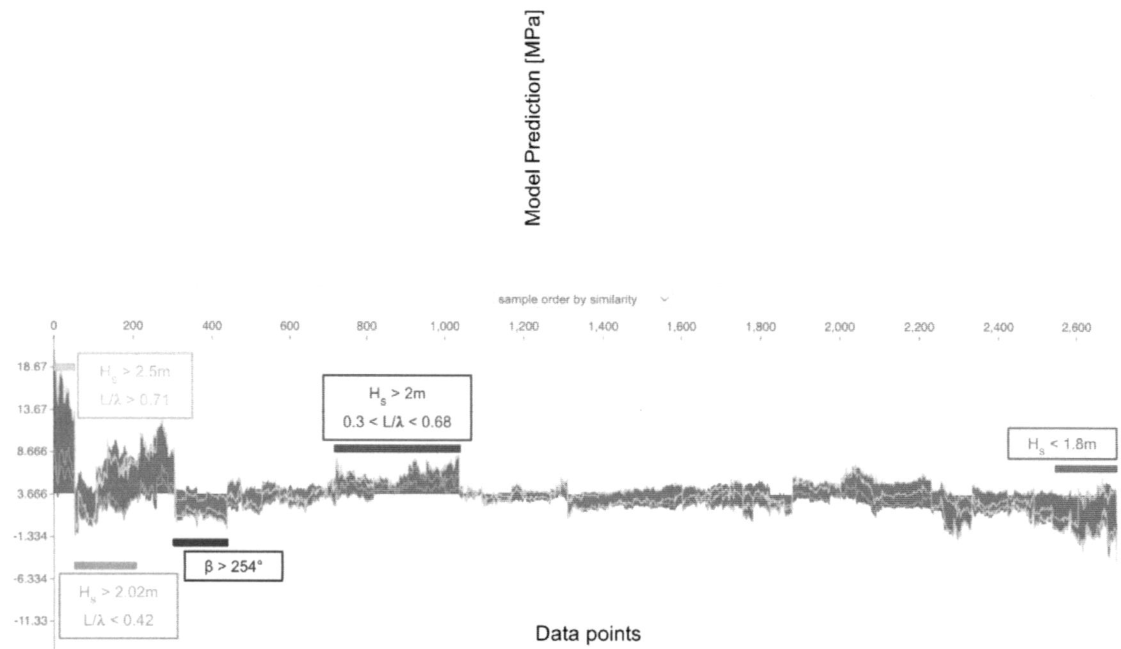

Figure 11: SHAP stacked force plot highlighting clusters of wave environments that produce similar effects on model prediction of maximum stress at SG01. Red areas represent wave environments that increase stress predictions and blue areas represent wave environments that reduce stress predictions.

Figure 13 presents a SHAP stacked force plot for predictions of maximum SG03 stress, though with data points along the horizontal axis sorted by H_S. Data points with similar effects on model predictions for SG03 are combined and averaged to obtain information about wave features' ranges that produce different stress predictions. The following is observed:

- The blue cluster on the far right has the highest predicted SG03 stresses (greater than 35 MPa) in the dataset. The average values of H_S, β and L/λ are 3.28 m, 143° and 0.84, respectively.
- The next clusters (green and orange) have the next highest predicted SG03 stresses, averaging 33.1 to 34.1 MPa. The corresponding average values of H_S, β and L/λ are 3.07 m, 124° and 0.68, respectively.

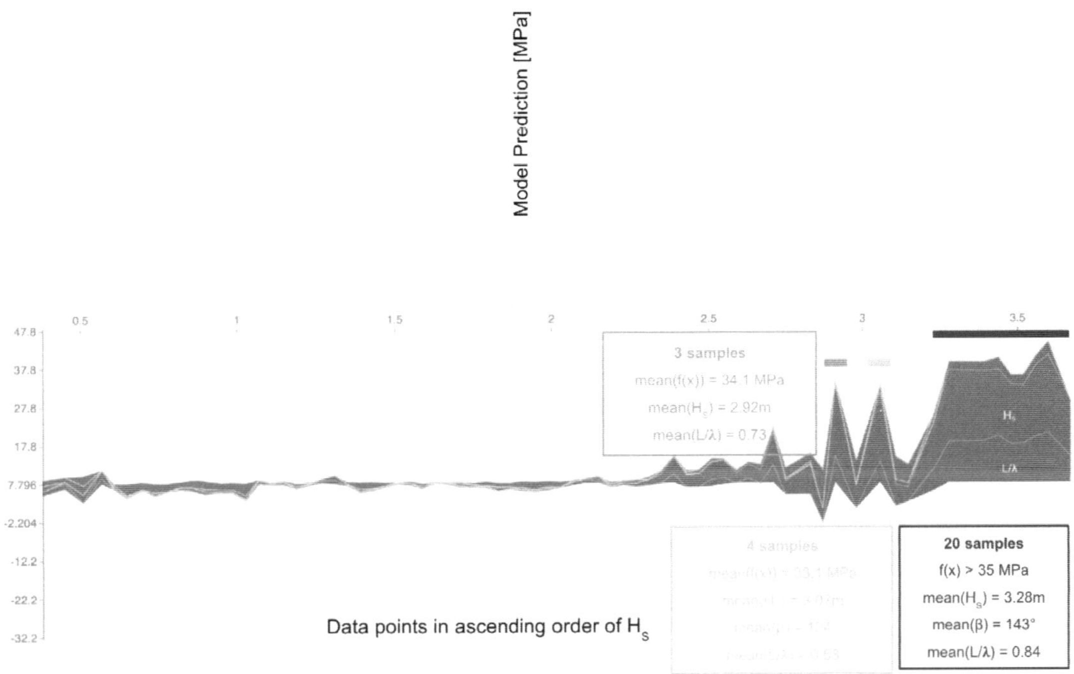

Figure 12: SHAP stacked force plot highlighting clusters of wave environments that produce similar effects on model prediction of maximum stress at SG03. Red areas represent wave environments that increase stress predictions and blue areas represent wave environments that reduce stress predictions.

4. DISCUSSION

The results demonstrate that the developed ML framework is able to identify and explain trends present in the *USCGC Stratton* dataset. Specifically, wave environments that produced maximum hull stress responses for SG01 and SG03 were identified. These wave environments featured H_S greater than 2.4 m and L/λ between approximately 0.8 and 1.

Relative to Hs and L/λ, β was shown to have the least significance on the predicted maximum stresses. β is used to calculate L/λ and thus shares some degree of correlation with L/λ's effects on model predictions. However, in heavier seas the crew may prefer to sail in head seas (rather than in beam or oblique seas), as suggested by Figure 4b. As such, it may be argued that the dataset is unbalanced due to the influence of the operator during the deployment. Alternatively, there is an implicit relationship between H_s and β. Nonetheless, this observation highlights the benefit of the ML framework, in that it implicitly takes into account the interdependences between the wave environment features, such as joint probabilities, and operational factors such as voluntary/involuntary speed reduction and heading changes [27].

Interestingly, the relative importance of the features to the predicted maximum stresses at the two strain gauge locations differs (refer to Figure 8a and 8b). The ratio L/λ has greater relative influence to the stresses at SG03 than those at SG01. The reason for the difference in the relationships between L/λ and the stresses is not clear. Both strain gauges are located at the same vertical height (01 deck), but SG03 is further forward and SG01 is closer to midships. From Figure 4a and 4b, there appears to be a stronger degree of non-linearity in the relationship between L/λ and the maximum stress at SG03 compared to that for SG01. This could be a whipping/flexural contribution, which would be relatively more important at the foreship. The range of stresses measured by SG03 are also larger than that at SG01. Thus, for the conditions that the *USCGC Stratton* encountered during the deployment, it is proposed that the sensitivity of the structure to L/λ at SG03 was greater than that at SG01. This hypothesis could be tested by examining additional deployment datasets.

The advantages of the developed ML framework are that it is flexible, capable of accounting for different feature variables and scales, and training the models is relatively quick in comparison to deep learning neural networks (for example, 30 s per model for the *USCGC Stratton* dataset). By improving the efficiency of naval structural analysis, including explanation of the association between wave environments and hull stresses experienced by a ship during its operation, this framework can assist in reducing the time and resources required to conduct fatigue life assessments and predict hull maintenance costs.

As the results are based on model predictions of stress, it is important to understand that SHAP explanations relate specifically to wave environments that affect model predictions, rather than true recordings of stress and speed. By appropriately training ML models and minimising prediction error, this study has obtained model predictions for the USCGC dataset that largely reflect the true data. Therefore, the identified wave environments that cause high model stress predictions are highly correlated to wave environments that cause high hull stress recordings. However, the models are unable to predict stress accurately for all elements of the dataset. As such, the identified wave environment patterns cannot be relied on without checking against the recorded wave environment values. This limitation is inherent when using ML predictions to represent true values.

As noted in Section 3.2, there is a level of scatter in the predictions. A possible reason for the scatter is that the heading angle is a non-linear feature in the ML model. As shown in Figure 4, the structural responses at the SG01 and SG03 are asymmetric and it was decided to preserve this characteristic. However, it is possible that the models will perform better by linearising the heading angle, by taking the cosine, and assuming symmetry.

The quality of insights that can be generated by the ML framework is dependent on the quality of the naval ship dataset and hindcast data. There may be an error due to the coupling of the ship data with wave hindcast data from the closest timestamp/frequency and location and differences in the values of wave parameters across different hindcast datasets [28]. The variability across datasets, as observed by Mogeke and Magoga [29], requires validity checking, such as comparison of similar variables, which was not within the scope of this study. Thus, the results generated in this project, and other studies utilising hindcast data, are dependent on the quality and availability of hindcast data for the given date and location. Further development of the framework would require quantification of these errors, like that observed in Hageman el al. [30] to determine the applicability of the predictions.

5. CONCLUSION

Driven by the need to efficiently assess the fatigue life of naval ships, and the opportunity presented by large naval performance datasets to apply 'data-driven' approaches, an exploratory ML framework for a full-scale dataset has been developed. The dataset has been acquired from the *USCGC Stratton* hull monitoring system enriched with wave hindcast data from CMEMS.

Three candidate supervised learning models were trained to predict the maximum stresses at two strain gauge locations based on the feature variables significant wave height, encountered wave angle, and ship length to wavelength ratio. The best performing model, the CatBoost model, was then assessed using the SHAP library to investigate the relative importance of the feature variables and to determine if patterns in wave environment parameters associated with relatively large structural stresses can be identified.

The results demonstrate that the developed ML framework is able to identify and explain trends that are present in the *USCGC Stratton* dataset. Wave environments that produced maximum hull stress responses at the strain gauge locations were identified. These wave environments featured H_S greater than 2.4 m and L/λ between approximately 0.8 and 1. The identified wave environment patterns were confirmed by the SHAP dependence and stacked force plots.

By appropriately training the chosen ML model (CatBoost) and minimising the prediction error, this study has obtained model predictions that largely reflect the measured data. Therefore, the identified wave environments that cause high model stress predictions are highly correlated to wave environments that cause high hull stress recordings. However, the models are unable to accurately predict the stress for all elements of the dataset. In addition, the quality of insights that can be generated by the ML framework is dependent on the quality of the naval ship dataset and hindcast data. To improve the performance and applicability of the models, it is recommended that linearisation of the ship encountered wave angle (as a feature variable) and wave hindcast data uncertainty be investigated.

Nonetheless, the presented data-driven approach serves as a proof of concept, demonstrating the use case for ML to assist in conducting fatigue life assessments of military ships. Further development of the ML framework may lead to efficiencies in conducting fatigue life assessments, allowing ship operators to better predict maintenance requirements.

6. ACKNOWLEDGEMENTS

The authors would like to acknowledge the contribution of the Fleet Ready Joint Industry Project (FReady.JIP) members including the United States Coast Guard, MARIN, BAE Systems, Bureau Veritas, DAMEN, Defence Research and Development Canada, Defence Science and Technology Group, Dutch Ministry of Defense, TNO, and TU Delft.

7. REFERENCES

1. DRUMMEN, I., HAGEMAN, R., AND STAMBAUGH, K., 'Structural fatigue life assessment and maintenance needs for a new class of US Coast Guard Cutters', *Life-Cycle of Engineering Systems: Emphasis on Sustainable Civil Infrastructure,* 2016.

2. STAMBAUGH, K., and KAMINSKI, M L., 'Ship Structure Fatigue and Life Cycle Risk Management Approaches', *Life-Cycle of Engineering Systems: Emphasis on Sustainable Civil Infrastructure*, 2016.

3. DONG, Y., GARBATOV, Y., and SOARES, C G., 'Review on uncertainties in fatigue loads and fatigue life of ships and offshore structures', *Ocean Engineering*, 2022.

4. ALPAYDIN, E., 'Introduction to machine learning', *MIT press*, 2020.

5. SCHOLCZ, T P., DÜZ, B., HAGEMAN, R B., and MAK, B., 'Consistency Assessment of Wave Directional Spectrum Predictions from Machine Learning Based Ship-As-a-Wave-Buoy Methods'. *International Conference on Offshore Mechanics and Arctic Engineering,* 2022.

6. DIEZ M., SERANI A., GAGGERO, M., and CAMPANA, E., 'Improving knowledge and forecasting of ship performance in waves via hybrid machine learning methods', *Proceedings of the 34th Symposium on Naval Hydrodynamics*, 2022.

7. KIM, J S., and JEONG, J S., 'Pattern recognition of ship navigational data using support vector machine', *International Journal of Fuzzy Logic and Intelligent Systems*, 2015.

8. GAO, M., and SHI, G Y., 'Ship-handling behavior pattern recognition using AIS sub-trajectory clustering analysis based on the T-SNE and spectral clustering algorithms', *Ocean Engineering*, 2020.

10. KOTSIANTIS, S B., ZAHARAKIS, I., and PINTELAS, P., 'Supervised machine learning: A review of classification techniques', *Emerging Artificial Intelligence Applications in Computer Engineering*, 2007.

11. HAFIZ, A M., HASSABALLAH, M., and BINBUSAYYIS, A., 'Formula-Driven Supervised Learning in Computer Vision: A Literature Survey', *Applied Sciences*, 2023.

12. MISHRA, P., 'Practical Explainable AI Using Python: Artificial Intelligence Model Explanations Using Python-based Libraries, Extensions, and Frameworks', *Apress,* 2022.

13. CHARBUTY, B., and ABDULAZEEZ, A., 'Classification based on decision tree algorithm for machine learning'. *Journal of Applied Science and Technology Trends*, 2021.

14. DO, T N., LENCA, P., LALLICH, S., and PHAM, N K., 'Classifying very-high-dimensional data with random forests of oblique decision trees', *Advances in Knowledge Discovery and Management,* 2010.

15. NATEKIN, A., and KNOLL, A., 'Gradient boosting machines, a tutorial', *Frontiers in Neurorobotics*, 2013.

16. GANJISAFFAR, Y., CARUANA, R., and LOPES, C., 'Bagging gradient-boosted trees for high precision, low variance ranking models', *Proceedings of the 34th international ACM SIGIR conference on Research and development in Information Retrieval*, 2011.

17. SHAPLEY, L S., 'Notes on the n-person game—ii: The value of an n-person game'. 1951.

18.. LUNDBERG, S M., and LEE, S I., 'A unified approach to interpreting model predictions', *Advances in Neural Information Processing Systems*, 2017.

19. STAMBAUGH, K., DRUMMEN, I., CLEARY, C., SHEINBERG, R., and KAMINSKI, M., 'Structural fatigue life assessment and sustainment implications for a new class of US coast guard cutters', *Transactions Society of Naval Architects and Marine Engineers*, 2014.

20. DRUMMEN, I., ROGERS, L., BENHAMOU, A., HAGEMAN, R., and STAMBAUGH, K., 'Hull Structure Monitoring of a New Class of US Coast Guard Cutters', *Proceedings of the ASNE Technology, Systems & Ships*, 2019.

21. COPERNICUS MARINE SERVICE, 'Copernicus Marine Data Store', https://data.marine.copernicus.eu/products, 2024.

22. MANSOUR, A., LIU, D., and PAULLING, J R., 'The principles of naval architecture series', *The Society of Naval Architects and Marine Engineers*. 2008

23. XGBoost Documentation - xgboost 2.0.0 documentation. *https://xgboost.readthedocs.io/en/stable/* (accessed: 5 September 2023), 2023.

24. CatBoost - Yandex Technologies. *https://yandex.com/dev/catboost/,* (Accessed: 11 September 2023), 2023.

25. THANAKI, J., 'Machine Learning Solutions: Expert Techniques to Tackle Complex Machine Learning Problems Using Python', *Packt Publishing*, 2018.

26. LUNDBERG, S., ERION, G., and LEE, S.., 'Consistent individualized feature attribution for tree ensembles', *arXiv*, 2018.

27. MAGOGA, T., OJEDA, R., CANNON, S., and THOMAS, G., 'Interdependencies between variables in fatigue analysis of a weight-optimised naval ship', *Procedia Structural Integrity*, 2019.

28. KAWAI, T., KAWAMURA, Y., OKADA, T., MITSUYUKI, T., and CHEN, X., 'Sea state estimation using monitoring data by convolutional neural network (CNN)', *Journal of Marine Science and Technology*, 2021.

29. MOGEKE, M., and MAGOGA, T., 'Towards improved understanding of naval ship structural performance via virtual hull monitoring', *Procedia Structural Integrity*, 2023.

30. HAGEMAN, R., DRUMMEN, I., THOMPSON, I., and STAMBAUGH, K., 'Fleet structural integrity through monitoring and data fusion', *Proceedings of the 15th International Symposium on Practical Design of Ships and other floating structures*, 2022

8. AUTHORS BIOGRAPHY

L McLeod is a student at Monash University, Australia. Mr McLeod's research interests include the application of supervised and unsupervised machine learning techniques to solve engineering problems.

M Herandi holds the current position of Associate Professor of Artificial Intelligence (AI) and Machine Learning at the Department of Electrical and Computer Systems Engineering, Faculty of Engineering, Monash University, Melbourne, Australia. He has contributed to designing the AI model and analysis of the results in this work. Mehrtash's previous experience includes extensive work in computer vision, machine learning, and AI.

T Magoga is the Discipline Lead Global Naval Structures at the Defence Science and Technology Group, Australia. Dr Magoga is responsible for leading capability development for improved assessment of naval structures. Dr Magoga's previous experience includes instrumented and virtual hull monitoring, and fatigue assessment of naval ships.

M Mogeke holds the current position of Researcher Global Naval Structures at the Defence Science and Technology Group, Australia. Mark is responsible for the development of data exploitation of naval ships for structural assessment. Mark's previous experience includes virtual hull monitoring and data science related to naval ships.

P Dennis holds the current position of Senior Researcher Naval Architecture and Platform Systems Analysis at the Defence Science and Technology Group, Australia. Dr Dennis is responsible for developing methods for the rapid assessment of whole-of-platform performance. Dr Dennis's previous experience includes modelling, simulation and data analysis for naval mechanical systems.

I Drummen holds the current position of Teamleader Hydro-Structural Services at MARIN, the Netherlands. He is a hydrostructual professional who received his education at TU Delft (MSc) and at the CeSOS Centre of Excellence in Trondheim Norway (PhD). For more than 15 years he has been working with at MARIN being among other responsible for running the Valid JIPs and FReady JIP.

R Hageman holds the current position of Senior Project Manager at MARINs Hydro-Structural Services team, the Netherlands. The Author is responsible for research coordination on hydro-structural topics. Dr. Hageman's previous experience includes various long-term monitoring campaigns onboard naval vessels and offshore units.

WARSHIP SURVIVABILITY NETWORKS

Dr V Capizzi, SYSTRA, STS, Australia

SUMMARY

This study delves into the optimization of survivability metrics. The term "survivability," synonymous with effectiveness and capability, is underscored by US legislation, including US CFR Title 10 Section 2366, which mandates realistic threat-based testing for surface ships and submarines.

Naval ships, considered as systems of systems, present intricate engineering challenges characterized by fragmented design information, diverse analytical systems, and varying levels of accuracy. Survivability is reliant on integrated systems and design cohesion.

Graph Theory proves essential in comprehending networked, distributed, and interdependent systems. By introducing visual nodal concepts and quantifying connectivity and interdependence, it becomes pivotal in shaping architectural features like robustness (survivability).

The application of "clustering" in physical systems analysis, driven by Graph and Percolation theory, emerges as a determinant of survivability. System topology tracks the evolution of interactions, especially during subsystem breakdowns.

Survivability, from system distribution analysis, depends on early architectural decisions, facilitating informed choices among alternatives. Robustness becomes evident, offering clarity to decision-makers seeking to optimize warship survivability. This paper contributes insights into warship survivability through a structured Network Theory analysis.

1. INTRODUCTION

Australia's strategic investment in warships prioritizes the procurement of vessels that are designed to endure the most challenging conditions encountered during wartime operations. This imperative imposes significant constraints on the budgetary framework. The concept of 'survivability,' which is tantamount to effectiveness and capability, is underscored by U.S. legislation, specifically CFR Title 10 Section 2366, which compels realistic threat-based evaluations for surface ships and submarines. In alignment with this directive, naval authorities are tasked with developing strategic business frameworks that necessitate judicious decision-making to realize desired outcomes. Concurrently, these frameworks are buttressed by analytical tools that substantiate decisions pertaining to Defence Survivability capability. Design addresses critical elements inherent to the lifecycle of any project or product. However, it does this in a silo approach. Defence requires systems to be integrated and optimized, but also executed in a manner that aligns with overarching strategic Defence goals.

Functional relationships play a crucial role in shaping the survivability of a ship's capability to fight and defend itself.

This research paper explores the use of Network Analysis, mathematically known as Graph Theory, to identify complex physical, functional, operational, and restorative relationships to enhance ship survivability. By focusing on survivability, this analysis helps to gain a clearer understanding of the relationships, their immediate connections, and their influence on the ship's architecture.

2. SURVIVABILITY

Survivability is a critical aspect of system design for warships, particularly for systems that are essential for war time mission success. Here are the key points that enshrine survivability:

- Mission Completion: Adequate Survivability architectures ensures that a warship is in the best ship configuration to complete its mission despite facing attacks, damage, setbacks or accidents.

- Essential Services: It involves maintaining essential services even when the system experiences successful intrusions.
- Timely Recovery: The ability to recover compromised services quickly after an intrusion is a vital component of survivability.
- Robustness: Overall, survivability is about the resilience of a system to withstand and bounce back from unforeseen challenges.

Survivable network systems also embody another essential characteristic; they need to ensure survivability in environments of unbounded networks with dynamic architectures. Generally, insufficient emphasis is placed on these survivability whole of system issues; the processes and techniques need to be geared for addressing integrated survivability, to provide adequate safety with associated threats. New concepts are moving towards a system engineering framework for integration of systems that are safe, achieve the mission, are fault tolerant, are safe for crew, as well as being dynamic, adaptable, and diverse.

The concept proposed herein uses Network Analysis as a framework, providing an integrated view of warship architectures.

The integration process presents an architecture that is co-existent with being physical, functional, operational, and restorative:

- Physical refers to the ship's physical architecture, enclosed by a hull, comprising decks and bulkheads, corridors, and component spaces.
- Functional refers to the systems requirement to meet the mission objectives, and their inter-system relationships.
- Operational refers to the set of operational requirements.
- Restorative refers to the ability to overcome system failures due to threats being realised.

In determining robustness of ship designs, generally for a surface combatant, the following constraints are considered, and proposed as a point of departure to set survivability objectives in this study, shown in Table 1.

Table 1. Post Event Capability Index

POST-IMPACT SYSTEM SURVIVAL CAPABILITY INDEX	MUST MAINTAIN CAPABILITY FOR:
1.0	To Fight and Win at Sea
0.75	To Move to Minimise Vulnerability and Be Out of Harm's Way
0.5	To Float, Be Out of Harm's Way and Not Be Vulnerable
0.20	To Abort Ship and Crew Be Safe

These constraints are useful starting points to determine the ships survivability outcomes.

3. MISSION CRITICAL

A ship can be generally considered to conduct four primary missions that cover the entire operational needs of the Navy. As identified in Table 2, these four missions all have objectives relating to the operational state at the time of conducting its primary missions. It is noted that the mission to fight and win at sea follows the kill chain for a single-shot engagement, see Figure 3.1. The mission states relate to the relationship between survivability (robustness), vulnerability and susceptibility, as shown in Figure 3.1. The vulnerability state can result in any missions 1 to 4, as defined in Table 2. The approach to be followed is to use a "network centrality" measure and subject a test case to peculation theory, as discussed in Section 5.3. Centrality is further explained below.

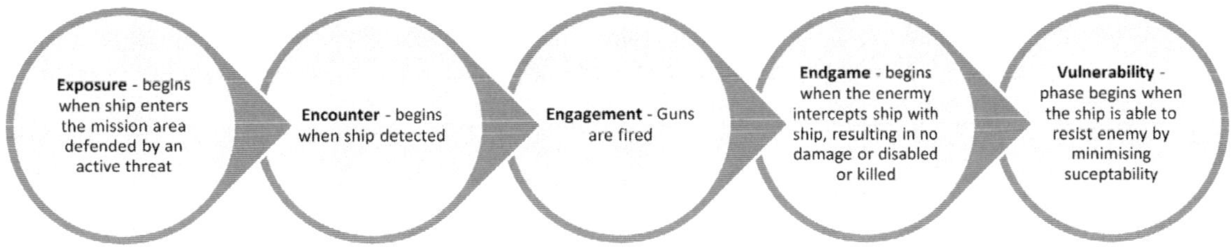

Figure 3.1: Kill Sequence

Table 2. Critical Systems for Mission Scenarios

MISSION CATEGORIES AND CRITICAL SYSTEMS			
1.To Fight and Win at Sea	**2. To Move to Minimise Vulnerability and Be Out of Harm's Way**	**3. To Float, Be Out of Harm's Way and Not Be Vulnerable**	**4. To Abort Ship and Crew Be Safe**
Structural And Hull Integrity,	Structural And Hull Integrity	Structural And Hull Integrity	Life Support
Stability, Control Surfaces,	Stability, Control Surfaces	Stability	Communications
The Bridge System	Navigation	Power Source	Human Machine Operations
Navigation	Propulsion	Life Support	Damage Control
Propulsion	Power Source	Self-Protection Systems	Vulnerability
Power Source	Fuel	Communications	
High Energy Weapons	Life Support	Damage Control	
Surveillance Systems, Radar	Self-Protection Systems	Human Machine Operations	
Combat Management Systems	Communications	Vulnerability	
Self-Protection Systems	Damage Control		
Fuel	Human Machine Operations		
Damage Control	Vulnerability		
Communications			
Life Support			
Human Machine Operations			
Vulnerability			

The loss of any of these components and their related location spaces or, any combination of, could render the ship unable to proceed with any of the primary missions.

An integrated view ties layers from the physical, functional, operational, and restorative.

- Physical layer representing ship spaces and locations; the ship platform and its spaces are fundamental for a ship to carry out its Mission to Fight. The considerations in this study relate to high level abstractions of hull and space.
- Functional layer representing the interactions between the systems from a functional perspective. Systems/components and their interfaces are the nerves of the Fight capability. This paper places higher risk to the interfaces as these have far reaching influences on system reliability.
- Operational layer represents the mission scenarios, i.e., Fight, Move and Float
- Restorative layer represents the actions to restore capability in the face of a system component failing due to damage; this comprises of reliability and damage control, human operations, and lifesaving.

4. PERCOLATION THEORY.

Robustness of networks against failure of their component parts is related to the study of percolation. The process that results from taking a network and removing some fraction of its vertices or edges, Percolation provides a natural model for studying the robustness of networks as it provides a framework for understanding connectivity in networks and has practical implications across various robust domains.

The mathematical framework is used to study the behaviour of connected components in random networks. It quantifies the threshold value of a local control parameter (such vertex occupation probability or edge activation probability) for which there exists a giant connected component (GCC).

Simply put, percolation describes the "behaviour" of a network when vertices or links are either taken away or added. It is a geometric phase transition, occurring when a "critical fraction" of vertices or edges are randomly added, or removed causing small, disconnected clusters to merge into significantly larger connected clusters, finally resulting in a giant connected component (GCC)

Note: In "site." percolation, randomly placed vertices while in "bond" percolation, refer to randomly placed edges.

The percolation threshold refers to the critical value of the control parameter that separates two regimes of network connectedness: of smaller clusters. When the network transitions to isolated clusters, then the component relational ties break. When applied to networks, it provides insights into the robustness of the system architecture. For this study, the assumption includes a threshold value of 0.5.

There are various methods to remove vertices from a network, for example in order of degree from highest to lowest; other methods will remove vertices by any centrality measure available. For the purposes of this study, we will introduce removal of edges by random activation. Based on this, we can quantify robustness by the R-index, which is defined by:

$$\mathrm{R} = 1/N \sum_{i=1}^{N} \sigma(\frac{i}{N})$$

N is the number of vertices
σ is the number of vertices removed.
The maximum value is (1/2) $(1 - 1/N)$
The minimum value is 1/N.
As a result, we introduce the Vulnerability Index (V-index), which quantifies a network's susceptibility to a specified pattern of vertex elimination, serving as the counterpart to the Robustness Measure.

$$V = \frac{1}{2} - R$$

5. WHAT IS NETWORK ANALYSIS?

In the development of any organisation involving people or systems, or assessing any social media applications, social/hierarchical functions appear that govern how the organisation operates. How the system/organisation operates is the subject of a common technological methodology known as Network Analysis.

As a mathematical tool, network analysis can be modelled as a series on linear equations with constraints to optimise outcomes.

5.1 WHAT IS A NETWORK?

A network refers to a visual structure representing a group of objects/people and relationships between them. Also known as a graph in mathematics, a network structure consists of vertices and edges. Here, vertices, represent notional objects, such as spaces or functions that require analysis while edges represent the relationships between those objects and the direction of their activity.

For example, if we are studying a social relationship between Facebook users, vertices are target users and edges are relationships such as friendships between users or group memberships. In Twitter, edges can be following-follower relationships.

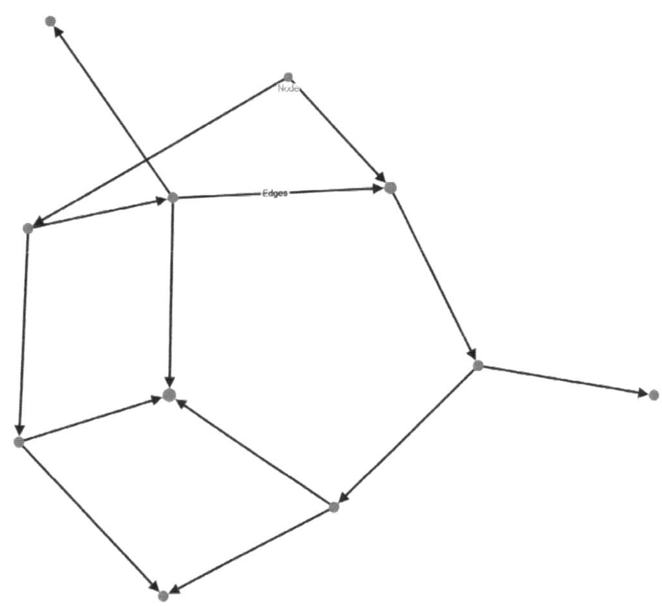

Figure 5.1: Network Representation

5.2 WHY NETWORK ANALYSIS?

Network Analysis is useful in many living application tasks, represented as shown in Figure 5.1. It helps us in providing a deep understanding of the structure of a relationship in organisational/social networks, a structure or process of change in natural phenomenon, or even the analysis of biological systems of organisms. This means of network representation adds greatly to the understanding the overall picture of the network's architecture. (Loem, 2021).

Network analysis plays a mainstream and important role in time series analysis, natural language processing, telecommunication network analysis, Machine Learning (Deep Learning) is also used in network analysis.
The use of the network of media, or social users, is used as an example. The analysis of this example network identifies:

- Identifying the most influential person or people in a group
- Defining similar characteristics of groups of participants
- Prediction of appropriate and suitable items for participants
- Identifying targets, groups/clusters, etc.

5.3 WHAT IS THE IMPORTANT SYSTEM COMPONENT?

A crucial application of network analysis is identifying the most important vertex(s) in a network. The methods used are called "centrality analysis". This refers to the task of identifying the most influential members, or the representatives of the group (Rodrigues, 2019).

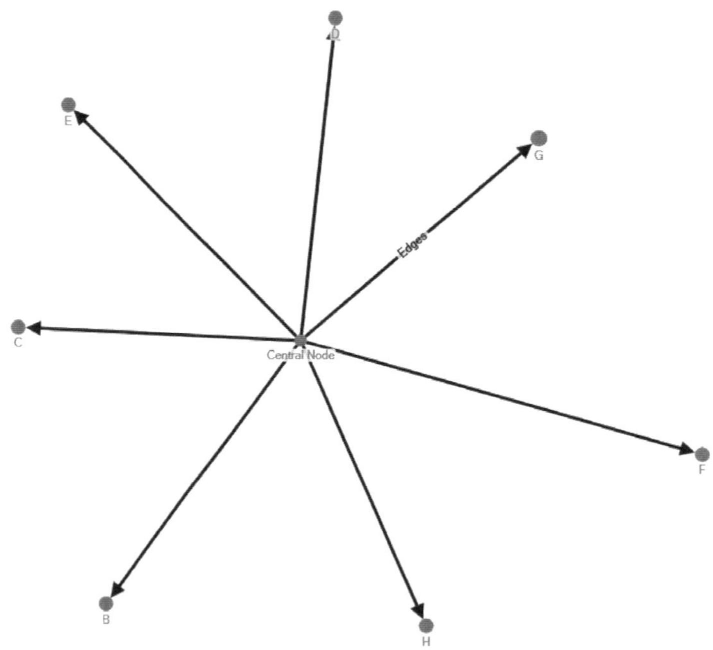

Figure 5.2: Network showing Most Influential System

The central vertex is the most important one in the Figure 5.2.

To define the most important vertex, specific definitions are required to identify several characteristics of important vertices for which there are several indicators used to measure the centrality of a vertex.

Degree centrality: The vertex(s) with a higher number of incoming edges, known as degrees, have higher degree centrality.

Eigenvector centrality: Supplementing the degree of one vertex, the centralities of neighbour vertices are considered. As a result, the eigenvector corresponding to the highest eigenvalue (of the mathematical entity, the adjacency matrix) represents the centrality of vertices in the network. Whereas degree centrality ranks a vertex as being important, if it is connected to many other vertices, the eigenvector centrality is based on the more subtle notion that a vertex should be viewed as important if it is linked to other vertices which are *themselves* important.

Betweenness Centrality: A different notion of centrality is betweenness that measures the extent to which a vertex lies on the shortest paths between pairs of other vertices in the network. A vertex with high betweenness centrality acts like a bridge between different parts of the network and can facilitate the flow of information or resources. Alternatively explained, it measures how many short paths between vertices in the network pass through a given vertex.

Closeness centrality: The length of the path from the i-th vertex to other vertices in the network is considered as the i-th vertex's closeness centrality. With this definition, for example, this centrality can be applied in the task of defining a suitable evacuation site in a city where the mean distance between a given vertex and all other vertices in the network [24,25]. To define closeness centrality which introduces the concept of a geodesic path in a network.

Clustering coefficient: measures how connected a vertex's *neighbours* are to one another. More generally, a clique or complete graph occurs when all vertices in a group are directly connected to each other. If vertex neighbours are all connected, there is a high clustering coefficient. If vertex neighbours are not really connected, then it results in a low clustering coefficient.

Adaptability: Adaptability indicates the systems flexibility in the changes of routing between nodes. Essentially, is a measure of redundancy evaluating how many nodes it takes for the system to work. A system component has a better chance of can be routed to two same systems (system and backup) requires two routes (interfaces) to consider it fails.

5.4 PAGE RANK

The PageRank algorithm, or Google algorithm was introduced by Lary Page, one of the founders of Google. For this paper, it can be thought of as an alternative centrality measure. It was first used to rank web pages in the Google search engine. Nowadays, it is more and more used in many different fields, for example in ranking users in social media etc... What is fascinating with the PageRank algorithm is how to start from a complex problem and end up with a very simple solution.

5.5 GROUPING AND CLUSTERING

One application of network analysis is the systems grouping. The purpose is to divide a network into groups of vertices (systems) that are similar in any specific features. Examples of this task are of defining groups of users in Damage Control who share common technical skills and domain experience, or to find groups of customers to offer specific skills, or to provide recommendation of systems architectures and the like, as presented in Figure 2.3 for a naval sustainment project.

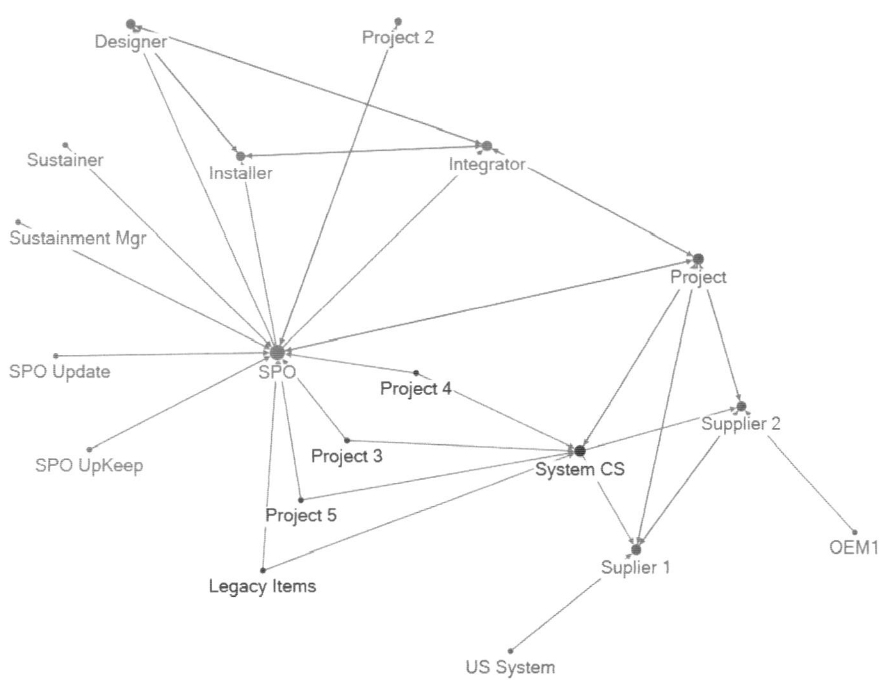

Figure 5.3: Three simple role clusters (Red, dark blue and Green) Based on In-degree metric and type for sustainment.

6. ASSESSMENT

The surface combatant architecture (SCA) has been modelled as a network and presented in Figure 2.4. This top-level view of the SCA displays the main constituents in the network and how they relate to vertices.

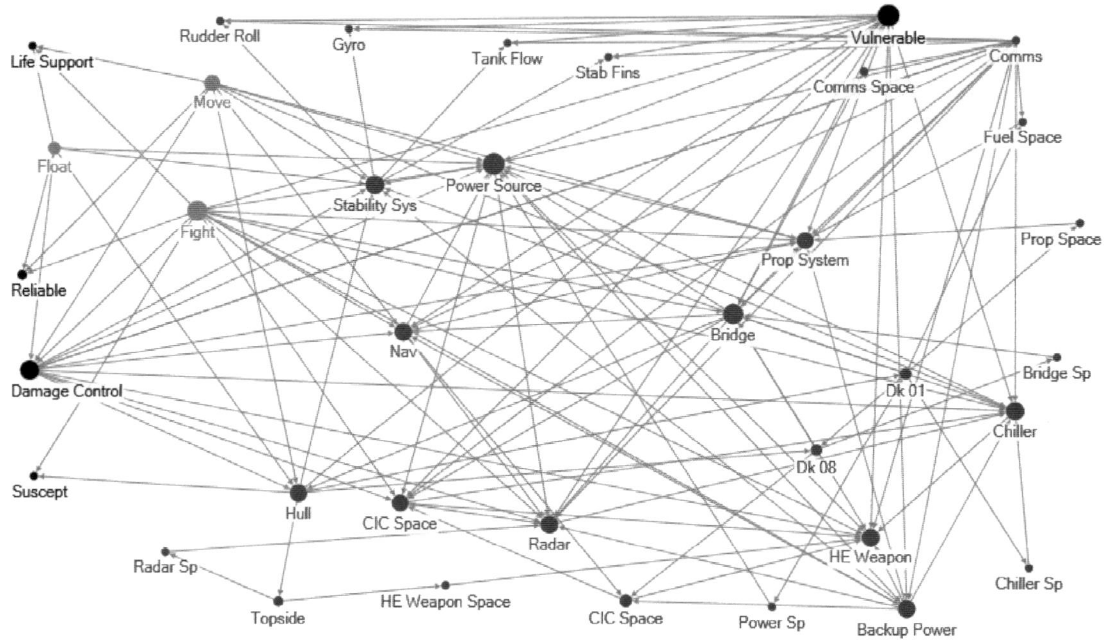

Figure 6.3: Notional Integrated Surface Combatant Baseline

The following features are identified:

- The colours of red, blue, purple represent mission, structure and location, and system components.
- The size of the vertex represents the vertex degree, i.e., the larger the number of relations, the larger the vertex.

Table 3. SCA Model Network Metrics

Graph Metric	Value
Graph Type	Directed
Vertices	36
Unique Edges	132
Edges With Duplicates	0
Total Edges	132
Self-Loops	0
Connected Components (SCA is Component)	1
Single-Vertex Connected Components	0
Maximum Vertices in a Connected Component	36
Maximum Edges in a Connected Component	132
Maximum Geodesic Distance (Diameter)	4
Average Geodesic Distance	1.998457
Graph Density	0.104761905

- The network has the following SCA model characteristics, shown in Table 3.
- The Mission vertices are the drivers for the SCA model.
- There are 132 unique edges: 36 individual vertices, spanning the physical, functional, operational, and restorative layered objectives. The SCA model is provided as an example of high-level modelling purposes only.
- The maximum geodesic distance is 4. The geodesic distance is the length (in terms of "vertex hops") of the shortest path between two vertices. This is an important discriminator of communications capability.
- The average geodesic distance is 2. This is the average length of the geodesic distances between all pairs of vertices. It gives a sense of how "close" vertices are from one another.
- The graph density is 0.1. The graph density is a number between 0 and 1 that indicates the percentage of possible edges that are connected. This is an important discriminator of the SCA. It indicates, based on the SCA model, how difficult communications and data operations occur in that environment. A measure of 0.1 indicates that only 10% of the SCA is connected. This measure is based on organisational constraints and functional properties. However, improving this number increases network value in relation to interdependency, adaptability, and connectivity.
- The metric designated adaptability shows the component that has the highest number of in-degree paths indicating redundancy in the component. The analysis assumes that paths are more important from a redundancy point of view because the component may be healthy, but the interfaces have failed.

Table 4 presents the vertex survivability various centrality outcomes for the SCA. Note that Page Rank has been included to provide an example of how Google uses this centrality metric for its business activities.

The centrality analysis shows that the most important systems are the Power Source, the Bridge, and the "Fight" function. This is based on the eigenvalue centrality analysis which indicates that it is the principal component, and neighbours of this vertex are as important. These systems are most critical and are also backed up by the Page Rank analysis.

Table 4. Example of SCA Centrality Results – Eigenvalue Sort

Vertex/Component	Degrees	In-Deg.	Out-Deg.	Betweenness Centrality	Closeness Centrality	Eigenvect Centrality	PageRank	Clustering Coefficient	Adaptability
Power Source	15	10	6	77.690	0.018	0.067	1.946	0.271	0.313
Bridge	13	6	11	225.663	0.019	0.059	2.295	0.162	0.188
Fight	13	1	13	77.280	0.017	0.056	1.622	0.269	0.031
Vulnerability	15	2	13	107.045	0.018	0.056	1.888	0.176	0.063
Communications		2	15	135.352	0.018	0.056	2.067	0.146	0.063
Damage Control	12	4	10	29.432	0.017	0.055	1.575	0.263	0.125
Chiller	11	11	1	54.180	0.017	0.054	1.499	0.318	0.344
Combat Inf Centre	9	8	5	27.285	0.017	0.053	1.466	0.288	0.250
Radar	10	9	2	63.157	0.016	0.051	1.400	0.345	0.281
Navigation	10	7	4	18.423	0.016	0.050	1.347	0.309	0.219
Backup Power	10	4	7	22.346	0.016	0.049	1.349	0.282	0.125
Propulsion System	9	10	1	67.969	0.016	0.046	1.429	0.245	0.313

7. ANALYSIS

Section 6 explains the SCA Model and provides the starting point for the survivability scenario to be implemented and how changes to the SCA architecture affect survivability and vulnerability.

Percolation

The SCA percolation scheme removes edges by a random number generator, each component system is subject to a unique random number for inclusion. If the random number is greater that the percolation thresh hold, assumed to be 0.5 or above, and is included as a constraint in the model. The SCA "scarred" model properties due to table 5. Percolation removes edges and is considered the worst case for this example. Table 5 shows the effect on the un-percolated system shown in Table 3.

Table 5. Percolation for ASC Architecture.

Graph Metric	Value
Graph Type	Directed
Vertices	34
Unique Edges	66
Edges With Duplicates	0
Total Edges	66
Self-Loops	1
Maximum Edges in a Connected Component	65
Maximum Geodesic Distance (Diameter)	6
Average Geodesic Distance	2.609174
Graph Density	0.057932264
Robustness/Survivability	0.4962

- Robustness has fallen to half of that exhibited in the SCA baseline, (assumed to be 1).
- There are now 66 unique edges: 34 individual vertices, spanning the physical, functional, operational, and restorative layered objectives. Note this analysis is for high level modelling purposes only.
- The maximum geodesic distance has decreased to 6.
- The average geodesic distance is 2.6 This is the average length of the geodesic distances between all pairs of vertices. It gives a sense of how "close" vertices are from one another.
- Robustness now has decreased by 50% of the SCA robustness baseline.
- The graph density is 0.05. The measure indicates that only 5% of the SCA edges are now connected.

Table 4 presents the vertex survivability various centrality outcomes for the SCA. Note that Page Rank has been included to provide an example of how Google uses this centrality metric for its business activities.

Table 6 shows the centrality measures for the percolated scenario. Table 7 shows that model properties have changed significantly.

Table 1. Example of Percolated System centrality Characteristic (Eigenvector Sort.)

Vertex	Deg.	In-Degree	Out-Degree	Betweenness Centrality	Closeness Centrality	Eigenvector Centrality	PageRank	Clustering Coefficient	Adaptability
Vulnerability	10	1	9	150.338	0.015	0.083	2.249	0.111	0.030
Power Source	10	6	4	256.738	0.017	0.079	2.340	0.089	0.182
Fight	8	0	8	142.933	0.016	0.072	1.835	0.143	0.000
Chiller	7	6	1	98.852	0.015	0.069	1.652	0.167	0.182
Combat Inf Centre	6	5	1	53.100	0.014	0.060	1.386	0.167	0.152
Navigation	6	5	1	80.771	0.014	0.058	1.368	0.167	0.152
Bridge	6	2	4	69.886	0.014	0.056	1.378	0.133	0.061
High Energy Weapon	6	6	0	150.800	0.014	0.052	1.504	0.133	0.182
Backup Power	5	2	3	18.433	0.013	0.051	1.139	0.250	0.061
Communications	7	1	7	156.357	0.015	0.047	1.723	0.024	0.030
Radar	5	4	1	73.452	0.013	0.042	1.248	0.100	0.121
Propulsion System	5	5	0	38.771	0.013	0.041	1.161	0.100	0.152
Damage Control	5	2	4	75.333	0.013	0.031	1.222	0.100	0.061
Rudder Roll	3	3	0	24.933	0.013	0.026	0.795	0.000	0.091
Stability System/Spaces	4	2	2	104.867	0.013	0.021	1.150	0.000	0.061
Move	5	0	5	78.267	0.012	0.021	1.361	0.100	0.000
Combat Inf Centre Space	2	0	2	2.700	0.011	0.019	0.559	0.000	0.000
Fuel Space	3	2	1	15.600	0.011	0.018	0.787	0.000	0.061
Power Space	2	1	1	8.300	0.011	0.016	0.580	0.000	0.030
Reliability	2	2	0	8.400	0.011	0.016	0.576	0.000	0.061
Hull	5	2	3	92.333	0.012	0.016	1.414	0.050	0.061

For the Fight mission, percolation has been applied for a typical single shot adversary on the ship with the result that edges/interfaces have been destroyed. Assuming a single and capable adversary, the mission is carried out while incurring system failures. (a removal of edges). Even though some edges are removed, system vertices may still be operational. The scenario shows that most components survive; deleted components include Deck 8 and the Propulsion system space, having "failed". This affects all mission operations.

The scenario involves percolation and its impact on the ship's system during the Fight mission. The key points:

Percolation application is to assess the ship's robustness against a single-shot adversary. A typical single-shot adversary targets edges/interfaces, resulting in mutual destruction. System failures occur which results in the removal of edges, thereby replicating leads to system failures.

However, some vertices survive despite edge removals, and so, some system vertices remain operational. As most components survive the attack, failed components are Deck 8 and the Propulsion system space. These failed and critical components affect mission operations.

The implications are:

- Resilience: The system shows resilience by keeping operational vertices even after edge removals.
- Critical Areas: The failure of Deck 8 and Propulsion system space highlights their critical role in mission success.

- Mitigation Strategies: Consider redundancy, backup systems, and rapid repair protocols to enhance system robustness.

In summary, percolation analysis offers valuable insights into the ship's vulnerability and informs strategies for improving its resilience during adversarial scenarios.

The percolation model results include:

- Fight mission system interfaces that have failed include Radar System, Bridge, Combat System Information Centre, Propulsion System Space, Damage Control
- For the "Move" mission, the following interface systems have been removed, Stability, System/Spaces and Propulsion System
- For the "Float" mission, the following interface systems have been removed:, Hull, Stability System/Spaces, Reliability and Damage Control.

Table 5 shows the most important systems in the SCA architecture are the Power Source, Chiller, and Vulnerability systems. This pattern of interface removals shows that the system after engagement may not meet the requirements of Fight, Move, or Float. The ship needs to be aborted.

Designing for Survivability

A practical means to improve the SCA Model is to examine the centrality measures to better the SCA model robustness. For example:

- Increasing the In-Degree measure, which is related to Adaptability, will provide added edges to increase access to systems. This increases overall system/component robustness and adaptability.
- As well as edge redundancy, component redundancy improves robustness.
- Finding the vertices that have low eigenvalue centrality measures can be improved by increasing neighbour importance by linking to other pathways.
- Increasing graph density will provide for added edge access.
- Validating those vertices with "out-degrees" of zero need to be bridged to alternative vertices so that it closes the gaps in the component make-up.
- Using the Closeness centrality measure, ensure that all nodes in the SCA can be traversed in the minimum number of "hops."
- Clustering measures show neighbours are connected. Ensure neighbours in themselves are connected to close neighbours.

8. CONCLUSIONS

The investigation of survivability for surface combatants is a critical aspect, especially during the preliminary design phases. The key aspects of this approach:

Objective and Context:

- The paper aims to assess the survivability of surface combatants.
- It introduces an approach commonly used in the Mathematical Sciences, which is now applied to engineering design practices.

Effective Timing:

- The proposed approach is most effective during the preliminary design phases of surface combatants.
- It supports systems engineering trade studies by providing insights into survivability.

Systems Architecture and Adaptability:

- The approach ensures that the systems architecture is designed to support integrated survivability.
- Designers can assess adaptability when the system is exposed to issues or threats.

Graph Theory and Connectivity:

- Graph Theory plays a crucial role in understanding networked, distributed, and interdependent systems.
- It quantifies connectivity and examines centrality measures to recognize deficiencies in the network.

Clustering and Survivability:

- The application of "clustering", informed by Graph and Percolation theory, determines a measure of survivability as a high-level integrated view.
- Clustering helps identify critical components and their interdependencies.

Integrated Robustness:

- Early architectural design decisions significantly impact survivability.
- Integrated robustness emerges as a key factor in ensuring a resilient system.

Optimizing Warship Survivability:

- Decision-makers can optimize warship survivability to reinforce capability and enhance safety of life at sea.
- This involves balancing trade-offs between various design parameters.

In summary, this research contributes valuable insights into warship survivability through a structured Graph Theory analysis. The approach provides a practical design tool for calculating parameters and metrics, aiding informed decision-making during the design process.

9. REFERENCES

1. BROWNLOW, et al., "A Multilayer Network Approach to Vulnerability Assessment for Early-Stage Naval Ship Design Programs", *Ocean Engineering Volume 225*, 2021.

2. MEAD, N., "Requirements Engineering for Survivable Systems", *Carnegie Mellon Software Engineering Institute*, 2003

3. IYER, S., et al., "Attack Robustness and Centrality of Complex Networks", *PLOS One*, 2013

4. LOKE, S., "Naval Survivability and Susceptibility reduction Study—Surface Ship", *Naval Graduate School Thesis*, 2012.

10. AUTHORS BIOGRAPHY

Dr Vincent Capizzi is the Technical Director for SYSTRA STS in Australia. With a career spanning over 30 years in STEM projects and activities in the Space and Defence domains. As a Systems Engineer, Vincent has worked in the UK, the US, and Spain, participating in extraordinary projects related to Space, Guided Weapons, Satellites, and Naval Defence acquisition. His main contributions include:

- Simulation-backed solutions development: Vincent combines simulation techniques to create innovative solutions.
- Model-Based Systems Engineering (MBSE): He leverages MBSE principles to enhance system design and development.
- Network Theory: Vincent's understanding of network theory informs his work in complex systems.

At SYSTRA, Vincent provides technical services to Defence projects and offers technical advice for business development activities, contributing to the STS Leadership Team. SYSTRA is a global engineering and consulting group specializing in security, transport, and mobility solutions.

The Royal Institution of Naval Architects
8-9 Northumberland Street
London, WC2N 5DA
United Kingdom

ISBN 979-8-3313-0375-4
9 798331 303754

CURRAN ASSOCIATES INC.
proceedings.com